Regional Conference on Science, Technology and Social Sciences (RCSTSS 2014)

Nor Azizah Yacob · Mesliza Mohamed
Megat Ahmad Kamal Megat Hanafiah
Editors

Regional Conference on Science, Technology and Social Sciences (RCSTSS 2014)

Science and Technology

Springer

Editors
Nor Azizah Yacob
Universiti Teknologi MARA
Pahang
Malaysia

Megat Ahmad Kamal Megat Hanafiah
Universiti Teknologi MARA
Pahang
Malaysia

Mesliza Mohamed
Universiti Teknologi MARA
Pahang
Malaysia

ISBN 978-981-10-0532-9 ISBN 978-981-10-0534-3 (eBook)
DOI 10.1007/978-981-10-0534-3

Library of Congress Control Number: 2016931603

Printed on acid-free paper

This Springer imprint is published by SpringerNature
The registered company is Springer Science+Business Media Singapore Pte Ltd.

Preface

The Regional Conference on Science, Technology and Social Sciences 2014 (RCSTSS 2014) is a biennial conference organized by Universiti Teknologi MARA (UiTM) Pahang. Showcasing recent advancements and trends in the three major academic disciplines, namely science, technology, and social sciences, RCSTSS 2014 facilitated knowledge sharing and networking among participants concerning new challenges in their fields. But more importantly, it also served as a platform to disseminate research findings and a catalyst to promote innovations in the development of the country as well as the region. More than 200 papers were presented by participants from various local and foreign universities and institutions of higher learning. Of these, 64 science and technology manuscripts have been selected to be included in this publication, namely architecture, biology, computer and information technology, engineering, environment and management, food science, forestry, health and medicine, mathematics and statistics, plantation and agrotechnology, physics, robotics and sport science. The papers included in this book have undergone a careful selection process to ensure that they meet the objectives of the conference. Hopefully, this publication will serve as a significant reference to academicians, researchers, and students who are pursuing further research in their respective fields.

Pahang, Malaysia

Nor Azizah Yacob
Mesliza Mohamed
Megat Ahmad Kamal Megat Hanafiah

Acknowledgements

In the name of Allah, the Most Gracious, the Most Merciful. Praise be upon him for giving us the courage, strength, and patience to complete the *Proceedings of the Regional Conference on Science, Technology and Social Sciences* (*RCSTSS 2014*), UiTM Pahang.

We would like to express our sincere gratitude to all RCSTSS 2014 committee members, participants and reviewers for their help and support. Many thanks also go to the Editorial Committee, Mohd Rozaidi Ismail, Aiza Harun, Shaikh Abdul Karim Yamani Zakaria, Nazirah Ramli, Neni Kartini Che Ramli, Mohd Kamal Azman Jusoh, Fairuz Shohaimay, Zurhana Mat Hussin, Faizan Abdul Jabar, Sharifah Norhuda Syed Wahid, Siti Noorul Ain Nor Azemi, Salimah Ahmad, and Siti Suhaila Harith, for their tireless effort in the preparation of this book.

Special thanks also go to the Rector of UiTM Pahang Prof. Dr. Jamaludin Kasim and the RCSTSS Chair, Associate Prof. Dr. Azhan Hashim for their encouragement and moral support. The editors would also like to thank Universiti Teknologi MARA Pahang for the technical and financial support that ensured the materialization of RCSTSS 2014.

Alhamdulillah. Thank you.

Contents

Part I Engineering

Part VI Forestry

Part XII Food Science

Part XIII Sustainable Development

Part XIV Physics

Part I
Engineering

Chapter 1
Strength Properties of Lightweight Foamed Concrete Incorporating Waste Paper Sludge Ash and Recycled Concrete Aggregate

Siti Shahidah Sharipudin, Ahmad Ruslan Mohd Ridzuan, Raja Nor Husna Raja Mohd Noor and Asmawati Che Hassan

Abstract A rapid growth in concrete industry is known to leave an enormous environmental footprint on planet. As the amount of cement being produced continues to rise, it results the high levels of carbon dioxide. In the meantime, a growing development in construction sector creates the sheer volumes of construction waste, mainly concrete waste. Thus, the present study aims at correlating the strength response to the contribution of using waste paper sludge ash (WPSA) and fine recycled concrete aggregate (FRCA) as an alternative of cement and natural fine aggregate substitute in the manufacturing of lightweight concrete (LWC). The different levels of WPSA and FRCA which are 0, 5, 10, 15, 20 and 30 % to cement and sand weight respectively were adopted. In this investigation, the densities of the foamed concrete used varied between 1400 and 1800 kg/m^3. The results found that the presence of proportion of WPSA had reduced the compressive strength of foamed concrete. It is revealed that the compressive strength of foamed concrete with addition of FRCA attained favourable strength at the optimum replacement of 5 and 15 % for density of 1400 and 1800 kg/m^3 respectively. It also showed that 1800 kg/m^3 concrete that contained a combined 20 % WPSA and 30 % FRCA recorded higher strength than that of control spec-

S.S. Sharipudin (✉) · A. Che Hassan
Faculty of Civil Engineering, Universiti Teknologi MARA, Masai, Johor, Malaysia
e-mail: sitishahidah85@yahoo.com

A. Che Hassan
e-mail: asmawati_chehasan@yahoo.com

A.R. Mohd Ridzuan
Faculty of Civil Engineering, Universiti Teknologi MARA, Shah Alam, Selangor, Malaysia
e-mail: ruslanridzuan@yahoo.co.uk

R.N.H. Raja Mohd Noor
Faculty of Civil Engineering, Universiti Teknologi MARA, Kepala Batas, Pulau Pinang, Malaysia
e-mail: rajanorhusna@ppinang.uitm.edu.my

N.A. Yacob et al. (eds.), *Regional Conference on Science, Technology and Social Sciences (RCSTSS 2014)*, DOI 10.1007/978-981-10-0534-3_1

imen. The results confirmed that the density and different replacement level of WPSA and FRCA have influenced in the strength of foamed concrete.

Keywords Compressive strength · Fine recycled concrete aggregate · Foamed concrete · Waste material · Waste paper sludge ash

1 Introduction

In recent years, the construction industry has shown significant interest in the use of lightweight concrete (LWC) particularly foamed concrete as a building material in construction application. The typical range of densities of foamed concrete is approximately 300–1850 kg/m^3 (Neville 2002). It has been successfully utilised since the ancient Roman times and it has gained its popularity due to its low density, superior thermal insulation, reduction of dead load, faster building rate and lower haulage cost (Bai et al. 2004). Nevertheless, it is stated that a lower density foamed concrete can achieve the strength similar to that of higher density foamed concrete by increasing cement content (Hamidah et al. 2005). It is known that cement is mainly used for the production of concrete that leads to the release of a significant amount of CO_2 and other greenhouse gas (GHGs). Therefore, this has challenged to look for sustainable solutions for future concrete construction. The solution of this issue is to use the supplementary cementitious material, such as silica fume, fly ash, rice husk ash, palm oil fuel ash and others to replace as much cement. However, the extensive use of industrial by-product in concrete production needs for the search a new material that comparable to those ashes.

Waste paper sludge ash (WPSA) from burning process of waste paper residue incinerator has been investigated in order to reduce the waste products resulted from an incineration process and turned this waste into useful building materials. According to Mozaffari et al. (2009), WPSA is possible reuse as a cement replacement in the production of concrete, since it can facilitate the hydration of hydraulic materials as well as undergo its own hydration. Previous attempts have also reported the use of WPSA as constituents in concrete in which this ash contained considerable amounts of alumino-siliceous material and could be used as a pozzolanic material (Bai et al. 2003; Chaipanich et al. 2005). However, utilisation of WPSA to replace part of cement is not well known in producing lightweight foamed concrete.

Concrete waste is another waste material which has been dumped by construction site until today. In Malaysia, it is estimated that concrete wastes recorded the highest amount of total recycled construction waste materials with 67.64 % (Begum et al. 2006). On another note, the reduction in available sources of natural aggregate is also affecting the production of concrete. Thus, recycling of concrete wastes as a source of aggregate is one of the few viable options to solve the increasing concrete waste that generated from the construction industry. Durmus et al. (2009) have proven that the coarse recycled concrete aggregate (CRCA) can be successfully used as a partial or full substitution of natural aggregates (NA) in the production of concrete. Although the consumption of RCA has become more popular, it is still fewer studies on the use of the smallest size fractions of the fine recycled concrete aggregate (FRCA) (i.e. <1.18 mm) in concrete product.

WPSA and FRCA have been found to be studied separately as a replacement for raw material to produce concrete. However, there is limited data available on the use of these two waste materials combined to produce lightweight foamed concrete. Thus, the strength properties of foamed concrete containing different level of WPSA and FRCA corresponding to cement and sand substitute are thoroughly studied.

2 Materials and Methods

Materials

Locally available ordinary Portland cement type I was used throughout this study in compliance with BS EN 197-1: 2000. The physical properties and chemical characterisation of OPC and WPSA are determined and summarised in Table 1. WPSA obtained from Malaysian Newsprint Industries Sdn Bhd (MNI), Pahang was used in the entire investigation. Silica sand with a fineness modulus and specific gravity of 2.46 and 2.67, respectively, was used in this study. FRCA was obtained from crushing waste concrete cubes in the laboratory. The crushed concrete was sieved into smallest size fraction of the FRCA (<1.18 mm). The fineness modulus and specific gravity of FRCA are 2.53 and 2.41, respectively. The particle size

Table 1 Physical properties and chemical composition of OPC and WPSA (Malaysian Newsprint Industries)

	OPC	WPSA
Physical properties		
Specific gravity	3.10	1.90
Fineness (% passing 90 μm sieve)	97.80	91.10
Chemical composition		
Silicon dioxide (SiO_2)	15.05	15.16
Aluminium oxide (Al_2O_3)	2.56	6.06
Ferric oxide (Fe_2O_3)	4.00	1.11
Titanium oxide (TiO_2)	0.12	0.45
Magnesium oxide (MgO)	1.27	2.00
Calcium oxide (CaO)	72.17	55.87
Sodium oxide (Na_2O)	0.08	0.19
Potasium oxide (K_2O)	0.41	0.34
Phosphorous oxide (P_2O_5)	0.06	0.48
Manganese oxide (MnO)	0.06	0.05
Sulphur trioxide (SO_3)	2.0	0.78
Loss of Ignition (LOI)	1.33	17.51

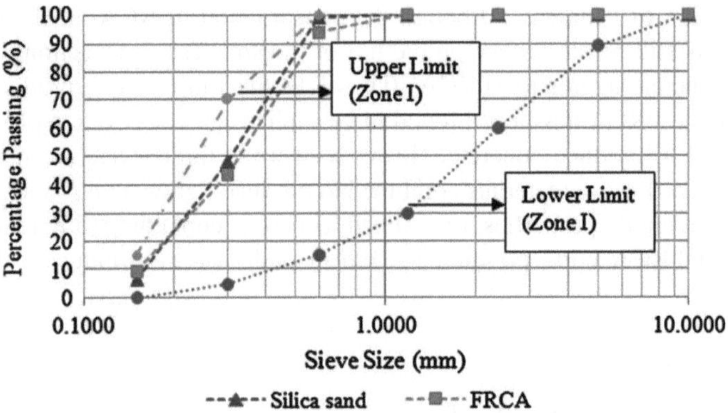

Fig. 1 Particle size distribution of silica sand and FRCA

distribution of silica sand and FRCA is illustrated in Fig. 1. The synthetic-based foaming agent was used to create air voids in the foamed concrete production.

Mix Proportions

In the present study, foamed concrete of 1400 and 1800 kg/m^3 were prepared. These densities are selected because the mix composition of paste determines the strength at the higher density in which the air voids are far apart to have an effect on the strength of foamed concrete (Ramamurthy et al. 2009). The mix proportions were prepared using a spreadsheet developed in-house to produce 1 m^3 of foamed concrete specimens. Otherwise, the sand-to-cement and water/binder ratios were set for all mixes as 1.5 and 0.6, respectively, which also can be calculated using that sheet. In this investigation, three sets of foamed concrete mixtures were prepared. For the first set, WPSA was employed as 0, 5, 10, 15, 20 and 30 % by weight replacements of the cement. In second set, FRCA was used to replace silica sand by masses 0, 5, 10, 15, 20 and 30 %. In addition, in final set of mixes, different percentages of WPSA were combined with various levels of FRCA to cement and sand content respectively. Foamed concrete mix consists of 100 % ordinary Portland cement (OPC) and 100 % sand was prepared for each density as control specimen.

Specimen Preparation

The preformed foam was prepared by mixing 900 ml of foaming agent with 17.1 l of water in a Portafoam generator. The base mix comprises of cement, WPSA, sand, FRCA, water and preformed foam was mixed in a drum mixer.

Testing of Specimens

In this study, cube specimens of 100 mm in size were prepared to determine the compressive strength of foamed concrete accordance to BS EN 12390-3: 2009.

3 Results and Discussion

Workability of Foamed Concrete

The flow table test for all foamed concrete mixes obtained in the present study was performed and the results are shown in Figs. 2, 3 and 4a, b respectively. As presented in Fig. 2, the flow of the WPSA foamed concrete mix with density of 1800 kg/m^3 exhibits a lesser spread value than that of the foamed concrete with a density of 1400 kg/m^3. This result can be explained that higher density of foamed concrete required higher amount of cement content, thereby resulting higher proportion of replacing WPSA in the mixes. It is obviously shown that the flow value decreases when WPSA was included in the foamed concrete mix. This is thought to be due to the increase in WPSA content from 5 to 30 % addition level. This finding is in line with research work by Mozaffari et al. (2006) who indicated that incorporation of WPSA into concrete mix experienced a relatively high water demand due to its porous nature. Similarly, the flow values reduced consistently with increasing of FRCA replacement that were represented in Fig. 3. The probable effect to the reduction in flowability is mainly attributed to the very high absorption

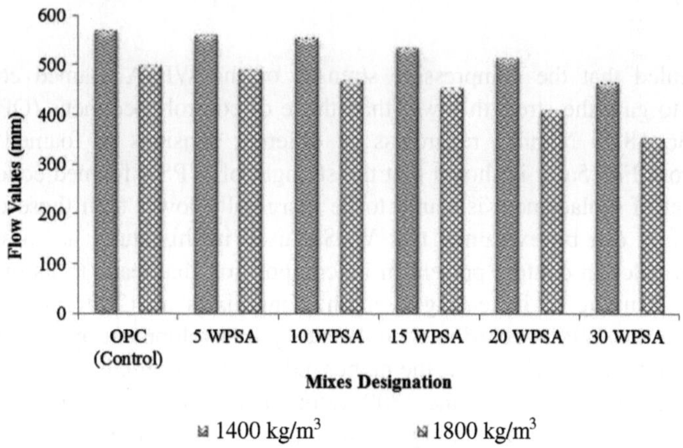

Fig. 2 The flow of spread (in mm) of WPSA foamed concrete for different density

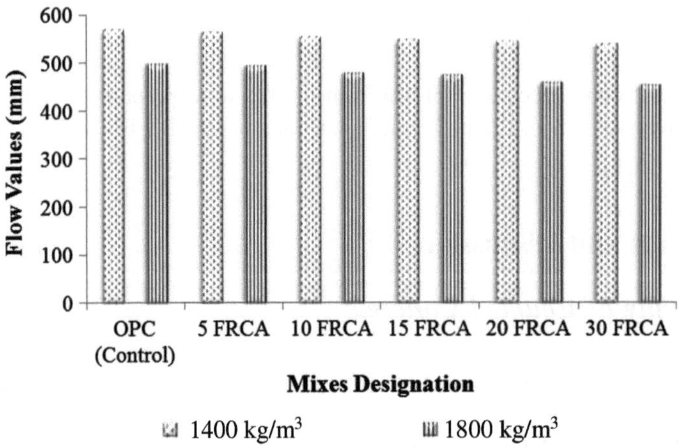

Fig. 3 The flow of spread (in mm) of FRCA foamed concrete for different density

capacity of FRCA and also its rough angular shape. As can be observed in Fig. 4a, b, the result portrays that the incorporation of WPSA and FRCA in concrete mixes provide flowability reduction as compared to plain foamed concrete. That result could be explained that the absorption of water by the WPSA and FRCA reduced the amount of water in the mixes. However, according to Neville (2002), a flow value of 400–650 mm was appropriate for all concrete mixes.

Influence of WPSA on Compressive Strength of Foamed Concrete

It is revealed that the compressive strength of the WPSA foamed concrete is observed to gain the strength lower than those of control specimens (OPC) in the range 2.36–18.60 N/mm^2, regardless of different densities of foamed concrete mixes. From Fig. 5a, it is shown that the strength of WPSA foamed concrete with 30 % level of replacement is found to be marginally lower than those of control specimens. It can be explained that WPSA used in this study has a significant amount of calcium content present in its composition that leads to its own hydration, thus resulting of increasing strength. Kinuthia et al. (2001) supported this explanation that WPSA contributes to the strength development by hydrating itself in the concrete mixes. However, the higher reduction in strength is noticed at 30 % WPSA replacement level for the 1800 kg/m^3 mixes as depicted in Fig. 5b. This result might be affected by the higher quantities of WPSA present in the concrete mixtures, as the lower cement content that disturbs the process of hydration within

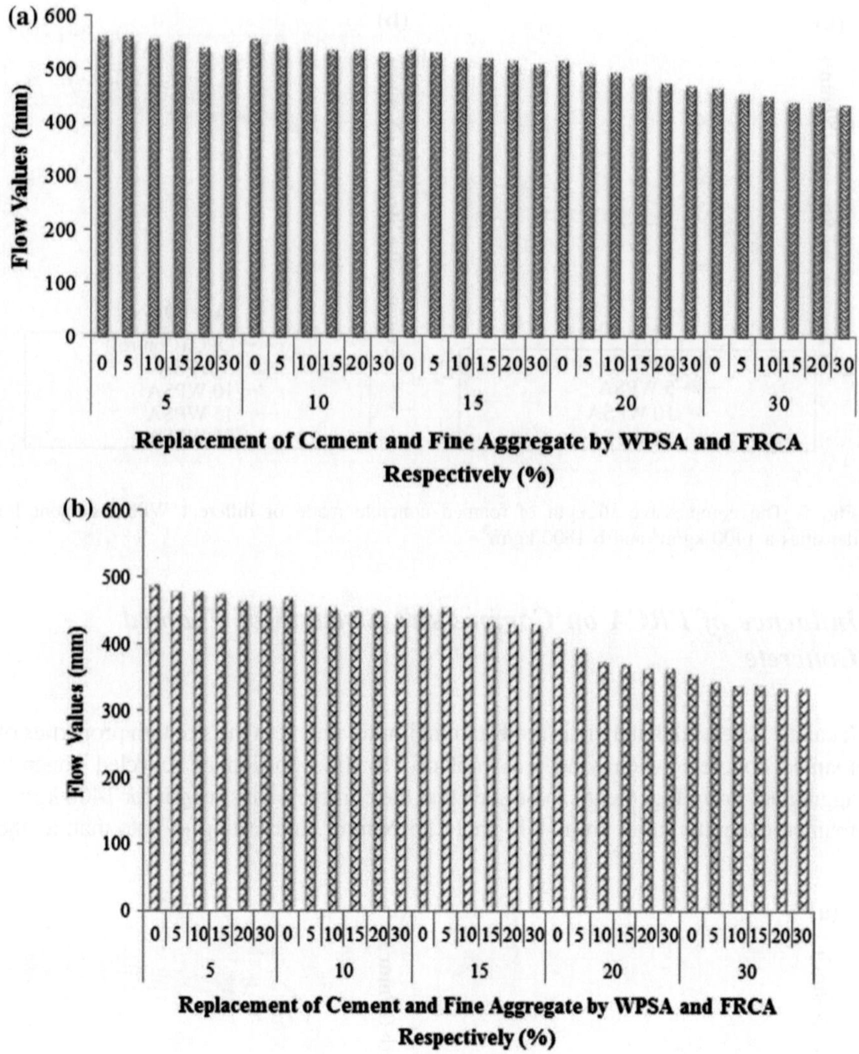

Fig. 4 The flow of spread (in mm) of combined WPSA and FRCA foamed concrete for different density **a** 1400 kg/m^3 and **b** 1800 kg/m^3

the foamed concrete matrix. Furthermore, WPSA derived from high carbon content which is indicated by loss of ignition (LOI) with 17.51 % compared to cement by the value of 1.33 %. Thus, the carbon content has absorbed water leading to reduction of water demand in the process of hydration which resulted in lower compressive strength.

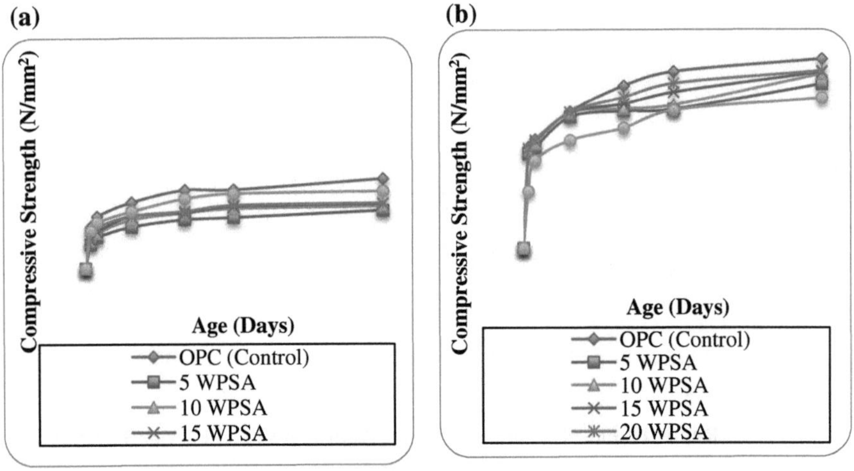

Fig. 5 The compressive strength of foamed concrete made of different WPSA content for densities **a** 1400 kg/m^3 and **b** 1800 kg/m^3

Influence of FRCA on Compressive Strength of Foamed Concrete

It can be observed that there is a substantial improvement in the strength properties of foamed concrete when sand was replaced by fine portion of recycled concrete aggregate. From Fig. 6a, it can be seen that the compressive strength for 1400 kg/m^3 foamed concrete varies from 4.13 to 11.25 N/mm^2. Interesting to note that, as the

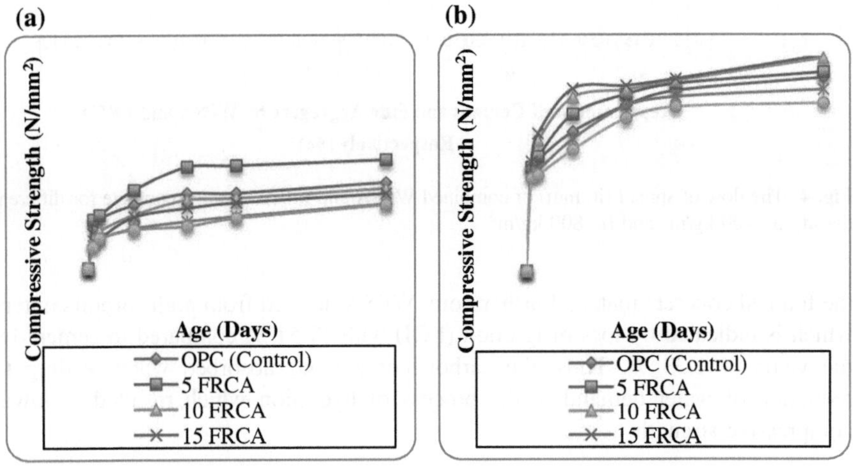

Fig. 6 The compressive strength of foamed concrete made of different FRCA content for densities **a** 1400 kg/m^3 and **b** 1800 kg/m^3

FRCA content was replaced at 5 % level, the compressive strength attained is higher than that of control specimens with the value of 5.05, 5.52, 8.12, 10.56, 10.69 and 11.25 N/mm^2 at 3, 7, 28, 60, 90 and 180 days, respectively. This result demonstrated that the inclusion of FRCA provided greater strength enhancement than mix without FRCA. However, the compressive strength decreased from 0.54 to 0.93 as compared to that of control when a higher replacement level namely 10–30 % was adopted. The trend depicted in Fig. 6b implies that foamed concrete is sensitive to the presence of FRCA with respect to curing time. It can be observed that the ratio of compressive strength at different replacement levels of FRCA marginally reduces as the density of the foamed concrete increases. According to Fig. 6b, the 1800 kg/m^3 foamed concrete exhibits the ratio of strength from 0.86 to 0.99 which is almost similar to control specimens as compared to the density of 1400 kg/m^3 concretes whose ratios are much lower with respect to similar series of FRCA employed. Based on the findings obtained, 15 % FRCA concrete mixes attained the highest compressive strength whose values are 11.21, 14.10, 18.70, 19.02, 19.71 and 22.12 N/mm^2 for 3, 7, 28, 60, 90 and 180 days, respectively. However, when further substitution level was over 20 %, the strength of 1800 kg/m^3 FRCA foamed concrete specimens tends to reduce marginally. It might be due to the presence of a large volume fraction of porosity governed by FRCA that affected water demand in producing the product of cement hydration. Besides, this situation is true according to another researcher that the replacement levels of FRCA were suggested to remain at or below 30 % of fine aggregates content (Obla et al. 2007).

Influence of Combined of WPSA and FRCA on Compressive Strength of Foamed Concrete

The result of the compressive strength of 1400 and 1800 kg/m^3 foamed concrete containing different levels of WPSA with 0, 5, 10, 15, 20 and 30 % incorporation of FRCA by 0, 5, 10, 15, 20 and 30 % to cement and sand content is presented in Figs. 7a–e and 8a–e, respectively. Obviously, it can be seen that the ratio in compressive strength to those of plain foamed concrete for WPSA and FRCA consistently decreases as replacement level of WPSA increases from 5 to 30 % with combination of FRCA. In this case, the strength of concrete mix composed of 30 % WPSA and 30 % FRCA is close and equivalent to those mixes without WPSA and FRCA addition for both densities. The graph depicts that the compressive strength seems to increase as the ratios of FRCA increased. These significant strength increases were due to the pozzolanic reactivity and presence of good interfacial bonding between paste and FRCA. This result was also confirmed by Berndt (2009) who claimed that the interfaces between old mortar and new paste remained intact to contribute to good interfacial bonding, thus resulting in higher strength. The mix incorporation of 10 % WPSA with the highest replacement level of 30 % FRCA as shown in Fig. 7b demonstrated the optimum compressive strength with the values

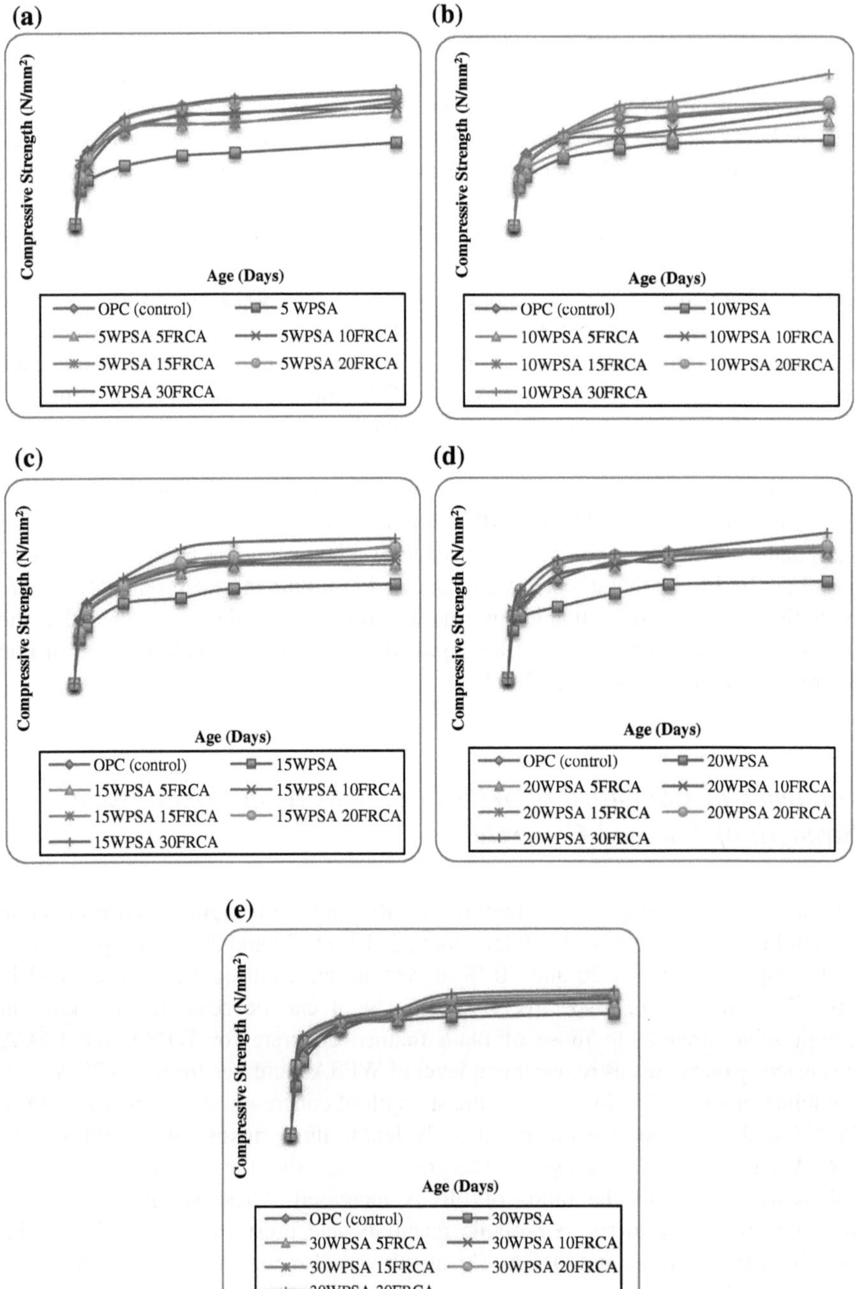

Fig. 7 **a–e** The compressive strength of 1400 kg/m^3 foamed concrete made of different WPSA and FRCA content

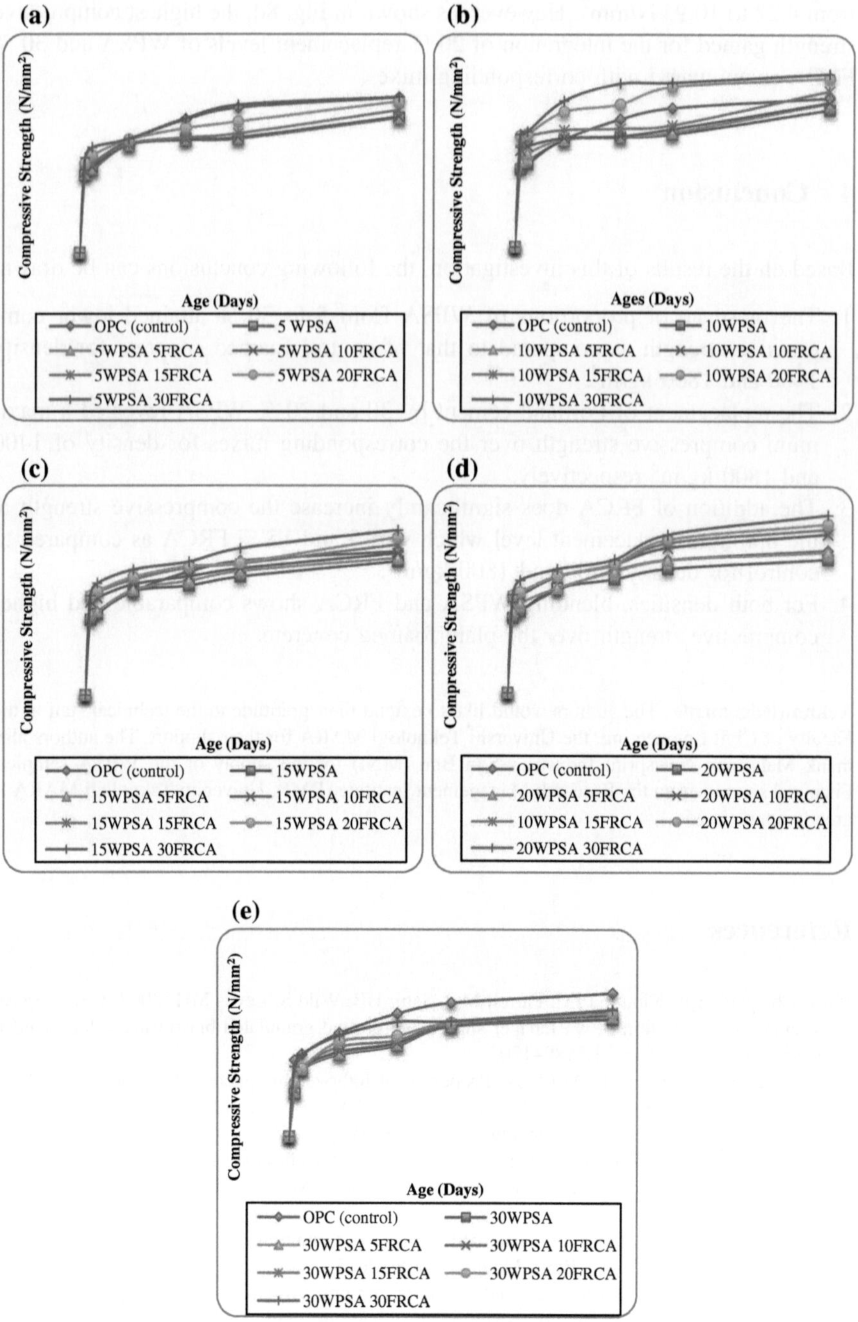

Fig. 8 **a–e** The compressive strength of 1800 kg/m^3 foamed concrete made of different WPSA and FRCA content

from 4.22 to 10.95 N/mm^2. However, as shown in Fig. 8d, the highest compressive strength gained for the integration of 20 % replacement levels of WPSA and 30 % FRCA as compared with corresponding mixes.

4 Conclusion

Based on the results of this investigation, the following conclusions can be drawn:

1. The presence of proportions of WPSA from 5 to 30 % attained lower compressive strength as compared to that of control foamed concrete for density 1400 and 1800 kg/m^3.
2. The replacement of Portland cement by 30 and 20 % WPSA revealed a maximum compressive strength over the corresponding mixes for density of 1400 and 1800 kg/m^3 respectively.
3. The addition of FRCA does significantly increase the compressive strength at the optimum replacement level which was 5 and 15 % FRCA as compared to control for density 1400 and 1800 kg/m^3.
4. For both densities, blending WPSA and FRCA shows comparable and higher compressive strength over the plain foamed concrete.

Acknowledgements The authors would like to extend their gratitude to the technical staff at the Faculty of Civil Engineering, the Universiti Teknologi MARA for their support. The authors also thank Malaysian Newsprint Industries Sdn Bhd (MNI) for the supply of the **WPSA** samples. Financial support from the Research Management Institute (RMI), Universiti Teknologi MARA is greatly appreciated.

References

Bai J, Chaipanich A, Kinuthia JM, Farrell MO, Sabir BB, Wild S, Lewis MH (2003) Compressive strength and hydration of wastepaper sludge ash-ground granulated blastfurnace slag blended paste. Cem Concr Res 33:1189–1202

Bai Y, Ibrahim R, Basheer PAM (2004) Properties of lightweight concrete manufactured with fly ash, furnace bottom ash, and lytag. In: Proceedings of the international workshop on sustainable development and concrete technology, Beijing, China

Begum RA, Siwar C, Pereira JJ, Jaafar AH (2006) A benefit-cost analysis on the economic feasibility of construction waste minimisation: the case of Malaysia. Resour Conserv Recy 86–98

Berndt ML (2009) Properties of sustainable concrete containing fly ash, slag, and recycled concrete aggregate. Constr Build Mater 23:2606–2613

BS EN 12390-3 (2009) Testing hardened concrete: compressive strength of test specimens

BS EN 197-1 (2000) Cement composition, specifications, and conformity criteria: common cements

Chaipanich A, Bai J, Farrell MO, Kinuthia JM, Sabir BB, Wild S (2005) Setting time and heat of hydration of wastepaper sludge ash-ground granulated blastfurnace slag blended pastes. In: 6th

international congress on global construction: ultimate concrete opportunities, cement combinations for durable concrete, University of Dundee
Durmus G, Simsek O, Dayi M (2009) The effect of coarse recycled concrete aggregate on concrete properties. J Fac Eng Archit Gazi Univ 24:183–189
Hamidah MS, Azmi I, Ruslan MRA, Kartini K (2005) Optimisation of foamed concrete mix of different sand-cement ratio and curing condition. In: Proceedings of the international conference on the use of foamed concrete in construction University of Dundee, Scotland, UK
Kinuthia JM, Farrell MO, Sabir BB, Wild S (2001) A preliminary study of the cementitious properties of wastepaper sludge ash—ground granulated blastfurnace slag (wsa:ggbs) blends. In: Proceedings of the international symposium, Recovery and Recycling of Paper
Mozaffari E, Kinuthia JM, Bai J, Wild S (2009) An investigation into the strength development of wastepaper sludge ash blended with ground granulated blastfurnace slag. Cem Concr Res 39:942–949
Mozaffari E, O'Farrell M, Kinuthia JM, Wild S (2006) Impoving strength development of wastepaper sludge ash by wet-milling. Cement Concr Compos 28:144–152
Neville AM (2002) Properties of concrete. Pearson Education Limited, London
Obla K, Kim H, Lobo C (2007) Crushed returned concrete as aggregate for new concrete. RMC Research and Education Foundation
Ramamurthy K, Nambiar EKK, Ranjani GIS (2009) A classification of studies on properties of foamed concrete. Cement Concr Compos 31:388–396

Chapter 2
Investigation of Cutting Edge Radius Effect in Macro-machining and Micro-machining

Juri Saedon, Noor Aniza Norrdin, Mohd Azman Yahaya, Mohd Shahir Kasim and NorHafiez Mohamad Nor

Abstract Chip formation is a dynamic process that is often nonlinear in nature. A chip may not form when the depth of cut is less than a minimum chip thickness. This paper presents an investigation of cutting edge radius effect in macro- and micro-machining of AISI D2 steel via simulation using ABAQUS software. Through the arbitrary Lagrangian–Eulerian FE modeling approach, the chip growth, chip formation, and cutting force were investigated under three criteria such as $a/r < 1$, $a/r > 1$, and $a/r = 1$. The results from this simulation can provide useful information for choosing reasonable cutting edge to improve surface integrity and prolong cutting tool life in macro- and micro-milling operation. It is found that the chip is formed at $a/r > 1$ while material extrusion performed under $a/r < 1$. The investigation on the cutting force found that value of a/r ratios greatly affects the cutting force. The cutting mechanism in micro-milling is similar to macro-milling due to the process undergoes both ploughing and shearing mechanism.

Keywords Chip formation · FEA · Cutting edge radius · Minimum chip thickness

---

J. Saedon · N.A. Norrdin (✉) · M.A. Yahaya · N. Mohamad Nor
Faculty of Mechanical Engineering, Universiti Teknologi MARA,
Shah Alam, Selangor, Malaysia
e-mail: nr.aniza@yahoo.com

J. Saedon
e-mail: juri41@yahoo.com

M.A. Yahaya
e-mail: yahayama@yahoo.com

N. Mohamad Nor
e-mail: norhafiez@yahoo.com

M.S. Kasim
Faculty of Manufacturing Engineering, Universiti Teknikal Malaysia,
Durian Tunggal, Malaysia
e-mail: shahir@utem.edu.my

N.A. Yacob et al. (eds.), *Regional Conference on Science, Technology and Social Sciences (RCSTSS 2014)*, DOI 10.1007/978-981-10-0534-3_2

1 Introduction

AISI D2 hardened tool steel is a high chromium, high carbon, tool and die steel with hardness in the range 54–62 HRC used for cold working operations. It has a high strength, very high resistance to cracking, and high resistance to softening and wear. Most recently, there are large amounts of work on hardened tool steel material toward the implementation of replication technology for mass production of miniaturization parts and components. But machining of hardened tool steel materials in manufacturing of moulds for replication technologies are mostly challenging. This is due to unpredictable tool life, premature of tool failure and differences in process mechanisms compared to macro-milling. Therefore, a good finite element model (FEM) simulation of machining AISI D2 would be very useful to optimize the machining process and led to reduce its costs, machining time saving, improved the quality and quantity.

In order to present the main differences in chip formation due to the scaling-down from macro- to micro-milling, a short review of the current state of the art in chip formation and minimum chip thickness in micro-milling is carried out. Then the FEM developed is presented and the results will be comparing to those found in the literature. The chip formation mechanism was studied quantitatively via FEM approach.

2 Comparison Between Micro-milling and Macro-milling

The effect of cutting edge radius was the significant contrast in cutting process between macro- and micro-machining methods. In macro-cutting, the cutting edge radius of carbides tools might be considered to be sharp as the uncut chip thickness is significantly larger than the cutting edge radius, see Fig. 1a. However, in micro-cutting, the uncut chip thickness is frequently smaller than the cutting edge

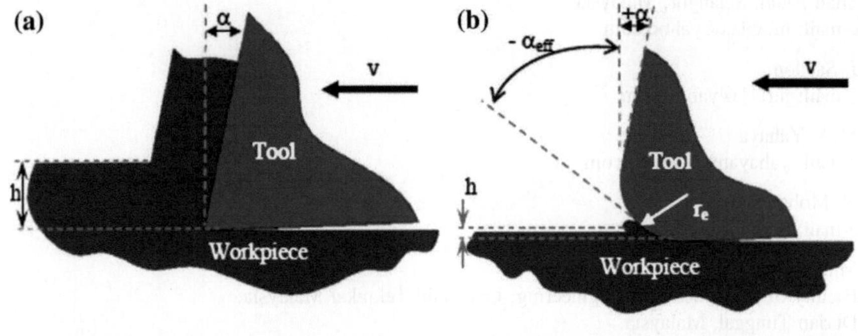

Fig. 1 Effect of cutting edge radius to chip thickness in macro-cutting (**a**) and micro-cutting (**b**) (Saedon et al. 2013)

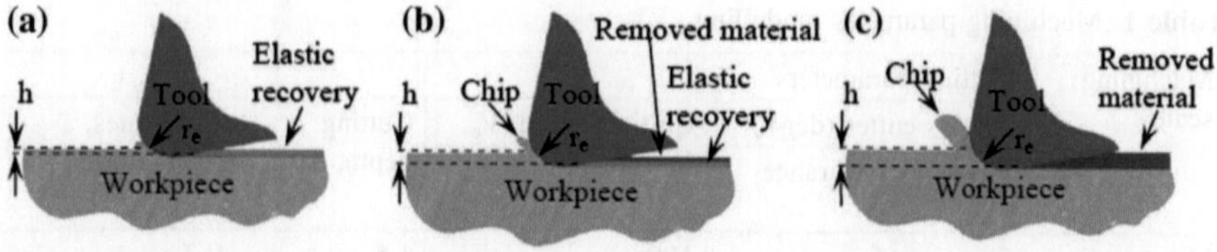

Fig. 2 Chip formation with respect to chip thickness in micro machining based on Saedon et al. (2013). **a** $h < h_{min}$, **b** $h \cong h_{min}$, **c** $h > h_{min}$

radius and the chip is structures in the zone of the cutting edge radius, see Fig. 1b. Thus, a profoundly negative rake angle exists and the apparent relative bluntness of the tool increases the specific cutting forces (Saedon 2011).

The minimum chip thickness h_{min} is characterized as the minimum undeformed thickness of the chip removed from a work surface at a cutting edge. In macro-machining, the depth of cut is generally greater than the cutting tool edge radius, and it is assumed that the cutting tools totally remove the surface of the workpiece and generate the chips. It is different in micro-machining where, the small depth of cut compared to the edge radius of the tool causes a large negative rake edge. The consequence of this is a process, which might be dominated more by rubbing and compression instead of cutting. This phenomenon causes a rough surface roughness and elastic recovery of the workpiece, increasing the possibilities tool damages to happen. Three different cases happen in micro-chip formation, as shown in Fig. 2.

At the point when the uncut chip thickness, *a*, is less than a critical minimum chip thickness, h_{min}, elastic deformation happens and the cutter does not remove any chip as shown in Fig. 2a. When the uncut chip thickness comes near to the minimum chip thickness, h_{min}, chips are form by shearing of the workpiece, with some elastic deformation even at this time still happening as shown in Fig. 2b. However when the uncut chip thickness increases further than the minimum chip thickness, h_{min}, the elastic deformation become lessen and chip is removed as shown in Fig. 2c. The minimum chip thickness phenomena can contribute to the increase of the cutting force due to the rising of slipping forces and ploughing of the machined surface.

3 Methodology—Finite Element Modelling (FEM)

In this study, ABAQUS/EXPLICIT software was used in macro-milling and micro-milling of AISI D2 tool steel. A two dimensional explicit dynamic algorithm coupled with Arbitrary Lagrangian–Eulerian (ALE) as a solution method was developed. The workpiece and cutter material for the both macro- and micro-scale model is AISI D2 tool steel and tungsten carbide, respectively. The workpiece is modelled as a rectangular block with a dimension of 30 mm long 15 mm thick for

Table 1 Machining parameter modelling

Machining scale	Cutting parameters				
	Angles cutter (deg)		Cutting speed, V_c (m/min)	Cutting depth, d	Feed rates, f (mm/tooth)
	Rake γ	Clearance α			
Macro scale	5	5	100	1.5	0.2
Micro scale	5	5	100	3	0.2

macro-scale model and while for micro-scale model is 50 μm long and 15 μm thick. The machining parameters: cutting speed, V_c (m/min), cutting depth, d, and feed rates, f, (mm/tooth) are shown in Table 1.

The cutting tool geometry is treated as perfectly rigid solids while workpiece is assumed to be deformable. The simulation was carried out with positive 5° rake angle and 5° clearance angle at different cutting tool edge radius, $r = 0.5, 1.5, 2.5$, and 5.7 mm for macro scale and $r = 1, 3, 5$ and 8 μm for micro scale. Various a/r ratios have been considered in order to study the influence of the cutting tool edge radius on the chip formation process. Four different cases have been simulated from $a/r = 3$, $a/r = 1$, $a/r = 0.6$ and $a/r = 0.2$. The workpiece is assumed homogeneous and plastic flow behaviour around the cutting edge was governed by Johnson–Cook material model of Eq. (1).

$$\sigma = (A + B\,\varepsilon_\rho^n)\,(1 + C\ln(\acute{\varepsilon}/\acute{\varepsilon}_o))\,(1 - (T - T_\gamma/T_m - T_\gamma)^m) \tag{1}$$

where σ is flow stress, ε_ρ and ε are strain and strain rate, ε_o is the reference stain rate $(1/s)$ and n, m, A, B and C are constant parameters for Johnson–Cook material model as shown in Table 2.

Table 2 FEM properties

Workpiece properties (AISI D2)	
Yield strength (Mpa)	1776
Hardening modulus (Mpa)	904
Strain rate sensitivity	0.012
Thermal softening coefficient	3.38
Hardening coefficient	0.312
Reference strain rate	1
Melting temperature (°)	1460
Density (kg m^{-3})	7750
Elastic modulus (Gpa)	200
Poisson's ratio	0.3
Specific heat (J kg^{-1} K^{-1})	485
Thermal expansion(10^{-6} °C)	10.4
Thermal conductivity (W/m K)	21

Total mesh generated for the micro scale model was 4791 of elements, while 4955 nodes have been set as the region of the chip formation analysis. It is different to the macro scale model where there was 5213 of elements and 5061 nodes. Figure 3 shows meshed and the predefined boundary conditions where the reference point was set at the cutting tool tip. The bottom and left side of the workpiece are fixed in vertical and horizontal direction. The displacement of tool for *y*-axis and rotation of tool for both *x*-axis and *y*-axis are fixed but the tool is permitted to move horizontally in *x*-axis at a cutting speed of 100 m/min.

The study can estimate the chip growth, chip formation, and cutting force at several tool radiuses, *r* while the cutting speed, V_c (m/min), feed rates, *f* (mm/tooth) and depth of cut, *d* were held constant. The values were estimated by using FEM as shown in Table 3.

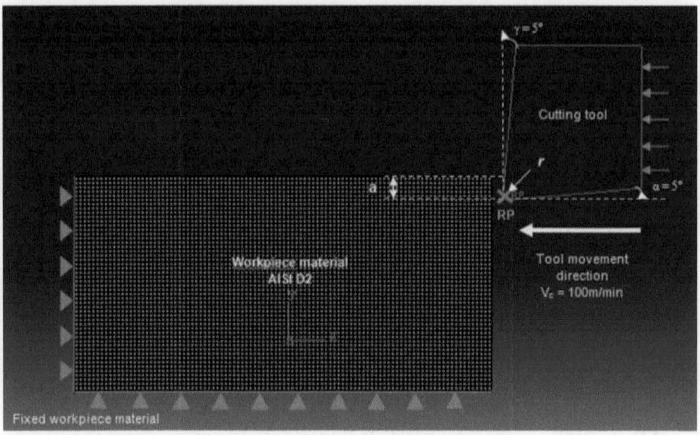

Fig. 3 FEM model with predefined boundary condition

Table 3 Machining process performed at several edge radiuses

Machining scale	Simulation numbers	Edge radiuses, *r* (mm, µm)	Cutting parameters		
			Cutting speed, V_c (m/min)	Cutting depth, *d* (mm)	Feed rates, *f* (mm/tooth)
Macro scale	S1	0.5	100	1.5	0.2
	S2	1.5			
	S3	2.5			
	S4	5.7			
Micro scale	S5	1	100	3	0.2
	S6	3			
	S7	5			
	S8	8			

4 Results and Discussion

Chip Formation

Figures 4, 5, 6, and 7 shows chip formation and material deformation for various *a*/*r* ratios values. A chip certainly formed when *a*/*r* = 3 and 1, however it is not for *a*/*r* = 0.6 and 0.2 for both macro- and micro-scale model. Therefore, it is possible to give judgment on the existence of a chip. The presence of a saw-tooth chip when *a*/*r* = 3, as observed in Fig. 4. This kind of chip is not found any more for smaller simulated *a*/*r* values as shown in Figs. 6 and 7.

A primary shear zone is able to be seen on Von Mises stress contours when *a*/*r* = 3, as in micro- and macro-cutting with a sharp tool. Chip forms easily within this area. As the material reaches its yield point, chips will forms from the parent

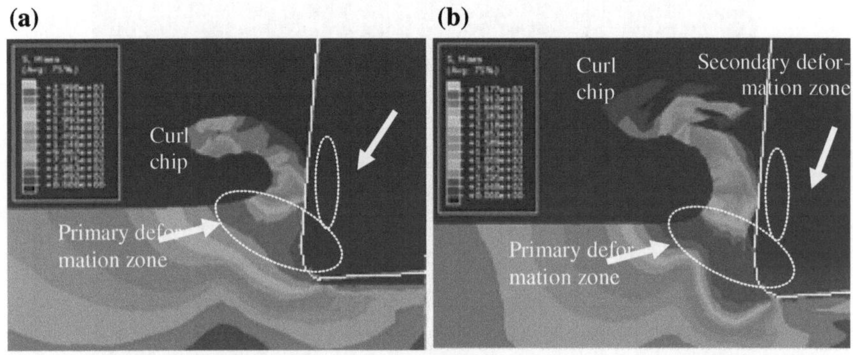

Fig. 4 Von Mises stress contours (10^3 Pa) during chip formation **a** macro-cutting and **b** micro-cutting at *a*/*r* = 3

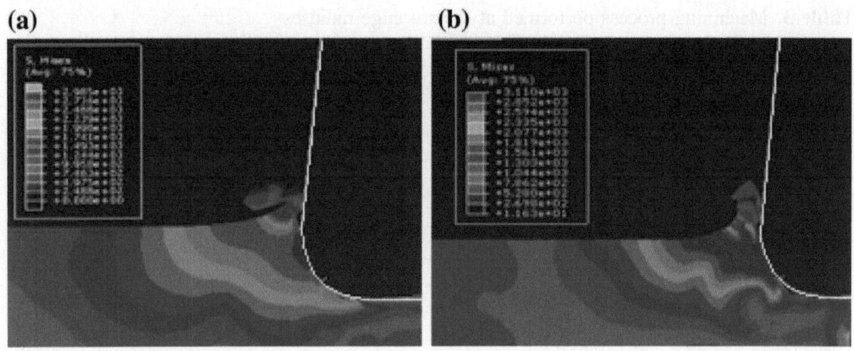

Fig. 5 Von Mises stress contours (10^3 Pa) during chip formation **a** macro-cutting and **b** micro-cutting at *a*/*r* = 1

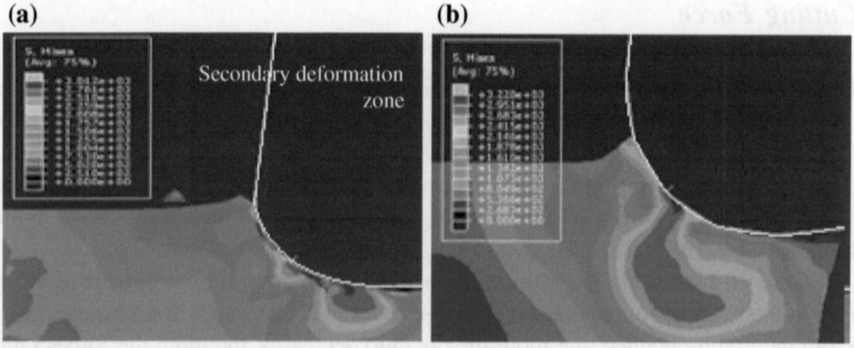

Fig. 6 Von Mises stress contours (10^3 Pa) during chip formation **a** macro-cutting and **b** micro-cutting at *a*/*r* = 0.6

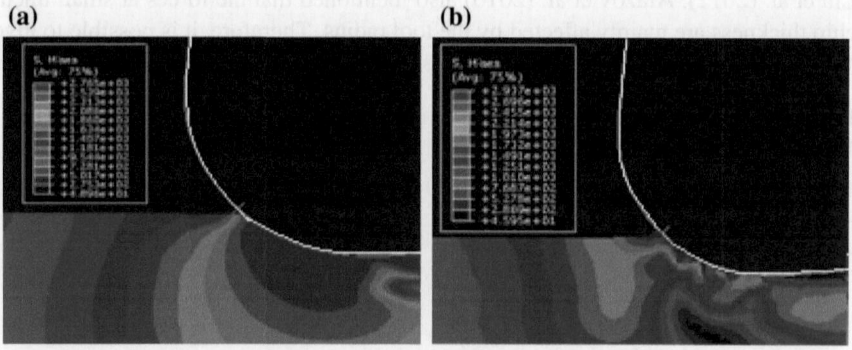

Fig. 7 Von Mises stress contours (10^3 Pa) during chip formation **a** macro-cutting and **b** micro-cutting at *a*/*r* = 0.3

material and slides along the primary shear plane, pushing material ahead of the cutting tool as in Fig. 4. A secondary shear zone occurs along the face of the cutting tool. In this zone, the friction raises temperature when the chip slides up the face of the cutting tool.

The more the *a*/*r* ratio decreases, the more the primary shear zone fades and it cannot be distinguish any longer from the *a*/*r* = 0.2 value. When *a*/*r* < 1, a primary shear zone can still be seen but the stress value seems to be smaller, also there are no secondary deformation zone found due to the reason that no chip is formed.

The results for micro-cutting are similar to those ALE model presented by Woon et al. (2008). The cutting tool edge radius and also the *a*/*r* value has great influence on the chip formation in micro-cutting. When looking to Figs. 4, 5, 6, and 7 the cutting mechanism in micro-milling is similar to macro-milling due to the reason that process undergoes both ploughing and shearing mechanism.

Cutting Force

Figures 8 and 9 show the cutting and feed forces for a 1.5 mm and 3 μm depth on cut. The cutting force and feed directions are determined at the reference point (RP). The cutting force is greater than the feed force in micro-cutting, which also observed in macro-cutting. The fluctuation evolution can be seen, as expected due to the saw-toothed chip where the slipping plane occurs and the up and down in the forces is observed.

Figure 10 represents the ratio between feed and cutting forces for the eight *a*/*r* simulated ratios, four ratios for macro-cutting and four ratios for micro-cutting. From the graph obtained, the more the *a*/*r* ratio decreases, the more the forces ratio (feed/cutting force) increases. This is due to the feed force becomes greater than the cutting force when the *a*/*r* ratio value becomes smaller. It is similar to both macro- and micro-cutting. These observations are similar to those made experimentally by Liu et al. (2012). Afazov et al. (2010) also mentioned that the forces at small uncut chip thickness are mainly affected by the tool radius. Therefore, it is possible to give

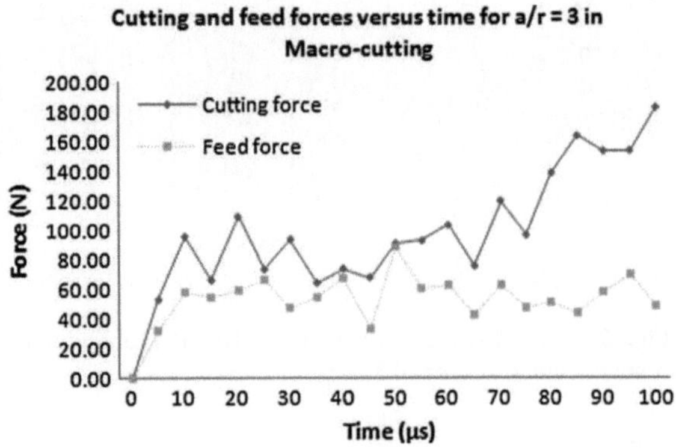

Fig. 8 Forces during macro-cutting when *a*/*r* = 3

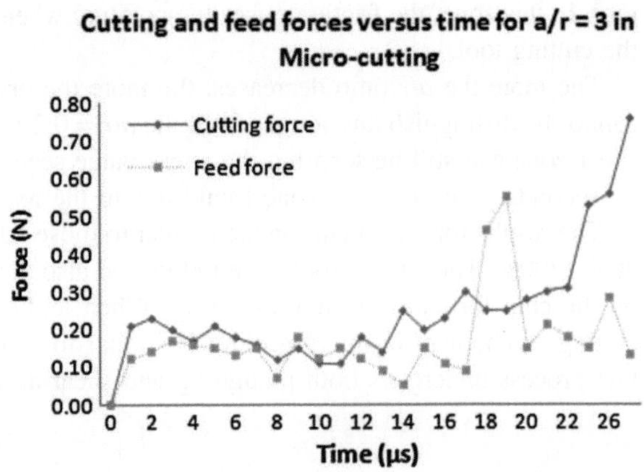

Fig. 9 Forces during micro-cutting when *a*/*r* = 3

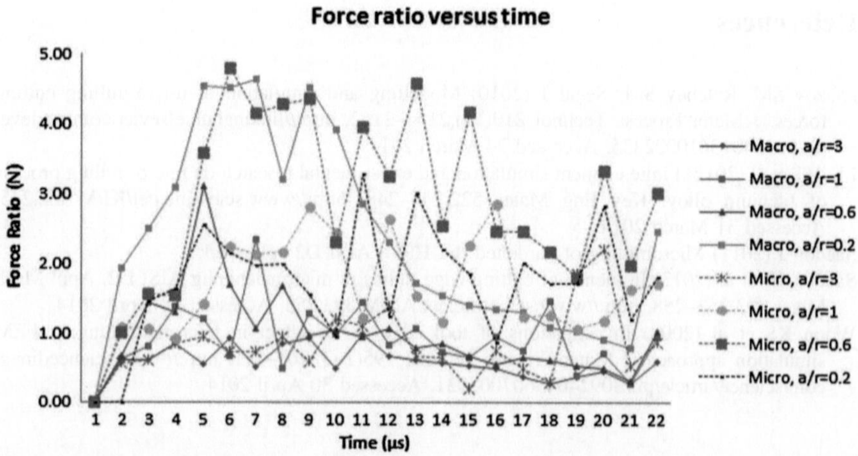

Fig. 10 Forces ratio during macro- and micro-cutting various *a*/*r*

judgment on the smaller value of *a*/*r* ratio creates a negative rake angle between the cutting tool and the workpiece which affects the chip formation and the cutting force, respectively.

5 Conclusion

In this work, the chip growth, chip formation and cutting force in the macro- and micro-machining process was investigated. The chip formation mechanism was studied quantitatively via FEM approach. Some important results were found as follows:

1. The transition from macro- to micro-cutting leads to some difference in chip morphologies when compared to orthogonal cutting. From the results, it is showed that the cutting mechanism in micro-milling is similar to macro-milling due to the process undergoes both ploughing and shearing mechanism. The macro- and micro cutting model showed that the chip formed perfectly at $a/r > 1$ and $a/r = 1$ while material extrusion performed under $a/r < 1$ ratio.
2. There are two zones for plastic deformation such as primary shear zone and secondary shear zone. From the observation, the primary shear zone appears to be most predominant and there are no secondary shear zone was found under $a/r < 1$ ratio due to no chip formed.
3. The smaller value of *a*/*r* ratio creates a negative rake angle between the cutting tool and the workpiece which affects the chip formation and the cutting force, respectively.

References

Afazov SM, Ratchev SM, Segal J (2010) Modelling and simulation of micro-milling cutting forces. J Mater Process Technol 210(15):2154–2162. http://linkinghub.elsevier.com/retrieve/pii/S0924013610002335. Accessed 24 March 2014

Liu HT et al (2012) Finite element simulation and experimental research on micro-milling process of titanium alloy. Key Eng Mater 522:245–248. http://www.scientific.net/KEM.522.245. Accessed 31 March 2014

Saedon J (2011) Micromilling of hardened (62 HRC) AISI D2 cold work

Saedon JB et al (2013) Influence of cutting edge radius in micromachining AISI D2. Appl Mech Mater 393:253–258. http://www.scientific.net/AMM.393.253. Accessed 28 April 2014

Woon KS et al (2008) Investigations of tool edge radius effect in micromachining: a FEM simulation approach. J Mater Process Technol 195(1–3):204–211. http://www.sciencedirect.com/science/article/pii/S0924013607005031. Accessed 30 April 2014

Chapter 3
Clay Stabilization Using OPC and Bottom Ash as Additives

Juhaizad Ahmad, Kamaruzzaman Mohamed, Abdul Samad Abdul Rahman, Mohd Ikmal Fazlan Rosli and Assrul Reedza Zulkifli

Abstract Clay can be found in many parts of Malaysia. It is very susceptible to settlement due to its low shear strength. Therefore, stabilization is needed especially for a wide construction area. Nowadays, the use of waste materials as construction material is crucial because this practise will help to protect the environment. One of the waste material that has low economical value and less explored by researches is bottom ash (BA). This material is produced from the electric power plant. This material has pozzolanic properties which is able to substitute cement in ground improvement. The objective of this study is to determine the unconfined compressive strength of clay after stabilization with the ordinary portland cement (OPC) and BA. The disturbed soil sample was taken from Pulau Indah, Klang, Malaysia. The basic and engineering properties tests were conducted on the clay samples. Moreover, 18 samples were prepared in cylinder form for unconfined compression test (UCT). The samples were then cured for 0, 7 and 14 days in order to see the strength development over time because the pozzolanic reaction is affected by time. Therefore, the strength of the soil mix will increase with the increase in cement content and curing time. Hence, this waste material can be used for ground improvement work in order to protect the environment.

Keywords Bottom ash · Cement replacement · Clay stabilization · Ground improvement

---

J. Ahmad (✉) · K. Mohamed · A.S. Abdul Rahman · M.I.F. Rosli · A.R. Zulkifli
Faculty of Civil Engineering, Universiti Teknologi MARA,
Shah Alam, Selangor, Malaysia
e-mail: juhaizad@salam.uitm.edu.my

K. Mohamed
e-mail: kzmohd@yahoo.com

A.S. Abdul Rahman
e-mail: kempass@yahoo.com

M.I.F. Rosli
e-mail: ikmal601@salam.uitm.edu.my

A.R. Zulkifli
e-mail: reedza03@gmail.com

N.A. Yacob et al. (eds.), *Regional Conference on Science, Technology and Social Sciences (RCSTSS 2014)*, DOI 10.1007/978-981-10-0534-3_3

1 Introduction

Clay is a natural material composed primarily of fine-grained minerals. Clay consists of tiny particles that show the properties of plastic and adhesion. Clay posses the small voids and pores so it is capable to retain water, so in this condition clay tends to expand and shrink which can lead to settlement. Clay can also be a problematic soil. When exposed to increment of water, it tends to be soft and becomes liquid. Soft clay has often caused difficulties in construction with its low strength and low stiffness. This has caused serious problems in geotechnical engineering because low strength soil leads to damage in building's foundation and cracks in road pavement.

When looking at the rapid growth development of infrastructures' facilities in Malaysia, it is impossible to avoid construction on clay. Clay consists of 20 % of the total soils in Peninsular Malaysia (Ahmad et al. 2011). It can be found on west and east coast of Peninsular Malaysia. This type of soil is normally classified as marine clay. This clay is originated from the floods in the ancient time. The sedimentation of seabed was very thick and can be up to 60 m in depth. That is about a 20 storey building's height. Hence, the construction over clay may experience bearing capacity failure and excessive settlement. Stabilization of soil using the cementitious material becomes optional to solve this problem.

Cementitious materials are several binding materials that may mix with water to form a plastic paste. OPC is used as a common cementitious binding agent. From the previous study, stabilization of soil using cement is one of the soil treatments that applied to improve soil plasticity and workability. Therefore, this research focuses on determination of the strength that can be produced by using BA as a part of the additive mixture. This will decrease the use of OPC on the stabilization of clay. By doing this more economical soil mix can be produced.

2 Literature Review

BA from electric power plant is estimated to increase in the developing country such as Malaysia. Above all, the limited area of landfill makes this situation worst. In the Peninsular Malaysia, there are four famous electric power plants which are located at Perak, Johor, Selangor and Negeri Sembilan, which uses coal as a material to generate electricity. One private sector is building a 2,100 MW capacity coal-fire power plant in Johor and another private sector group is building a 1,700 MW capacity plant in Negeri Sembilan. TNB's electric power plant in Perak, which is using coal, began its operations in September 2002 with a capacity of 2100 MW (3 × 700 MW). This coal was taken directly from Sarawak. TNB has also been importing high quality coal from Indonesia, Australia, the United States, Canada and China. TNB uses 1.5 million tons of coal each year. To control fly ash and dust hovering in the atmosphere recipient's electrostatics were built and used for trapping 99 % of ash and dust (TNB 2013).

Coal BA are physically coarse, porous, glassy, granular, greyish and incombustible materials that are collected from the bottom of furnaces that burn coal. The type of BA produced depends on the type of furnace and also the sources of coal. From the burning process of coal, 80 % of product becomes fly ash and the remaining 20 % of product is BA.

BA is collected at the bottom of the combustion chamber in a water-filled hopper which is then removed by means of high-pressure water jets and conveyed by sluiceways to a decanting basin for dewatering followed by stockpiling and possibly crushing. Chemical composition of the BA was determined by Bai et al. (2010) and are tabulated in Table 1. In general, BA was composed of 54.8 and 28.5 % of silica and alumina, respectively. These two constituents can be found in OPC. The chemical properties of BA are almost the same as OPC. Hence, it can function as cement in concrete mix.

BA has the capability to promote the pozzolanic reaction due to high percentages of Si and Al contents (Filiponi et al. 2002). From the research done by Cheriaf et al. (1999), the high amount of amorphous and highly reactive silica in BA also exhibit the pozzolanic reaction.

Generally, the strength of BA and the cement mixture can be defined by its activity index. According to Cheriaf et al. (1999), the activity ndex will increase with the time steadily. The previous study showed that in 7 days of curing the activity index increased to 0.72 while in 90 days of curing the activity index will achieve 0.97. This improvement is given by a chemical reaction that happened between BA and OPC. This will be a good additive for soil in order to increase the strength.

Table 1 Chemical properties of bottom ash and OPC (Bai et al. 2010)

Chemical composition	Bottom ash weight (%)	Cement OPC weight (%)
Silica, SiO_2	54.8	20.6
Alumina, Al_2O_3	28.5	5.7
Iron oxide, Fe_2O_3	8.49	2.9
Titania, TiO_2	2.71	–
Magnesia, MgO	0.35	1.8
Calcium Oxide, CaO	4.2	63.6
Na_2O	0.08	0.12
K_2O	0.45	0.75
P_2O_5	0.28	–
SO_3	–	3.2
Cl	–	0.01
Loss in ignition 1000 °C	2.46	1.5

3 Methodology

The main materials used in this research are clay, BA and OPC. Clay was collected from Pulau Indah, Klang and BA was collected from the coal power plant located near the Port Dickson. OPC is a normal cement that is available easily.

Clay was taken as disturbed sample by hand dug using shovel at the depth of 4 m. The clay was stored in plastic bag to preserve the natural moisture content. The clay was oven-dried at the temperature of 105 °C for 24 h to determine its moisture content. Particle size distribution for clay was conducted using sieving test in accordance of BS 1377-1:1990 (BSI 1990).

Portland cement is defined as hydraulic cement that hardens when it reacts with water and also forms a water resistant product. Besides, water is an important substance for any work that is used by concrete. If water is combining with a cementitious material, it will form a cement paste by the process of hydration. For this research, a good source for water is necessary in order to make sure that no other substance might affect the hydration of cement. Thus, tap water was used as it is free from any particles and contaminants.

The BA was extracted at the bottom of combustion chamber by using high-pressure water jets. It was then transported to decanting basin for the process of dewatering. Later it was stored and it was crushed if the size of the particles was not easily transported out from the store. Before conducting the characteristic and engineering tests, the BA was dried for 24 h in oven with a temperature of 105 °C to determine its moisture content.

After it is oven-dried for 24 h, the BA was sieved and the BA sample that retained on 425 μm sieve was collected. Later, the mixture was produced by the predefined proportions of OPC, BA and clay as tabulated in Table 2. The mixture was calculated from the total weight of clay. 1000 g of clay is needed for each sample, so the weight of the admixtures, which are OPC and BA, was calculated from the weight of clay. The clay sample used for this study was also oven-dried for 24 h prior to mixing. The OPC was not oven-dried since it is a mixture of various chemical constituents which are sensitive to temperature changes. It is not advisable to oven-dry OPC because it will disturb the hydration process of cement.

The mixing process of the clay and additives was conducted in a stainless steel tray after which the sample was poured in the cylindrical mould. This cylinder mould was the same as the one that is used in standard proctor test. The mixture was compacted in the standard proctor test equipment. In order to form the size of 100 mm diameter sample the tube sampler was pushed into the soil. The soil in the tube sampler was then extracted and the soil was cut into three pieces. Each of the soil samples were stored in the plastic container which was capable to preserve the moisture content during the curing time. The curing time was 0, 7 and 14 days. Each sample was tested according to the specified curing time.

Table 2 Sample name with percentage of admixtures

Sample no.	Sample name	Curing period (Day)	OPC (%)	BA (%)
CO	COA	0	0	0
	COB	7		
	COC	14		
C1	C1A	0	20	0
	C1B	7		
	C1C	14		
C2	C2A	0	18	2
	C2B	7		
	C2C	14		
C3	C3A	0	16	4
	C3B	7		
	C3C	14		
C4	C4A	0	12	8
	C4B	7		
	C4C	14		
C5	C5A	0	4	16
	C5B	7		
	C5C	14		

4 Experimental Results and Discussion

Physical properties and classification tests were conducted in order to identify the clay soil. These properties were moisture content, specific gravity, liquid limit and particle size distribution. It was observed that the colour of the soil was dark grey. It was very cohesive and could be remoulded by hand. This soil showed the high plasticity behaviour. The summary of the soil physical properties is shown in Table 3. The results from Kobayashi et al. (1990) were used as a reference.

The physical properties of the BA were also determined to identify its properties. It was found that the moisture content of BA was 15.71 % while the specific gravity was 2.14. This is almost the same as reported by Cheriaf et al. (1999) which were 15.8 and 2.0–2.25 for moisture content and specific gravity, respectively.

UCT was conducted to determine the unconfined compressive strength of the samples. The UCT was selected because this test is easy to handle and gives a faster result than the other strength tests. 18 samples were tested using UCT machine. The size of the sample is 50 mm diameter and 100 mm height.

The parameters such as deviator stress and axial strain were obtained from UCT results. These two values were plotted and it gave the strength relationship over time. There are two main reasons that can effect the strength of these samples. The first one is the proportion of OPC and BA as the additives, and the second factor is

Table 3 The soil physical properties tests

Properties	Results from this research	Range value Kobayashi et al. (1990)
Moisture content (%)	49.35	40–125
Size distribution (%)	36.0 clay	25–28 clay
	43.77 silt	15–60 silt
	20.23 sand	0–45 sand
Soil description	Clayey silty	–
Liquid limit	49.66	50–130
Plastic limit	29.92	20–70
Specific gravity	2.48	2.45–2.7
Max dry density (mg/m^3)	2.1	–
Optimum moisture content	24	–

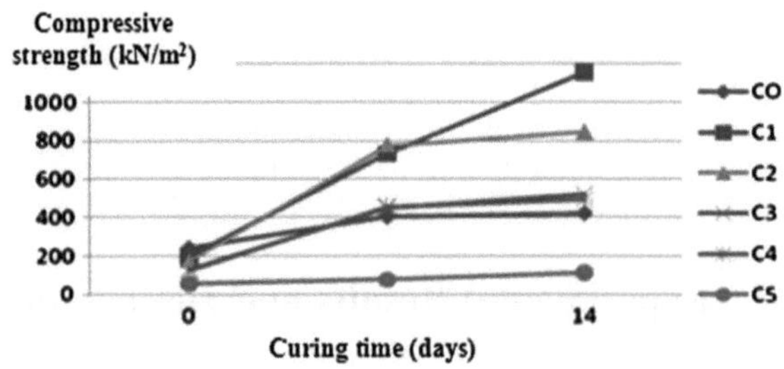

Fig. 1 The strength development over time

the curing period, whereby, longer curing period will cause the sample to attain higher strength as shown in Fig. 1.

It was observed that the optimum percentage of BA needed to stabilize the clay with the additive of cement is 2 % from the weight of the clay while the cement content is 18 %. The behaviour of clay with BA and cement stabilization is affected greatly by the percentage of cement. The greater the percentage of cement, the higher strength can be achieved. However, the function of BA as cement reducer is still significant. It can be seen that the strength of clay is almost the same for sample C1 (20 % cement alone) and C2 (18 % cement + 2 % BA) after 7 days, but the strength was changed dramatically with the increase in curing time.

5 Conclusion

In conclusion, this study has shown that portland cement consumption can be reduced by using BA as pozzolanic agent. This can lead to eco-friendly construction by using this waste as construction material. The finding from this research can be beneficial to engineers. This has created a new knowledge on the suitability of the BA as a substance in ground improvement work. This research can help engineers to fully utilize the waste products that are disposed from the power plants.

Furthermore with the increasing demand of cement and high cost of construction, the cost can be decreased with the supplementary cementitious material like BA. Hazardous material originated from BA may contain toxic elements. By reusing it as an additive it can minimize this problem. The result obtained from this research shows that BA has also the ability to be additive in the stabilization of clay, thus protecting the environment from hazards.

The chemical properties of OPC and BA show similar oxide composition with high amorphous silica that leads an ability to ignite the pozzolanic reaction. This is a good way to protect the environment and create a sustainable development.

This research can be extended by studying the chemical reaction that happened during pozzolanic reaction. It can be done by studying the images from X-ray diffraction analysis (XRD). Above all, the pozzolanic reaction of BA is expected to increase by using finer particles (use of nanoparticles). This is due to the increase in the specific surface of the particles that might accelerate the process.

Acknowledgements The authors would like to express their sincere gratitude to Research Management Institute (RMI, UiTM) and the Faculty of Civil Engineering, UiTM Shah Alam for providing equipments and financial support for this research. It was funded under Excellent Fund UiTM 600-RMI/ST/DANA 5/3/Dst (14/2009). They would like to thank undergraduate students Hafiz bin Mohamad Muzaffar, Khif Farhan bin Abd Rahman and Masnor Hashim bin Mohamad for their valuable contributions in this study.

References

Ahmad J, Md.-Zain NH, Ashaari Y, Abdul-Rahman AS (2011) Lateral movement and settlement of sandwiched soft soil using physical model. In: 2011 IEEE colloquium on humanities, science and engineering research (CHUSER 2011), 5–6 Dec 2011, Penang, p 71

Bai Y, Ibrahim R, Basheer M (2010) Properties of lightweight concrete manufactured with fly ash, furnace bottom ash, and lytag. In: International workshop on sustainable development and concrete technology. Accessed 24 April 2013

BSI (1990) BS 1377-1:1990—methods of test for soils for civil engineering purposes. General requirements and sample preparation. British Standards Institution, United Kingdom

Cheriaf M, Cavalante J, Rocha J, Pera J (1999) Pozzolanic properties of pulverized coal combustion bottom ash. Cem Concr Res 29:1389–1390

Filiponi P, Poletini A, Pomi R, Sirini P (2002) Physical and mechanical properties of cement based product containing incineration bottom ash. Waste Manag 23:145–156

Kobayashi Y, Todo H, Weerasinghe WAY, Chandra P (1990) Comparison of coastal clay found in Singapore, Malaysia and Indonesia. In: 10th Southeast Asian geotechnical conference, 16–20 April 1990, Taipei, Taiwan

Tenaga Nasional Berhad (2013) The coal-fired power plant in Malaysia. http://www.tnb.com.my. Accessed 24 August 2013

Chapter 4
On the Use of Spectral Feature Fusions for Enhanced Performance of Malaysian English Accents Classification

Mohd Ali Yusnita, Murugesa Pandiyan Paulraj, Sazali Yaacob, Abu Bakar Shahriman, Rihana Yusuf and Shahilah Nordin

Abstract Accent problem is a current issue that degrades the intelligibility and performance of speech recognition (ASR) systems. Despite English accents have been extensively researched in the United States, Britain, Australia, China, India, and Singapore, the study of Malaysian English (MalE) is still at infancy. There is till date, very limited evidence to corroborate how ethnically diverse accents in MalE of its three main ethnics can be identified from their speech signals. Most studies about MalE tackles issues from the view point of attitudinal studies and making use of human perceptual analysis. Instead, this paper presents experimental methods by means of acoustical analysis and machine learning techniques. In order to enhance the performance of accent classifier to classify the Malay, Chinese, and Indian accents this paper proposes fusion techniques of popularly known mel-frequency cepstral coefficients (MFCC) and linear prediction coefficients (LPC) with formants termed here as spectral feature fusions (SFFs). In these SFFs feature extractors, the main spectral features are fused with five usable formants and the extracted features are used to model K-nearest neighbors and artificial neural networks (ANN). Using independent test samples technique, gender-dependent accent classifiers were evaluated. Experimental results showed that the proposed SFFs surpassed the baseline features by 7.8 and 3.9 % increment of the classification rates for the LPC-formants and MFCC-formants fusions, respectively. The highest accuracies yielded for the fusion of MFCC and formants were 96.4 and

M.A. Yusnita (✉) · R. Yusuf · S. Nordin
Faculty of Electrical Engineering, Universiti Teknologi MARA, Kepala Batas, Pulau Pinang, Malaysia
e-mail: yusnita082@ppinang.uitm.edu.my

M.P. Paulraj · A.B. Shahriman
School of Mechatronic Engineering, Universiti Malaysia Perlis, Arau, Perlis, Malaysia
e-mail: paul@unimap.edu.my

S. Yaacob
Universiti Kuala Lumpur Malaysian Spanish Institute, Kuala Lumpur, Kedah, Malaysia
e-mail: sazali.yaacob@unikl.edu.my

N.A. Yacob et al. (eds.), *Regional Conference on Science, Technology and Social Sciences (RCSTSS 2014)*, DOI 10.1007/978-981-10-0534-3_4

92.5 % on the male and female datasets. Speaking of LPC-formants fusion, the results were also promising, i.e., 92.6 and 88.8 % on the male and female datasets, respectively.

Keywords Accent classification · Formants · Linear prediction coefficients · Mel-frequency cepstral coefficients · Malaysian english

1 Introduction

There is a large volume of published studies on automatic classification of native and/or foreign accents in American English (Arslan 1996; Sangwan and Hansen 2012), British English (Hanani et al. 2011), and Australian English (Nguyen et al. 2010). Although it is evident that the use of accent-specific database improves the recognition rate of automatic speech recognition (ASR) system, the success rates vary among languages and accent types. Malaysian English (MalE) is one form of non-native English varieties that is now known to the world but its uniformity as the national spoken language is being debated by many scholars as it is also influenced by the local languages at different degree due to implication of localized ethnic speech diversity (Gill 1993; Lowenberg 1992; Nair-Venugopal 2000; Pillai et al. 2010). At present, there is a lack of research to corroborate how ethnically diverse accents in MalE of its three main ethnics can be identified from speech signals rather than investigation through perceptual and attitudinal studies. In the ASR market product nowadays, the conventional way is to treat MalE as a uniform English variety (Basem and Tan 2011), but apparently this notion is debatable (Phoon 2010).

The aim of this paper is to present an experimental methods by means of acoustical analysis and machine learning techniques to identify a speaker accent that is caused by ethnicity for MalE speech. Previously, many good techniques have been proposed to extract speech features for accent classification such as mel-frequency cepstral coefficients (MFCC) and perceptual linear prediction (Arslan and Hansen 1996; Diederich and Pedersen 2008; Dupont et al. 2005; Fohr and Illina 2007; Humphries et al. 1996; Picone 1993; Pitton et al. 1996; Rabiee and Setayeshi 2010), linear prediction coefficients (LPC) and formant frequencies (formants) analysis (Deshpande et al. 2005; Ghesquiere and Compernolle 2002; Ghorshi et al. 2008; Teixeira et al. 1996; Vieru et al. 2011). Formants were proven very useful to mark accent traits, however, using solely formants in the past study (Arslan and Hansen 1996, 1997) was not good enough to gain high accuracy due to its dynamic range of intra-group speakers. Researchers started to use formants as additive features so as to improve the accuracy. Yusnita et al. (2011) found that, using formants alone resulted in poor results of accuracy between 45 and 60 % only. Following this, Yusnita et al. (2013a) found that the first five formants played an important role to identify accents of MalE regardless of gender. In this paper, to enhance the performance of automatic accent classification (AAC), spectral feature fusions (SFFs) of formants with LPC and likewise, formants with MFCC with minimal orders of the baseline features are proposed that can no longer be improved by merely increasing the order. Two classification algorithms, namely, K-nearest neighbors (KNN) and artificial neural networks (ANN) are proposed as the modeling methods.

The remaining sections have the following content. Section 2 elaborates about speech database and speech features extractors for AAC of MalE accented speech. The findings are discussed in Sect. 3. Finally, Sect. 4 concludes with the important findings and investigation of this paper.

2 Methodology

Speech Database and Experimental Setup

For the purpose of this research, a new MalE accents database was developed using experimental setup as shown in Fig. 1. Several recording sessions were conducted to elicit speech from 51 male speakers and 52 female speakers of three main ethnics i.e. the Malays, Chinese, and Indians. The tasks consisted of two modes of speaking, namely, isolated words (IWs) and sentences (STs) using reading-speaking style. The source for speech materials was taken from (Yusnita et al. 2013a, b). The data collection consisted of 9265 samples of IWs-speech and 5172 samples of STs-speech recorded from 103 students of Universiti Malaysia Perlis (UniMAP).

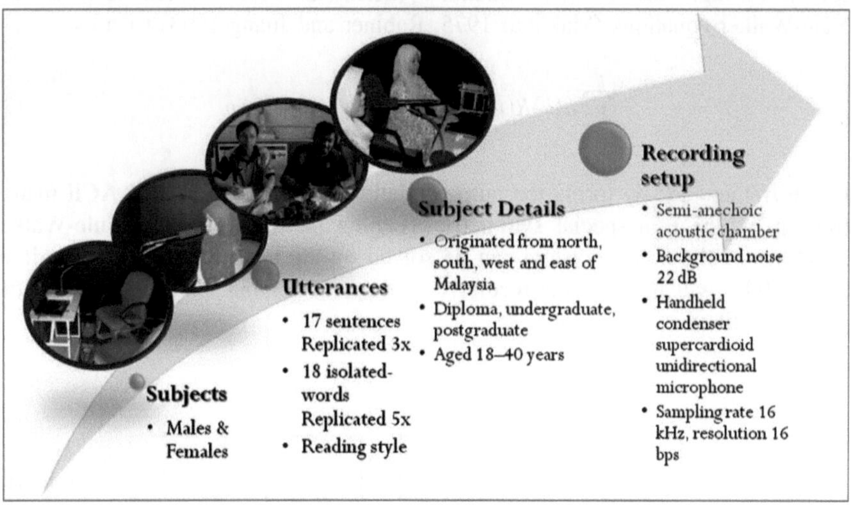

Fig. 1 Speech recording setup and activities

Spectral Feature Fusions as the Proposed Feature Extractors

A block diagram showing the steps involve in extracting MFCC is portrayed in Fig. 2. The working principle of MFCC processor is based on a set of filter banks constructed from several bandpass filters, in a form of triangular-shaped window functions (Davis and Mermelstein 1980). For better human perception this frequency range is warped using, a new scale known as Mel-scale, the known variation of the human ear's critical bandwidths with frequency (Picone 1993). The cepstral coefficients of the Mel-scale filter banks (Chew et al. 2011) can be computed as in Eq. (1) by summing all the product of Fourier transform derived log-energy output of individual bandpass filter and discrete cosine transform (DCT).

$$C_m = \sum_{k=1}^{N} E_k \cos[m(k - 0.5)\pi/N] \tag{1}$$

where variables C_m and E_k represent the mth cepstral coefficient (cepstrum) and the kth log-energy, respectively. N is the number of filters in the filter banks and the number of cepstrum takes in this order, i.e., $m = 1, 2, 3, \ldots, M$.

While MFCC signal processor mimics human auditory system, the principle of LPC extractor is based on human vocal tract characteristics of speech production system (Furui 2001; Makhoul 1975). The spectrum of speech signal is represented as a pth order autoregressive system which is a special case of all-pole infinite impulse response (IIR) filter. This IIR filter coefficients $a(k)$ can be solved using the least squares method as a result of the minimization of the mean or total squared error with respect to each of the coefficients. The least squares problem can be solved through the normal equations (Makhoul 1975) which lead to the Yule-Walker equations (Makhoul 1975; Rabiner and Juang 1993) such as

$$\sum_{k=1}^{p} a(k)R(i-k) = -r(i), \quad 1 \le i \le p \tag{2}$$

where $r(i)$ and $R(i\text{–}k)$ forms the autocorrelation function (ACF) and ACF matrix (p-by-p) which is a special symmetric Toeplitz matrix. Then, the Yule-Walker equations of the p-by-p system can be solved by the Levinson–Durbin algorithm (Garg 2003) which is a direct recursive method for solving the coefficients of the prediction filter.

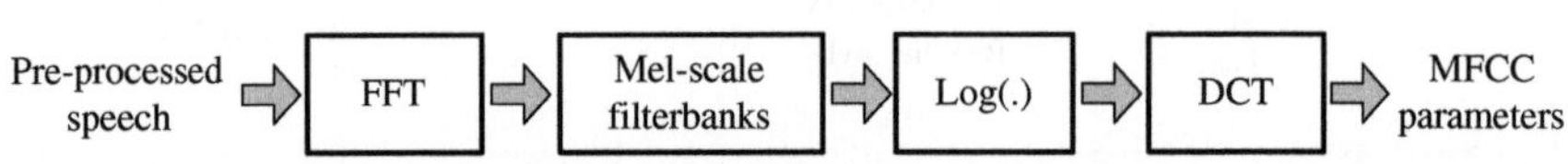

Fig. 2 Block diagram of MFCC feature extraction

Another important accent features which can be directly derived from the spectrum of LPC are the formants using root-solving technique (Vallabha and Tuller 2002). They are good representative of accent discrimination (Deshpande et al. 2005; Liu and Fung 1999; Rabiee and Setayeshi 2010), because the position of each formant is shifted according to the tongue movement, thus indicating a speaker's accent. Figure 3 shows the block diagram for extracting LPC and formant features.

In this work, in order to enhance the formulation of speech features for accent classification, SFFs of MFCC-formants and LPC-formants are proposed as in Fig. 4. The speech signal is initially preprocessed such as normalization, preemphasis using first order FIR filter of emphasis coefficient of 0.9375, frame blocking into 32 ms short segments and 50 % overlapping and windowing using Hamming function prior to feature extraction. Since MFCC and LPC parameters are of different measurements from formants, the joint SFFs features require normalization before modeling task to avoid underrating smaller valued features.

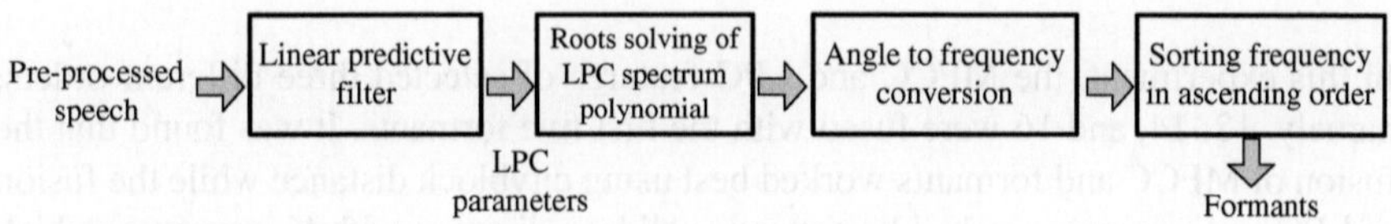

Fig. 3 Block diagram of LPC and formant frequencies extraction

Fig. 4 The process flow for SFFs feature extraction in automatic accent classification (AAC) system

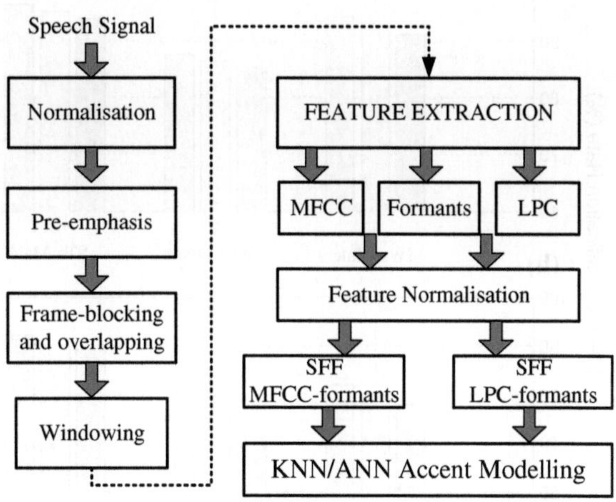

3 Results and Discussion

This section aims to investigate the ability of KNN and ANN classification models to classify accent accurately based on SFFs of two mixtures, i.e., MFCC-formant and LPC-formant extracted from two modes of speech, namely, IWs and STs of male and female speakers. The formulation of the models using these two different learning algorithms for AAC entail eight different accent models for the IWs and STs modes respectively. The results, particularly mean classification rates (CRs) are reported using independent test samples of 60–40 ratio of training and testing datasets for models using KNN. For ANN models, the standard practice is to divide the whole data into 60 % training dataset, 15 % validation dataset and the remaining 25 % is assigned as testing dataset (Fahlman 1988; Looney 1997). To average out any sources of randomness in the learning methods, ten runs per experiment were conducted for the same architecture in case of any parameter change.

Spectral Feature Fusions Using K-Nearest Neighbors

In this experiment, the MFCC and LPC features of selected three different orders, namely, 12, 14, and 16 were fused with the first five formants. It was found that the fusion of MFCC and formants worked best using cityblock distance while the fusion of LPC and formants worked best using euclidean distance with K-parameter which is set to 2. Figure 5 shows the average performance of these SFFs feature sets across

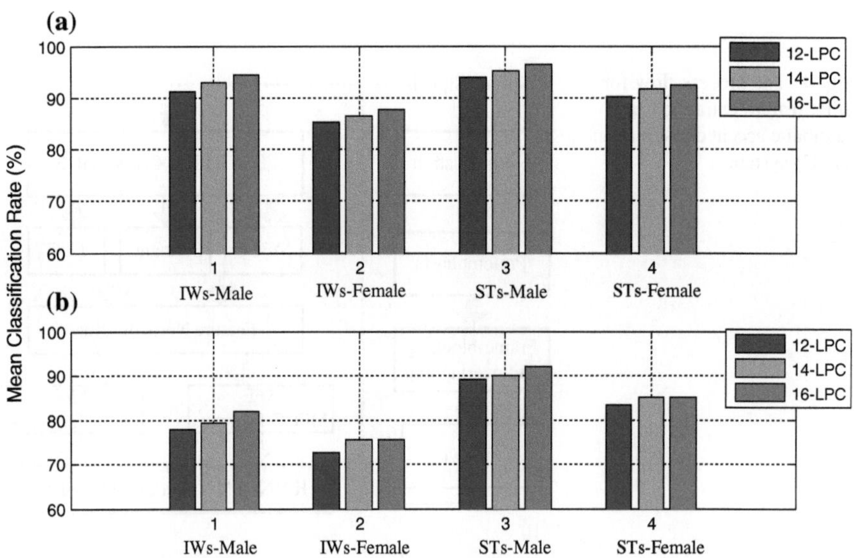

Fig. 5 Performance comparison of AAC based on KNN models using SFFs of **a** MFCC-formants and **b** LPC-formants for different orders of the baseline features

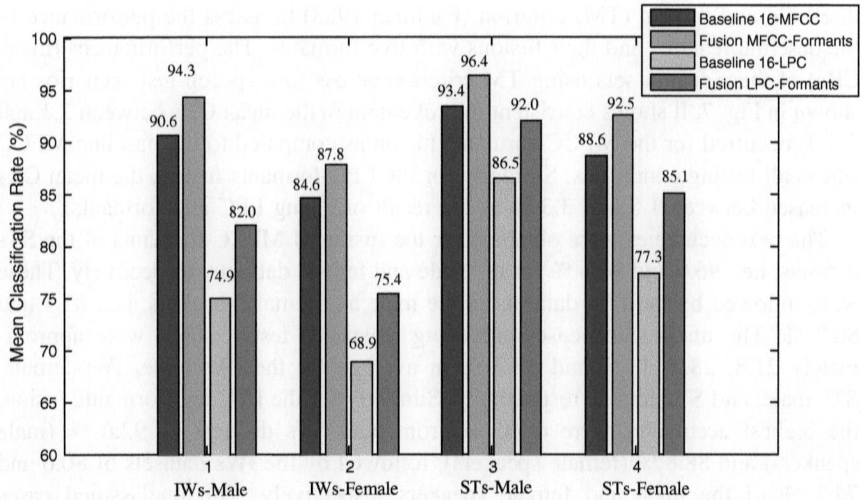

Fig. 6 Performance comparison of AAC based on KNN models between the baseline features of MFCC and LPC and the SFFs features

different orders. Referring to the results, it could be noticed that the performance increased in accordance with additional number of the baseline features of MFCC and LPC. The highest accuracies yielded for the fusion of MFCC-formants occurred for both male datasets, i.e., 96.4 % (STs-speech) and 94.3 % (IWs-speech) followed by the female datasets of 92.5 % (STs-speech), and 87.8 % (IWs-speech). Speaking of the fusion of LPC-formants, the highest accuracies were recorded from both STs datasets, i.e., 92.0 % (male speakers) and 85.1 % (female speakers), followed by the IWs datasets of 82.0 and 75.4 % of the male and female speakers respectively.

In order to compare how good the performance of SFFs features to that of their baseline features, Fig. 6 shows the comparison between different baseline features, i.e., MFCC and LPC of the 16th order together with their respective fusions with five formants (F_1 to F_5) for all speech test scenarios. It is worth noting from the results that by fusing the baseline features with five formants, a great improvement in the CRs have been achieved. Speaking of the CRs improvement generally the fusion of LPC-formants benefited more than that of the fusion of MFCC-formants. The bar chart shows that additional CRs between 5.5 and 7.8 % occurred for the LPC-formants fusion as compared to the MFCC-formants fusion, i.e., from 3.01 to 3.9 % with respect to their baseline performances.

Spectral Feature Fusions Using Artificial Neural Networks

The testing procedures for these SFFs of MFCC-formants and LPC-formants for the 16th order were repeated using feedforward multilayer ANN by employing threshold and margin (TM) criterion (Fahlman 1988) to assess the performance of the baseline features and their fusions with five formants. The performances (mean CRs) of these fusion sets using TM criterion across four speech test scenarios are shown in Fig. 7. It shows consistent improvement in the mean CRs between 2.1 and 2.9 % occurred for the MFCC-formants fusion as compared to the baseline MFCC across all testing conditions. Similarly for the LPC-formants fusion, the mean CRs increased between 1.8 and 3.3 % as the result of fusing LPC and formants.

The best accuracies were obtained for the fusion of MFCC-formants of the STs datasets, i.e., 96.0 and 90.6 % for the male and female datasets, respectively. These were followed by the IWs datasets of the male and female speakers, i.e., 87.1 and 80.2 %. The unclassified cases according to speech test scenario were approximately 21.8, 28.9, 11.5, and 18.2 % in average for the IWs-male, IWs-female, STs-male, and STs-female respectively. Similarly for the LPC and formants fusion, the highest accuracies were obtained from both STs datasets of 92.6 % (male speakers) and 88.8 % (female speakers), followed by the IWs datasets of 80.0 and 74.3 % of the male and female speakers respectively. The unclassified cases according to speech test scenario were approximately 28.9, 34.0, 13.2, and 18.1 % in average for the IWs-male, IWs-female, STs-male and STs-female respectively. Despite better performance had been achieved using fusion techniques, features extracted from the IWs-female dataset using LPC were still poor indicated by the accuracy of detection of less than 80 %.

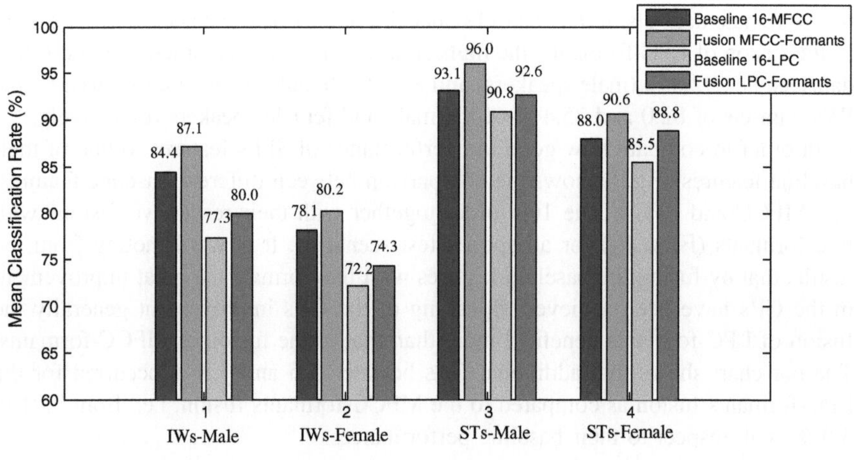

Fig. 7 Performance comparison of AAC based on ANN models between the baseline features of MFCC and LPC and the SFFs features

4 Conclusion

This paper has considered investigation of SFFs features of MFCC-formants and LPC-formants for enhanced performance of AAC for MalE accents database. The uniformity of MalE as spoken by Malaysians is debatable, thus requires experimental methods to identify accents of three majority ethnics namely, the Malay, Chinese, and Indian. The experimental results based on KNN and ANN classification models gained very promising success on the use of the proposed SFFs feature extractors. The resulted additional mean CRs of 5.5–7.8 % were yielded for the LPC-formants fusion while that of 3.01–3.9 % for the MFCC-formants fusion as compared to their respective baseline performances. The highest accuracies yielded for the fusion of MFCC-formants were 96.4 and 92.5 % for the male and female datasets respectively. The other SFF extractor, LPC-formants yielded the highest accuracies of 92.6 and 88.8 % of the male and female datasets respectively.

References

Arslan LM (1996). Foreign accent classification in American English. Ph.D, Duke University

Arslan LM, Hansen JHL (1996) Language accent classification in American english. Speech Commun 18(4):353–367

Arslan LM, Hansen JHL (1997) Frequency characteristics of foreign accented speech. Paper presented at the 1997 IEEE international conference on acoustics. Speech Signal Process, 21–24 April 1997

Basem HAA, Tan TP (2011) Non-native accent pronunciation modeling in automatic speech recognition. Paper presented at the 2011 international conference on asian language processing (IALP), Penang, Malaysia, 15–17 Nov 2011

Chew LW, Seng KP, Ang LM, Ramakonar V, Gnanasegaran A (2011) Audio-emotion recognition system using parallel classifiers and audio feature analyzer. Paper presented at the third international conference on computational intelligence. Model Simul, Langkawi, Malaysia, 20–22 Sept 2011

Davis S, Mermelstein P (1980) Comparison of parametric representations for monosyllabic word recognition in continuously spoken sentences. IEEE Trans Acoust Speech Signal Process 28 (4):357–366

Deshpande S, Chikkerur S, Govindaraju V (2005) Accent classification in speech. Paper presented at the fourth IEEE workshop on automatic identification advanced technologies, 17–18 Oct 2005

Diederich J, Pedersen C (2008) Accent in speech samples: support vector machines for classification and rule extraction. In: Diederich J (ed) Rule extraction from support vector machines, vol 80. Springer, Berlin, pp 205–226

Dupont S, Ris C, Deroo O, Poitoux S (2005) Feature extraction and acoustic modeling: an approach for improved generalization across languages and accents. Paper presented at the 2005 IEEE workshop onautomatic speech recognition and understanding, San Juan, 27–27 Nov 2005

Fahlman SE (1988) An empirical study of learning speed in back-propagation networks. Carnegie Mellon University

Fohr D, Illina I (2007) Text-independent foreign accent classification using statistical methods. Paper presented at the 2007 IEEE international conference on signal processing and communications, 24–27 Nov 2007

Furui S (2001) Digital speech processing, synthesis, and recognition, vol 7. CRC

Garg M (2003) Linear prediction algorithms. Bombay, India: Indian Institute of Technology (IIT), pp 1–15

Ghesquiere PJ, Compernolle DV (2002). Flemish accent identification based on formant and duration features. Paper presented at the The 2002 IEEE international conference on acoustics. Speech Signal Process (ICASSP) Orlando, FL, United States, 13–17 May 2002

Ghorshi S, Vaseghi S, Yan Q (2008) Cross-entropic comparison of formants of British, Australian and American english accents. Speech Commun 50(7):564–579

Gill SK (1993) Standards and pedagogical norms for teaching english in Malaysia. World Englishes 12(2):223–238

Hanani A, Russell M, Carey MJ (2011) Speech-based identification of social groups in a single accent of British English by humans and computers. Paper presented at the 2011 IEEE international conference on acoustics, Speech Signal Process

Humphries JJ, Woodland PC, Pearce D (1996) Using accent-specific pronunciation modelling for robust speech recognition. Paper presented at the fourth international conference on spoken language Philadelphia, PA, USA, 3–6 Oct 1996

Liu WK, Fung P (1999) Fast accent identification and accented speech recognition. Paper presented at the IEEE international conference on acoustics, Speech Signal Process, Phoenix, AZ, USA 15–19 March 1999

Looney CG (1997) Pattern recognition using neural network: theory and algorithm for engineers and scientists. Oxford University Press, New York

Lowenberg P (1992) The marking of ethnicity in Malaysian english literature: nativization and its functions. World Englishes 11(2/3):251–258

Makhoul J (1975) Linear prediction: a tutorial review. Proc IEEE 63(4):561–580

Nair-Venugopal S (2000) English, identity and the Malaysian workplace. World Englishes, vol 19. Blackwell Publishers Ltd, Oxford, pp 205–213

Nguyen P, Tran D, Huang X, Sharma D (2010) Australian accent-based speaker classification. Paper presented at the third international conference on knowledge discovery and data mining, Phuket, Thailand, 9–10 Jan 2010

Phoon HS (2010) The phonological development of Malaysian english speaking chinese children: a normative study. Doctor of Philosophy, University of Canterbury. Communication Disorders, Christchurch, New Zealand. http://hdl.handle.net/10092/4336

Picone JW (1993) Signal modeling techniques in speech recognition. Proc IEEE 81(9):1215–1247. doi:10.1109/5.237532

Pillai S, Mohd Don Z, Knowles G, Tang J (2010) Malaysian English: an instrumental analysis of vowel contrasts. World Englishes 29(2):159–172. doi:10.1111/j.1467-971X.2010.01636.x

Pitton JW, Kuansan W, Biing-Hwang J (1996) Time-frequency analysis and auditory modeling for automatic recognition of speech. Proc IEEE 84(9):1199–1215

Rabiee A, Setayeshi S (2010) Persian accents identification using an adaptive neural network. Paper presented at the second international workshop on education technology and computer science, Wuhan, Hubei, China, 6–7 March 2010

Rabiner L, Juang BH (1993) Fundamentals of speech recognition, vol 103. Prentice Hall, Englewood Cliffs, New Jersey

Sangwan A, Hansen JHL (2012) Automatic analysis of Mandarin accented English using phonological features. Speech Commun 54(1):40–54

Teixeira C, Trancoso I, Serralheiro A (1996) Accent identification. Paper presented at the fourth international conference on spoken language, Philadelphia, PA, 3–6 Oct 1996

Vallabha GK, Tuller B (2002) Systematic errors in the formant analysis of steady-state vowels. Speech Commun 38(1–2):141–160

Vieru B, de Mareüil PB, Adda-Decker M (2011) Characterisation and identification of non-native French accents. Speech Commun 53(3):292–310

Yusnita MA, Paulraj MP, Yaacob S, Shahriman AB, Saidatul A (2011) Malaysian english accents identification using LPC and formant analysis. Paper presented at the 2011 IEEE international conference on control system, Computing and Engineering (ICCSCE), Penang, Malaysia, 25–27 Nov 2011

Yusnita MA, Paulraj MP, Yaacob S, Fadzilah MN, Shahriman AB (2013a) Acoustic analysis of formants across genders and ethnical accents in Malaysian english using ANOVA. Procedia Engineering 64(2013):385–394. doi:10.1016/j.proeng.2013.09.111

Yusnita MA, Paulraj MP, Yaacob S, Yusuf R, Shahriman AB (2013b) Analysis of accent-sensitive words in multi-resolution mel-frequency cepstral coefficients for classification of accents in Malaysian english. Int J Automot Mech Eng 7(2013):1053–1073

Chapter 5
Robustness Analysis of Feature Extractors for Ethnic Identification of Malaysian English Accents Database

Mohd Ali Yusnita, Murugesa Pandiyan Paulraj, Sazali Yaacob, Abu Bakar Shahriman, Rihana Yusuf and Mokhtar Nor Fadzilah

Abstract Accent is fascinating human speech behavior that can be used to mark personal identity and social characteristics of its bearer. However, it also has potential to bias social interaction which includes prestige, job competency and ethnic discrimination. Albeit many successful methods have been deployed in the past to identify a speaker accent, the success rates are most likely database-dependent. This study aims to inquire about identification of Malaysian English (MalE) accents caused by ethnic diversities in this country. Robustness analysis was conducted using seven noisiness levels by corrupting the speech signals with additive white Gaussian noise (AWGN) to investigate the performance of four different schemes of feature extractors under clean and noisy conditions. These methods are filter bank analysis consists of mel-frequency cepstral coefficients (MFCC) and a new set of formulated features named as descriptors of mel-bands spectral energy (MBSE). Principle component analysis (PCA) was utilized to transform to another new features called PCA-MBSE. Second, vocal tract analysis consists of linear prediction coefficients (LPC) and formant frequencies (formants). Third, hybrid analysis consists of discrete wavelet transform (DWT) and LPC. The last scheme is fusions of spectral features (SFFs) of MFCC with formants and LPC with formants. Experimental results showed that SFFs techniques possess more sturdy noise resistivity than MFCC, LPC, MBSE, and DWT-derived LPC features. Similarly, PCA-transformed MBSE was just moderately affected as compared to the original features. While PCA-MBSE only caused a performance drop of 15 % in

M.A. Yusnita (✉) · R. Yusuf · M.N. Fadzilah
Faculty of Electrical Engineering, Universiti Teknologi MARA,
Kepala Batas, Pulau Pinang, Malaysia
e-mail: yusnita082@ppinang.uitm.edu.my

M.P. Paulraj · A.B. Shahriman
School of Mechatronic Engineering, Universiti Malaysia Perlis, Arau, Perlis, Malaysia
e-mail: paul@unimap.edu.my

S. Yaacob
Universiti Kuala Lumpur Malaysian Spanish Institute, Kuala Lumpur, Kedah, Malaysia
e-mail: sazali.yaacob@unikl.edu.my

N.A. Yacob et al. (eds.), *Regional Conference on Science, Technology and Social Sciences (RCSTSS 2014)*, DOI 10.1007/978-981-10-0534-3_5

average and the SFFs were just slightly affected by the AWGN from 8 to 13 % drop, the percentage drop of other feature sets were fairly above 30 %.

Keywords Ethnic identification · Discrete wavelet transform · Linear prediction coefficients · Melfrequency cepstral cefficients · Malaysian English

1 Introduction

The fact that Malaysian English (MalE) accents are new, problem in automatic accent classification (AAC) field motivates this research work. Accent is fascinating human speech behavior that can be used to mark personal identity and social characteristics of its bearer. According to Ahmed et al. (2013), accent is a language behavior that indicates personal and social characteristics of a speaker and has potential to bias social interaction which includes prestige, job competency, and ethnic discrimination. Thus, it is useful in the investigation of communicative phenomenon where significant social decision-making is required. The emergence of accents or sub-variants of English in Malaysia is obvious due to the scenario of multi-ethnic, multi-cultural and multi-lingual society comprises of 28.3 million population of 50.1 % Malays, 22.6 % Chinese, 6.7 % Indians, and others (Hassan 2012) complicates the structure of English spoken by its residents. It is thus a challenge to inquire about identification of MalE accents caused by ethnic diversities in this country.

Over the past and recent years, many good techniques have been proposed to obtain either high accuracy or robust performance. However, very few reported to achieve both qualities in the accent features. Prevalent methods of speech analysis such as mel-frequency cepstral coefficients (MFCC) as reported in (Arslan and Hansen 1996; Nguyen et al. 2010; Rabiee and Setayeshi 2010; Vergyri et al. 2010), normally provides good accuracy but suffer from deterioration issues under noisy conditions (Nhat and Lee 2004; Tufekci and Gowdy 2000), while linear prediction coefficients (LPC) despite its simple and robust implementation, are not as accurate as the earlier mentioned feature extractor (Tanabian and Goubran 2005; Teixeira et al. 1996). Formant frequencies are robust features (Ghorshi et al. 2008; Mustafa and Bruce 2006) for accent classification, however, using these features alone limits the accuracy that can be achieved due to their dynamic properties (Arslan 1996). Another problem with the existing method relates to the limited precision of fixed window size using short-time Fourier transform (STFT). Previous efforts to modify and improve these prevalent techniques have been made (Yusnita et al. 2013a, c, 2012a, b) that directly dealt with AAC of MalE speech. This paper is an extension effort to analyse feature extractors that possess both robust and salient acoustic features by corrupting speech signals with additive white Gaussian noise (AWGN) to investigate their various performances under clean and noisy conditions.

The remaining sections have the following direction. Section 2 elaborates about speech database and multiple feature extraction methods in AAC system. The results are discussed in Sect. 3. Finally, Sect. 4 concludes the important findings and investigation of this paper.

2 Methodology

Speech Database and Experimental Setup

For the purpose of this research, a new MalE accents database was developed by conducting a series of recording sessions in a semi-anechoic acoustic chamber using a handheld condenser, supercardioid and unidirectional microphone using a laptop computer sound card and MATLAB program. The recorded background noise level in the chamber was 22 dB. The sampling rate and bit resolution were set to 16 kHz and 16 bps for normal high quality used in automatic speech recognition applications. Accented speech were elicited from 51 male volunteers and 52 female volunteers of three main ethnics, i.e., the Malays, Chinese and Indians. The tasks consisted of continuous speech mode in the form of sentences (STs) using reading speaking style. The source of speech materials can be referred from (Yusnita et al. 2013b). Each sentence was replicated three times for each speaker to increase the number of samples per speaker and to increase precision and provide an estimate of the experimental error per speech and per speaker. The data collection consisted of 5172 samples of STs speech recorded from 103 students of Universiti Malaysia Perlis (UniMAP). Their age were between 18–40 years and they originated from various north, south, west and east regions of the country and as such they were also influenced by their regional accents.

Feature Extractors for Robust Ethnic Identification

In this work, in order to investigate the robustness quality of feature extractors for accent identification based on ethnic diversity, four different schemes of feature extractors are proposed as in Fig. 1 as a complete AAC system. In the pre-processing stage, speech signal is initially normalised, pre-emphasized using first order FIR filter of emphasis coefficient of 0.9375, frame-blocked into 32 ms short segments with 50 % overlapping and windowing using Hamming function prior to feature extraction. The second stage within the dotted line consists of four different schemes of feature extractors to model the K-nearest neighbors (KNN) classifier namely, MFCC and a new set of formulated features named as descriptors of mel-bands spectral energy (MBSE). Previous works (Yusnita et al. 2013a, e) are referred for MFCC and MBSE theoretical background. Principle component analysis (PCA) was utilized to transform to another new features called PCA-MBSE. This method is described in (Yusnita et al. 2013c). Secondly, vocal tract analysis consists of LPC and formant frequencies (formants). The theoretical

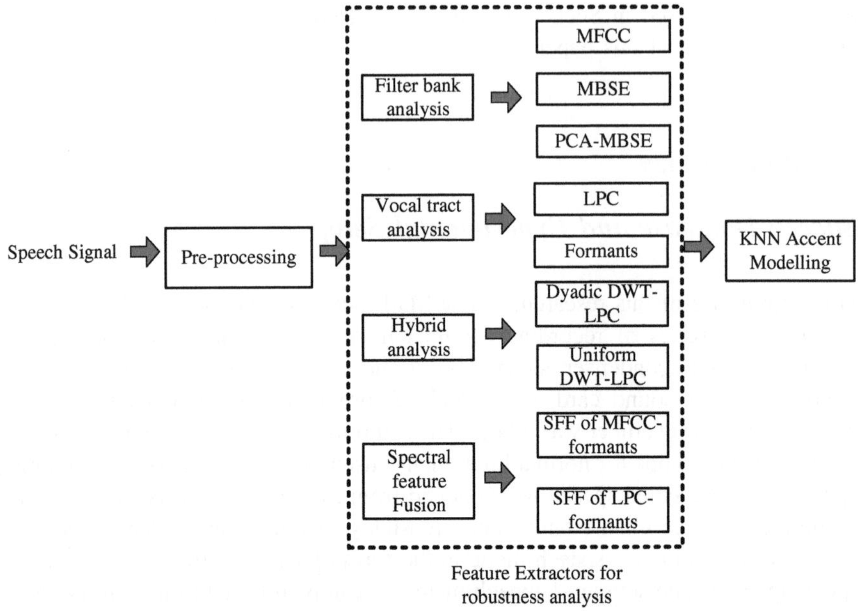

Fig. 1 A block diagram showing multiple feature extraction methods for robust automatic accent classification (AAC) system

background of LPC and formants is reported in (Yusnita et al. 2013d, 2011). Thirdly, hybrid analysis consists of discrete wavelet transform (DWT) and LPC. The DWT-derived LPC consists of two methods of extracting the LPC coefficients namely dyadic manner and uniform manner. These hybrid methodologies can be referred from Yusnita et al. (2012a). The last scheme is fusion of spectral features (SFFs) of MFCC with formants and LPC with formants. This method is elucidated in the other paper of this conference.

3 Results and Discussion

This section compares the performance of various formulated accent feature extractors under clean and noisy conditions using KNN classifier. In order to test the robustness quality, seven levels of noisiness are presented here. For each feature extraction method, the test dataset which constituted of 40 % of the overall audio volume was corrupted with AWGN to stimulate as background noise in the real environment. The train dataset was as clean as originally recorded. The performances of different feature vectors under different level of signal-to-noise ratio (SNR) of 35, 30, 25, 20, 15, 10 dB and down to 5 dB were tested. Comparison of susceptibility to noise levels for two speech test scenarios which are the combinations of STs-Male and STs-Female are reported. All results reported are the mean and maximum (max) values of classification rates (CRs) of ten trials.

Robustness Comparison for Noisy Conditions

Comparison of susceptibility to noise levels for two speech test scenarios are depicted in Figs. 2 and 3 in terms of degradation of mean CRs. Generally speaking, from observations made of these two figures, all types of the tested feature vectors were quite less susceptible to AWGN under noisy conditions between 25 and 35 dB. For more noisy conditions of below 25 dB, some feature vectors experienced significant performance drop while others showed robust performance. For instance, the baseline features such as MFCC, LPC, and MBSE and also both of the proposed DWT-derived LPC features were highly affected when corrupted with high level of AWGN. On the other hand, the PCA-transformed MBSE was just moderately affected. The results also suggested that the SFFs of MFCC-formants and LPC-formants were of robust quality features as the performance only started to drop noticeably at the SNR of 15 dB by approximately 0.5–1.2 % for the STs speech with respect to the SNR of 35 dB.

Table 1 compares the percentage of drop in the performance of mean CR when the SNR level were set between 25 and 5 dB using different feature extractors. It can be inferred that the degradation was above 30 % for all baseline features and the DWT-derived LPC features with the largest drop using the dyadic DWT-LPC by averagely 43 %. While, the PCA-MBSE features only dropped by 15 % in average

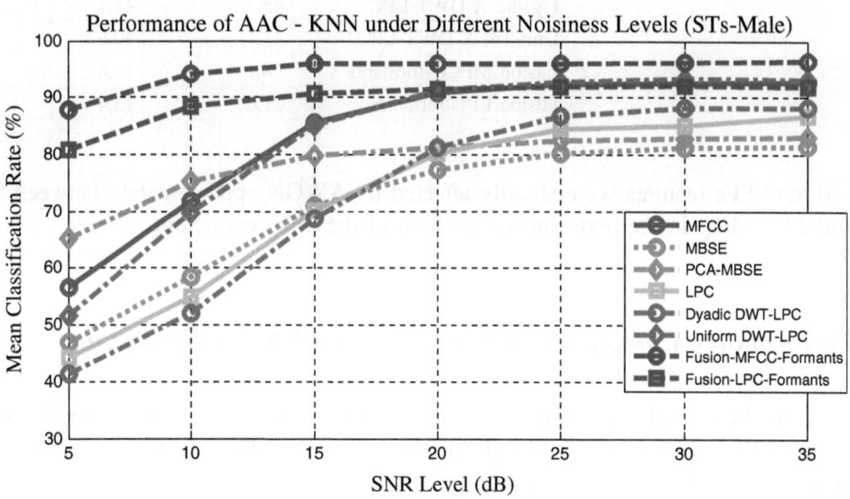

Fig. 2 Robustness performance of various feature extractors under different noisiness levels for STs speech of male speakers

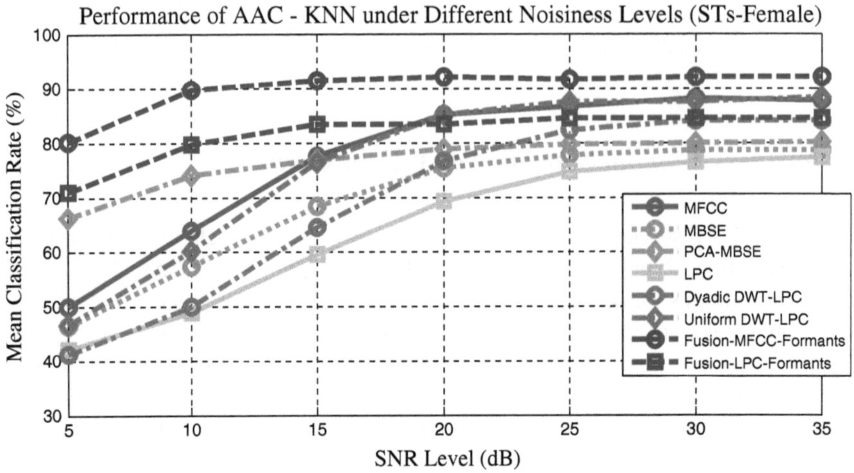

Fig. 3 Robustness performance of various feature extractors under different noisiness levels for STs speech of female speakers

Table 1 Performance drop percentage in mean CR for different SNR levels

Feature vector	Mean CR drop from 25 to 5 dB (%)	
	STs-Male	STs-Female
MFCC	35.7	36.8
MBSE	33.1	31.3
PCA-MBSE	17.1	13.6
LPC	40.2	32.7
Dyadic X DWT-LPC	45.3	41.1
Uniform X DWT-LPC	41.1	40.9
Fusion MFCC-formants	8.1	11.6
Fusion LPC-formants	11.0	13.4

and the SFFs features were slightly affected by AWGN approximately between 8 and 13 % drop in the performance for both of the SFFs extractors.

Performance Comparison Under Clean and Noisy Conditions

Next, the best result (max CRs) of eight feature vectors of ten runs of simulation under clean and three selected levels of noisiness are compared in Figs. 4 and 5 for both speech test scenarios, respectively.

These results suggested that for the AAC developed for MalE speakers of three main ethnics, namely the Malay, Chinese and Indian, the best feature extraction

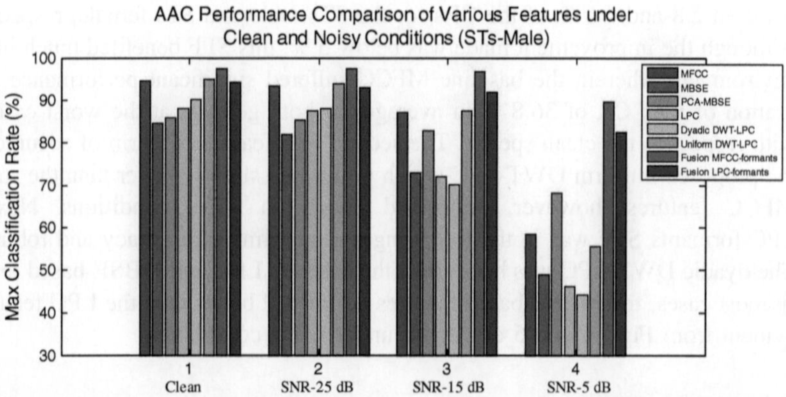

Speech Condition	MFCC	MBSE	PCA-MBSE	LPC	Dyadic DWT-LPC	Uniform DWT-LPC	Fusion MFCC-formants	Fusion LPC-formants
Clean	94.6	84.6	85.8	88.1	90.2	95.0	97.4	94.3
25 dB	93.3	81.9	85.2	87.5	88.1	94.0	97.5	93.1
15 dB	87.3	73.0	82.8	72.1	70.1	87.5	96.7	91.9
5 dB	58.4	48.9	68.2	46.2	44.2	55.6	89.6	82.4

Fig. 4 Performance (max CR) of different feature extractors under clean and different SNR for STs speech of male speakers

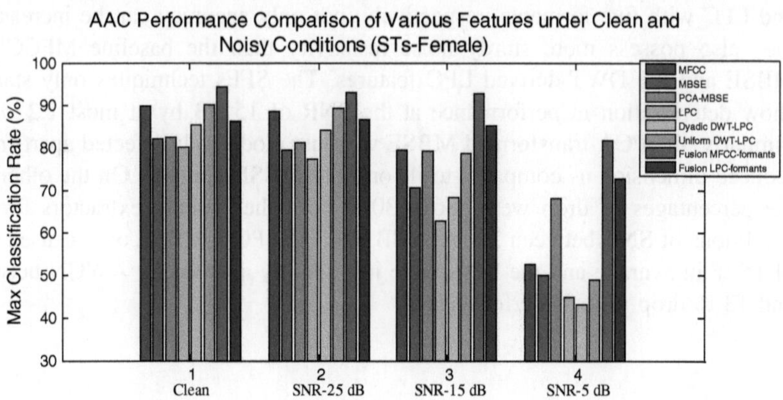

Speech Condition	MFCC	MBSE	PCA-MBSE	LPC	Dyadic DWT-LPC	Uniform DWT-LPC	Fusion MFCC-formants	Fusion LPC-formants
Clean	89.9	82.1	82.5	80.1	85.4	90.2	94.5	86.4
25 dB	88.7	79.5	81.3	77.3	84.2	89.3	93.0	86.6
15 dB	79.4	70.6	79.3	62.8	68.3	78.8	92.8	85.2
5 dB	52.6	50.1	68.2	44.9	43.1	49.0	81.8	72.7

Fig. 5 Performance (max CR) of different feature extractors under clean and different SNR for STs speech of female speakers

method that accomplished the highest accuracy rate was the mixture of MFCC-formants SFF which occurred for both speech test scenarios. The proposed MFCC-formants fusion outperformed the baseline MFCC by approximately between 2.8 and 4.6 % of the CRs for the STs-male and STs-female, respectively. Although the improvement made was below 5 %, this SFF benefited much in noisy environment wherein the baseline MFCC suffered significant performance deterioration of max CR of 36.8 % in average for both genders at the worst case SNR with respect to the clean speech. The second best features in term of accuracy was the proposed uniform DWT-LPC which performed slightly better than the standard MFCC features, however, performed poorly in noisy conditions. Next, the LPC-formants SFF was in the following line in terms of accuracy and robustness. The dyadic DWT-LPC was better than the baseline LPC and MBSE-based features. In most cases, the MBSE-based features performed better than the LPC features as evident from Figs. 4 and 5 especially under noisy conditions.

4 Conclusion

This paper has considered the study of robustness quality of eight feature extractors using four different schemes to model KNN classifier for yielding the most accurate and robust features performance in identifying the ethnic group from MalE accents database. Robustness analysis using speech corrupted with seven different levels of AWGN has proven that by fusing the powerful baselines features such as MFCC and LPC with five formants, respectively, not only accuracy can be increased but they also possess more sturdy noise resistivity than the baseline MFCC, LPC, MBSE and the DWT-derived LPC features. The SFFs techniques only started to show deterioration in performance at the SNR of 15 dB by at most 1.2 % only. Similarly, the PCA-transformed MBSE was just moderately affected apart from its reduced dimension as compared to its original MBSE features. On the other hand, the percentages of drop were above 30 % for other feature extractors for noisy conditions of SNR between 25 and 5 dB while the PCA-MBSE only caused a drop of 15 % in average and the SFFs were just slightly affected by AWGN between 8 and 13 % drop in their performances.

References

Ahmed ZT, Abdullah AN, Heng CS (2013) The role of accent and ethnicity in the professional and academic context. Int J Appl Linguist Engl Lit 2(5):249–258

Arslan LM (1996) Foreign accent classification in American English. Duke University, Durham

Arslan LM, Hansen JHL (1996) Language accent classification in American English. Speech Commun 18(4):353–367

Ghorshi S, Vaseghi S, Yan Q (2008) Cross-entropic comparison of formants of British, Australian and American English accents. Speech Commun 50(7):564–579

Hassan AR (February, 2012) Monthly statistical bulletin Malaysia

Mustafa K, Bruce IC (2006) Robust formant tracking for continuous speech with speaker variability. IEEE Trans Audio Speech Lang Process 14(2):435–444

Nguyen P, Tran D, Huang X, Sharma D (2010) Australian accent-based speaker classification. In: Paper presented at the third international conference on knowledge discovery and data mining, 9–10 Jan 2010, Phuket, Thailand

Nhat VDM, Lee S (2004) PCA-based human auditory filter bank for speech recognition. In: Paper presented at the 2004 international conference on signal processing and communications (SPCOM'04), 11–14 Dec, 2004, Bangalore, India

Rabiee A, Setayeshi S (2010) Persian accents identification using an adaptive neural network. In: Paper presented at the second international workshop on education technology and computer science, 6–7 March 2010, Wuhan, Hubei, China

Tanabian MM, Goubran RA (2005). Speech accent identification with vocal tract variation trajectory tracking using neural networks. In: Paper presented at the computational intelligence for homeland security and personal safety, 31 March–April 1 2005, proceedings of the 2005 IEEE international conference on CIHSPS 2005

Teixeira C, Trancoso I, Serralheiro A (1996) Accent identification. In: Paper presented at the fourth international conference on spoken language, 3–6 Oct 1996, Philadelphia, PA

Tufekci Z, Gowdy JN (2000) Feature extraction using discrete wavelet transform for speech recognition. In: Paper presented at the Proceedings of the IEEE Southeastcon 2000, 7–9 April, 2000, Nashville, TN

Vergyri D, Lamel L, Gauvain JL (2010) Automatic speech recognition of multiple accented english data. In: Paper presented at the 11th annual conference of the international speech communication association: spoken language processing for all, 26–30 Sept 2010, INTERSPEECH 2010, Makuhari, Chiba, Japan

Yusnita MA, Paulraj MP, Sazali Y, Shahriman AB, Fadzilah MN (2013a) Statistical band selection for descriptors of MBSE and MFCC-based features for accent classification of Malaysian English. Int J Electr Electron Syst Res 6(2013):31–46

Yusnita MA, Paulraj MP, Yaacob S, Fadzilah MN, Shahriman AB (2013b) Acoustic analysis of formants across genders and ethnical accents in Malaysian English using ANOVA. Procedia Eng 64(2013):385–394

Yusnita MA, Paulraj MP, Yaacob S, Shahriman AB (2012a) Classification of speaker accent using hybrid DWT-LPC features and K-nearest neighbors in ethnically diverse Malaysian English. In: Paper presented at the 2012 IEEE Symposium on Computer Applications and Industrial Electronics (ISCAIE), 3–4 Dec 2012, Kota Kinabalu

Yusnita MA, Paulraj MP, Yaacob S, Shahriman AB (2013c) Feature space reduction in ethnically diverse Malaysian English accents classification. In: Paper presented at the 2013 7th International Conference on Intelligent Systems and Control (ISCO), 4–5 Jan 2013, Coimbatore, Tamilnadu

Yusnita MA, Paulraj MP, Yaacob S, Shahriman AB, Mokhtar NF (2013d) Statistical formant descriptors with linear predictive coefficients for accent classification. In: Paper presented at the 8th IEEE Conference on Industrial Electronics and Applications (ICIEA) Melbourne, 19–21 June 2013, Australia

Yusnita MA, Paulraj MP, Yaacob S, Shahriman AB, Nataraj SK (2012b) Speaker accent recognition through statistical descriptors of mel-bands spectral energy and neural network model. In: Paper presented at the 2012 IEEE Conference on Sustainable Utilization and Development in Engineering and Technology (STUDENT), 6–9 Oct 2012, Kuala Lumpur

Yusnita MA, Paulraj MP, Yaacob S, Shahriman AB, Saidatul A (2011) Malaysian English accents identification using LPC and formant analysis. In: Paper presented at the 2011 IEEE International Conference on Control System, Computing and Engineering (ICCSCE), 25–27 Nov 2011, Penang, Malaysia

Yusnita MA, Paulraj MP, Yaacob S, Yusuf R, Shahriman AB (2013c) Analysis of accent-sensitive words in multi-resolution mel-frequency cepstral coefficients for classification of accents in Malaysian English. Int J Automot Mech Eng (IJAME) 7(2013):1053–1073

Chapter 6
Effect of Heat Treatment on Timbers—A Review

Nur Ilya Farhana Md. Noh, Zakiah Ahmad and Nur Kamaliah Mustaffa

Abstract Heat treatment is one of the environment-friendly ways to treat timber that will lead to the improvement of timber natural quality and equip the timber with new properties by modifying the timber. It is an eco-friendly and alternative treatment method that will modify the properties of timber by using high temperature instead of using chemical preservatives. This paper presents the overview of heat treatment effects on physical, chemical and mechanical properties in the case of heat treated timbers research. It is learned that at temperature over 150 °C, the physical, mechanical and chemical properties of the wood gradually get altered permanently. The research would be essential for the purpose of finding a better alternative to treat the timber to replace the chemical treatment, Copper Chromium Arsenic (CCA) which is not environment friendly. Research had shown that heat treatment on softwood as internal purpose especially in furniture application, hence there is a need to expand this treatment as an alternative method on hardwood timbers to be used in construction industry.

Keywords Heat treatment · Environmental friendly · Timbers · Properties · High temperature

N.I.F. Md. Noh (✉) · N.K. Mustaffa
Faculty of Civil Engineering, Universiti Teknologi MARA,
Shah Alam, Selangor, Malaysia
e-mail: nurilyafarhana@yahoo.com

N.K. Mustaffa
e-mail: nurkamaliah@salam.uitm.edu.my

Z. Ahmad
Institute of Infrastructure Engineering and Sustainable Management,
Universiti Teknologi MARA, Shah Alam, Selangor, Malaysia
e-mail: zakiah@salam.uitm.edu.my

N.A. Yacob et al. (eds.), *Regional Conference on Science, Technology and Social Sciences (RCSTSS 2014)*, DOI 10.1007/978-981-10-0534-3_6

1 Introduction

Heat treatment is the more environmental way to treat timber. It is an eco-friendly and alternative treatment method that will modify the properties of timber by using high level of temperature (Heräjärvi 2009). Heat treatment will lead to the improvement of the timber's natural quality and equip the timber with new properties by modifying the timber. In Malaysia, heat treatment has been applied to timber used for furniture application and also for the softwood timber. There is a need to expand this technology for hardwood timbers to be used for structural applications. Timbers used for furniture are timbers of low density as compared to timbers suitable for structural applications. Hence the heat treatment for higher density hardwood timber may require different temperatures and pressures and these can impose different effects on the physical, chemical and mechanical properties of the timbers. Therefore, this paper aims to review the effects of heat treatment on physical, chemical and mechanical properties in the case of heat treated timbers research.

The most commonly method of treatment for timber is by using chemical preservatives known as Copper Chromium Arsenic (CCA) (Hardin and Beckermann 2005). It is a world most widely used wood preservatives including Malaysia. It has been proven that CCA can help to improve timber properties and can also help to prolong the life of a timber species from few years to 30 years or more (Tong 2005). Due the usage of chemical preservatives which are toxic, there are many issues arise related to CCA. One of the issues is related to human where CCA can give a threat to their health. EC Scientific Committee on Toxicity, Eco-toxicity, and the Environment (CSTEE) in year 2002 had announced that CCA is both genotoxic and carcinogenic. Based on these issues, many countries have put restriction and banned on the usage of CCA for treating timbers such as Canada, US, Japan, Denmark, Switzerland, Vietnam and Indonesia (Read 2003). However, Malaysia still allows CCA to be used for treatment. In light of this issue, alternative treatment method needs to be explored and made available in Malaysia. One of the possible method of treatment is heat treatment which is by using heat rather than chemicals. Heat treatment is the most environment-friendly way to treat a timber species that will be an alternative way to replace CCA. This method of treatment will alter the substrate of a timber species by using high level of temperature.

2 Literature Review

Heat Treatment

Heat treatment is one of a method of treatment to alter and improve the properties of timbers by using a high-level temperature which also finally can equip the timbers with new properties (Sundqvist 2004). It is an environment-friendly way to improve and alter the timber's structures and components ("Heat Treatment of Timbers" 2005). It differs from Copper Chromium Arsenic (CCA) which is widely used in Malaysia, heat treatment is an eco-friendly way of treating timber because there is no chemicals or preservatives used during its process. Based on the previous researches, heat treatment may affect the physical, mechanical and chemical properties of timbers. Therefore, the next section reviewed the effects of heat treatment on physical, chemical and mechanical properties of timbers.

According to Militz (2002), the heat treatment process involves exposing wood to an elevated temperatures ranging from 160 to 260 °C in an atmosphere with low oxygen content to avoid the wood from burning due to the high temperature. The temperature and duration for heat treatment generally vary from 180 to 280 °C and 15 min–24 h, respectively, depending on the process, wood species, sample size, moisture content of the sample, the desired mechanical properties, resistance to biological attack and dimensional stability of the final product. The extent of the change in timber properties during heat treatment is determined by the method of thermal modification, the wood species and its characteristic properties, the initial moisture content of the wood, the surrounding atmosphere, the treatment temperature and time. It is reported that temperature has a stronger effect on timber properties than the amount of time the timber was exposed to heat. Treating wood at lower temperature for longer time periods does not produce the desired properties. Syrjanen and Oy (2000) found that temperature over 150 °C alter the physical and chemical properties of wood gradually.

Heat treatment reduces certain mechanical properties, but the dimensional stability and the biological durability of wood increases through heat treatment. Yildiz (2002) said that heat treatment also resulted changes in the physical properties of the wood including reduces shrinkage and swelling, low equilibrium moisture content, enhanced weather resistance, a decorative dark colour and better decay resistance. Therefore, heat treated wood is an eco-friendly alternative to impregnate wood materials, and heat treated wood can be used for garden, kitchen, and sauna furniture, cladding on wooden buildings, bathroom cabinets, floor material, ceilings, inner and outer bricks, doors, and window joinery, and a variety of other outdoor and indoor wood applications.

Based on Petrissans (2003), most of earlier experiments showed that heat treatment reduces the equilibrium moisture content (EMC), and thickness swelling and slow down water absorption (WA) and wet ability of wood. However, heat treatment actually increased the WA of pine sapwood heat treated at 170, 190, 210 ° C. Rousset et al. (2004), said that heat treatment significantly reduces the ability of bound water to migrate through the wood sample due to the chemical modification of the cell wall, but the permeability value did not change significantly.

Effects of Heat Treatment on Physical Properties

Physical properties of timber are related to the natural characteristics of the timber such as colour, smell, moisture content, durability, density and dimensional stability ("Timber Structures Design 8.1." 2008). Treatment or modification on timber could affect the properties of timber. Heat treatment also will modify and give effect to properties of timber including physical properties of timber. According to "Heat Treatment of Timbers" (2005), heat treatment will affect one of the physical properties of beech timber which is colour. Treating timber using high level of temperature within some duration will affect its appearance (Vernois 2001). The colour of beech that being treated using temperature of 200 °C up to eight (8) hours had caused it to become dark. Figure 1 shows the chronology of colour changes from the first hour until the final hour of heating.

Based on ("Heat Treatment of Timbers" 2005), the researcher only used one temperature level and had concluded that the colour changes due to the duration of heat treatment. A proposed study needs to be done on the effect of heat treatment on Malaysian tropical hardwood timber. The proposed study will be analysed as the temperatures and durations vary to verify either temperature level or duration that actually effect the colour changes. This study also has been proposed to determine either the same effect on colour changes will occur on hardwood timber or not. Besides that, another physical properties that are affected by heat treatment are movement and stability of the timber, which are referring to swelling and shrinkage of timber. Heat treatment will cause a decrement of timber movement which means improve the stability of timber. "Heat Treatment of Timbers" (2005) also had found that beech timber that been treated at temperature of 200 °C within 8 h reduced 50 % of its swelling movement. Figure 2 shows the relationship of swelling percentage and numbers of days after the heat treatment that had been done on beech timber.

Stability European Hophornbeam wood also had been improved after treating by heating it at 120, 150 and 180 °C for 2, 6 and 10 h (Korkut et al. 2009). The stability has been highly improved when there was a higher increment percentage of swelling reduction when the timber was heated at 180 °C at longer durations. Table 1 shows the increment percentage of swelling reduction for this timber after heat treatment at each temperature and duration.

Based on this research, it showed that stability of the European Hophornbeam had also been improved at almost 50 %. Another research had been done by using the same temperature level and durations on different species of timber to analyse

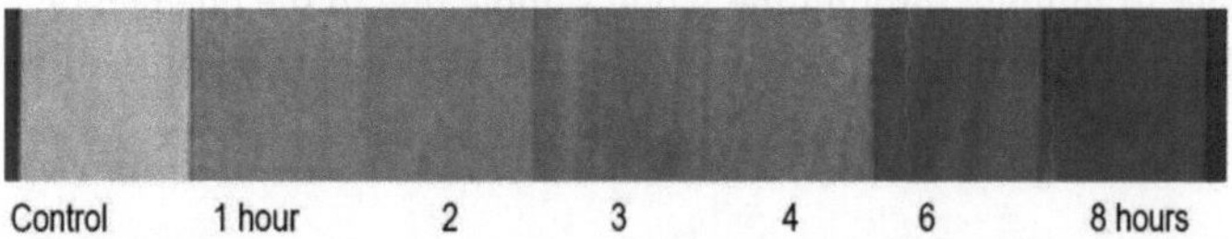

Fig. 1 Effect of heat treatment on physical properties of beech timber (*colour*) ("Heat Treatment of Timbers" 2005)

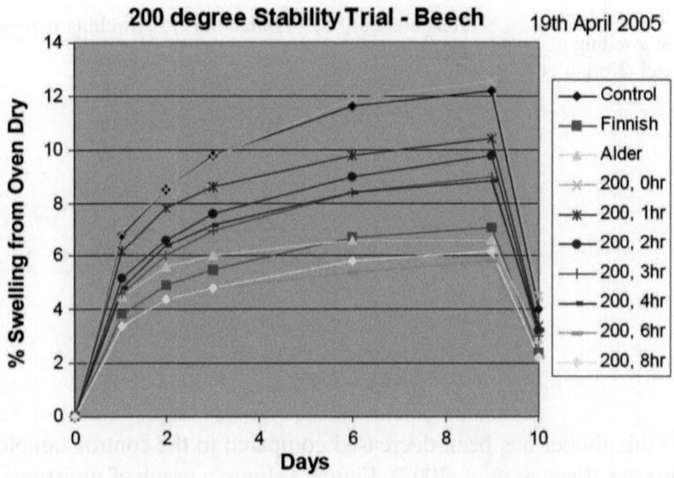

Fig. 2 A graph of swelling percentage versus days after heat treatment ("Heat Treatment of Timbers" 2005)

Table 1 Effect of heat treatment on swelling of European Hophornbeam (Korkut et al. 2009)

Temperature (°C)	Duration (h)	Swelling percentage (%)
120	2	26.062
	6	27.123
	10	29.784
150	2	30.517
	6	31.634
	10	34.485
180	2	36.896
	6	37.928
	10	48.710

its stability after treatment. The species was Turkish Hazel. The results showed that the percentage of swelling reduction only increased up to 20.29 % as in Table 2 (Korkut et al. 2008).

From Table 2, it can be seen the properties of timber varies even though the same level of temperatures and durations have been applied for the treatment. Therefore, it is the intention of this proposed study to determine the effect of heat treatment on Malaysian tropical hardwood timber. The temperature level above 200 °C will be used to investigate either there is improvement on timber stability when temperature above 180 °C been applied.

Another physical property that will be affected by heat treatment is moisture content of timber. A heat treatment on eucalypt wood had been done at the temperature level of 170–200 °C within 2–24 h. The result shows that the moisture

Table 2 Effect of heat treatment on swelling of Turkish Hazel (Korkut et al. 2008)

Temperature (°C)	Duration (h)	Swelling percentage (%)
120	2	0.87
	6	1.02
	10	7.08
150	2	8.44
	6	9.49
	10	12.67
180	2	14.67
	6	15.63
	10	20.29

content of this timber has been decreased compared to the control sample after the heating process (Esteves et al. 2007). Figure 3 shows a graph of moisture content of eucalypt wood after been treated by heat treatment.

Based on Fig. 3, it shows that the moisture content is decreasing corresponding to mass loss. In this proposed study, the heat treatment will be done on Malaysian tropical hardwood timber to investigate either it will give the same effect on moisture content as on softwood. This is important as the hardwood in this study is treated to be used as structural material, it is important to determine either its moisture content level will decrease or not. This is because materials with lower moisture content will equip with higher strength compared to materials with high moisture content ("Timber Structures Design 8.1." 2008). There will be dramatical changes in timber properties when its moisture content is below the point of fibre saturation point (FSP), which is occurred at moisture content of 25–35 % ("Timber Structures" 2008).

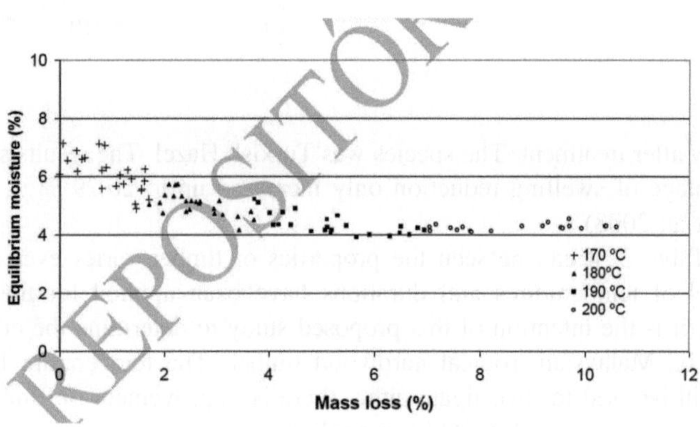

Fig. 3 Moisture content versus mass loss of eucalypt wood after heat treatment (Esteves et al. 2007)

To achieve the moisture content lower than FSP, there is a need to remove the free water from the cell lumen of timber. The green timber that had been dried will first lose the free water in the cell lumen (Köhler 2007). Figure 4 shows the location of free water in cell lumen and bound water in cell wall of a timber cell.

In this proposed study, the heat treatment will be used as a method of treatment to dry out the green Malaysian tropical hardwood timber in various durations and temperature levels to achieve moisture content below of FSP so that the mechanical properties in terms of strength also will be improved. The reason for applying different temperatures and durations is to find out what is the right temperature level with what duration to achieve the moisture content below the FSP level. It is important so that the effect of heat treatment on mechanical properties in terms of strength can be investigated either it is related to the moisture content level or not for each species of hardwood timber samples. The relationship between duration, temperature and the properties of timber will be tabulated in a smart table accordingly.

Effects of Heat Treatment on Mechanical Properties

Other than physical properties, previous researches also had found that heat treatment also affected mechanical properties of a timber. Mechanical properties can be defined as material characteristics that have been exposed to any external applied forced such as strength and durability ("Timber Structures Design 8.1." 2008). Bending strength and compression strength are classified as mechanical properties.

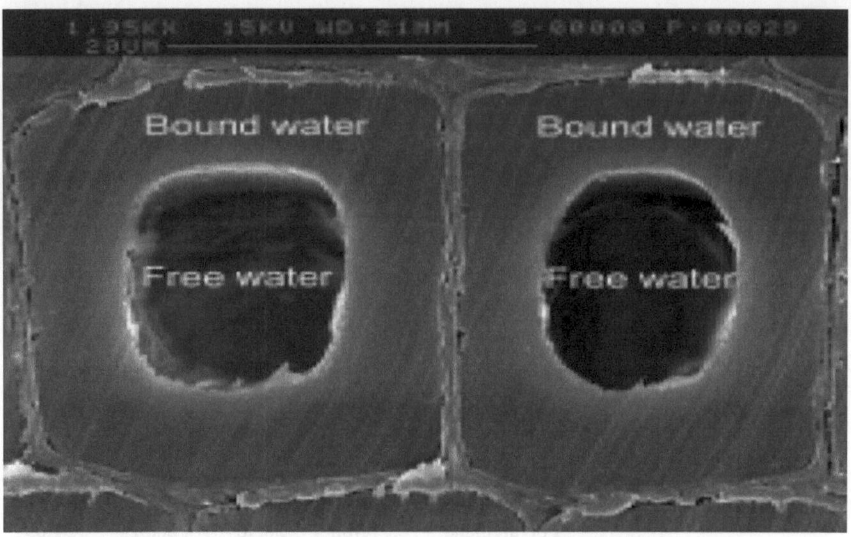

Fig. 4 Free water and bound water location in a timber cell (Köhler 2007)

A heat treatment on laminated black pine (*Pinus nigra*) had caused the positive impact to its bending and compression strength (Ordu et al. 2013). The laminated black pine had been heated at temperature 100–150 °C for 4 h. Both bending and compression strength are affected positively by the heating process. The higher improvement of these strengths occurred at higher level of temperature applied in this research with same duration. Tables 3 and 4 show the result of bending and compression strength of its timber samples before and after it had been treated at two different levels of heat within 4 h with two different adhesives respectively.

Both of these tables show that during treatment there are positive increments of strength after it been heated. This finding is supported by a study of heat treatment on laminated beech (*Fagus orientalis L.*) wood which they suggested the treatment had improved the bending and compression strength of this wood (Wood and Altinok 2010). Tables 5 and 6 show the tabulated data of these strengths for each of temperature used.

Both of these previous researches found positive impact on mechanical properties in terms of strength of laminated timber. As in this proposed study, the investigation of strength after heat treatment will be done on fresh wood itself without lamination. This proposed study will determine the effect of heat at temperature above 200 °C and different durations to learn which temperature and duration of treatment on fresh hardwood timber will give the best effect on its bending and compression strength. Besides of strength, heat treatment also had helped to improve mechanical strength in terms of durability (Viitaniemi 2001). Durability of timber is its ability to resist fungi and decay attack (Muthike 2008). Japanese cheddar had achieved improvement in durability after it had been treated by heat at temperature 105–150 °C for 6–72 h. The durability had been increased when the samples have been heated at temperature above 135 °C. Longer duration

Table 3 Bending strength of laminated black pine (Ordu et al. 2013)

Adhesives	The value of bending strength (N/mm^2)			
		Control	100 °C	150 °C
Polyvinyl acetate	Min	57.013	59.555	61.282
	Max	62.412	66.604	66.855
Polyurethane	Min	54.523	56.829	59.384
	Max	60.231	62.89	65.361

Table 4 Compression strength of laminated black pine (Ordu et al. 2013)

Adhesives	The value of compression strength (N/mm^2)			
		Control	100 °C	150 °C
Polyvinyl acetate	Min	22.16	23.83	29.76
	Max	27.48	27.85	33.53
Polyurethane	Min	26.96	28.52	31.63
	Max	32.21	32.68	36.36

Table 5 Bending strength of laminated black pine in N/mm^2 (Wood and Altinok 2010)

Samples	Polyvinyl acetate			Polyurethane		
	Control	100 °C	150 °C	Control	100 °C	150 °C
1	67.26	70.48	74.30	59.73	62.57	64.67
2	66.91	68.58	74.72	61.40	63.70	63.41
3	68.68	69.73	71.81	61.44	66.72	67.47
4	65.97	69.03	75.00	56.84	61.41	66.26
5	70.75	66.69	73.54	63.36	67.55	60.65
6	63.78	72.88	78.95	59.44	64.39	65.70
7	66.96	66.08	73.86	60.12	63.37	65.57
8	64.85	68.38	72.33	60.90	61.67	66.32
9	64.94	70.43	74.26	62.59	64.19	64.85
10	68.59	69.75	74.11	58.46	63.07	65.12

Table 6 Compression strength of laminated black pine in N/mm^2 (Wood and Altinok 2010)

Samples	Polyvinyl acetate			Polyurethane		
	Control	100 °C	150 °C	Control	100 °C	150 °C
1	30.42	34.72	35.14	23.38	29.46	34.12
2	28.12	36.24	38.56	24.12	31.68	36.71
3	29.98	35.19	35.05	24.33	30.04	35.11
4	31.26	33.25	37.15	23.62	32.78	35.48
5	32.10	35.24	38.80	22.63	30.85	33.22
6	28.80	32.21	39.16	21.48	28.84	34.06
7	30.72	33.16	36.08	21.02	32.74	35.63
8	29.33	35.12	36.03	22.50	32.45	36.85
9	33.21	32.64	37.98	23.26	33.09	35.79
10	31.31	34.46	36.82	21.62	32.47	37.96

with high temperature of treatment will improve the timber resistance against fungus (Momohara 2003). In this proposed study, the treatment duration and temperature vary to achieve moisture content below FSP level. Fungal decay can be reduced when the timber achieves moisture content less than 20 % which is below the FSP level. It shows that lower moisture content will equip the timber with new and improved properties. Wood is a hygroscopic material where it will absorb and loss moisture in humid and dry condition accordingly (Schott 2006). Heat that been applied on timber will cause a reduction in hygroscopic and low level of moisture content. This condition occurred due to the modification of timber cell. In this proposed study, the effect of heat treatment on chemical properties of Malaysian tropical hardwood timber will be identified.

Effects of Heat Treatment on Chemical Properties

Chemical properties of timber will also affect by heat treatment process. Timber are hygroscopic materials where its cell consists of three main polymers known as lignin, cellulose and hemicelluloses that comes with reactive hydroxyl group (Schott 2006; Tjeerdsma 2001). When heat is applied, this reactive hydroxyl group will react and affect the chemical properties of timber, in which Maritime pine, spruce, beech and poplar wood treated by heat treatment. When heat had been applied, the hygroscopic had been reduced and attack the timber polymers known as lignin and hemicelluloses (Kamdem et al. 2002). The researchers found that the content of lignin will be higher compared to the lignin content of untreated timber. The increment of lignin content will cause a reduction of hemicelluloses where in this condition it will lead to low moisture content.

Once the timber has been heated, hemicelluloses will be the first cell polymer that will be attacked and experience dehydration while celluloses are hardly been attacked by heat treatment (Esteves and Pereira 2009). The hydroxyl group will be reduced when the heat attack the hemicelluloses and cause less water absorption of the timber cell wall. This chronology will finally leads to reduction of moisture content level (Korkut et al. 2008). In this proposed study, a few tests will be done on timber to analyse these cell polymers movement and content so that the right temperature and duration of heat treatment can be identified where the lowest hemicelluloses content and the highest lignin content can be achieved to get to the low moisture content.

3 Conclusion

Heat treatment research is important for the purpose of finding a smarter and better alternative to treat material for construction works. The heat treatment might give an acceptable effect on the physical properties such as colour, equilibrium moisture content, water absorption and dimensional stability of timber species in strength grouping SG2 to SG4 as well as SG5 (for comparison) and to the mechanical properties (bending and tensile strength) and also the chemical properties. The quality of timber SG2 to SG4 shall be improved in eco-friendly way, thus providing a better and sustainable alternative for construction material. Furthermore, new data for treatment of tropical hardwood timbers could be acquired from the research. This will allow timbers in the future to be treated without using chemicals and become materially more appropriate in many applications especially in construction and structural use.

Acknowledgments This research is financially supported by the Research Acculturation Grant Scheme, Ministry of Education (MOE) Malaysia 600-RMI/RAGS 5/3 (53/2012) and Universiti Teknologi Mara (UiTM) whom both are greatly acknowledged.

References

Esteves BM, Pereira HM (2009) Wood modification by heat treatment: a review 4(1965):370–404

Esteves B, Domingos I, Pereira H (2007) http://www.freepatentsonline.com/article/ForestProducts-Journal/160106273.html, 57:47–52

Handbook 1 (2008) Timber structures

Hardin RA, Beckermann C (2005) Simulation of heat treatment distortion (3). Heat treatment of timber. (2005). Calu technical notes, (July)

Heräjärvi H (2009) Effect of drying technology on aspen wood properties, 43

Kamdem DP, Pizzi A, Jermannaud A (2002) Durability of heat treated wood. Karadeniz Technical University

Köhler J (2007) Reliability of timber structures. Institute of Structural Engineering Swiss Federal Institute of Technology

Korkut DS, Korkut S, Bekar I, Budakç M, Dilik T (2008) The effects of heat treatment on the physical properties and surface roughness of Turkish hazel (*Corylus colurna L.*) Wood 1772–1783

Korkut S, Alma MH, Elyildirim YK (2009) The effects of heat treatment on physical and technological properties and surface roughness of European Hophornbeam (*Ostrya carpinifolia scop.*) wood 8(20):5316–5327

Militz (2002) Thermal treatment of wood: European processes and their background. In: Cardiff-Wales: IRG/WP02-40241 33rd Annual Meeting

Muthike G (2008) Low cost timber drying method for sawyers, merchants and other users

Momohara I (2003) Effect of high-temperature treatment on wood durability against the brown rot fungus, fomitopsis palustris, and the termite, coptotermes formosanus, pp 284–287

Ordu M, Altinok M, Atilgan A, Ozalp M, Peker H (2013) The effects of heat treatment on some mechanical properties of laminated black pine (*Pinus nigra*)

Petrissans (2003) Wettability of heat treated wood, Holzforchung, pp 301–307

Read D (2003) Report on copper, chromium and arsenic (CCA) treated timber

Rousset et al (2004) Modification of mass transfer properties in polar wood (*P. robusta*) by a heat treatment at high temperature, Holz als Roh-und Werkstoff, pp 377–386

Schott BW (2006) Bamboo in the laboratory

Sundqvist B (2004) Colour changes and acid formation, pp 1402–1544

Syrjänen T, Oy K (2000) Production and classification of heat treated wood in Finland

Timber structures design 8.1. (2008). DelDOT, (April), 1–30

Tjeerdsma B (2001) Heat treatment of wood by the plato-process

Tong HL (2005) Treatment of timber for trusses. Malaysian Wood Preserving Association (MWPA)

Vernois DM (2001) Heat treatment of wood in France—state of the art

Viitaniemi P (2001) Heat treatment of woof—better durability without chemicals

Wood FOL, Altinok M (2010) The effects of heat treatment on some mechanical properties of laminated beech

Yildiz (2002) Physical, mechanical, technological and chemical properties of beech and spruce wood treated by heating. PhD dissertation

Chapter 7
Driver Behaviour and Compliance at Signalised Intersection

Li-Sian Tey, Muhamad Kasimi Abd Khalil and Fairus Azwan Azizan

Abstract In Malaysia, many studies have been conducted concerning highway and traffic safety but there has been no proper study involving traffic safety at signalised intersection. In order to maintain the safety, it is important to identify the behaviour of the driver that leads to an unsafe traffic conduct. This paper studied on the evaluation of driver behaviour and compliance at signalised intersection. Parameters adopted are driver approaching speed towards the signalised intersection, driver compliance rate to red light signal and driver responses at dilemma zone during amber light signal. Field data collection was conducted using video recording, measuring the driver responses at signalised intersection. The measuring location is an intersection categorised as four-leg intersections which consist of two (2) lanes for each approach, located below the Butterworth–Kulim Expressway (BKE) flyover connecting three (3) main roads among Projek Lebuhraya Usahasama Berhad (PLUS) expressway, BKE and Jalan Permatang Tok Kandu in Penang. The data of driver compliance were collected during peak hour and off-peak hour by counting the number of vehicles that comply or not comply with the red light signal. It was found that the average driver approaching speed during off-peak hour is higher than during peak hour. The average driver approaching speed resulted around 13.49 m/s during the weekend and it was found slightly the same as during the off-peak hour. As for the driver compliances, it shows that the number of drivers who did not comply with the red light signal increased as the volume of the vehicles increased. This proved that at average, driver compliances to the regulation of signalised intersection are very poor during peak hour which provide a better understanding in term of the driver responses towards existing system at the studied signalised intersection.

L.-S. Tey · M.K.A. Khalil · F.A. Azizan (✉)
Faculty of Civil Engineering, Universiti Teknologi MARA,
Permatang Pauh, Pulau Pinang, Malaysia
e-mail: fairusaz1@ppinang.uitm.edu.my

L.-S. Tey
e-mail: lstey@ppinang.uitm.edu.my

M.K.A. Khalil
e-mail: kasimi90@yahoo.com

N.A. Yacob et al. (eds.), *Regional Conference on Science, Technology and Social Sciences (RCSTSS 2014)*, DOI 10.1007/978-981-10-0534-3_7

Keywords Driver compliance · Red light signal · Signalised intersection

1 Introduction

An intersection is a shared space that is used by more than one approach at a time. An intersection may be three-way, a T junction/T intersection or fork, four-way or a crossroads, five-way or more depending on the number of road segments (arms) that come together at the intersection. It may often be controlled by traffic lights. A traffic light indicates which traffic is allowed to proceed at any particular time. A signalised control device is more often used at the intersection in order to smooth the traffic control.

Driver behaviour is considered as one of the factors that can contribute to the traffic accidents (Shope 2006). Behaviour describes our observable actions on how we conduct ourselves and driving is a complex activity that requires multi-level skilled behaviour. Yet much of our driving behaviour is automatic, developed from experiences and requiring minimal allocations of attention (Murcotts 1969).

Traffic light is an alternative device that can be applied as rule to curb traffic accident at intersection. It works in the typical sequence of colour phases namely red, amber and green. The green light allows traffic to proceed in the direction denoted, if it is safe to do so. The yellow or amber light denoting prepares to stop short of the intersection, if it is safe to do so, whereas the red signal prohibits any traffic from crossing the stop line.

The term of driver compliance is one of the parameters commonly used to test driver behaviour at signalised intersection. Drivers are grouped into several categories, such as conservative, normal and aggressive, according to their behaviour at the signalised intersection namely, their responses to the existence of a dilemma or option zone, their initial approaching speed and the non-compliance when facing the red or yellow light signal.

Studying driver behaviour at signalised intersection provides additional knowledge and beneficial information that can lead to proper measure taking. Therefore, it is important to identify the driver behaviour and compliance at a signalised intersection, driver compliance rate to red signal light and driver responses at dilemma zone during amber signal light. The area of study was carried out at a signalised intersection located at Permatang Tok Kandu, Permatang Pauh, Pulau Pinang.

2 Driver Behaviour

Bjorklund and Berg (2005) stated that in many road intersections, formal traffic regulation is regulating driver behaviour. Drivers based on informal regulation depend on road design in addition to the formal regulation and on other drivers' behaviour. Other signal such as difference in vehicle size and speed and relative positions of vehicles in the intersection are of essential in real life. However, based on a self-report survey which describe a different crossing situation and the possibilities of different cues in a driving situation questionnaire resulted in the drivers having limitation to decide their judgement either to yield or not to yield (Bjorklund and Berg 2005).

As part of Panagiotis (2006) study, the drivers can be classified according to their behaviour. The behaviour of drivers at signalised intersection can be determined according to certain criteria, where the first refers to the approaching speed, as compared to the posted speed limit and classifies drivers into two groups: aggressive and non-aggressive. Some factors that define aggressive driving the risky situations are overtaking, heavy braking, low headway, large speed differentials, conflicting movements and near misses (Bonsall et al. 2005). The aggressive driving behaviour at signalised intersection can be determined in terms of speeding and of ignoring yellow or red light. Drivers are always practice aggressive behaviour that is associated with non-compliance of traffic ethic and eventually reduces road safety levels at signalised intersection.

Dilemma Zone

Dilemma zone theory is when a driver approaching a signalised intersection at high speeds and the light turns yellow, the dilemma zone is the space from the intersection to the point on the road where it may be difficult for the driver to discern whether they should cross the yellow light or brake to be safe (Merrill Scott 2008). Dilemma zone divided into two types which is Type I dilemma zone and Type II dilemma zone. These two types of dilemma zone situations may arise for drivers faced with a yellow indication on the approach to a signalised intersection. This is known as the option zone.

Type I dilemma zones are associated with signal timing for clearance intervals. The Type I dilemma zone were described and first referenced by Gazis et al. (1960) as the possibility that a vehicle presented with a yellow indication while approaching a signalised intersection will, be unable to safely cross the intersection or stop prior to the stop line and it is depend on the physical parameters of the situation. Type II dilemma zone is defined as a segment of road on the approach to a signalised intersection where, if occupied by a driver presented with the circular yellow indication, is likely to result in a driver having difficulty deciding to stop at the stop line or proceed through the intersection (David et al. 2012).

3 Methodology

In this study, there were a few data of driver behaviour was collected by video recording, namely, driver approaching speed towards the signalised intersection, driver compliance rate to red signal light and driver responses at dilemma zone during amber signal light. The selected study site was carefully investigated so that the camera was set on tripod stand and installed at study site was hidden from drivers' attention, which may have affected their driving behaviour.

The area of the data collection is at Permatang Tok Kandu, Permatang Pauh. It is located below the Butterworth–Kulim Expressway (BKE) flyover connecting the three main roads among Projek Lebuhraya Usahasama Berhad (PLUS) Highway, BKE and Jalan Permatang Tok Kandu. This intersection is categorised as four-leg intersections and consists of two lanes for each approach as shown in Figs. 1 and 2, respectively.

Driver behavioural and compliance data was collected during Monday on 24th March 2014 at 10:15 A.M–12:15 P.M, Thursday on 27th March 2014 at 7:30 A.M–9:30 A.M, Friday on 28th March 2014 at 8:00 A.M–10:00 A.M, Wednesday on 9th April 2014 at 12:00 P.M–2:00 P.M, Friday on 11th April 2014 at 7:30 A.M–9:30 A.M, Monday on 14th April 2014 at 4:30 P.M–6:30 P.M, Tuesday on 15th April 2014 at 4:30 P.M–6:30 P.M, Thursday on 17th April 2014 at 7:30 A.M–9:30 A.M, and Saturday on 19th April 2014 at 7:45 A.M–9:45 A.M.

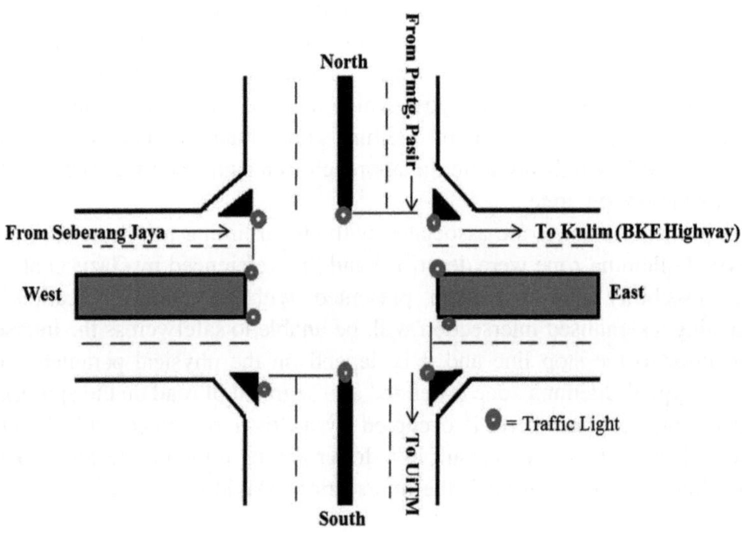

Fig. 1 The schematic diagram of study site layout plan

Fig. 2 Study site

Approaching Speed

Traffic was recorded on the intersection approaches using video camera installed at suitable location approaching the signalised intersection. The video was recorded for 2 h. The approaching speed profile of the subject as it approached and stopped at the signalised intersections was retrieved. The subject is refers to the first vehicle approaching the intersection from more than 180 m that was uninhibited by other vehicles after the red light had been activated. The times when the red light is activated and the subject moved from 180 m until it reached the stop line were recorded. The distance interval of 25 m was also considered as a distance division.

Driver Compliance Rate to Red Signal Light

This location was different from data collection location for driver approaching speed and it provides a view of vehicles approaching the monitored signalised intersection. The distance of video camera installation from stop line is 100 m. For the purpose of this study, data on the red light compliance at the study site were limited to those vehicles that approach stop line 0.5 s or more after onset of the red signal. To being counted as red light compliance, a vehicle's rear tyres must had been positioned behind the crosswalk or stop line prior to entering on red light. The data of driver compliance were collected during peak hour and off-peak hour by counting the number of vehicles comply or not comply with the red signal light.

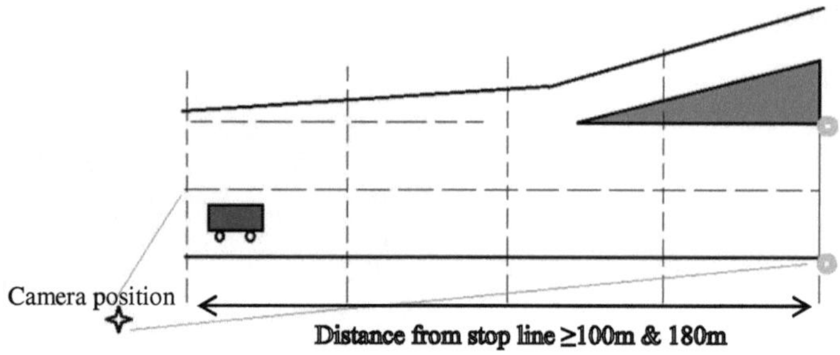

Fig. 3 Schematic diagram of a video-based data collection system

Driver Responses at Dilemma Zone During Amber Signal Light

To capture the driver responses at dilemma zone during amber signal light, this study focused on collecting the acceleration/deceleration rates of drivers before and after a yellow phase and the distance and probability of vehicle stopping to the stop-line when drivers perceive the commencement of amber signal light. In order to measure the above information reliably, this study adopted a video-based data collection system, as shown in Fig. 3, which has the functions of precisely tracking each individual vehicle trapped in the yellow phase and computing the exact distance and time for a vehicle to reach the stop-line from the start of a yellow phase.

4 Results and Discussions

The data collection was conducted during peak hour and off-peak hour during weekdays as well as a single data during weekend because the traffic volume is constant during weekend. The chosen of peak hour and off-peak hour is based on the volume of vehicle approaching the signalised intersection during that time. Usually, the peak hour for this roadway is during lunch time (11:00 A.M–2:00 P.M), evening peak hour (4:00 P.M–7:00 P.M) while the off-peak hour is besides the peak hour period. The data was taken at least more than one data set for comparison and validation.

Determination of the Driver Approaching Speed Towards the Signalised Intersection

Figure 4 shows the comparison of the results between peak hour and off-peak hour for the driver approaching speed data during peak hour (10:15 A.M–12:15 P.M)

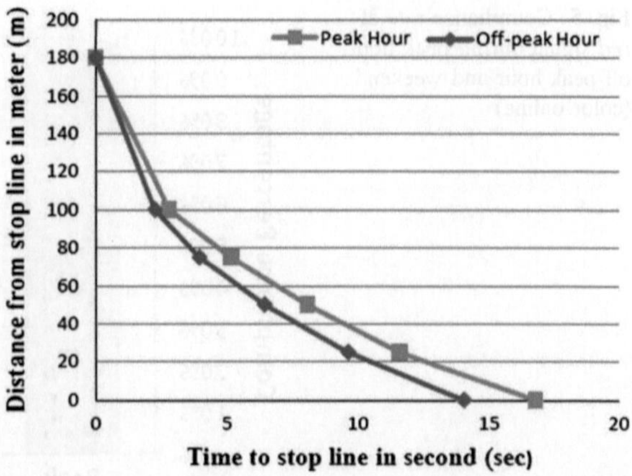

Fig. 4 Comparing results between peak hour and off-peak hour

and off-peak hour (7:30 A.M–9:30 A.M) for the speed comparison. There were 45 and 46 times red light signal within 2 h for the peak hour data and the off-peak hour data respectively. It was found that the speed reduction during peak hour is more than during off-peak hour and it depend on the longer time taken by driver to reach the stop line. Based on the graph, the average time to reach the stop line for the peak hour and off-peak hour are around 16.78 and 14.07 s, respectively. The average speed to reach the stop line for the peak hour and off-peak hour are around 10.73 and 12.79 m/sec, respectively. In general, drivers reduced their approaching speed later during off-peak as compared to peak hour. The traffic volume during certain time may become one of the main factors that can contribute to driver approaching speed towards signalised intersection. The more increment in the traffic volume the more decrease of the driver approaching speed towards signalised intersection. The tendency of driver to slow down later influenced by several factors such as when the driver knew that they were still far to approaching the leading vehicle. Some drivers increased their speed at the onset of the amber signal light and tried to across the signalised intersection safely.

Determination of Driver Compliance Rate to Red Signal Light

The result in Fig. 5 shows the result of comply and non-comply rate at red signal. It shows that the number of non-compliance during off-peak hour (34 %) is slightly less compared to the peak hour (47 %). While much higher compliance (84 %) was observed during weekend. These outcomes may possibly due to high traffic volume on the road or 'time pressure' of drivers for work. The driver behaviour during off-peak hour is under control depending on situation at that time which is more relax without interruption from any traffic congestion. Although there were also some driver still tend to ignore the red signal light, perhaps their action is caused by their own attitude which always break the traffic regulations. On the other hand,

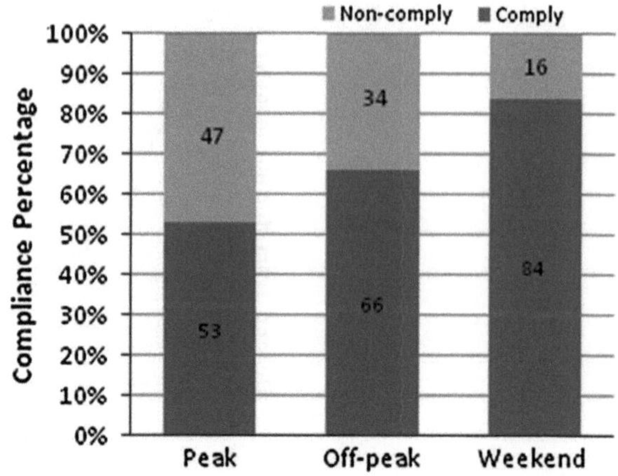

Fig. 5 Compliance rate at *red* signal during peak hour, off-peak hour and weekend (color online)

based on the point that the majority of the non-compliance driver tend to happen during peak hour, it indicates that the volume of vehicle is the main factor that can influence the driver behaviour.

Determination of Driver Responses at Dilemma Zone During Amber Signal Light

Figure 6 indicated that during peak hour, the probability of vehicle stopping from the distance 100–75 m is 100 %, while from the distance 75–50 m is approximately 94 %. The probability of vehicle stopping from the distance 50–25 and 25 m to the stop line dropped to approximately 80 and 22 %, respectively. During off-peak hour the probability of vehicle stopping from the distance 100–75, 75–50, 50–25 and

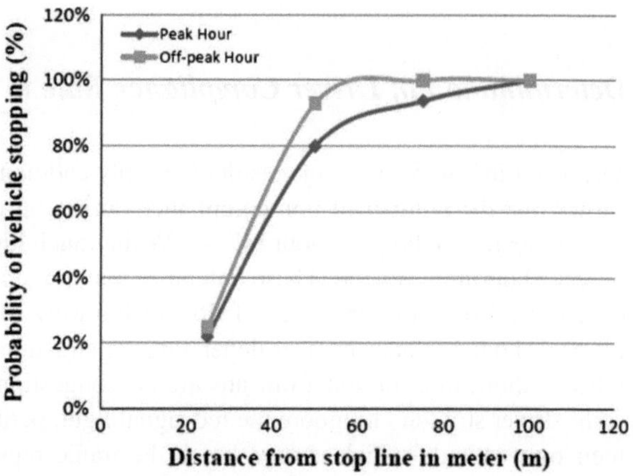

Fig. 6 Comparison of vehicle stopping probability percentage at certain interval from stop line during amber signal light

25 m to the stop line is 100, 100, 93 and 25 %, respectively (similar results were observed during weekend). The results indicate that the probability of driver to stop at amber signal light is higher before the distance of 50 m from the stop line.

Thus, the predicted dilemma zone interval exists between 50–25 m where 90 % of drivers stopped (Hesham et al. 2008) or as measured from the onset of the yellow indication, there were 3 s or more back from the stop line, 85 % of drivers would stop (Chang et al. 1985).

5 Conclusion and Recommendation

Driver approaching speed towards the signalised intersection, the variations of data taken were affected by several factors, such as traffic volume, the driver rushing ('time pressure') behaviour and the driver does not alert with the red signal light timing system. The driver compliance rate to red signal light shows that the driver behaviour towards the regulation of the signalised system influenced by traffic volume. The higher the traffic volume the higher number of driver not complied with the red signal light. Driver responses at dilemma zone during amber signal light, based on the Type II dilemma zone boundaries established by applying the 85 % of driver stopped proposed by Chang et al. (1985) and dilemma zone boundaries (exist in a distances where 10–90 % of vehicle stop) proposed by Hesham et al. (2008), the predicted dilemma zone region exists between 50–25 m interval where the percentage at that interval is between 80 and 93 % for both peak hour and off-peak hour respectively. It was obvious that driver compliance and awareness to the traffic rules at signalised intersection is more likely to be neglected during peak hour. The study has also produced a better understanding of the drivers' responses towards the existing systems at the studied signalised intersection. In the future, more driver behaviour and compliance studies should be conducted at signalised intersections for bigger sample size. In addition, further evaluation on driver behaviour and compliance can also be carried out by using other data collection methods and new technologies such as driving simulator for comparison of alternative traffic control systems.

Acknowledgements The authors would like to thank the anonymous reviewers for their valuable comments and suggestions and most importantly to the Faculty of Civil Engineering, Universiti Teknologi MARA, Pulau Pinang for the support in providing the opportunity to present this paper.

References

Bjorklund GM, Berg LA (2005) Driver behaviour in intersections: formal and informal traffic rules. Dalarna University, 781 88 Borlange, Sweden, Transportation Research Part F 8, p 239–253

Bonsall P, Liu R, Young W (2005) Modelling safety-related driving behaviour, impact of parameter values. Transp Res Part A 39:425–444

Chang MS, Messer CJ, Santiago AJ (1985) Timing traffic signal change intervals based on driver behaviour. Transp Res Rec 1027:20–30

David SH, Wang H, AK Michael Jr, Ni D, Derek M (2012) Fuzzy sets to describe driver behaviour in the dilemma zone of high-speed signalised intersections. Transp Res Part F 15:132–143

Gazis DC, Herman R, Maradudin A (1960) The problem with the amber signal light in traffic flow. Oper Res 8:112–132

Rakha H, El-Shawarby I, Amer A, Setti J, Inman V, Davis G (2008) Modelling driver behaviour at the onset of a yellow phase transition. Final report (VPI-2006-06), Department of Civil Engineering, Virginia Polytechnic Institute and State University

Scott M (2008) What is a dilemma zone. http://wiki.answers.com/Q/What_is_a_dilemma_zone?#slide=4

Murcotts (1969) Driver behaviour. http://www.murcotts.edu.au/resources/driver-behaviour-training/driver-behaviour/

Papaioannou P (2006) Driver behaviour, dilemma zone and safety effects at urban signalised intersections in Greece. Aristotle University of Thessaloniki, Department of Civil Engineering, Transport Section, Thessaloniki GR-541, 24

Shope JT (2006) Influences on youthful driving behaviour and their potential for guiding interventions to reduce crashes. Inj Prev 12:i9–i14. doi:10.1136/ip.2006.011874

Chapter 8
Enhancement of Thermophilic (*Geobacillus stearothermophilus*) Cement–Sand Mortar Properties

Raden Maizatul Aimi Mohd Azam, Hamidah Mohd Saman, Kartini Kamaruddin and Noor Hana Hussain

Abstract This study describes the incorporation method of *Geobacillus stearothermophilus* (ATCC 12978) inside the cement–sand mortar. Different cell concentrations of *Geobacillus stearothermophilus* were incorporated and the improvement in the properties of cement–sand mortar has been observed. Based on the study conducted, the involvement of *Geobacillus stearothermophilus* (ATCC 12978) has a positive impact to improve the performance of cement–sand mortar in terms of compressive strength and water absorption characteristic as compared to the controlled mortar. The incorporation of *Geobacillus stearothermophilus* has catalyst the occurrence of wollastonite ($CaSiO_3$) that has proven capable to the enhanced concrete properties by modifying concrete pore structures. This study shows that the optimum concentration gives the most enhanced properties of the resulted cement–sand mortar as 1×10^9 cfu/ml. The current work demonstrates that the direct involvement of *Geobacillus stearothermophilus* in mortar has a great potential to enhance concrete performance in such natural way.

Keywords Cell concentration · Cement–sand mortar · Growth curve · Microorganisms · *Geobacillus stearothermophilus*

R.M.A. Mohd Azam (✉) · H. Mohd Saman · K. Kamaruddin
Faculty of Civil Engineering, Universiti Teknologi MARA,
Shah Alam, Selangor, Malaysia
e-mail: raden.amy@gmail.com

H. Mohd Saman
e-mail: hamid929@salam.uitm.edu.my

K. Kamaruddin
e-mail: kartini@salam.uitm.edu.my

N.H. Hussain
Faculty of Applied Sciences, Universiti Teknologi MARA,
Shah Alam, Selangor, Malaysia
e-mail: noorhana@salam.uitm.edu.my

N.A. Yacob et al. (eds.), *Regional Conference on Science, Technology and Social Sciences (RCSTSS 2014)*, DOI 10.1007/978-981-10-0534-3_8

1 Introduction

Biomineralisation process occurs due to microbial mineral precipitation seeking researchers interest in finding sustainable yet natural method to improve the behaviour of concrete (Ghosh et al. 2006, 2009; De Muynck et al. 2008; Van Tittelboom et al. 2010; Jonkers et al. 2010; Achal et al. 2011; Siddique and Chahal 2011; Chahal et al. 2012; Xu and Yao 2014; Wang et al. 2014a, b; Kunal et al. 2014). Achal et al. (2013) stated that the biomineralisation process produces calcium carbonate or also known as calcite ($CaCO_3$) by common metabolic activities such as photosynthesis, sulphate reduction and urea hydrolysis. Calcite is the most widespread and common minerals on the earth. It is constituted at 4 % of the earth crust by weight (Achal and Pan 2014) making it suitable to be explored as an alternative natural enhancer in concrete and cementitious material properties.

Previous studies on the ureolytic bacteria, as their capabilities in producing urease enzyme that leads to biomineralisation process has shown promising results in sustainable concrete repair methodology (Ghosh et al. 2006, 2009; De Muynck et al. 2008; Van Tittelboom et al. 2010; Jonkers et al. 2010; Achal et al. 2011; Siddique and Chahal 2011; Chahal et al. 2012; Xu and Yao 2014; Wang et al. 2014a, b; Kunal et al. 2014). Some bacterial species such as *Bacillus cohnii* (Jonkers and Schlangen 2007; Jonkers et al. 2010), *Bacillus pasteurii/Spocarcina pasteurii* (Jonkers and Schlangen 2007; Maheswaran et al. 2014; Chidara et al. 2014), *Bacillus sphaericus* (Wang et al. 2010; Jagadeesha Kumar et al. 2013) and *Bacillus subtilis* (Seshagiri Rao et al. 2013) have proven their ability in enhancing concrete performance.

Most of the bacteria that have been used come from the genus *Bacilli* that are able to transform to endospore by the nutrient deprived of surrounding environment. The longevity, tenacity and persistence of endospore (Arden Jones et al. 1979; Nicholson et al. 2002) become the main factors in determining the suitable bacteria to be incorporated within the cement matrices.

With this background in view, an attempt was made to examine the effect of incorporating thermophilic type of *Geobacillus stearothermophilus* in the mortar. The *G. stearothermophilus* was chosen due to the needs of searching the bacteria that capable to resist extreme environmental stresses due to harsh concrete environment. Only with those stresses, the pinnacle of evolution that represents bacterial spores will occur (Arden Jones et al. 1979; Nicholson et al. 2002; Sapto et al. 2010).

This study using living cells (vegetative cells) instead of spores which in its active state to be incorporated in the cement–sand mortar to create a harsh environment for the bacteria. The present study is also aimed to investigate the existence of biomineralisation product derive from *G. stearothermophilus* in its vegetative cells into cementitious material. Different cell concentrations of bacteria were counted to measure the optimum growth and optimum concentrations of bacteria to be used in mortar. The initiative to use *G. stearothermophilus* would be beneficial in varying capabilities of bacteria in enhancing concrete properties in such natural ways.

The result of this experiment will draw either positive or negative outcome towards cementitious properties throughout the addition of thermophile *Geobacillus stearothermophilus* (ATCC 12978). For record, there has been no attempt so far to include *Geobacillus stearothermophilus* (ATCC 12978) in cement mortar leaving all possibilities to be studied of. The main objective of this experiment is to study the effect on direct involvement of the anaerobic *Geobacillus stearothermophilus* (ATCC 12978) in improvise the characteristics of cement–sand mortar.

2 Materials and Methods

Materials

Ordinary Portland Cement (OPC) was used in this study. The OPC fineness is recorded as 98 % its particle passing 90 μm. The fine aggregate used in this study is natural river sand, locally available, clean and well graded. The specific gravity of sand is 2.67. In order to reduce other effects on the results, the distilled water was used as mixing water in preparing the mortar.

Culture Collections

Geobacillus stearothermophilus (ATCC 12978) was purchased from All Eights (M) Sdn. Bhd., Malaysia. It is a gram positive thermophilic (heat loving) bacteria characterized by an inner cell membrane and a thick cell wall. *Geobacillus stearothermophilus* is a rod shaped anaerobe found in thermophilic habitats like thermal vents. The bacteria grows in temperature ranges from 35 to 65 °C (Sapto et al. 2010).

Growth Curve

The bacteria was cultured anaerobically in a nutrient broth placed in shaker incubator at 65 °C of temperature. To ensure the pure culture of the bacteria was used throughout the experimental work, the 24 h culture being examined under microscopic examination. The growth curve of the bacteria was carried up by monitoring of the growth through viability counting on the plate for several hours (Ninganagouda et al. 2014). The optical density of the cultures was measured by spectrophotometry at a wavelength of 600 nm (Domínguez et al. 2001). The OD

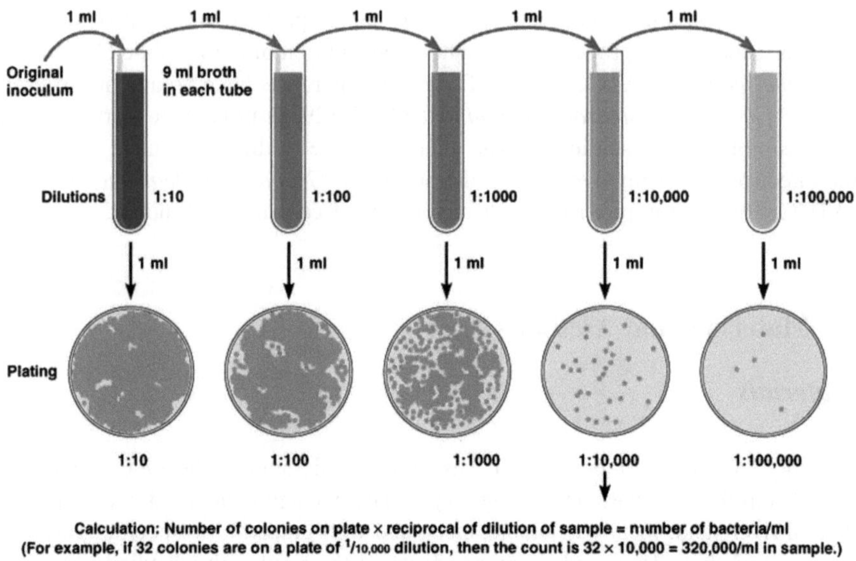

Fig. 1 Serial dilution of bacteria (Photo by Elizabeth H., accessed on 17/09/2014)

reading being taken every 0.5 h for first four (4) hours followed by one (1) hours for each interval until the OD reading drops indicating the death phase of the bacteria. The taken reading of OD were plotted and analyzed. However, OD reading only indicating the growth pattern or curve in meet requirement as standard growth curve which included four (4) different phases of bacteria growth namely lag phase, exponential growth phase, stationary phase and death phase.

To obtain the cell number based on the growth curve derived by microorganism, the plate count or also know as viable count method was used. Using same interval period of OD reading method as mentioned above, the bacteria were transferred aseptically into the diluent for serial dilution process as shown in Fig. 1. The serial dilution has been performed in order to having a higher dilution of microorganisms so that the bacteria was countable. Every 0.1 ml of bacteria taken from each dilution into the plate and incubation process took 24 h before count the growth of bacteria on the plate.

The calculation of the cell number was calculated by Eq. (1).

$$\text{cfu/ml} = (N_c \times D_f)/V_i \tag{1}$$

where

N_c Number of colonies
D_f Dilution factor
V_i Volume of inoculation.

Biochemical Test

Two types of biochemical tests were adopted which are determination of urease production and calcium content. The tests were performed to determine whether the bacteria is suitable to be incorporated inside the cement–sand mortar producing biomineralisation product in enhancing mortar performance.

Urease production The Christensen media was used for the detection of rapid urease activity of any urease-positive microorganism. The slant containing the Christensen media with droplets of 24 h culture *Geobacillus stearothermophilus* were incubated at 65 °C for 48 h. The presence of urease producing microorganisms indicates by the change colour of media from yellowish to deep red. The amount of urease produced in the medium is proportioned to the intensity of colour developed in the media (Sapto et al. 2010).

Calcium carbonate precipitation *Geobacillus stearothermophilus* was cultured in modified nutrient broth with other components like urea, yeast extract and calcium lactate ($C_6H_{10}CaO_6$). The aqueous ammonia oxalate added into 10 drops of 1 OD cells of media for alkalisation. With blank broth as controlled, the spectrophotometer of 600 nm was used to confirm the OD of the media.

Preparation of Cement–Sand Mortar Cubes

The cement–sand mortar mix was prepared for testing using parameters as followed:

The specific gravity of fine aggregate was 2.65
The specific gravity of cement was 3.15
A (cement:sand) mix ratio were 1:3 at w/c 0.5 was adopted
The dimensions of the cube were 50 mm × 50 mm × 50 mm.

Different concentrations ranged from 1×10^3 to 1×10^{11} cfu/ml of cell concentrations for every ml mixing water were mixed with cement and fine aggregates. The cement–sand mortar mix then was poured to the moulds and was compacted using vibrator table. After 24 h of casting, the demoulding took place and all the mortar specimens were undergone water curing until testing at 3, 7, 28 and 60 days.

Testing of Prepared Cement–Sand Mortar Cubes

Two (2) tests were conducted which are compressive strength and water absorption tests. Scanning Electron Microscopy (SEM) was also performed to examine the microstructure of the cement–sand mortar samples incorporated with *Geobacillus stearothermophilus*.

Compressive Strength Microbial cement–sand mortar undergone compressive strength test to evaluate the imprint of *G. stearothermophilus* direct incorporated inside the cement–sand mortar. Five (5) concentrations of bacteria were tested and compared with normal controlled mortar. The cement–sand mortar specimens were tested and its compressive strength as methods prescribed based on ASTM C1586—05 (2011) and ASTM C780.

Water Absorption The 50 mm × 50 mm × 50 mm cement–sand mortar cube were taken out from curing tank and subjected to water absorption test (based on ASTM C642—13) at age 3, 7, 28 and 60 days. These specimens were then oven dried for 72 h at the temperature 105 °C until the constant mass was obtained. After that the specimen was kept in vacuum desiccator for 24 h before weighed. This weight was noted as the dry weight (W_1) of the cement–sand mortar. For the weight changes, the specimens were immersed completely in water and the specimens were weighed every 30 min for 4 h. Then, this weight was noted as the wet weight (W_2) of the cubes. The saturated water absorption (Wang et al. 2014b) was calculated by Eq. (2).

$$W_s = (W_w - W_d)/W_d \times 100\% \tag{2}$$

where

W_s saturated water absorption ratio
W_d dry weight of the specimen
W_w wet weight of the water saturated specimens.

SEM Examination and EDX Analysis

The morphology and composition of cement–sand mortar surface were investigated using Scanning Electron Microscopy (SEM). The broken samples from compressive test were collected and dried before coated by gold sputter equipment at the Faculty of Applied Sciences, UiTM Shah Alam. Using the same machine, Energy-Dispersive X-ray Spectroscopy (EDX) was also performed to examine the element composition of a sample.

3 Results

Growth Curve

Figure 2 shows the pure culture of *Geobacillus stearothermophilus* with its rod shape and almost similar size without the existence of spore intact. The pure culture is an individual cell that will divide and multiplying each cell to form a discrete

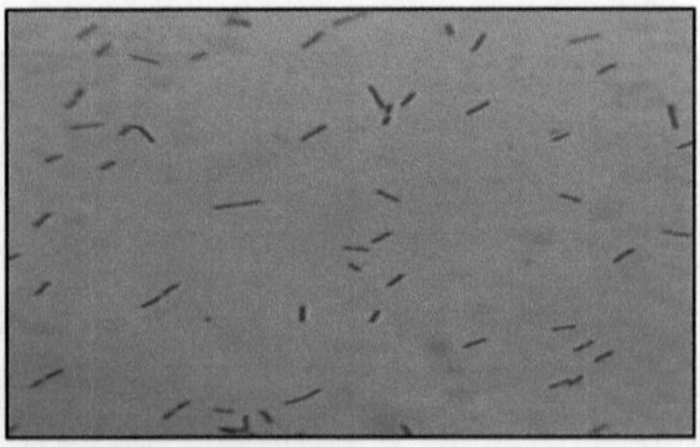

Fig. 2 Shape of *Geobacillus stearothermophilus* with (100×) magnification as seen under microscope

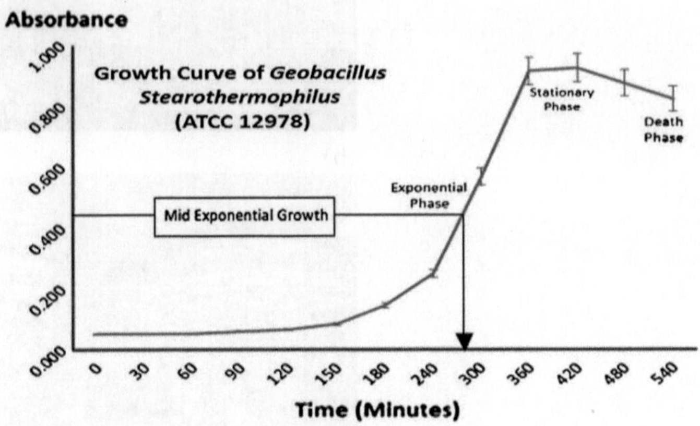

Fig. 3 Growth curve of *G. stearothermophilus*

colony. Figure 3 shows the growth curve of *Geobacillus stearothermophilus* (ATCC 12978). Based on multiple examinations and calculations formula on Eq. 1, the plate count of *Geobacillus stearothermophilus* (ATCC 12978) during its mid exponential growth was found to be 3.75×10^{11} cfu/ml.

Biochemical Test

Urease production After 48 h incubation of Christensen media agar with *Geobacillus stearothermophilus*, the colour changed from yellow to red colour (Fig. 4a) indicated the ability of bacteria to produce urease enzyme. The microorganisms that capable producing urea will induce precipitation and trapping calcium ions (Wang et al. 2014b).

Fig. 4 **a** Christensen media without bacteria (*left*); *Geobacillus stearothermophilus* in Christensen media after incubation for 48 h (*right*). **b** The presence of calcium by *Geobacillus stearothermophilus* by white precipitation (*right*) as compared to control (*left*)

Calcium carbonate precipitation The ability of bacteria to induce carbonate precipitation confirms by the presence of calcium carbonate in the medium. It is recognised through white precipitation based on reaction with Ammonium oxalate. Figure 4b shows the presence of calcium by white precipitation of the media.

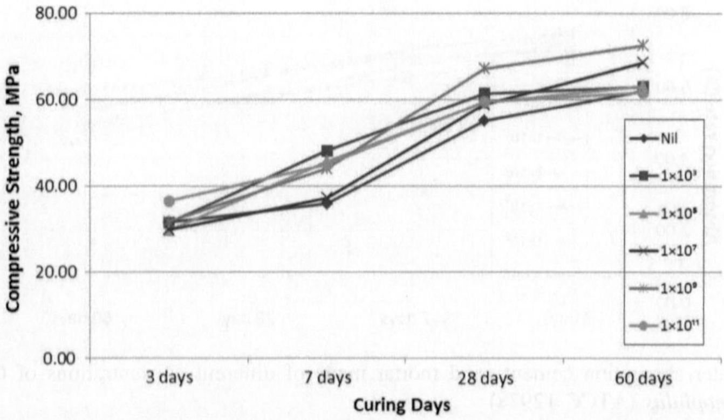

Fig. 5 Compressive strength of cement–sand mortar incorporated with different concentration of *Geobacillus stearothermophilus* (ATCC 12978)

Compressive Strength of Microbial Cement–Sand Mortar

Compressive strengths of cement–sand mortar cubes were tested to determine the efficiency of improvement by bacterial cells *Geobacillus stearothermophilus*. Figure 5 summarize the results of compressive strength and the variation of strength given with the optimum bacterial concentration that charts highest strength is 1×10^9 cfu/ml. It is revealed that the addition of *G. stearothermophilus* into the cement–sand mortar mix has improved the strength characteristics of the cement–sand mortar.

Water Absorption of Microbial Cement–Sand Mortar

The water absorption for five (5) series of microbial cement–sand mortar with normal mortar as control is shown in Fig. 6. As shown in Fig. 6, the addition of *Geobacillus stearothermophilus* had a positive effect on decreasing the capillary water absorption towards cement–sand mortar. The lowest water absorption has been achieved by inclusion of *Geobacillus stearothermophilus* at the concentration of 1×10^9 cfu/ml. The water absorption result portrays a good relationship with the improved strength of cement–sand mortar due to the inclusion of bacteria. The absorption rate decreased as the concrete matured due to the product hydration of cement particles hydration in mortar with the aid of bacteria that filled up the spaces.

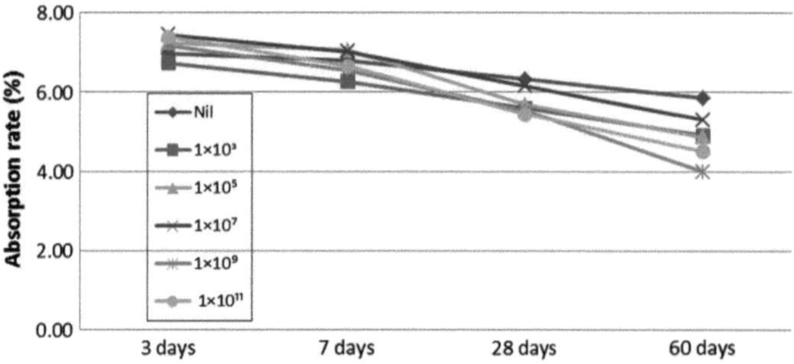

Fig. 6 Water absorption cement–sand mortar made of different concentrations of *Geobacillus stearothermophilus* (ATCC 12978)

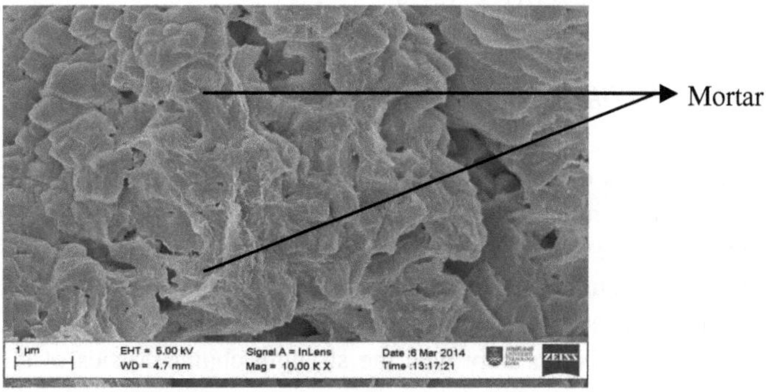

Fig. 7 SEM image on normal controlled mortar for 28 days

SEM Analysis and EDX Examination

The broken sample for 28 days for control cement–sand mortar and 1×10^9 cfu/ml was examined under Scanning Electron Microscopy (SEM). Direct incorporation of bacteria into the cement–sand mortar alters the microstructure and the morphology of bacteria and clearly can be seen through the image captured. Figure 7 shows the SEM micrograph of the control cement–sand mortar while Fig. 8 shows the EDX analysis of the sample. Figure 9 shows the evidence of the alteration in the microstructure with the addition of *Geobacillus stearothermophilus*. The analysis for EDX with sample added of *Geobacillus stearothermophilus* for 28 days is shown in Fig. 10. It is found that inclusion of G. *stearothermophilus* has precipitated calcite within the cement matrix. Bacteria was found has an intimate interaction between cement–sand mortar and the production of calcite results in between.

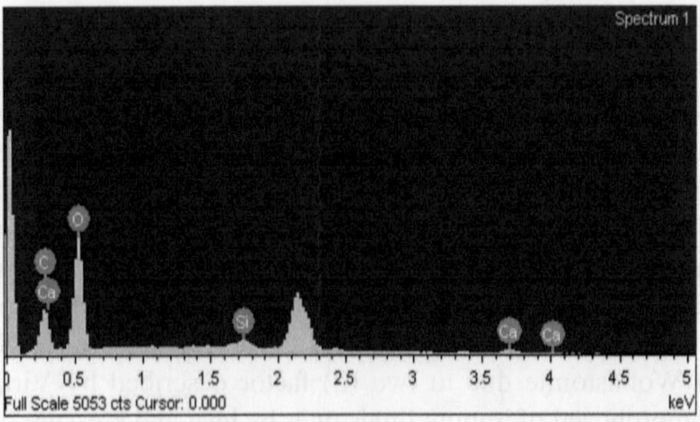

Element	Weight%	Atomic%
C K	10.21	17.49
O K	46.24	59.46
Si K	3.14	2.30
Ca K	40.41	20.75
Totals	100.00	

Fig. 8 The EDX results of sample control cement–sand mortar

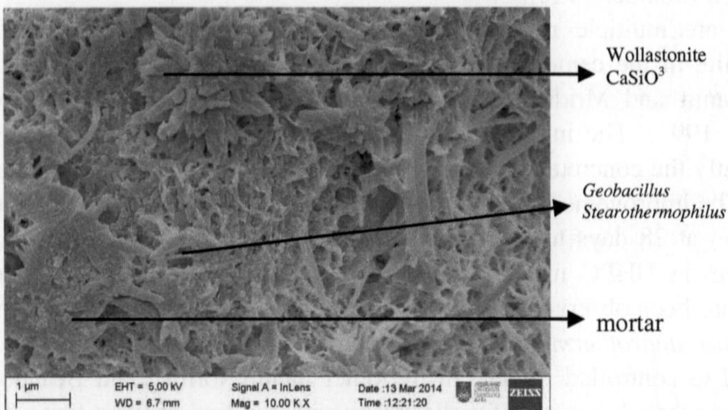

Fig. 9 SEM image on microbed cement–sand mortar (with addition of *Geobacillus stearothermophilus*) concentration of 1×10^9 cfu/ml for 28 days

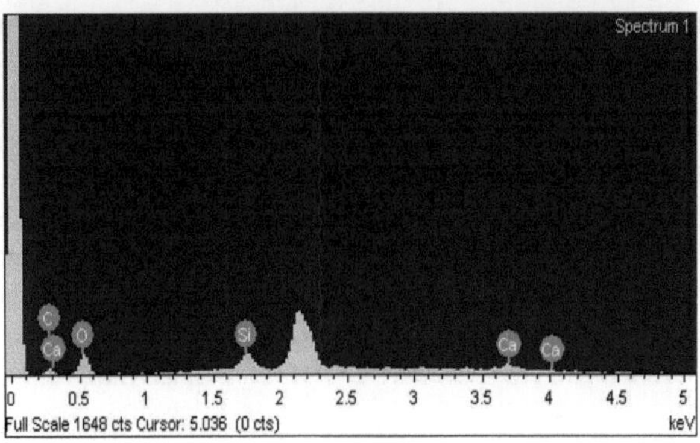

Element	Weight%	Atomic%
C K	2.41	6.13
O K	13.94	26.63
Si K	10.59	11.52
Ca K	73.06	55.71
Totals	100.00	

Fig. 10 The element results of microbed cement–sand mortar (with addition of *Geobacillus stearothermophilus*) concentration of 1×10^9 cfu/ml for 28 days

4 Discussion

The optimum concentration of *G. stearothermophilus* that has proven in this study is 1×10^9 cfu/ml giving a positive impact for both compressive strength and water absorption characteristics. The improvement made on compressive strength is 15 % and water absorption 31 %, respectively, is corresponding to the control samples. The SEM EDX analysis also proved that the cement–sand mortar induced minimum calcite but high wollastonite ($CaSiO_3$) as compared to controlled mortar. This can be seen through combination of single element that forms $CaSiO_3$ which are Ca, Si and O. The occurrence of Wollastonite due to two (2) factor described by Virta (2001), which are, (1) metamorphosed of impure limestones by heat and pressure or (2) silica bearing fluids are introduced into calcareous sediments during metamorphic processes. In both cases, calcite reacts with silica to produce wollastonite and carbon dioxide.

There are multiple researches previously conducted on the contribution of wollastonite in the cementitious material (Soliman and Nehdi 2014; Kalla et al. 2013; Manjit and Mridul 2007; Mathur et al. 2007; Virta 2001; Norman and Beaudoin 1991). The incorporation of wollastonite in the concrete has increased significantly the concrete mechanical properties as reported by Soliman and Nehdi (2014). The improvement in compressive strength (28–35 %) and flexural strength (36–42 %) at 28 days to 56 days has discovered by the addition of wollastonite microfibers in UHPC mixtures. A positive effect on the early-age compressive strength has been observed. This is similarity implicated by higher concentration of *Geobacillus stearothermophilus* in this study resulting early strength at 3 days as compared to controlled mortar. In the other hand, Norman and Beaudoin (1991) demonstrate the advantages of wollastonite with addition of silica fume. The results are discussed in terms of pore structure parameters and physical characteristics as determined by the mercury porosimetry and the helium gas pycnometry measurements. Based on the results of Mercury porosimetry, it is indicated that the total porosity in the cement–silica fume-wollastonite composite is in the range of 15.4–18.5 % and the corrected total porosity is in the range of 15.5–21.5 %. Porosity in both data sets increases gradually with the amount of wollastonite micro-fibres (2–15 % by volume).

In summary, the positive potential of *Geobacillus stearothermophilus* that has been demonstrated in this study provides an interesting concept of its direct involvement. The capability of *Geobacillus stearothermophilus* in cementitious material by modifying the microstructure hence improving its pore properties and increase the natural occurrence of wollastonite become an interesting field to be explored further in near future.

5 Conclusion

From the results and discussion, several conclusions can be drawn as follows:

1. It is deduced that *Geobacillus stearothermophilus* improve the compressive strength of the resulted mortar by 14 % as compared to the control one.
2. The higher wollastonite detected reduced the water absorption rate by 31 % of the microbed cement–sand mortar specimens.
3. High concentration of the *Geobacillus stearothermophilus* induces higher amount wollastonite and contributes to early strength of microbed cement–sand mortar.
4. The percentage of wollastonite is found to increase in the microbed cement–sand mortar as compared to control mortar specimens.

Acknowledgement The authors would like to acknowledge the financial support from Fundamental Research Grant Scheme (600-RMI/ST/FRGS 5/3/Fst 28/2011), Ministry of Higher Education (MOHE), and the management of Research Monitoring Unit (RMI), Universiti Teknologi MARA (UiTM). Special appreciation for Faculty of Applied Sciences, the UiTM for providing the facilities in the microbiologist fields and Faculty of Civil engineering, UiTM for providing the facilities to carry out testing on mortar and for the technical support.

References

Achal V, Pan X, Özyurt N (2011) Improved strength and durability of fly ash-amended concrete by microbial calcite precipitation. Ecol Eng 37:554–559

Achal V, Pan X (2014) Influence of calcium sources on microbially induced calcium carbonate precipitation by Bacillus sp. CR2. App Biochem Biotechnol 173:307–317

Achal V, Mukerjee A, Sudhakara Reddy M (2013) Biogenic treatment improves the durability and remediates the cracks of concrete structures. Constr Build Mater 48:1–5

Arden Jones MP, McCarthy AJ, Cross T (1979) Taxonomic and serological studies on Micropolyspora faeni and Micropolyspora strains from soil bearing the specific epithet rectivirgula. J Gen Microbiol 115:343–354

Beaudoin JJ, Low MP (1991) Mechanical properties of high performance cement binders reinforced with wollastonite micro-fibres

Chahal N, Siddique R, Rajor A (2012) Influence of bacteria on the compressive strength, water absorption and rapid chloride permeability of fly ash concrete. Constr Build Mater 28:351–356

Chidara R, Nagulagama R, Yadav S (2014) Achievement of early compressive strength in concrete using *Sporosarcina pasteurii* bacteria as an admixture. Adv Civil Eng

De Muynck W, Cox K, Belie ND, Verstraete W (2008) Bacterial carbonate precipitation as an alternative surface treatment for concrete. Constr Build Mater 22:875–885

Domínguez MC, de la Rosa M, Victoria Borobio M (2001) Application of a spectrophotometric method for the determination of post-antibiotic effect and comparison with viable counts in Agar. J Antimicrobial Chemotherapy 47:391–398

Ghosh P, Mandal S, Chattopadhyay BD (2006) Effect of addition of microorganism on the strength of concrete. Indian Conc J 45–48

Ghosh S, Biswas M, Chattopadhyay BD, Mandal S (2009) Microbial activity on the microstructure of bacteria modified mortar. Cement Concr Compos 31:93–98

Jagadeesha Kumar BG, Prabhakara R, Pushpa H (2013) Effect of bacterial calcite precipitation on compressive strength of mortar cubes. Int J Eng Adv Technol (IJEAT) 2

Jonker HM, Schlangen AE (2007) Crack repair by concrete-immobilized bacteria. In: Proceedings of the first international conference on self healing materials, Noordwijk Aan Zee, The Netherlands, pp 1–7

Jonkers HM, Thijssen A, Muyzer G, Copuroglu O, Schlangen E (2010) Application of bacteria as self-healing agent for the development of sustainable concrete. Ecol Eng 36:230–235

Kalla P, Misra A, Gupta RC, Csetenyi L, Gahlot V, Arora A (2013) Mechanical and durability studies on concrete containing wollastonite-fly ash combination. Constr Build Mater 40:1142–1150

Kunal, Siddique R, Rajor A (2014) Influence of bacterial treated cement kiln dust on the properties of concrete. Constr Build Mater 52:42–51

Maheswaran S, Rama Chandra Murthy A, Bhuvaneshwari B, Ramesh Kumar V, Palani GS, Iyer NR, Krishnamoorthy S, Sandhya S (2014) Strength improvement studies using new type wild strain Bacillus cereus on cement mortar. Curr Sci 106:50–58

Manjit S, Mridul G (2007) Durability of cementing binders based on fly ash and other wastes. Construc Building Mater 21:2012–2016

Mathur RM, AK Misra, Goel P (2007) Influence of wollastonite on mechanical properties of concrete. J Sci Ind Res 66:1029–1034

Nicholson W, Fajardo-Cavazos P, Rebeil R, Slieman T, Riesenman P, Law J, Xue Y (2002) Bacterial endospores and their significance in stress resistance. Antonie Van Leeuwenhoek 81:27–32

Ninganagouda S, Rathod V, Singh D, Hiremath J, Singh AK, Mathew J, Ul-Haq M (2014) Growth kinetics and mechanistic action of reactive oxygen species released by silver nanoparticles from Aspergillus niger on Escherichia coli. Biomed Res Int 2014:9

Sapto RA, Prima Endang T (2010) Isolation of thermostabl α-amylase from local thermophilic bacteria for liquefaction. In: Proceedings of the third international conference on mathematics and natural sciences (ICMNS 2010), pp 342–354

Seshagiri Rao MV, Srinivasa Reddy V, Hafsa M, Veena P, Anusha P (2013) Bioengineered concrete—a sustainable self-healing construction material. Res J Of Eng Sci 2:45–51

Siddique R, Chahal NK (2011) Effect of ureolytic bacteria on concrete properties. Constr Build Mater 25:3791–3801

Soliman AM, Nehdi ML (2014) Effects of shrinkage reducing admixture and wollastonite microfiber on early-age behavior of ultra-high performance concrete. Cement Concr Compos 46:81–89

Van Tittelboom K, De Belie N, De Muynck W, Verstraete W (2010) Use of bacteria to repair cracks in concrete. Cem Concr Res 40:157–166

Virta RL (2001) Wollastonite. U.S. geological survey minerals yearbook, 002, 83.1–83.2

Wang JY, De Belie N, Verstraete W (2010) Potential of applying bacteria to heal cracks in concrete. In: Proceedings of the second international conference on sustainable construction materials and technologies, pp 1–12

Wang JY, Snoeck D, Van Vlierberghe S, Verstraete W, De Belie N (2014a) Application of hydrogel encapsulated carbonate precipitating bacteria for approaching a realistic self-healing in concrete. Constr Build Mater 68:110–119

Wang JY, Soens H, Verstraete W, De Belie N (2014b) Self-healing concrete by use of microencapsulated bacterial spores. Cem Concr Res 56:139–152

Xu J, Yao W (2014) Multiscale mechanical quantification of self-healing concrete incorporating non-ureolytic bacteria-based healing agent. Cem Concr Res 64:1–10

Chapter 9
A Congestion Control Optimal Model Design Perspectives of Non-safety Applications on Vehicular Ad Hoc Networks: A Survey of Requirements, Protocols, and Challenges

Shamsul Jamel Elias, Mohd Nazri Mohd Warip, Badlishah Ahmad and Aznor Abd Halim

Abstract The issues of congestion control on non-safety applications in vehicular ad hoc networks (VANETs) have received considerable crucial attention in the field of intelligent transport systems (ITS). The critical challenges of ITS applications are in the events of intermittent connectivity and data collisions. Another challenge is the event-driven effect on both safety and non-safety applications that are critically relied on delay and reliability. So far, the development of protocol stacks for VANETs implies the WAVE protocol other IEEE 802.11p MAC protocol. Based on IEEE 802.11p MAC protocol standard in the 5.9 GHz band (other similar protocols such as C2C-CC for European and CALM from ISO) in which adopt of vehicle manufacturers of Information Communication Technology (ICT) to address the safety systems, emergency (breakdown, SOS), traffic control, environmental, navigational, and tracking. Little attention has been given to non-safety applications, for instance, chat, twitter, SMS, internet services, video-on-the-go, etc. Besides, roadside-vehicle communications (RVC) caters for information concerning repair notifications, remote diagnostics, context information, navigation information, and alerts system. In this survey, we present a framework that proposes a guideline and recommendation for the congestion control mechanism perspectives in VANETs. In particular, researches are centered on the investigation of the

S.J. Elias (✉)
Faculty of Computer and Mathematical Sciences,
Universiti Teknologi MARA, Merbok, Kedah, Malaysia
e-mail: shamsulje@kedah.uitm.edu.my

M.N. Mohd Warip · B. Ahmad · A. Abd Halim
School of Computer and Communication Engineering,
Universiti Malaysia Perlis, Arau, Perlis, Malaysia
e-mail: nazriwarip@unimap.edu.my

B. Ahmad
e-mail: badli@unimap.edu.my

N.A. Yacob et al. (eds.), *Regional Conference on Science, Technology and Social Sciences (RCSTSS 2014)*, DOI 10.1007/978-981-10-0534-3_9

mobility model of congestion control algorithms. Finally, this survey contributes to appropriate knowledge and challenges in legacy congestion control of VANETs and toward the future research development.

Keywords Non-safety applications · WAVE protocol · Inter-vehicle communications (IVC) · Roadside-vehicle communications (RVC)

1 Introduction

The vehicular ad hoc network (VANET) is a progressively indispensable area in mobile ad hoc networks (MANETs). Recently, researchers have shown an increased interest in vehicular communications. According to definition provided by authors in (Hartenstein and Laberteaux 2008), the VANETs comprise vehicle-to-vehicle (V2V) and vehicle-to-infrastructure (V2I) communications based on wireless local area network technologies. The distinctive set of candidate application makes VANET a unique area of wireless communication (Hartenstein and Laberteaux 2008) as follows:

1. collision warning, traffic-jam or accident, and local traffic information for drivers;
2. Internet resources (accessing the Internet, licensed spectrum, rechargeable power source),
3. the environment of wireless topology (e.g., vehicular traffic flow patterns, privacy concerns).

The Federal Communication Commission (FCC) formulated VANETs for wireless spectrum frequency prior to establish a channel in 2003 for communication, the dedicated short range communications (DSRC) service. The 5.850–5.925 GHz bands are used for public safety and private applications of non-safety and comfort applications for DSRC communication service (He et al. 2010).

In order to communicate using VANETs, the IEEE is working on the IEEE 802.11p (WAVE—Vehicular Environments Wireless Access) and 802.11 protocol standard for DSRC. The reliability and low latency requirements for non-safety applications are suitable for VANET environments, such as high vehicular mobility and faster topology changes. The DSRC was designed using a multichannel system. The FCC divided the spectrum into seven channels, each with 10–20 MHz, in which six were identified as service channels (SCHs) and one as a control channel (CCH). The CCH channel is used for safety messages while non-safety services (WAVE-mode short messages) are expected to go through the other six service channels available (Bouassida and Shawky 2008; Park et al. 2013a). Figure 1 explained the breakdown of control channel (CCH) and service channels (SCHs) derived from IEEE 1609.1/.2/.3/.4. (Morgan 2010). As specified in Fig. 1, we will focus on the congestion control of service channel control (SCH). Basically, there are two categories of SCH applications: real-time applications and non-real-time applications.

In essence, Fig. 2 presents a classification framework of congestion control mechanism implemented in VANET topology. There are five key domains in congestion control comprising applications, rebroadcast routing scheme, power control transmitting, and scheduling algorithm along with congestion control and control.

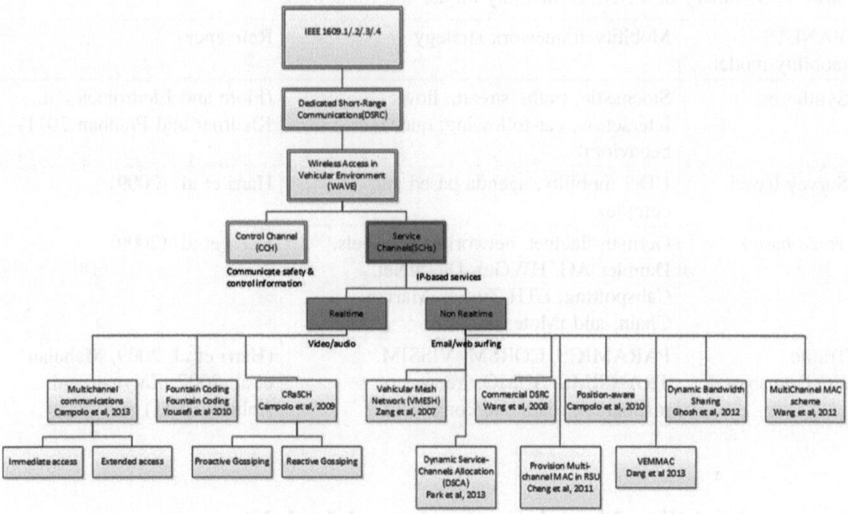

Fig. 1 Topology for IEEE 1609.1/.2/.3/.4 (Morgan 2010)

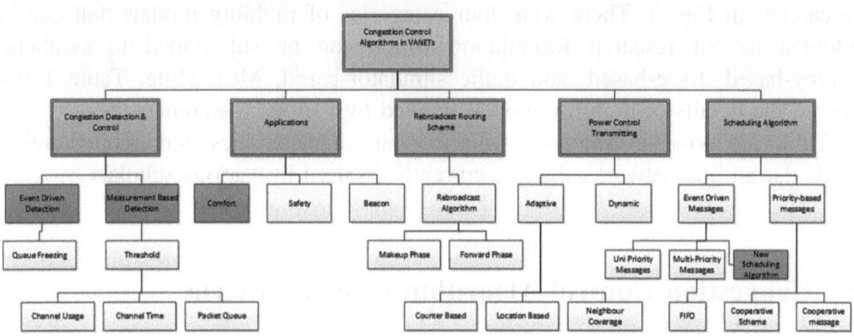

Fig. 2 Taxonomy of VANETs general topology with congestion control mechanism

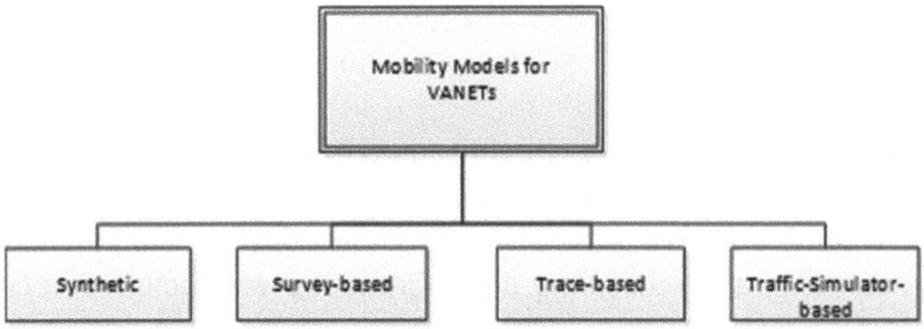

Fig. 3 Topology for mobility models based on Harri et al. (2009)

Table 1 Summary of VANETs mobility model and framework

VANETS mobility model	Mobility framework strategy	References
Synthesis	Stochastic, traffic stream, flows interaction, car-following, queue, behavioral	(Fiore and Elettronica n.d.; Khairnar and Pradhan 2011)
Survey-based	UDel mobility, agenda-based and complex	Harri et al. (2009)
Trace based	German fleetnet, networks on wheels, Daimler AG, HWGui, DieselNet, Cabspotting, ETH Zurich, Markov Chain, and iMotes	Harri et al. (2009)
Traffic simulator-based	PARAMICS, CORSM, VISSIM, TRANSIMS, SUMO, freeway, manhattan, road, and construction	(Harri et al. 2009; Mahajan et al. 2007; Zhiyuan and Jinhong 2011)

2 The Mobility Model Concepts and Insights

The research framework will be derived from the mobility models proposed by researcher in Fig. 3. There were four categories of mobility models that can be adopted in our research formulation which can be subdivided to synthetic, survey-based, trace-based, and traffic simulator-based. Meanwhile, Table 1 discussed the details of mobility models applied by various researchers.

Table 2 provides a summary of applications, characteristics, and architectures of vehicular ad hoc networks that are currently focused by various scholars.

3 Congestion Control Algorithm Design Criteria

The foundation challenges for designing congestion control algorithm can be adopted from the process proposed by Darus and Bakar (2011). Some modification has been suggested by author in carrying out this research. As illustrated in the

Table 2 Categorization of VANETs' applications derived from Cunha et al. (n.d)

Application	Characteristic	Architecture	Location awareness	Time awareness	Communication technology	Protocols	Challenges	Application examples
Safety	Delay	V2V, V2I	Yes	Yes	DSCR/RFID/Bluetooth/WiFi	Reliability	Reduce latency	Collision alert, intersection collision
Efficiency	Availability	V2V, V2I	Yes	Yes	DSCR	Real-time and reliability	Service availability	Traffic flow, accident, and road condition
Comfort	Reliability	V2I	Yes	No	WiMAX/WiFi/3G/4G/LTE	Real-time	Support on demand	Free parking space, music downloading
Interactive entertainment	Connectivity and availability	V2V-V2I	No	Yes	WiMAX/WiFi/3G/LTE	Unicast	Keep synchronization	Games, Internet and data synchronization
Non interactive entertainment	Delay tolerant	I2V	No	No	WiMAX/WiFi/3G/LTE	Data dissemination	Appropriate throughput	Video download and data synchronization
Urban sensing	Mobility	V2V-V2I	Yes	Yes	DSCR/WiMAX/WiFi/3G/LTE	Data collection	Security issue	Photographs, road condition

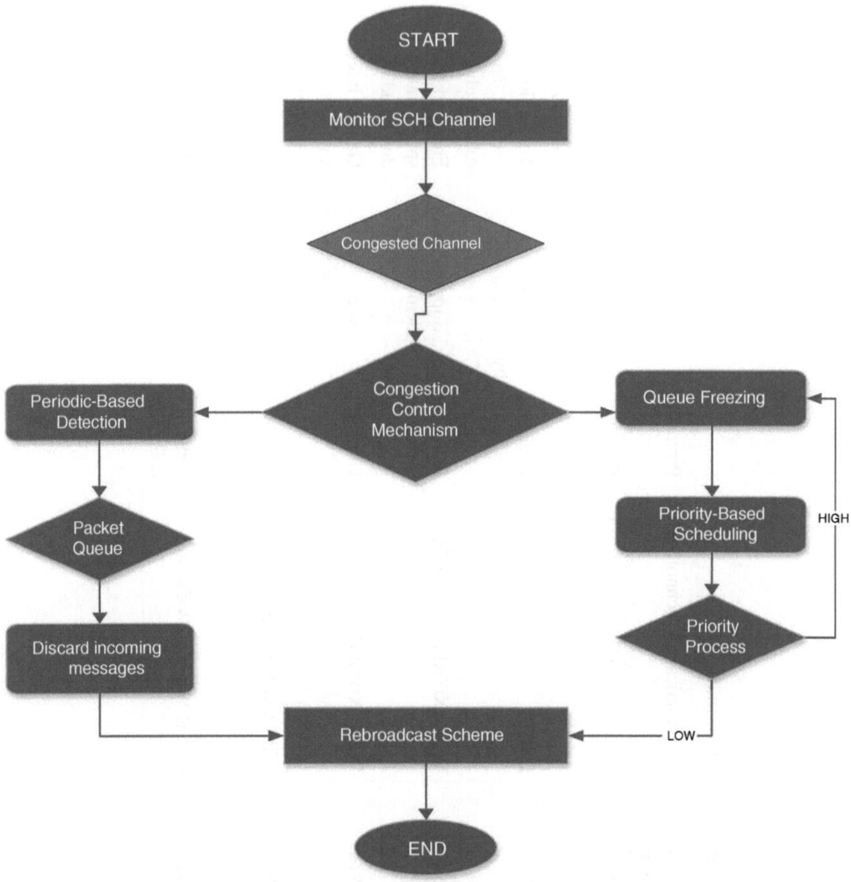

Fig. 4 VANETs flow chart algorithm adopted from Darus and Bakar (2011)

flowchart of Fig. 4, the congestion control algorithm process can be broken into measurement-based detection and queue freezing techniques which monitor all the data going through the SCH channel. Zang et al. (2007) proposed the novel VMESH MAC protocol for enhancing the performance of non-safety applications in vehicular environments based on the WAVE infrastructure. The proposed MAC protocol generates a distributed beaconing scheme and a reservation-based channel access (DRP) on SCH to increase the channel access efficiency. They also investigated the performance of the VMESH MAC protocol in more realistic scenarios with stochastic simulations and more realistic mobility, channel, and traffic models. The literature in Campolo et al. (2009) introduced CRaSCH, a cooperative scheme in 802.11p/WAVE-based vehicular networks for service-channels reservation. The proposed scheme targets vehicle-to-infrastructure and vehicle-to-vehicle communications, in which nodes, either roadside or on-board units, and acting as WAVE providers, choose a service channel for their initialized WBSS. CRaSCH (cooperative reservation of SCH) a gossip-based reservation mechanism relies on cooperation among nearby providers. Specifically, they proposed two approaches:

1. Proactive gossiping: every provider advertises its own SCH and SCHs are reserved by nearby users or providers
2. Reactive gossiping: every provider spreads out the perceived SCH status information to nearby users or providers.

Furthermore, research efforts are required to investigate the additional countermeasures taken by providers or users to protect data transmissions in the SCH interval against collisions due to the hidden nodes.

Cheng et al. (2011) considered a set of simple and easy-to-deploy data rate adaptation policies that rely on position information during the SCH interval in order to select the data rate for packet transmission. As a future work, the researchers plan to enhance the proposed solution to work in urban scenarios in which time-varying multipath channel conditions and higher vehicle density could affect performance. Moreover, the researchers have enhanced the solution with functionalities that are already partially available in the literature, to differentiate the causes of frame losses and failures due to collisions from failures caused by channel errors/weak signals, in order to avoid performance degradation due to rate under selection when the link conditions are good and not saturated.

Amadeo et al. (2009) studied a novel MAC and SCH allocation scheme to guarantee the QoS of non-safety services and the reliability of safety services in the RSU-assisted VANETs. We anticipate vehicle density for improving the non-safety service time. Accordingly, Fig. 4 explained the research methodology on congestion control algorithm adopted from previous scholar that focused on CCH instead of SCH. The studies have shown the improvement in QoS performance for non-safety services by applying a MAC-dedicated multichannel allocation scheme based on channel throughput analysis for centralized RSU in VANETs.

Wang et al. (2012) addressed a variable CCH interval (VCI) multichannel medium access control (MAC) scheme, which can adjust the length ratio between the CCH and SCHs. In addition, the scheme introduces a multichannel coordination mechanism to provide contention-free access of SCHs. Markov modeling is applied in optimizing the intervals range based on the traffic condition. The proposed scheme shown in IEEE 1609.4 MAC significantly enhanced the throughput of SCHs and reduced the transmission delay of the service packets.

The proposed taxonomy in Table 3 was derived from the characteristic of congestion control algorithms and functionality of each algorithm in VANETs. Sattari et al. (2012) concluded that proactive algorithm was one of the best congestion control algorithms for VANETs which used transmission power and packet generation rate control at the same time, based on dynamic carrier sense threshold thus differentiating the priorities of the packets types. Campolo and Molinaro (2013) studied the use of multiple channels in vehicular networks. The analysis commenced with the design challenges unique to the vehicular environment that need to be addressed in order to make decisions concerning the adoption,

Table 3 Summary of VANETs' congestion control techniques (Sattari et al. 2012)

Class	Approach	Packet rate	Utility function	Power control	Access priority	Carrier sense	Smart rebroadcast
Proactive	Vehicle-to-vehicle for cooperative collision warning (VCCW)	✓	×	×	×	×	×
	Utility-based packet forwarding and congestion control (UBPFCC)	×	✓	×	✓	×	×
	Dynamic priority-based scheduling (DPBS)	×	✓	×	✓	×	×
	Cross-layer congestion control (Cross-layer CC)	×	×	✓	✓	×	✓
	Broadcast reception rates and effects of priority access (BRR-EPA)	×	×	✓	✓	×	×
	Application-based congestion control (ABCC)	✓	×	✓	×	×	×
	D-FPAV—a fully distributed and localized algorithm	×	×	✓	✓	✓	×
Reactive	Power or rate based congestion control	✓	×	✓	×	✓	×
Hybrid	Power and rate combined congestion control	✓	×	✓	×	✓	×
	Concepts and framework for congestion control (CF for CC)	×	×	×	×	×	✓
	Adaptive inter-vehicle communication control (AICC)	✓	✓	✓	×	×	×

Yes ✓
No ×

adaptation, and improvement of the multichannel architecture proposed by the standardization bodies. Provisioning the safety-critical and commercial services on the road is crucial to vehicular ad hoc networks. Multiple channels are assigned in the 5 GHz spectrum to support these services.

Through the dynamic service channel allocation (DSCA) method, the throughput could be maximized by assigning different service channels to the users (Park et al. 2013b). The solution for multiple service channel assignment is based on some vehicular environment variables that include BERs, access categories, and traffic and road conditions applying a single transceiver. The realistic parameters have been used from CISCO traffic analysis in NS-3 simulation. The average throughput is extensively evaluated under various channel conditions and vehicle densities. However, non-safety applications can affect the network efficiency by exchanging traffic information using four different access categories as mentioned earlier. Hong (2013) proposed the enhanced multichannel MAC for VANETs, where exchange of non-safety messages is possible during CCH and SCH interval. The simulation result done by this researcher has shown that the VEMMAC protocol outperforms the IEEE 1609.4 in terms of aggregate throughput and average delay.

4 Conclusion

In this paper, we have presented a literature survey on congestion control of VANETs. The research contributions that we have reviewed in this paper exhibit the potential and limitations of congestion control implemented in service control channel (SCH) for vehicular ad hoc networks. We first compared each of the technologies and methods realized by various researchers. Then, we investigated and proposed a framework for congestion control algorithm for SCH applications mainly for non-safety applications. The congestion control approach is one of the solid solutions to reduce congestion in wireless communication channel. We have highlighted the algorithm for the non-safety messages, queues, and mechanism to monitor the channel communications based on defined threshold.

To provide a reliability performance for non-safety messages, we intend to develop new rebroadcast schemes and techniques. The congestion control approach is expected to reduce the channel loading thus meeting the QoS requirements for wireless network performance. In future research, we plan to do the test bed and evaluate the performance of our proposed congestion control technique using OMNet++ integrated simulation tools.

References

Amadeo M, Campolo C, Molinaro A, Ruggeri G (2009) A WAVE-compliant MAC protocol to support vehicle-to-infrastructure non-safety applications

Bouassida MS, Shawky M (2008) On the congestion control within VANET. In: 2008 1st IFIP wireless days, pp 1–5. doi:10.1109/WD.2008.4812915

Campolo C, Cortese A, Molinaro A (2009) CRaSCH : a cooperative scheme for service channel reservation in 802.11p/WAVE vehicular ad hoc networks

Campolo C, Molinaro A (2013) Multichannel communications in vehicular Ad Hoc networks: a survey. IEEE Commun Mag 51(5):158–169. doi:10.1109/MCOM.2013.6515061

Cheng N, Lu N, Wang P, Wang X, Liu F (2011) A QoS-provision multi-channel MAC in RSU-assisted vehicular networks (poster). IEEE Veh Netw Conf (VNC) 2011:193–197. doi:10.1109/VNC.2011.6117142

Cunha F, Villas L, Viana A, Loureiro AAF (n.d.) Data communication in VANETs : a survey, challenges and applications 1–12

Darus MY, Bakar KA (2011) Congestion control framework for emergency messages in VANETs. Control 2(3):643–646

Fiore M, Elettronica D (n.d.) Understanding vehicular mobility in network simulation

Harri J, Filali F, Bonnet C (2009) Mobility models for vehicular ad hoc networks: a survey and taxonomy. IEEE Commun Surv Tutorials 11(4):19–41. doi:10.1109/SURV.2009.090403

Hartenstein H, Laberteaux K (2008) A tutorial survey on vehicular ad hoc networks 164–171

He J, Chen H, Chen T, Cheng W (2010) Adaptive congestion control for DSRC vehicle networks. IEEE Commun Lett 14(2):127–129. doi:10.1109/LCOMM.2010.02.092002

Hong CS (2013) An enhanced multi-channel MAC for vehicular ad hoc networks. IEEE Wirel Commun Netw Conf (WCNC) 2013:351–355. doi:10.1109/WCNC.2013.6554589

Khairnar VD, Pradhan SN (2011) Mobility models for vehicular ad-hoc network simulation. IEEE Symp Comput Inf 2011:460–465. doi:10.1109/ISCI.2011.5958959

Mahajan A, Potnis N, Gopalan K, Wang A (2007) Modeling VANET deployment in urban settings, New York, pp 151–158

Morgan YL (2010) Notes on DSRC & WAVE standards suite: its architecture, design, and characteristics. IEEE Commun Surv Tutorials 12(4):504–518. doi:10.1109/SURV.2010.033010.00024

Park S, Chang Y, Khan F, Copeland JA (2013a) Dynamic service-channels allocation (DSCA) in vehicular ad-hoc networks. In: 2013 IEEE 10th consumer communications and networking conference (CCNC), pp 351–357. doi:10.1109/CCNC.2013.6488469

Park S, Chang Y, Khan F, Copeland JA (2013b) Dynamic service-channels allocation (DSCA) in vehicular ad-hoc networks. In: 2013 IEEE 10th consumer communications and networking conference (CCNC), pp 351–357. doi:10.1109/CCNC.2013.6488469

Sattari MRJ, Noor RM, Keshavarz H (2012) A taxonomy for congestion control algorithms in vehicular ad hoc networks. In: 2012 IEEE international conference on communication, networks and satellite (ComNetSat), pp 44–49. doi:10.1109/ComNetSat.2012.6380774

Wang Q, Leng S, Fu H, Zhang Y, Member S (2012) An IEEE 802.11p-based multichannel mac scheme with channel coordination for vehicular. Ad Hoc Netw 13(2):449–458

Zang Y, Stibor L, Walke B, Reumerman H-J, Barroso A (2007) A novel MAC protocol for throughput sensitive applications in vehicular environments. In: 2007 IEEE 65th vehicular technology conference—VTC2007-Spring, pp 2580–2584. doi:10.1109/VETECS.2007.531

Zhiyuan L, Jinhong H (2011) Framework of real VANET simulation research. In: 2011 third international conference on multimedia information networking and security (201012), pp 136–140. doi:10.1109/MINES.2011.15

Chapter 10
Pollutant Removal Efficiency for Wetland (PREWet) Stormwater Quality Model Performance for Constructed Wetlands in Tropical Climate

Nur Asmaliza Mohd Noor, Asmidar Alias, Kamisah Ariffin and Lariyah Mohd Sidek

Abstract Rapid development has increased impervious areas which have largely contributed to the occurrence of flash floods and deterioration of the water quality. Constructed wetland, a component under the best management practice, has been used as an alternative for water quality enhancement in Malaysia. However, water quality is a major issue in evaluating the effectiveness of constructed wetland. Hence, concerted efforts are needed to establish water quality modeling for constructed wetland. The objective of this study is to assess the treatment performance of constructed wetland using the pollutant removal efficiency for wetland (PREWet) as the stormwater quality model. The parameters involved are biochemical oxygen demand (BOD), total phosphorus (TP), total nitrogen (TN), and total suspended solid (TSS). In order to achieve the objective of the study, water quality sampling was conducted in 2012 at a small scale constructed wetland, located at the Humid Tropic Centre, Kuala Lumpur. The removal rate value was identified and the results indicated 0.32 day^{-1}, 0.4 and 0.3 m/s, 0.05 for BOD, TP, TSS, TN, respectively.

N.A. Mohd Noor (✉) · A. Alias
Faculty of Civil Engineering, Universiti Teknologi MARA,
Jengka, Pahang, Malaysia
e-mail: nurasmaliza@pahang.uitm.edu.my

A. Alias
e-mail: asmidar@pahang.uitm.edu.my

K. Ariffin
Academy of Language Studies, Universiti Teknologi MARA,
Jengka, Pahang, Malaysia
e-mail: kamisah@pahang.uitm.edu.my

L. Mohd Sidek
Civil Engineering Department, College of Engineering,
Universiti Tenaga Nasional, Bandar Baru Bangi, Selangor, Malaysia
e-mail: lariyah@uniten.edu.my

N.A. Yacob et al. (eds.), *Regional Conference on Science, Technology and Social Sciences (RCSTSS 2014)*, DOI 10.1007/978-981-10-0534-3_10

The results from the water quality sampling were then used for model calibration and verification to assess the effectiveness of the model. The findings from this study would provide a better understanding on the performance of constructed wetland as stormwater improvement among the researchers and stormwater engineers.

Keywords Constructed wetland · Water quality · Stormwater quality model · Pollutant removal efficiency

1 Introduction

The urbanization process has changed the land use pattern and the urban structure. Urban development has added impervious surfaces, increased runoff, decreased infiltration, ground water recharge, and has negative effects on the local water quality and quantity balance. A large portion of former forest and agricultural areas have been cleared and replaced by concrete buildings, roads and drainage systems which are impermeable. The problem of silting in rivers and drains, together with high rainfall intensity, has caused flash floods. Flash floods not only can cause problems to commuters who are usually caught in traffic jams but at times, to the residents of some settlements who have to evacuate to relief centres. In addition, the phenomenon of these flash floods has become more common as the surface runoff from developed urban and industrial areas has fully increased the peak discharges entering the rivers with shortened time of detention to peak flow (Abdullah 2007). In order to overcome flash flood problems, the Department of Irrigation and Drainage (DID) in Malaysia has taken a step by introducing the Stormwater Management Manual for Malaysia (MSMA). MSMA is a guideline published by the DID in 2000 and was revised in the MSMA 2nd Edition in 2012 to treat the long-term nationwide directions and needs in ensuring sustainable urban drainage systems are fully utilized (DID 2012).

Constructed wetland is one of the chapters under MSMA. It can be used for stormwater quality improvement and domestic wastewater. The successful application of constructed wetlands for wastewater has led to the exploration of the treatment for different sources, such as stormwater, industrial, agricultural, urban, airport runoff, and acid mine drainage (Kadlec and Knight 1996). The constructed wetlands have been evolved recently to the hybrid system with the application on the treatment of various industrial wastewater (Jing et al. 2001; Oovel et al. 2007; Kato et al. 2010).

Water quality has become a significant issue in constructed wetland. Stormwater quality modeling is conducted to simulate the key water quality processes in catchment areas and treatment system. The practice has become an important part of the urban stormwater quality planning and design stage (Trepel et al. 2000). Water quality models are usually classified according to model complexity, type of receiving water, and the water quality parameters that the model can predict. If more parameters needed to be considered, the model is, then, categorized as a complex model (Trepel et al. 2000). There are a few water quality models that can be used to estimate the amount for concentration at the outlet of constructed wetland, such as MUSIC, PREWet, XP-SWMM, WMod, and Aqualm. PREWet had been chosen due to its user-friendly model. It has no additional cost and can be applied with minimal input data estimating the amount of water quality improvement provided by wetlands. PREWet is an acronym for pollutant removal efficiency for wetland (PREWet). It is a screening level mathematical model for estimating the amount of pollutant removal efficiency (RE) provided by wetlands, where RE is stated in percentage, between 0 and 100 % (Mark 1997). The PREWet model assumes steady-state conditions and, either fully mixed or one-dimensional, longitudinally varying concentration (Mark 1997). These simplifying assumptions allow for rapid model implementation with minimal input data requirements. Parameters that can be measured using the PREWet model are BOD, TSS, coliform bacteria, TN, and TP (Mark 1997).

The water quality modeling in constructed wetlands are widely used in temperate climate countries compared to tropical climate countries, especially in Malaysia. Prior studies conducted by a few researchers in Malaysia have focused on macrophytes, flow interaction, and sedimentation process but none of them has examined water quality on constructed wetlands. Thus, this indicates that there are knowledge gaps in regard to the effectiveness of constructed wetland treatment performance under tropical climate.

The main objective of this study is to assess the treatment performance of constructed wetland using the PREWet as the stormwater quality model involving the BOD, TP, TN, and TSS parameters. Specifically, this study evaluates the performance of constructed wetland under tropical climate.

2 Methodology

Site Description

The Constructed Wetland at Humid Tropic Centre (HTC) Kuala Lumpur has been selected as the site location of this study. The site is located at the Humid Tropic Centre (HTC) Jalan Redang, Kuala Lumpur which is near to the Drainage Irrigation Department Headquarters, Kuala Lumpur. The design criteria of constructed wetland are 1.2 day for the retention time and 0.084 m^3/s for the inflow rate. The emergent plants that are planted in the HTC constructed wetland are *Hanguana malayana, Phragmites karka, Lepironia articulata*, and *Typha latifolia*. Figure 1 shows the location of the constructed wetland at the HTC.

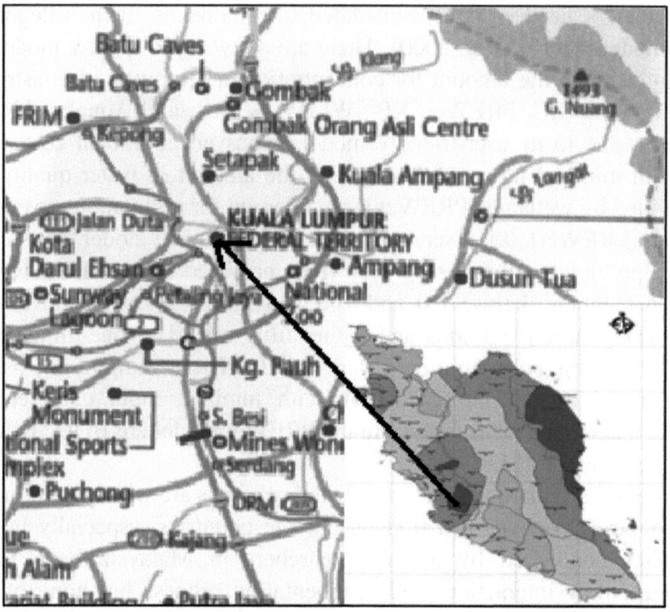

Fig. 1 Location of constructed wetland at Humid Tropic Center, Kuala Lumpur

Water Quality Sampling

Water quality sampling was carried out through the events from 24th April 2012 until 15th August 2012. Samples were collected at the inlet and the outlet of the constructed wetland. The parameters measured were TSS, TP, TN, and BOD. The

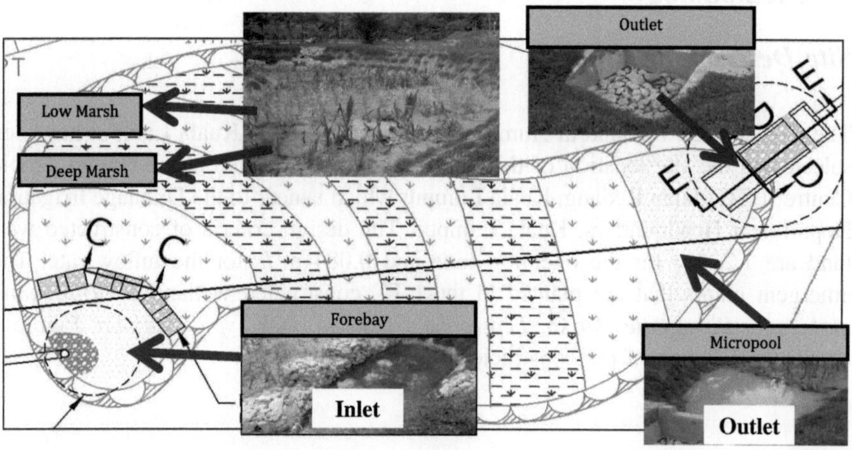

Fig. 2 Location of water quality sampling

sampling and testing procedures were conducted in accordance to the *Standard Method for Examination of Water and Wastewater 20th Edition.* Figure 2 shows the location of the sampling at the inlet and outlet of the constructed wetland.

Water Quality Simulation

The PREWet simulation, in order to generate the simulated concentration at the outlet of constructed wetland, required the input data on the wetland morphology, such as length, width, depth, and amount of inflowing water. Then, for the specific removal processes, different decay, TP to fraction of particulate to total phosphorus ratio, ratio of nitrate to TN, and settling rates were chosen. The detail process of PREWet simulation is shown in Fig. 3.

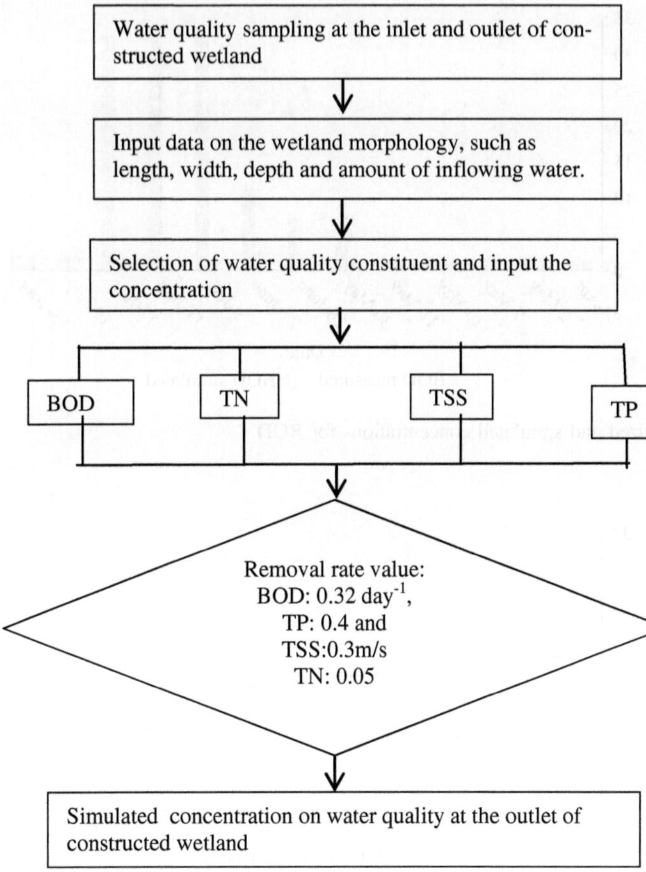

Fig. 3 Flow chart of the PREWet simulation

3 Results and Discussion

This study examined the comparison between observed and measured concentrations for the parameters, such as BOD, TP, TN, and TSS for constructed wetland under tropical climate. Figure 4 shows that the value of BOD produced by PREWet is almost similar to the measured BOD. However, certain samples show different values. This may be due to the existence of external elements influencing the removal of BOD which were not calculated in this model. Meanwhile, the correlation between measured and simulated concentration was conducted as shown in Fig. 5. The findings indicate that the R^2 is 0.1271. The study conducted by Mohd Noor (2010), on the other hand, found that the correlation between measured and simulated was around 0.525.

Figure 6 shows the measured and simulated concentration for TP. The simulation by PREWet produced around 88 % higher concentration compared to the

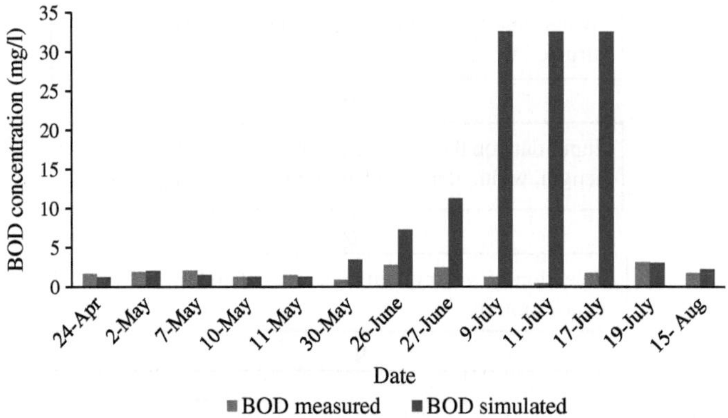

Fig. 4 Measured and simulated concentrations for BOD

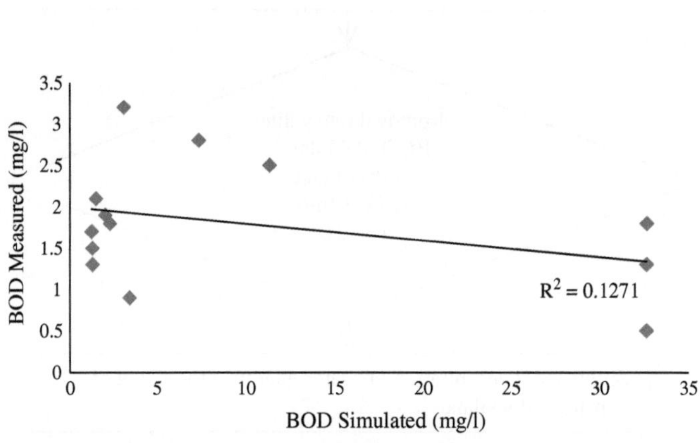

Fig. 5 Correlation between measured and simulated for BOD

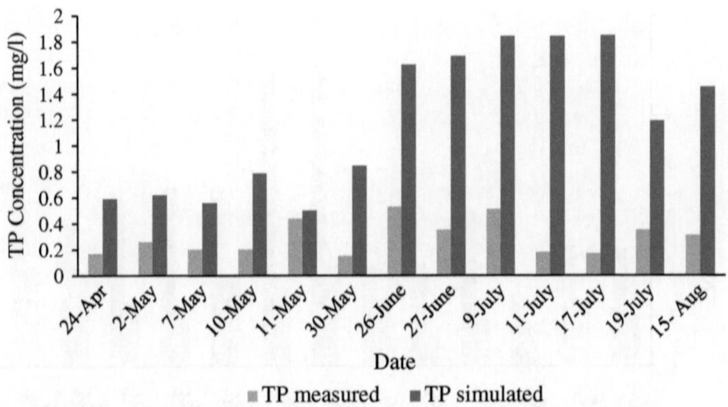

Fig. 6 Measured and simulated concentrations for TP

measured concentrations. Such higher simulation concentration may be contributed by some biological effects that are not considered by the model. Furthermore, the correlation between measured and simulated concentrations was conducted to evaluate the model performance and this is represented in Fig. 7 with the result as 0.0868. However, Mohd Noor's (2010) study on the PREWet model showed the correlation value as 0.479.

Figure 8 shows the measured and simulated concentration for TN. The simulated concentrations had predicted TN at higher concentration compared with measured TN due to the consideration of nitrogen removal processes but in actual condition it is expected to lose by burial in the sediments. The correlation between simulated and measured concentrations had been conducted and the R^2 was around 0.474. Figure 9 shows the correlation between simulated and measured concentrations for TN.

Figure 10 demonstrates the measured and simulated concentrations for TSS. The finding indicates that the measured and simulated concentrations produced almost

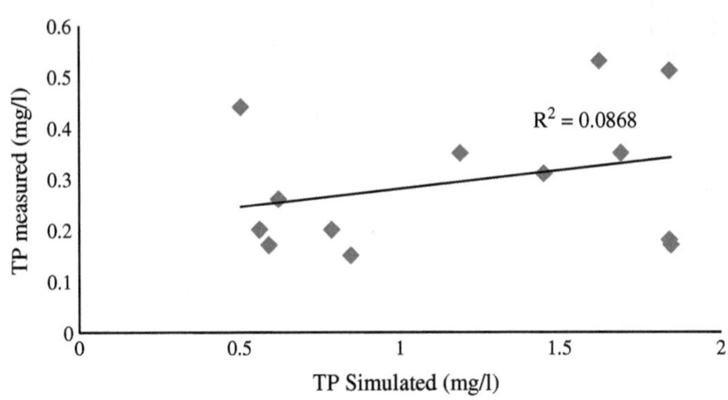

Fig. 7 Correlation between measured and simulated for TP

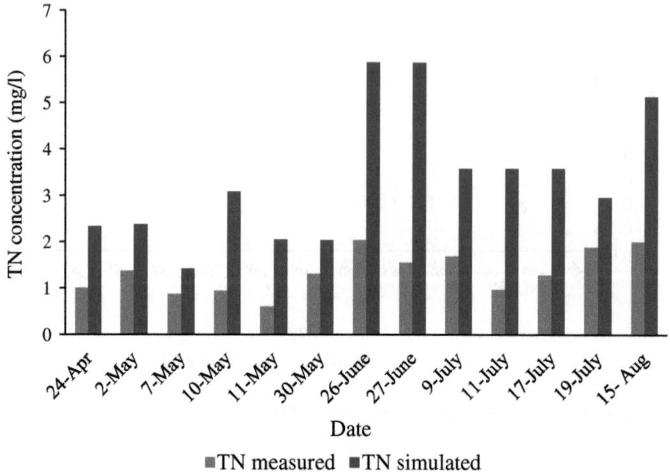

Fig. 8 Measured and simulated concentrations for TN

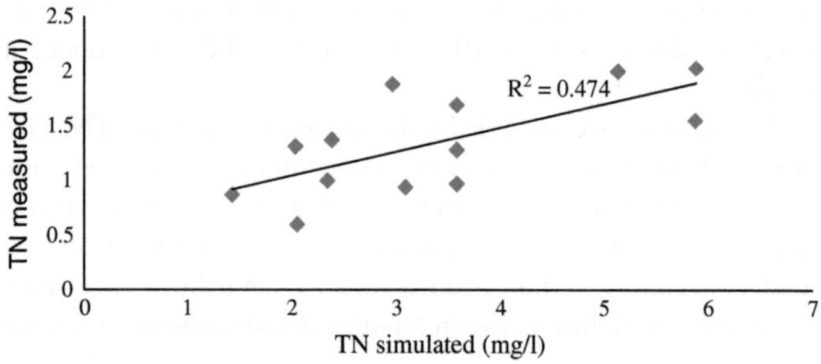

Fig. 9 Correlation between measured and simulated for TN

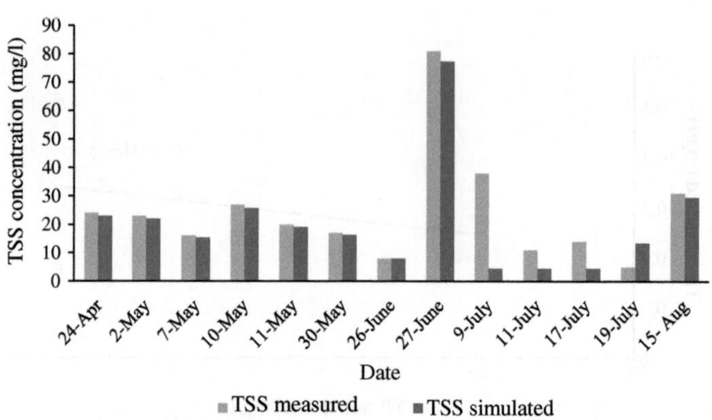

Fig. 10 Measured and simulated concentrations for TSS

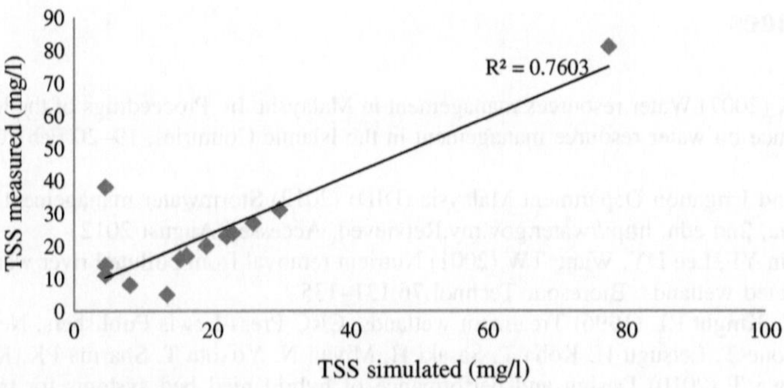

Fig. 11 Correlation between measured and simulated for TSS

similar results. However, some samples indicate that the inconsistency, which is due to the water quality, is often judged by the TSS concentration without distinction to the type. Meanwhile, the correlation between measured and simulated concentration indicates the correlation value as around 0.763. This is shown in Fig. 11. Mohd Noor (2010), in a similar study, obtained the correlation value as around 0.067.

4 Conclusion

The treatment performance of constructed wetland for BOD, TP, TN, and TSS were assessed by using the PREWet with the variation of R^2 correlation values. The use of PREWet model is an alternative to evaluate performance of constructed wetland under tropical climate. Given R^2 for acceptable values for TN and TSS removal, PREWet can be proposed as a model in predicting water quality for constructed wetland. However, further study in water quality modeling in constructed wetland should be properly conducted to incorporate major parameters for removal rates factors as constructed wetland in the tropical climate area is relatively new for discussion. The efficiency of stormwater quality model can provide for better understanding on constructed wetland performance to help in decision making by researchers and stormwater engineers.

Acknowledgments The authors would like to express their gratitude to the Universiti Teknologi MARA and Universiti Tenaga Nasional for their financial support.

References

Abdullah K (2007) Water resources management in Malaysia. In: Proceedings of the international conference on water resource management in the Islamic Countries, 19–20 Feb 2007, Tehran Iran

Drainage and Irrigation Department Malaysia (DID) (2012) Stormwater management manual for Malaysia, 2nd edn. http://water.gov.my.Retvieved. Accessed August 2012

Jing SR, Lin YF, Lee DY, Wang TW (2001) Nutrient removal from polluted river water by using constructed wetlands. Bioresour Technol 76:131–135

Kadlec RH, Knight RL (1996) Treatment wetlands. CRC Press-Lewis Publishers, New York

Kato K, Inoue T, Letsugu H, Koba T, Sasaki H, Miyaji N, Yokota T, Sharma PK, Kitagawa K, Nagasawa T (2010) Design and performance of hybrid reed bed systems for treating high content wastewater in the cold climate. In: Masi F, Nivala J (eds) Proceeding of the 12th international conference on wetland systems for water pollution control, Lisbon, Portugal, pp 295–305

Mark SD (1997) Screening level techniques for estimating pollutant removal by wetland. WRP Technical Note WQ-EV-5.1 (March 1997)

Mohd Noor NA (2010) Water quality modeling of constructed wetlands using PREWET model: a case study BIOECODS, USM. Unpublished master's thesis, Universiti Sains Malaysia

Oovel M, Tooming A, Mauring T, Mander U (2007) Schoolhouse wastewater purification in a LWA-filled hybrid constructed wetland in Estonia. Ecol Eng 29:17–26

Trepel M, Michele DO, Luigi DC, Marcel DW, Silvia O, Luca P, Jasper P, Nico MP, Tiemo T, Giuseppe B, Winfrid K, Sven EJ (2000) Model for wetland planning, design and management. Ecosys Bd 8:93–137

Chapter 11
An Idealized Model of Meandering Tidal River

Wei-Koon Lee and Irma Noorazurah Mohamad

Abstract Study on tidal river provides understanding to the dynamic interactions between river discharge, tidal fluctuation and other physical processes in the estuarine environment. However, the generalized approach is typically limited to a straight channel in one-dimension, although varying depth and bed profile may be included. The geometry of lowland meander, meanwhile, contains critical lateral component which is not taken into consideration in such simplification. Idealized representation of meander geometry is non-trivial. Harmonic analysis using Fourier Series model requires no less than 20 constituents to reproduce river plan form realistically. On the other hand, sine-generated curve from theory of minimum variance, though requires only few parameters, is applicable only to short symmetrical meandering reach. In this paper, we introduce a reference axis described by a polynomial function of appropriate power, combined with the sine-generated curve to produce good simplified planform approximation of meandering lowland reaches. The number of parameters requires is less and can more readily be interpreted in physical sense. The approach produces an idealized two-dimensional (2D) model of tidal river, which can be applied across a wide range of estuary-river system.

Keywords Estuary-river system · Idealized 2D-model · River meander · Sine-Generated curve · Tidal river

W.-K. Lee (✉)
Fluvial and River Engineering Dynamics Research Centre (FRiEnD),
Institute of Infrastructure Engineering and Sustainability Management (IIESM),
Universiti Teknologi MARA, Shah Alam, Selangor, Malaysia
e-mail: leewei994@salam.uitm.edu.my

I.N. Mohamad
Faculty of Civil Engineering, Universiti Teknologi MARA,
Shah Alam, Selangor, Malaysia
e-mail: irma1095@salam.uitm.edu.my

N.A. Yacob et al. (eds.), *Regional Conference on Science, Technology and Social Sciences (RCSTSS 2014)*, DOI 10.1007/978-981-10-0534-3_11

1 Introduction

Fresh water supply from the river, the fertile land on the flood plain, and the rich biodiversity of the estuary, together present themselves as a heaven of life sustenance. It is thus not surprising that most early civilizations were built near river mouths and continue to grow and expand. However, as human settlements began to mushroomed, the intricate physical processes in the river and the estuary began to be felt. We have come to know of the cyclic sea water level fluctuation, the seasonal variation in river discharge, and the destructive power of water. We have learnt that the dynamic morphology of the river and the estuary evolves through time, and its natural equilibrium is altered as human interference was imposed.

Our increased understanding of the river and estuary system has prompted us to continuously devise and implement various measures to safeguard the valuable real estate that was erected on these locations. Nevertheless, the battle which long began is yet to be won. One notable example is our wrestle with flood inundation. Modern landscape has significantly encroached upon the floodplain and is popularly sited along the coastal and river waterfront. In particular, natural rivers have in many cases, been lined and trained to make way for development. Alteration of river channels inevitably affects the natural river flow. This, coupled with increased surface runoff due to the reduction of permeable surfaces as a consequent of rampant development, have led to larger and more frequent urban flood disasters (Konrad 2003). In lowland rivers, pronounced tidal backwater effect poses additional threat especially when occurs in conjuction with high riverine discharge (Mohamad and Lee 2014). The recurrence of river bank overflow under these extreme condition has increasingly becoming a major concern with the anticipated sea level rise (DID 2000).

Flood risk of coastal inundation and lowland rivers is thus a subject of continual interest, and is closely related to the interactions of river and the estuary, another popular research area. Study on river-estuary system falls into two generally approaches: a realistic two-dimensional (2D) model, or a generalized simplified model in one-dimension (1D). For engineering purposes, computational models of estuary and river hydrodynamics are usually based on real geometry, bathymetry, and environmental forcings (e.g. Mahdi 2000; Mustafa 2007). The results obtained are thus specific only to the location of study and by no means can be extended to other river-estuary systems. On the other hand, conceptual investigation typically conducted on idealized models which appropriately represent the physical characteristics with considerable simplifications (e.g. Hetland and Geyer 2004; John et al. 2010) offer more generalized solutions but have so far been limited to linear one-dimensional models. Cai et al. (2014), for example, showed that simulation in a prismatic tidal channel suggests the dimensionless estuary shape is amongst the key parameters which control the estuarine hydrodynamics. Nonetheless, the estuary shape is represented only by a simple dimensionless parameter which characterize the effect of cross-sectional area and depth convergence along the linear longitudinal axis assumed.

Essentially, for research on river-estuary system, question arises as to whether it is possible to formulate an idealized model which permits results generalization, but still admits the ubiquitous lateral flow effects as observed in river meander. In this paper, we attempt to address the issue by proposing an idealized 2D tidal river model, incorporating the lowland meander features.

2 Study Area

We consider Lower Klang River (LKR) in the state of Selangor, Malaysia. Klang River basin is one of the biggest river basins in Selangor with a total catchment area of 1,288 km^2. Klang River has an estimated total length of 120 km, originating from Ulu Gombak Forest Reserved area, and fed by 13 major tributaries. The last major tributary is the Damansara River. Klang River exits on the west coast at Port Klang, where the tide is semi-diurnal and the estuary is well-mixed. LKR is located in a flat and low lying plain. The river bed slope is mild at the Klang-Damansara confluence point ($\sim$ 1/2300) and is near flat (1/7000) when the river meets the sea. Meanwhile, the river width increases from 50 to 300 m between the two locations.

River meander is a natural features which result from complex interactions of numerous physical processes as basin runoff travels from upstream catchment downvalley towards the sea. Meander geometry is typically studied using series approach to account for its quasi-randomness (e.g. Swamee et al. 2003), or analyzing the underlying regularity (e.g. Langbein and Leopold 1966). In the latter, meanders are typically idealized and approximated by generalized functions. This is typically true for most stable river form where the variability of erosion and deposition rate over time is minimized.

In the analysis that follows, we focus on LKR stretching from Port Klang estuary up to Taman Sri Muda, Shah Alam (Fig. 1) where the effect of tidal backwater is

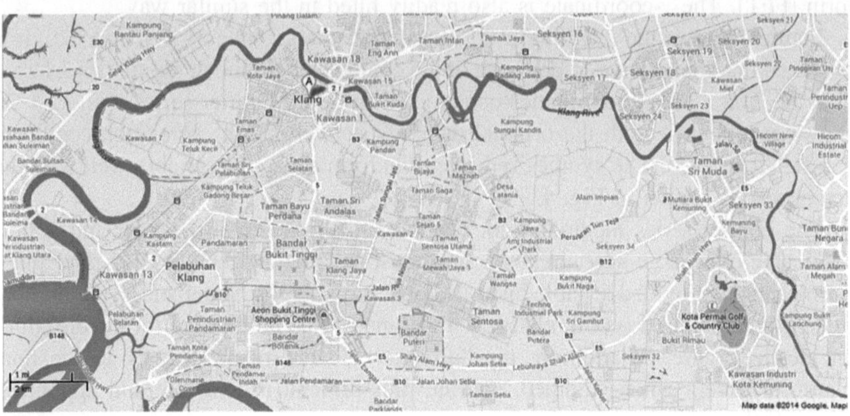

Fig. 1 Lower Klang River. *source* Google Map

most felt. For the purpose of this study, the (x, y) coordinates of LKR along its river centreline is discretized from Google Map. Analysis and idealized reconstruction of the river planform is presented in Sects. 3 and 4, respectively.

3 LKR Planform Description

Fourier Series Model

A widely used approach to approximate the geometry of river meander is Fourier Series model. The (x_i, y_i), $i = 1, 2, 3, \ldots, N$ coordinates of the river course can be treated as arbitrary functions of the index i. Swamee et al. (2003) showed that good approximation of the river planform can be obtained using a maximum of 20 harmonics, and convergence of the series improved by taking a linear function as the reference axis, where the first x_0 and last x_N data points coincide with the original data set.

Following their approach, we write the modified x-coordinate in the form of:

$$\xi_i = x_i - x_0 - (x_N - x_0)\frac{i}{N}, \tag{1}$$

where the coeffcients A_k and B_k of the Fourier series

$$\xi_i = \sum_{k=1}^{k_L} A_k \sin\frac{\pi k i}{N} + B_k \cos\frac{\pi k i}{N}, \tag{2}$$

with k_L number of harmonics is readily derived using standard fast-fourier transform (FFT). The y-coordinate is also readily fitted in the similar way.

Figure 2 shows the magnitudes of the coefficients for index $k = 1$ up to 20. It is observed that the relative importance of the coefficients diminish rapidly for k larger than 12. The x and y coordinates constructed from the Fourier series with $k_L = 20$ shows excellent fit with the original data, respectively, with correlation coefficent $R > 0.999$ (Fig. 3). The synthesized river planform agrees reasonably well with the actual one as shown in Fig. 4.

The Fourier series method is not restricted to symmetrical meander and is thus applicable over large spatial extent. It is most suitable for the study of river planform changes provided reliable historical data is available. Nonetheless, the number of constants required is still relatively high, and can be prohibitive for the design of an idealized case. In addition, the harmonics do not relate directly to the final planform except when summed, and are thus difficult to evaluate and applied in conceptual model.

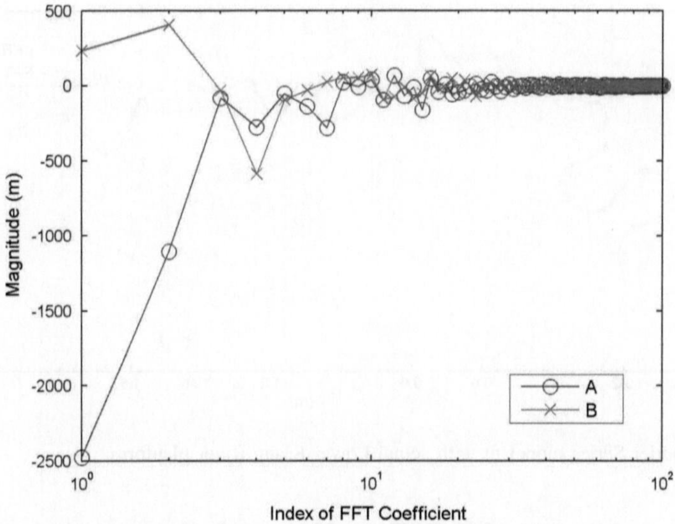

Fig. 2 Magnitudes of FFT coefficients for Lower Klang River

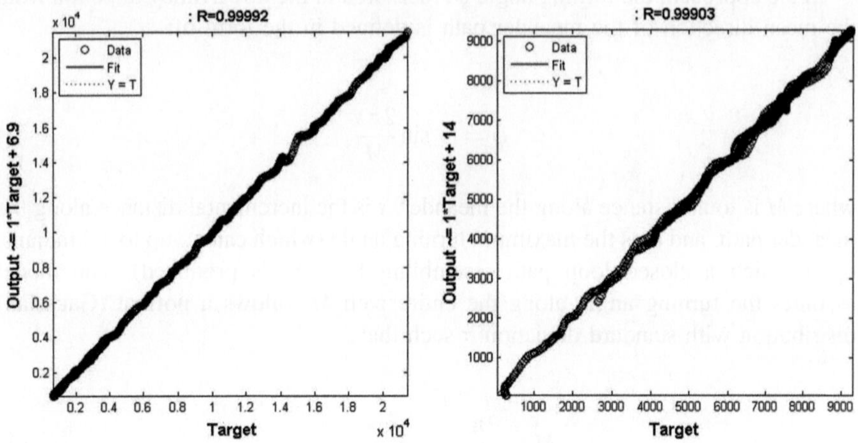

Fig. 3 FFT fit for x (*left*) and y (*right*) coordinates using index value, $k_L = 20$

Sine-Generated Curve

Using the theory of minimum variance, Langbein and Leopold (1966) showed that a symmetrical meander path can be approximated by a 'sine-generated curve', i.e. the channel direction changes is a sinusoidal function of distance along the reference axis. The resulting planform is not a sine curve, but characterized by more realistic rounded horseshoe loops. The theory was verified on large number of cases

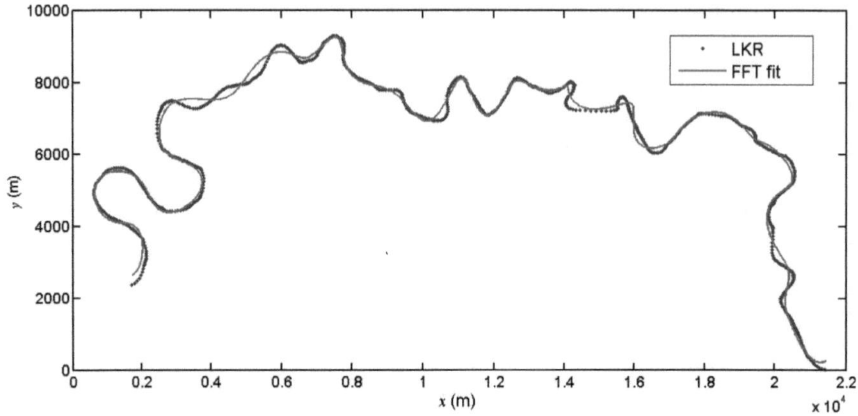

Fig. 4 Fourier Series model fit with actual Lower Klang River planform

(Langbein and Leopold 1966; Williams 1986), producing good planform approximation for cases where the ratio of bend curvature to channel width is in the range of 2–3.

In the approach, the turning angle ϕ, measured in the downvalley direction from the mean direction of the meander path is defined in the form of:

$$\phi = \omega \sin \frac{2\pi s}{M}, \tag{3}$$

where M is total distance along the meander, s is the incremental distance along the meander path, and ω is the maximum turning angle (which can go up to 2.2 radians, upon which a closed loop path resembling Fig. '8' is produced). The model assumes the turning angle along the entire path M follows a normal (Gaussian) distribution with standard deviation σ such that

$$M = \frac{2\pi}{\sigma} \frac{\omega}{\sqrt{2(1 - \cos \omega)}}. \tag{4}$$

Unfortunately, natural meanders do not maintain geometrical symmetry over long distances, and thus the method is only applicable to selected or segmented river each where a linear reference axis may be assumed. We postulate that by adopting a suitable reference axis, this method can be extended to more general cases. Using a 2-degree polynomial function as the reference axis for LKR (Fig. 5), we describe the river planform using a set of transformed coordinate (x', y') measured with respect to the reference axis such that the meander can be represented as undulation about a linear axis. Analysis of the downvalley turning angles of the modified meander path for the downstream end of LKR yields ω = 1.9 radians, and

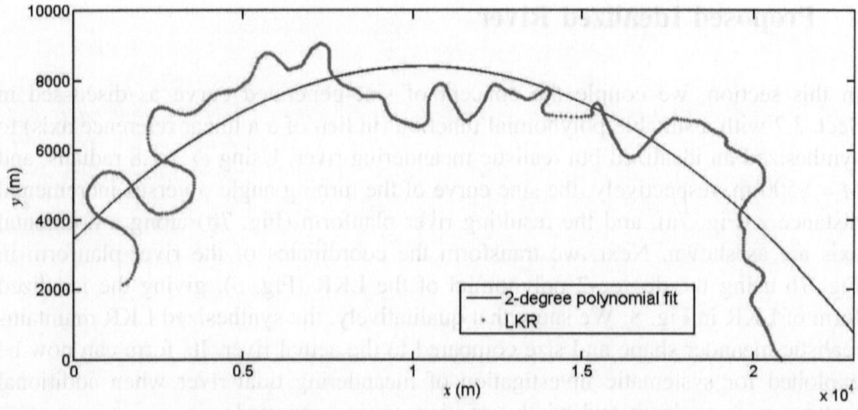

Fig. 5 LKR fitted with 2-degree polynomial

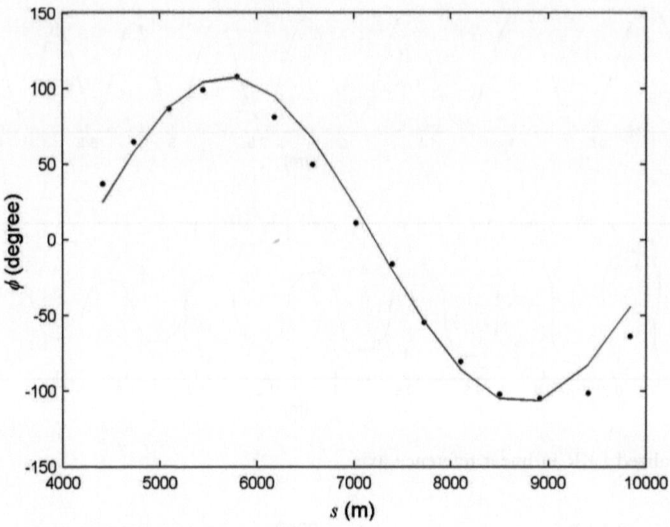

Fig. 6 Downvalley turning angles of LKR along the channel distance agrees well with the sinusoidal function (Eq. 3)

$M = 6000$ m, respectively. We show that the turning angles is found to approximate the sinusoidal function in Eq. 3 well (Fig. 6).

We conclude herein that the concept of sine-generated curve introduced by Langbein and Leopold (1966) adequately describes the changes of downvalley turning angle of a meandering river. The method requires only two physical quantities (ω and M) which are easy to interpret and can be readily extracted from available river data. It is thus more attractive compared to the Fourier series method to be used to design the planform of an idealized meandering river.

4 Proposed Idealized River

In this section, we couple the concept of sine-generated curve as discussed in Sect. 3.2 with a suitable polynomial function (in lieu of a a linear reference axis) to synthesized an idealized but realistic meandering river. Using $\omega = 1.8$ radians, and $M = 8500$ m, respectively, the sine curve of the turning angle ϕ versus incremental distance s (Fig. 7a), and the resulting river planform (Fig. 7b) along a horizontal axis are as shown. Next, we transform the coordinates of the river planform in Fig. 7b using the degree-2 polynomial of the LKR (Fig. 5), giving the idealized form of LKR in Fig. 8. We show that qualitatively, the synthesized LKR maintains realistic meander shape and size compared to the actual river. Its form can now be exploited for systematic investigation of meandering tidal river when additional features such as depth and width variation are incorporated.

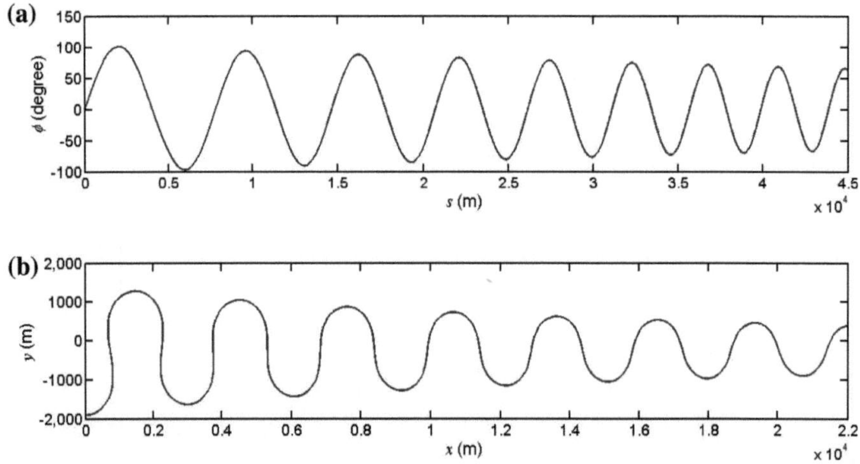

Fig. 7 Idealized LKR in linear reference axis

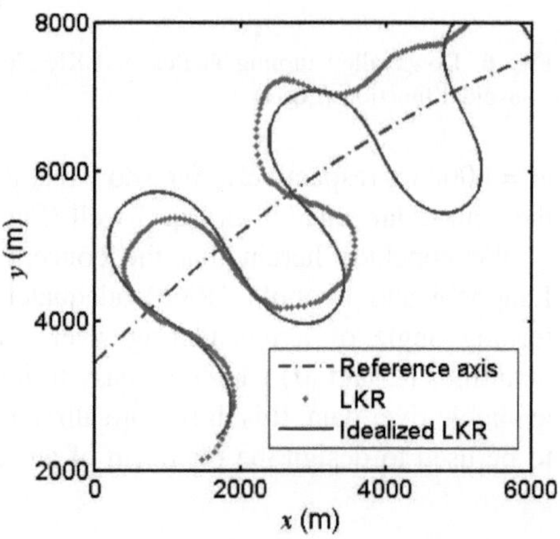

Fig. 8 Idealized LKR in polynomial reference axis

5 Conclusions

In this paper, we presented the development of an idealized meandering tidal river model using the Lower Klang River as a test case. We show that the combination of Langbein and Leopold's sine-generated curve and a suitable polynomial function of approriate power can be used to synthesize a realistic meandering river planform using only small number of phyical quantities which are easy to interpret and approximate. The resulting idealized model can be useful in systematic parametric investigation of estuary-river system without any loss of two-dimensionality and lateral flow effect. Our future works include to develop the numerical model of the proposed idealized tidal river, to simulate the behaviour of estaury-river system under varying external forcing, and to compare and contrast the dimensionless results with laboratory observations.

Acknowledgements The authors gratefully acknowledge the support provided by the Ministry of Higher Education Malaysia (FRGS/1/2012/TK03/UITM/03/6), and Research Management Institute (RMI) of Universiti Teknologi MARA.

References

Cai H, Savenije HHG, Toffolon M (2014) Linking the river to the estuary: influence of river discharge on tidal damping. Hydrol Earth Syst Sci 18:287–304

Department of Irrigation and Drainage (DID) (2000) Manual Saliran Mesra Alam, Chapter 46: Lowland, Tidal & Small Island Drainage, 1st edn

Hetland RD, Geyer WR (2004) An idealized study of the structure of long, partially mixed estuaries. J Phys Oceanogr 34:2677–2691

John ZS, Lu LF, Liu YN (2010) The hydrodynamics of an idealized estuarine plume along a straight coast: a numerical study. Environ Model Assess 15:487–502

Konrad CP (2003) Effects of urban development on floods. U.S. Geological Survey Fact Sheet 076-03, 4 p. Accessed on 04 July 2014. http://pubs.usgs.gov/fs/fs07603/

Langbein WB, Leopold LB (1966) River meanders—theory of minimum variance. Physiographic and Hydraulic Studies of Rivers: Geological Survey Professional Paper 422-H. US Department of the Interior

Mahdi MA (2000) Tidal effects on the flow and channel geometry of klang river estuary. Unpublished master dissertation, Universiti Pertanian Malaysia, Serdang, Malaysia

Mohamad IN, Lee WK (2014) Numerical and laboratory approach to the prediction of flood stage in lowland tidal river. International Conference on Applied Physics and Engineering (ICAPE), Penang, 17–18 Sept 2014, UiTM PP

Mustafa Z (2007) Effects of tides and surface runoff on channel geometry. Unpublished master dissertation, Universiti Teknologi Malaysia, Johor, Malaysia

Swamee PK, Parkash B, Thomas JV, Singh S (2003) Changes in channel pattern of River Ganga between Mustafabad and Rajmahal, Gangetic plains since 18th century. Int J Sedim Res 18 (3):219–231

Williams GP (1986) River meanders and channel size. J Hydrol 88:147–164

Chapter 12
Track Cyclist Performance Monitoring System Using Wireless Sensor Network

Sukhairi Sudin, Ali Yeon Md Shakaff, Fezri Aziz, Fathinul Syahir Ahmad Saad, Ammar Zakaria and Ahmad Faizal Salleh

Abstract The right training programs are an important factor to increase the cycling performance among the professional track cyclist. Over the years, the cyclist performance was based on the feedback from bicycle's kinematics and physiological condition. The advancement in sensor technologies allows the optimization of the training program; by combining both information from the cyclist's physiological condition and kinematic data from the bicycle. The physiological conditions such as heart rate variability (HRV) and forehead temperate can be combined with bicycle kinematic data such as speed and distance to provide accurate assessment of the track cyclist's condition and training program intensity. A system that combines data from physiological signal and bicycle kinematic has been developed for this purpose. Wearable physiological body sensors and bicycle kinematic sensors are deployed using wireless sensor network (WSN). HRV provide using photoplethysmography (PPG) technique that capture signal from cyclist's finger, which provide 3 % error rate refer to heart rate belt. Data handling and communication was developed based on Zigbee protocol whereby the WSN centralized base-station was supported by two repeater node which was used to extend signal coverage in Velodrome to prevent data losses. With two repeater nodes and adjustment on the routing protocol, the packet drops were reduced from 46 to 3 %. The propagation study was carried out in the Velodrome with environment temperature range from 28 to 30 °C and humidity was observed at 85 %. The optimization of network topology by considering the connectivity among the wireless nodes is crucial in order to reduce data losses.

Keywords Cyclist · Photoplethysmography · Performance · WSN · Zigbee

---

S. Sudin (✉) · A.Y. Md Shakaff · F. Aziz · F.S. Ahmad Saad · A. Zakaria
Centre of Excellence for Advanced Sensor Technology,
Universiti Malaysia Perlis, Arau, Perlis, Malaysia
e-mail: sukhairi_sudin@yahoo.com

S. Sudin · A.Y. Md Shakaff · F. Aziz · F.S. Ahmad Saad · A. Zakaria · A.F. Salleh
School of Mechatronic Engineering, Universiti Malaysia Perlis, Arau, Perlis, Malaysia

N.A. Yacob et al. (eds.), *Regional Conference on Science, Technology and Social Sciences (RCSTSS 2014)*, DOI 10.1007/978-981-10-0534-3_12

1 Introduction

Nowadays, the development of technology in sport field rapidly increases. In any sports event, an athlete cannot be better just with normal training without feedback about the current condition during training that shows it performance. This feedback is important to monitor athlete condition while reschedules their training ingredient to expand their performance for maximum and prevent from injury (Okamoto et al. 2014). Athlete physiological data provide such valuable information about their condition and more meaningful while combining with a little data from bicycle kinematic.

Physiological information such as heart rate variability (HRV) can be used as an earlier detection of any disease especially related to heart. Normal person with symptom like dizziness, presyncope, palpitation, angina, nausea and syncope related to heart and it can be monitored by heart rate (Anh and Tao 2010). Abnormalities in heart electrical waveform pattern that can be monitored by heart rate during training, games and normal daily life can turn to be a predictor of any diseases and sudden death (Wang et al. 2006). HRV normally estimated from the electrocardiogram (ECG) which sensor need to attached on human chest but it not suitable for condition that required too much movement (Lei et al. 2011). Photoplethysmography (PPG) technique used in this system to estimate HRV from cyclist's finger which less movement compared to others body part during cycling. PPG technique measure light reflection from blood microvascular inside the finger (Jihyoung et al. 2013).

Another physiological factor that takes into account is body temperature which it is a basic part to determine human condition and it's very important to keep the human body temperature between 36.12 and 37.5 °C. Slightly changes of human core body temperature can affect body system and human comfort level. Increasing body temperature can drive cyclist's to heat stroke (Jones 2006). Core body temperature usually measured using infrared thermocouple that placed inside the ear. Forehead temperature can be measured as representative to core body temperature which it 98.9 % in average similar to core body temperature (Malkinson 2002).

Easiest way to determined cyclist's performance is by measuring his cycling speed. Cycling performance always describe by shortest time cyclist's travelling in certain distance which it related to how fast cyclist's can cycle and maintain it. This normally displayed in cycling speed for certain duration. Bicycle is a moving vehicle for which wireless communication very important to transmit data from physiological sensor node and bicycle kinematic sensor node to base for real-time monitoring, record and analysis (Mu-Huo et al. 2008). Propagation study provide sufficient design for Xbee series 2 using a Zigbee protocol network with single base and two router for signal extended purpose. Data received at base displayed in a user interface panel for monitoring while recording for further analysis by coach and sports physician.

This paper proposed a wireless physiological and bicycle kinematic monitoring system for track cyclist. The Zigbee protocol provides low powered wireless

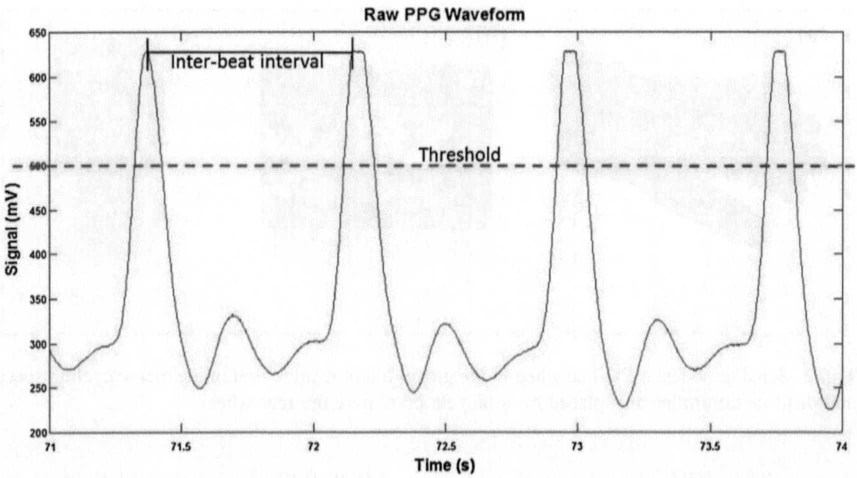

Fig. 1 PPG signal waveform

network with a greater transceiver range and minimum data packet loss that's covered whole Velodrome. Real time physiological and kinematic data from cyclist such as cyclist's heart rate, body temperature and bicycle speed displayed and recorded for comparison and analysis by a physician. In this paper, heart rate for cyclist extracted from cyclist's finger, which more easy to wear in glove and body temperature directly measured from helmet which easy and comfort to wear.

2 Experimental Method

The development of the monitoring system for cyclist's in track or Velodrom combined various sensing elements and method to provide different data that state about the cyclist's condition and performance. All these techniques and method tested inside the lab test bed and real field test on Velodrome with real national level athletes (Fig. 1).

Sensing Unit

Heart rate variability (HRV) and forehead temperature sensing element placed inside the cyclist's glove and helmet like illustrated in Fig. 2a, b. HRV based on photoplethysmography (PPG) is totally different from others method like ECG and bipolar ECG that use electrode to attach to human body to detect HRV. PPG concept was based on light absorption or reflectance on human skin especially on

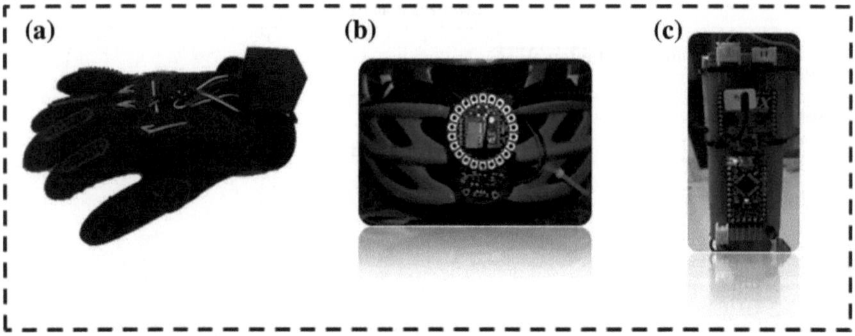

Fig. 2 Sensing nodes: **a** PPG attached to the glove; **b** temperature unit on helmet; **c** cycling speed and distance controller unit placed on a bicycle body near the rear wheel

fingers. This HRV detector designed using APDS-9007 low powered light photo sensor that received a green light reflected from a cyclist's finger. Green led produced 530 nm wavelength turn to be most suitable wavelength to react with blood melanin inside microvascular bed of tissue which provides better waveform (Jihyoung et al. 2013). Low pass filter with 2 Hz cutoff frequency applied to raw PPG signal to remove unwanted signal before it's been amplifed. This conditioning or acquisition circuit built on a single board with 10 mm diameter for fit under the finger skin. A Nova microcontroller with Arduino Leonardo bootloader was chosen as a controller due to its small size and low operated power consumption. Nova powered by 90 mAh Polymer Lithium Ion Battery that can last for 2 h direct usage.

Comparative method that used in peak finding process functioned by comparing each sampled signal received to find highest value that declared as peak as illustrated in Fig. 1. Time on each peak stored to find differences between each of them before calculating HRV using (2). HRV for each individual are exactly not same and there are many formulas to calculate maximum HRV for different categories. However, for standard users, Karvonan formula (3) still can be used to determine maximum HRV based on age.

$$T_i = t_b - t_{b-1} \tag{1}$$

$$HRV = T_i \times 60 \tag{2}$$

$$HRV_{\max} = 220 - Age \tag{3}$$

where T_i is time interval between peak to peak in second, t_b is time at current peak and t_{b-1} is time at previous peak. By using this comparative method with interbeat interval, HRV can be known much faster.

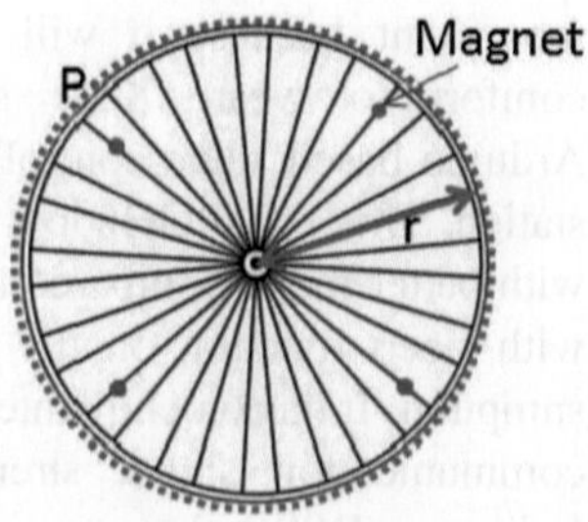

Fig. 3 Hall effect magnet place on bicycle rear wheel

Body skin temperature measured by using 406b contact type thermal probe series, which medical class temperature probe with 0.02 °C sensitivity level. Thermal probe attached inside the helmet at the front area which contact directly to cyclist's forehead. A quarter Wheatstone bridge designed to increase sensing accuracy before temperature raw data send to Arduino Pro Mini for converting from ADC to degree Celsius. Signal sampled every 2 ms and transmitted an average temperature every 2 s to reduce power usage. With sleep mode on Xbee series 2 and 800 mAh Polymer Lithium Ion Battery, the temperature sensor node can last for approximately 17 h.

Four magnets placed on a bicycle wheel like illustrated in Fig. 3 used to provide hall effect that will detect by the magnetic hall effect sensor that placed on the bicycle body near to wheel like illustrated in Fig. 2c to calculate bicycle speed and distance that have been travelled by cyclists. Track bicycle tyre perimeter approximately 2 m, which mean 1 tyre rotation equal to 2 m travelled distance. Bicycle tyre perimeter divides quarter by four magnets that each magnet will represent 0.5 m. Travelled distance can be calculated using Eq. (5) and cycling speed calculated based on Eq. (6) that implemented inside microcontroller.

$$T_r = t_m - t_{m-1} \tag{4}$$

$$Dist(m) = M_c \times 0.5 \tag{5}$$

$$Speed(kmj) = 3.6/T_r \tag{6}$$

where T_r is time between magnetic sensor triggered in second, t_m is time at current magnetic sensor triggered and t_{m-1} is time at previous magnetic sensor triggered. While M_c is the magnetic hall effect sensor triggered count.

Wireless Communication

Wearable device not complete without wireless communication on it. Wireless very important to apply onto a wearable device, especially that one is involved in movement because it will reduce wiring complexity and increase degrees of comfort to wear. Xbee series 2 with Zigbee protocol attached to each Arduino-based microcontroller as end device to transmit processed data to base station. This is 2.4 GHz low cost and low powered wireless communication device with better range compared to ANT wireless protocol devices. This module comes with sleep function on the end device side which useful to reduce power consumption. It has on chip antenna to strengthen transmitted and receive signal. Xbee communication signal strength measured based on Receive Signal Strength Indicator (RSSI) that provide in Pulse Width Modulation (PWM) output. This module operated at 12 MHz with 2400 total counts on PWM referred to 200 μs period. PWM reading in decimal from the analog port can convert to signal strength percentage using (8).

$$PWM = (41 \times RSSI) - 5928 \quad (7)$$

$$S(\%) = \left(\frac{PWM}{2400}\right) \times 100 \quad (8)$$

where *PWM* is the count of on cycle, *RSSI* is decimal value read through analog pin on the microcontroller that represent duty cycle in microsecond and *S* is signal strength in percentage.

Data packet from the end device at sensor node will try to transmit data directly to the coordinator at a base and this will reduce the transmitting time, but if the connection between the end device and coordinator failed, end device will transmit to the router as a signal repeater to pass data to the coordinator. Figure 4 illustrates a wireless network connection between end device, router and coordinator. This network connection reduces data packet drop and each packet assigned with count header to avoid data duplicated and identify packet drop.

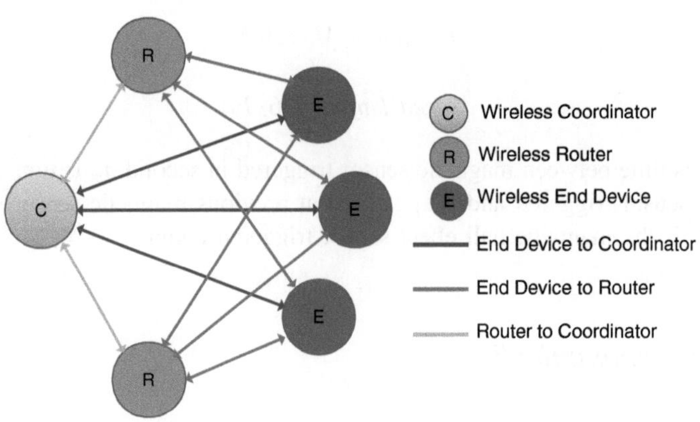

Fig. 4 Wireless network connection

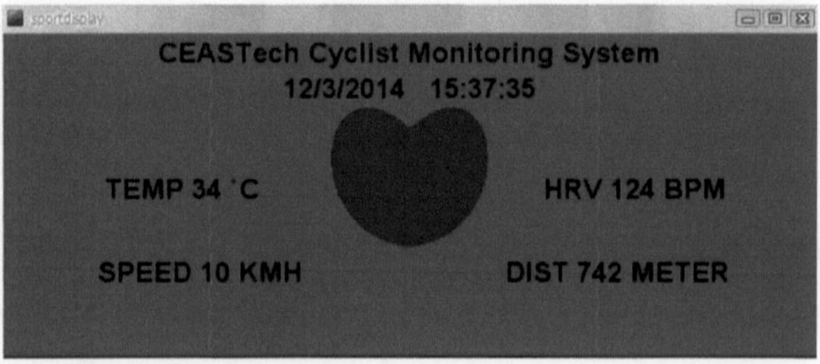

Fig. 5 User interface monitoring panel

Interface

Processing with Java-based was used to develop a graphical user interface which contains only useful readable information about HRV. Beside providing a user interface like illustrated in Fig. 5, processing also recorded incoming data into Excel for further action if there any abnormalities on data. Processing is one open source platform that can be used like a plain sketch that we can write or draw anything we want using Java language program. On this user interface, data displayed refresh every 10 ms.

3 Results

The system was applied to national junior cyclist to test on the field. Comparison between current on market Sigma PC10.11 heart rate belt with developed HRV glove based on PPG method, shows that there are small differences in HRV reading with 3 % average error rate as illustrated in Fig. 6.

Xbee series 2 with 1.5 m height point to point connection range inside Velodrome up to 50 m. After reaching 50 m, transmitted packet data start to loss due to the weak signal connection between these two nodes. The connection range test result shown in Fig. 7a. Two repeaters attached to the current system with distance from each router node to base and to end device not exceed 40 m like illustrated in Fig. 7b to reduce data packet drop as minimum as possible. With additional two repeaters, data packet loss reduced to 3 % from 46 % without repeater.

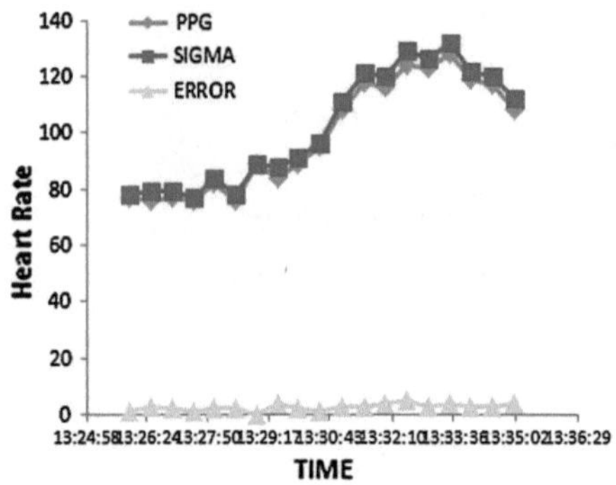

TIME	PPG	SIGMA	EFFICIENCY
13:26:00	77	78	98.7179
13:26:30	76	79	96.2025
13:27:00	77	79	97.4684
13:27:30	76	77	98.7013
13:28:00	82	84	97.6190
13:28:30	76	78	97.4359
13:29:00	89	89	100.0000
13:29:30	84	88	95.4545
13:30:00	89	91	97.8022
13:30:30	95	96	98.9583
13:31:00	108	111	97.2973
13:31:30	118	121	97.5207
13:32:00	116	120	96.6667
13:32:30	124	129	96.1240
13:33:00	123	126	97.6190
13:33:30	128	132	96.9697
13:34:00	119	122	97.5410
13:34:30	117	120	97.5000
13:35:00	108	112	96.4286
Average Accuracy			97.4751

Fig. 6 Comparison HRV between developed system with on market system

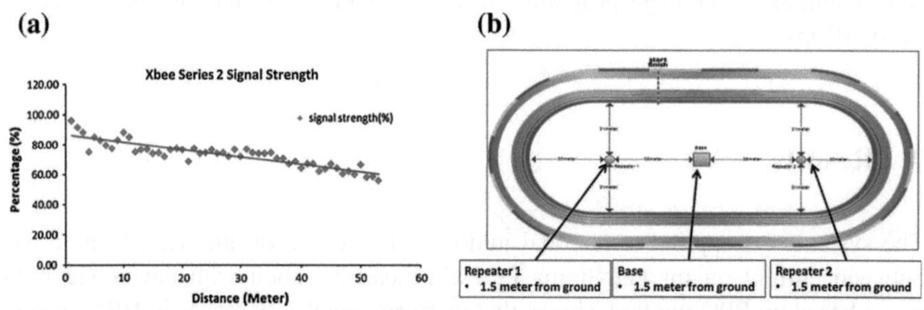

Fig. 7 **a** Xbee series 2 signal strength test; **b** repeater and base in Velodrome

4 Conclusion

Developed systems to monitor cyclist's on the Velodrom track has high reliability and tested can provide wide wireless communication inside the Velodrome with minimum data packet loss. PPG concept proves that it's suitable to capture cyclist's HRV during cycling and in more easy to wear as a glove. The system provides real-time monitoring data while same data will recorded for further analysis by a sports physician.

Acknowledgements This project is funded by the Sport Division on Ministry of Education Malaysia under Sport Research Grant Scheme. This research also has collaborated with the National Sport Council (ISN).

References

Anh D, Tao W (2010) Bandage-size non-ECG heart rate monitor using ZigBee wireless link. In: 2010 international conference on bioinformatics and biomedical technology (ICBBT)

Jihyoung L, Matsumura K et al (2013) Comparison between red, green and blue light reflection photoplethysmography for heart rate monitoring during motion. In: 2013 35th annual international conference of the IEEE engineering in medicine and biology society (EMBC)

Jones WD (2006) Taking body temperature, inside out [body temperature monitoring]. IEEE Spectr 43(1):13–15

Lei S, Chen W et al (2011) Study of algorithm for heart rate detection based on bipolar motion ECG. In: 2011 third international conference on measuring technology and mechatronics automation (ICMTMA)

Malkinson TJ (2002) Skin temperature response during cycle ergometry. In: 2002 IEEE CCECE Canadian conference on electrical and computer engineering

Mu-Huo C, Li-Chung C et al (2008) A real-time heart-rate estimator from steel textile ECG sensors in a wireless vital wearing system. In: 2008 ICBBE the 2nd international conference on bioinformatics and biomedical engineering

Okamoto S, Tsujioka T et al (2014) Design of wireless waist-mounted vital sensor node for athletes—performance evaluation of microcontrollers suitable for signal processing of ECG signal at waist part. In: 2014 IEEE topical conference on biomedical wireless technologies, networks, and sensing systems (BioWireleSS)

Wang L, Su SW et al (2006) Time constant of heart rate recovery after low level exercise as a useful measure of cardiovascular fitness. In: 2006 EMBS '06 28th annual international conference of the IEEE engineering in medicine and biology society

Chapter 13
Evaluation of Material Properties of Cold-formed Steel Channel Section with Different Thickness

Mohd Syahrul Hisyam Mohd Sani, Fadhluhartini Muftah, Ahmad Rasidi Osman, Mohd Azran Razlan and Cher Siang Tan

Abstract Cold-formed steel channel (CFSC) section is a popular material that used broadly in construction, especially in the roof truss system and non-structural elements. CFSC with a variety of sizes, shapes, thicknesses and grades is becoming admired because it offers several advantages such as lightweight, easy fabrication, fast installation, no formwork, termite resistance and corrosion resistance. The main objective of the study is to investigate the CFSC material properties with a variety of thicknesses. When the CFSC is utilised as in construction structural material, the CFSC material properties must be evaluated for further work, especially in design, failure capacity information and modelling intention. The thickness of the CFSC is important in study to check the failure mode and mechanical properties of the material. The material properties of CFSC are examined by using tensile coupon test specimens and universal tensile testing machine. The CFSC that be used in the study is CFSC 1.0 and CFSC 1.2 with the thickness of 1.0 and 1.2 mm respectively for grade G450. The CFSC is selected by having intermediate end stiffener and lipped. The specimen of tensile coupon test is located on flange, web, upper corner

M.S.H. Mohd Sani (✉) · F. Muftah · A.R. Osman · M.A. Razlan
Faculty of Civil Engineering, Universiti Teknologi MARA,
Jengka, Pahang, Malaysia
e-mail: msyahrul210@pahang.uitm.edu.my

F. Muftah
e-mail: fadhluhartini@gmail.com

A.R. Osman
e-mail: rasidi@pahang.uitm.edu.my

M.A. Razlan
e-mail: azran@pahang.uitm.edu.my

M.S.H. Mohd Sani · F. Muftah · C.S. Tan
Faculty of Civil Engineering, Universiti Teknologi Malaysia,
Johor Bahru, Johor, Malaysia
e-mail: tcsiang@utm.my

N.A. Yacob et al. (eds.), *Regional Conference on Science, Technology and Social Sciences (RCSTSS 2014)*, DOI 10.1007/978-981-10-0534-3_13

and bottom corner of overall section. From the results, the yield stress, elastic modulus and ultimate stress of the CFSC 1.2 is found higher than of CFSC 1.0.

Keywords Material properties · Cold-formed steel · Channel section

1 Introduction

Material properties are the important part in the investigation of the performance of steel structural element and also must be considered in designing criteria. The material properties of the section are affected when the temperature is rising or drop. The main criteria that have been tested in material properties are yield load, yield stress, ultimate load, ultimate stress, stress–strain relationship, load-extension relationship, modulus of elasticity, Poisson ratio and ductility. The minimum required for ultimate tensile stress is listed in the AISI Specification range from 290 to 690 MPa. Although, the effects of cold work condition of the mechanical properties depend on the factor of *fu/fy* ratios. Wang and Young (2014) were reported the material testing is the important thing that has been done before design the cold-formed steel channel beam with stiffeners. They were testing the material properties by using two coupon specimens for seven different members. Uzzaman et al. (2013) have tested and determined the material properties of CFSC with web hole, according to British Standard and the coupon location were placed from the web in the longitudinal direction. Dinis et al. (2014) has reported the elastic modulus on the web coupon specimen, G550 and G500 of cold-formed steel rack-section column is higher when comparing with the flange coupon specimen. All research includes beam, column and other application of CFSC were started with the material testing to determine the yield stress and elastic modulus.

Generally, the mechanical properties of cold-formed sections are different from other form, shape and size. The reason is the cold-forming usually increases the yield and tensile stress, and simultaneously also decrease the ductility. The increment of tensile stress is smaller when comparing with the yield stress. The material properties

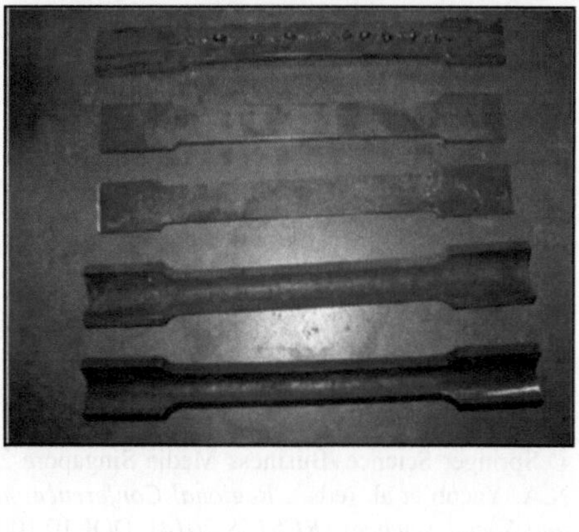

Fig. 1 Flat and corner tensile coupon specimens before testing (Hu et al. 2011)

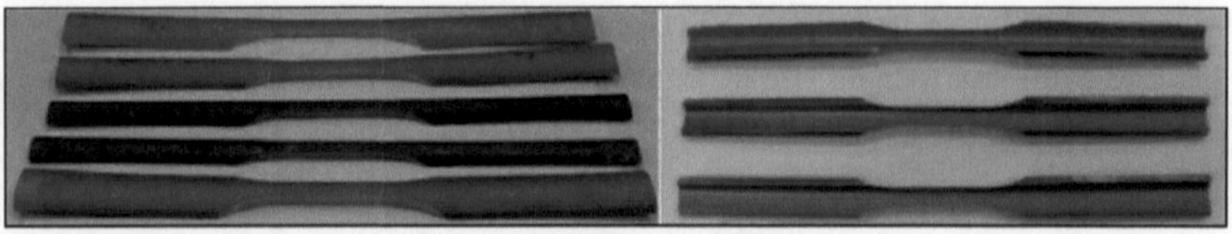

Fig. 2 Necked corner coupon specimens (Afshan et al. 2013)

Fig. 3 The location of *flat* and *corner coupon* test (Zhang and Young 2012)

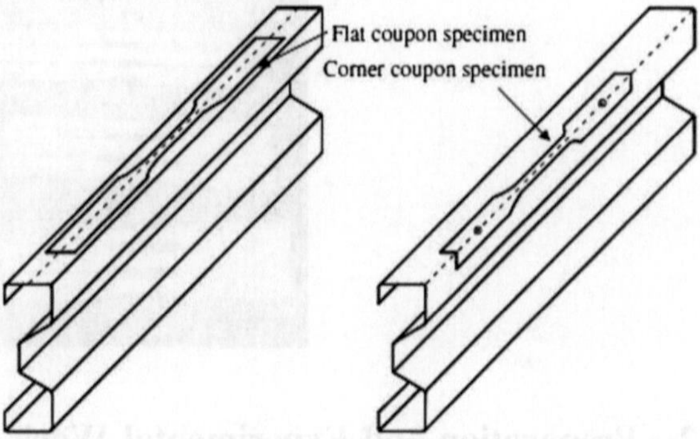

of the corner section are the higher result compared to flat section. Corner coupon specimen is produced to determine the strength enhancement due to cold-forming purpose and process (Afshan et al. 2013). Figures 1 and 2 shows the examples of flat and corner coupon specimens before testing. Zhang and Young (2012) has shown in Fig. 3 the location of flat and corner coupon specimens for checking the material properties of cold-formed steel I-shaped open section. Flat and corner coupon specimen was produced to determine the material properties of materials when without cold-forming process and also with the cold-forming process. This is important to know the relationship of the material properties as any part of cold-forming either at lipped or any part of curving by cold-forming without temperature rise.

There is a lot of standard and code of practice for the preparation of coupon specimens such as American Standard (ASTM E8-04 2004), Australian Standard (AS 1391 2007), Eurocode (BS EN 10002-1 2001) and Indian Standard (IS 1608 2005). In the study, the coupon specimen for flat and corner was followed the ASTM E8-04. The main objective of the study is to provide the material properties that have been broadly sold in the industry market for roof truss system and other application. The information about the material properties is very important and will be used in the design and modelling work. Besides, the material that is being used in construction must be checked with their properties for safety purpose.

Table 1 The dimension of the two types of CFSC in the study

Specimen	Breadth, B (mm)	Depth, D (mm)	Thickness, t (mm)	B/t	D/t	D/B
CFSC 1.0	38	75	1.00	38.00	75.00	1.97
CFSC 1.2	38	75	1.20	31.67	62.50	

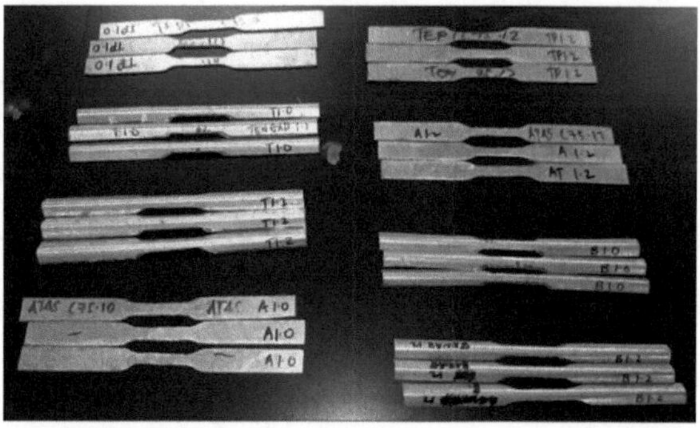

Fig. 4 Flat and corner coupon specimens

2 Preparation and Experimental Work

Two types of Cold-formed Steel Channel (CFSC) were used in this study and the description is listed in Table 1. Ratio of dimension between both the sections is also tabulated in Table 1. The CFSC is with the intermediate end stiffener and lipped that generally used in a roof truss system. Grade cold-formed steel that be used in the study is G450. The specimens of coupon either flat or corner were cut from the CFSC section. It was cut at 4 locations from the cross-section and noted for flange, web, upper corner, and the bottom corner as shown in Fig. 4. The flat and corner coupon specimens were extracted in the longitudinal direction of the column testing height.

The tensile coupon test was prepared and tested for tensile tests until failing circumstances or broken by using Universal Tensile Machine (UTM) with maximum load of 100 kN under displacement control rate of 1 mm/min. The extensometer was used to determine the strain of the specimen under tensile load. For checking the chemical composition or element, the GDS testing has been utilised to investigate the metal-based properties of CFSC. The sample was occupied randomly in the CFSC section in square shape with size 20 mm. The galvanized coating was removed prior the test to get the material base metal by using the sand paper and thinner liquid.

Table 2 The percentage of chemical composition of the cold-formed steel

Chemical compositions	Percentage		
	CFSC	CFS G550 (Chen and Young 2007)	CFS G450 (Chen and Young 2007)
Iron (Fe)	99.0	99.6	99.6
Carbon (C)	0.0538	0.055	0.055
Manganese (Mn)	0.175	0.20	0.21
Phosphorus (P)	0.0312	0.007	0.008
Sulfur (S)	0.00858	0.015	0.014
Silicon (Si)	0.0157	<0.005	0.005
Copper (Cu)	0.0906	0.009	0.009
Nickel (Ni_2)	0.317	0.026	0.027
Chromium (Cr)	0.0213	0.013	0.011
Vanadium (V)	0.0192	<0.003	<0.003
Molybdenum (Mo)	0.00772	0.002	0.002
Titanium (Ti)	0.0401	<0.003	<0.003
Aluminum (Al)	0.0182	0.037	0.038
Niobium (Nb)	0.00712	0.001	0.001
Zirconium (Zr)	0.232	–	–
Boron (B)	0.00263	–	–
Antimony (Sb)	0.0167	–	–
Cobalt (Co)	0.00990	–	–
Tin (Sn)	0.091	0.002	0.003
Lead (Pb)	0.0116	–	–

3 Experimental Results and Discussion

Chemical Properties

Chemical properties of the CFSC were tested by using Glow Discharge Spectrometer (GDS) and the result of testing was tabulated in Table 2. It showed that the CFSC that be used is contributing 99.0 % for iron in the total amount of chemical. The chemical composition of CFSC is compared with the CFS grade G550 and G450 that be used by Chen and Young (2007) in their research. The percentage difference between the CFSC and G550 and G450 is about 0.60 % amount of iron in the specimen. This is shown that the specimen that be tested is purely an iron-based. Besides, the chemical composition of Tin, Nickel, Phosphorus and Aluminum was reported a high percentage of difference between these specimens is 97.8, 91.8, 77.5 and 50.8 %, respectively. Nickel in CFS could help the improvement of the fracture toughness and corrosion resistance. Thus, the CFSC is more fracture toughness when compared with CFS G550 and G450. The Phosphorus is produced, its role and responsibility in reducing the ductility of the material. The content of Carbon in this material is 0.05 % and it is important to produce strength of the material. In addition, Chromium, Columbium, Copper, and Molybdenum also act as corrosion resistance of the material.

Material Properties

Flat Coupon Tests. The flats coupon specimens were occupied from the centre of the face at flange and web. The location of the flat coupon specimens was taken from the cold-formed steel channel as shown in Fig. 5. The calibrated extensometer with 50 mm gauge length was located into the flat coupon specimen. This extensometer was used to investigate the strain for the period of testing. All the testing was used a 100 kN capacity Universal Testing Machine as shown in Fig. 6.

Material properties such as yield stress, ultimate stress, elastic modulus, and extension at ultimate load are investigated by using a tensile coupon test. The result of the material properties of cold-formed steel channel with different thickness is

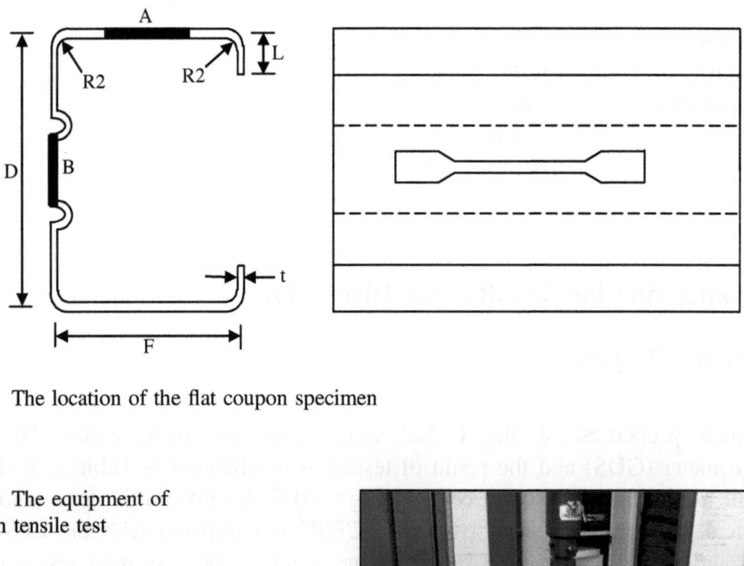

Fig. 5 The location of the flat coupon specimen

Fig. 6 The equipment of coupon tensile test

shown in Table 3. Predictably, the ultimate load, yield load, ultimate stress and yield stress of CFSC with thickness, 1.2 mm was shown higher than 1.0 mm. The percentage different of ultimate load of flange and web is 2.85 and 0.63 % in thickness, 1.0 and 1.2 mm, respectively. The ultimate stress of flange is smaller

Table 3 Material properties of flange and web coupon specimen of CFSC with different thickness

Parameter/property	1.0 mm thickness		1.2 mm thickness	
	A, flange	B, web	A, flange	B, web
Ultimate load (N)	3771.24	3881.85	4119.79	4093.86
Yield load (N)	3639.15	3771.40	4093.86	3903.97
fu, ultimate stress (MPa)	471.41	485.23	514.97	506.11
fy, yield stress (MPa)	454.89	471.43	511.73	488.00
fyf, yield stress factory	450.00	450.00	450.00	450.00
Modulus of elasticity (GPa)	203.72	203.91	205.97	206.63
Extension on ultimate load (mm)	1.3900	0.8585	1.5918	1.0585
fu/fy	1.036	1.029	1.006	1.037
fu/fyf	1.048	1.078	1.144	1.125

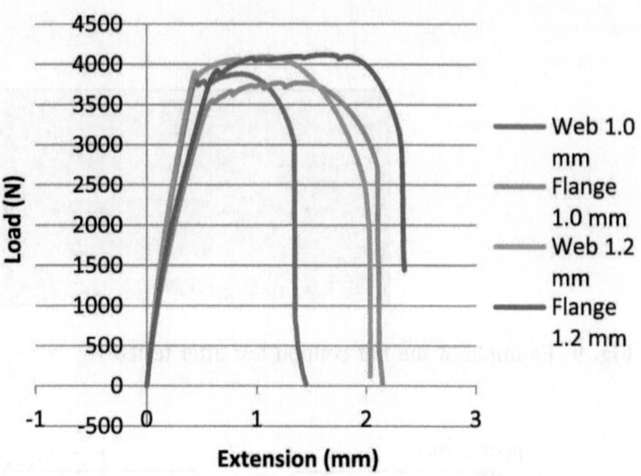

Fig. 7 The load versus extension curve for flat coupon test

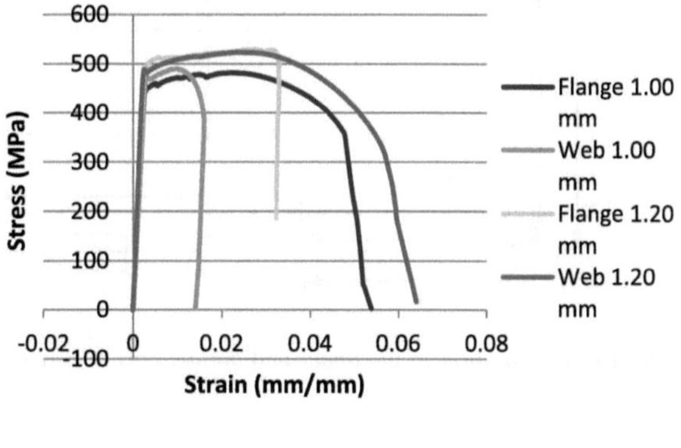

Fig. 8 The stress–strain curve of flat coupon test

when compared with the web for 1.0 mm thickness, but rather to the thickness, 1.2 mm. The flat coupon specimen is very important to ensure the specimen in exactly straight and flat. If the flat coupon specimen is not in exactly flat, it will be affecting the reading of the stress because the specimen is having an initial load or initial stress. The specimen will be failed on quickly and rapidly when subjected to tensile load. The process of cutting to get the flat coupon specimen will be produced the not flat or straight specimen.

The extension for ultimate load for the web is smaller than the flange for both thicknesses. The ratio of ultimate stress per yield stress is recorded in Table 3. The ratio of *fu/fyf* is reported more than *fu/fy* and it is meant that this material is safe to be used. This is because the yield stress of experiment is illustrated over the limit of factory data. The load versus extension and stress–strain curve is shown in Figs. 7 and 8, respectively. Figure 9 shows the example of the flat coupon specimen after testing.

Corner Coupon Tests. The corner coupon specimens were located from the corner at upper and bottom corner of the CFSC. The location of the upper corner coupon specimens was taken from the CFSC as shown in Fig. 10. The calibrated extensometer with 50 mm gauge length was located into the corner coupon specimen. This extensometer was used to investigate the strain for the period of testing.

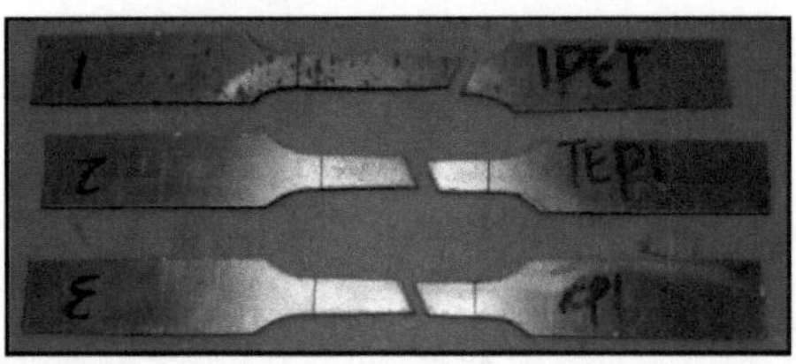

Fig. 9 Example of the flat coupon test after tested

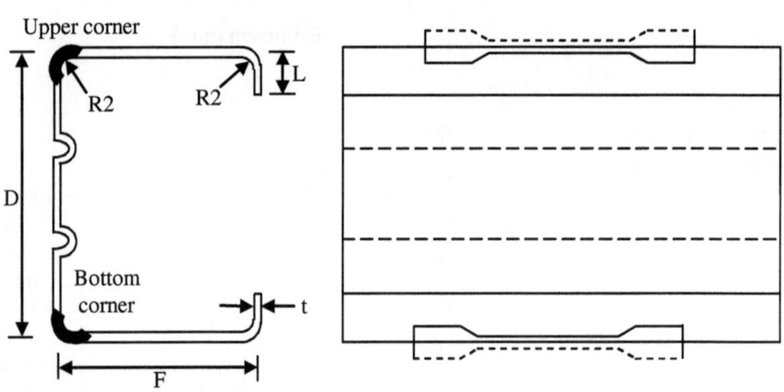

Fig. 10 The location of the *upper* and *bottom corner* coupon specimen

Fig. 11 The testing of corner coupon test

Table 4 Material properties of corner coupon specimen

Parameter/property	1.0 mm thickness		1.2 mm thickness	
	Upper corner	Bottom corner	Upper corner	Bottom corner
Ultimate load (N)	5992.41	4689.95	6321.41	5140.00
fu, ultimate stress (MPa)	704.99	689.70	650.18	629.98
fy, yield stress (MPa)	465.15	619.44	288.65	474.34
fyf, yield stress factory	450.00	450.00	450.00	450.00
Modulus of elasticity (GPa)	100.16	107.64	95.73	104.84
Extension on ultimate load (mm)	2.9785	1.7484	3.2251	1.9684
fu/fy	1.516	1.113	2.250	1.328
fu/fyf	1.567	1.533	1.445	1.400

All the testing was used a 100 kN capacity Universal Testing Machine and hot-rolled rod for gripping process as illustrated in Fig. 11.

Material properties such as yield stress and ultimate stress are investigated by using a tensile coupon test. The result of the corner coupon test was shown in Table 4. As expected, the ultimate load is increased with increasing the thickness of the CFSC. The percentage different of ultimate stress between upper corner and

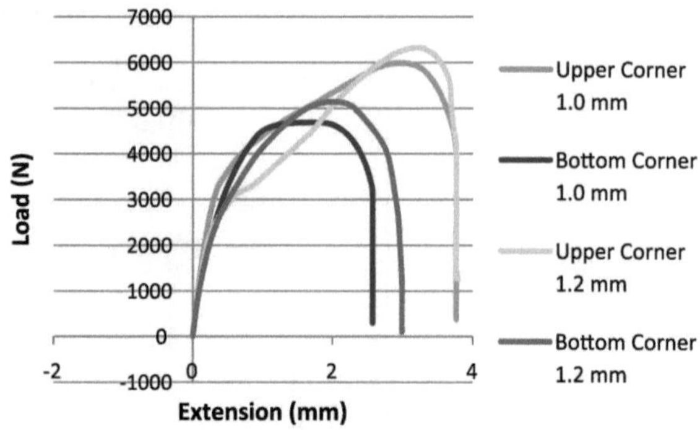

Fig. 12 The load-extension curve of corner coupon specimen

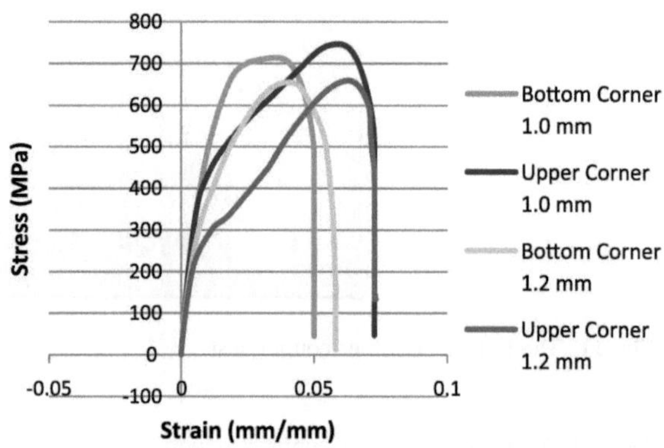

Fig. 13 The stress–strain curve of corner coupon specimen

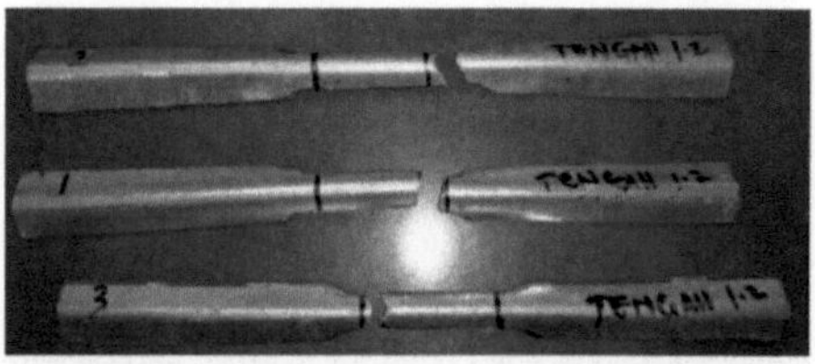

Fig. 14 Example of the corner coupon specimen after testing

bottom corner CFSC 1.0 and 1.2 mm thickness is 7.77 and 8.65 %, respectively. The upper corner for both specimens significantly showed a high value of ultimate load and stress when compared with bottom corner. The ratio of *fu/fy* is generally depends on the pattern of the graph and value of the yield stress. The ratio of *fu/fyf* for a corner coupon specimen is reported more than *fu/fy* and its intended this material is safe to be used. Figures 12 and 13 are utilised the load-extension curve and stress−strain curve. The line pattern of the graph is showed same line for upper and bottom corner although in different thickness. Figure 14 shows the corner coupon specimen after the testing process.

The upper corner with 1.2 mm thickness is shown inaccurate value because the phenomena of a small split in the testing. The split problem is the major issue that must be considered in investigating the result. This is because for the corner coupon test, the rounded hot-rolled bar is placed in the holder on the corner coupon specimen. The attachment between the hot-rolled bar and CFSC is utilised the weld connection. The problem is not same as flat coupon specimen that only tighten on the holding grips. In addition, corner coupon specimen is difficult to provide the best and reasonable result due to uncertainties of cutting size, cutting method, method of testing, and grip holder. Corner coupon also showed the difficulties in measuring the cross-section area for tensile load. The specification and standard also still give a limited information and guidance to solve the uncertainties. Therefore, many researchers came out with their own additional method to get the accurate result. As mentioned by (Afshan 2013), the result of the corner coupon specimen is lead to different results because the researchers have used different methods of testing.

4 Comparison Study

The comparison of the material properties was investigated according to grade 450 MPa, grade 550 MPa and the corner coupon type that shown in Table 5. All specimens show the ratio of *fu/fy* in the range of 1.00–1.05. The elastic modulus of all specimens is recorded in range 200.00–210 GPa and normally noted as steel material. The percentage differs of ultimate stress between CFSC 1.0 and CFSC 1.2 with G550 (Wei and Jihong 2012) is 29.16 and 24.82 %. With the different grade, but same thickness, the CFSC 1.0 is noted below 30.00 %. 10.56 and 5.07 % is reported for comparison ultimate stress between G450 (Kankanamge and

Table 5 Comparison of material properties with other study

Material	Coupon types	Thickness (mm)	*fy* (MPa)	*fu* (MPa)	*fu/fy*	E (GPa)
CFSC 1.0	Flat	1.00	471.43	485.23	1.029	203.91
CFSC 1.2	Flat	1.20	511.73	514.97	1.006	205.97
G550 (Wei and Jihong 2012)	Flat	1.00	669.00	685.00	1.024	218.80
G450 (Kankanamge and Mahendran 2011)	Flat	1.90	515.25	542.50	1.053	201.40
G450 (Kankanamge and Mahendran 2011)	Flat	1.50	536.25	561.80	1.048	209.24
G450 (Lam et al. 2006)	Flat	1.60	515.00	537.00	1.043	208.00
G450 (Lam et al. 2006)	Corner	1.60	566.00	583.00	1.030	208.00
G450 (Chen and Young 2007)	Flat	1.90	524.00	551.00	1.052	203.00

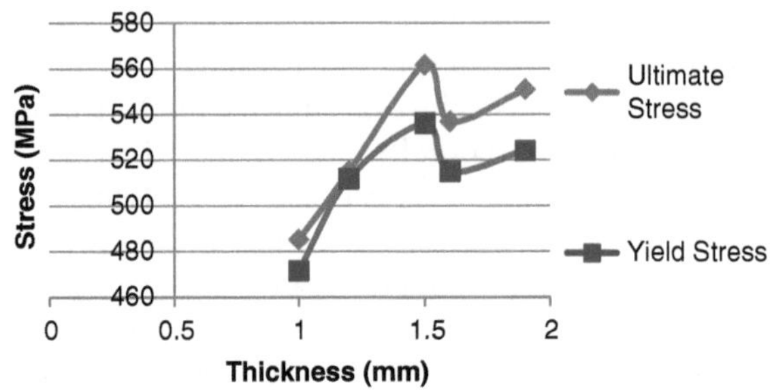

Fig. 15 The comparison graph of the different thickness

Mahendran 2011) with CFSC 1.0 and CFSC 1.2, respectively. Meanwhile, the percentage reduction of G450 (Lam et al. 2006) corner coupon with CFSC 1.0 and CFSC 1.2 is 16.77 and 11.67 %. The corner coupon specimen is normally produced high value of yield and ultimate stress when compare with the flat coupon specimen. This is because the production of the material is by cold-forming on ambient temperature could give the quick strain-hardening process.

According to Table 5, the comparison of the ultimate stress of grade 450 for different thickness is demonstrated in Fig. 15. Even though, the grade is similar and with thick specimen, but is generally not proposed the high value of ultimate stress. The value of the ultimate stress is established also according to their shape of the specimen and the testing procedure. The material properties of the specimen depend on the grade, thickness, chemical properties and testing method. The result of the G450 (Kankanamge and Mahendran 2011) for thickness, 1.50 is reported high value of ultimate stress when compared with the others specimen. It can be noted that the supplier of the material gives a high stress material and good quality of the iron-based material. The minimum yield stress for this specimen is 450 MPa and the percentage difference between the minimum yield stress and experimental yield stress is about 16.08 %. This is shown that the material properties of the specimens are becoming over design and adduce the non economical prospect. If compared with other specimen, the range of the increment between minimum yield stress and experimental yield stress is 5–15 %.

5 Conclusion

A series of experimental works are carried out to determine the material properties of cold-formed steel channel sections (CFSC) with different thicknesses. From the test result, some conclusions can be drawn

1. The ultimate and yield stress of the CFSC for flat coupon specimen with thickness, 1.2 mm is higher than thickness, 1.0 mm. The ultimate and yield stress increases with increasing of the thickness of CFSC.

2. All yield stress for the flat coupon specimen is reported more than the minimum requirement of minimum yield stresses that established by the steel factory. Therefore, the quality of the CFSC can be considered to achieve the prescribed quality and cannot be doubted to use in the construction industry.
3. The ultimate stress is recorded below 10 % difference between 1.0 and 1.2 mm thickness for upper and bottom corner coupon specimen.

Further study with experimental is recommended to identify the relationship of material properties with different grade. In addition, the study of factor that affected the material properties of CFSC must be investigated such as varies with temperature.

Acknowledgement The authors gratefully acknowledge the financial support from the Universiti Teknologi Mara Pahang under Dana Kecemerlangan Grant. Thanks also extend to Faculty of Civil Engineering and Faculty of Mechanical Engineering for providing machinery and equipment. Special thanks are extended to the lecturer and technician of UiTM Pahang and UTM Johor Bharu for their help during the experimental program.

References

Afshan S, Rossi B, Gardner L (2013) Strength enhancements in cold-formed structural sections-Part 1: material testing. J Constr Steel Res 83:177–188

AS 1391 (2007) Metallic materials—tensile testing at ambient temperature

ASTM E8-04 (2004) Standard test methods for tension testing of metallic materials

BS EN 10002-1 (2001) Tensile testing of metallic materials. Method of test at ambient temperature

Chen J, Young B (2007) Experimental investigation of cold-formed steel material at elevated temperatures. Thin-Walled Struct 45:96–110

Dinis PB, Young B, Camotim D (2014) Local-distortional interaction in cold-formed steel rack-section columns. Thin-Walled Struct 81:185–194

Hu S-D, Ye B, Li L-X (2011) Materials properties of thick-wall cold-rolled welded tube with a rectangular or square hollow section. Constr Build Mater 25:2683–2689

IS 1608 (2005) Metallic materials—tensile testing at ambient temperature

Kankanamge ND, Mahendran M (2011) Mechanical properties of cold-formed steels at elevated temperature. Thin-Walled Struct 49:26–44

Lam SSE, Chung KF, Wang XP (2006) Load-carrying capacities of cold-formed steel cut stub columns with lipped C-section. Thin-Walled Struct 44:1077–1083

Uzzaman A, Lim JBP, Nash D, Rhodes J, Young B (2013) Effect of offset web holes on web crippling strength of cold-formed steel channel sections under end-two-flange loading condition. Thin-Walled Struct 65:34–48

Wang L, Young B (2014) Design of cold-formed steel channels with stiffened webs subjected to bending. Thin-Walled Struct 85:81–92

Wei C, Jihong Y (2012) Mechanical properties of G550 cold-formed steel under transient and steady state conditions. J Constr Steel Res 73:1–11

Zhang, Young B (2012) Compression tests of cold-formed steel I-shaped open sections with edge and web stiffeners. Thin-Walled Struct 52:1–11

Chapter 14
Characterization and Development of Geldart's Fluidizing Velocity Profile of Sand Particles for the Application in Fluidized Bed Combustor (FBC)

Ahmmad Shukrie, Shahrani Anuar and Azri Alias

Abstract The objective of this study is to fully utilize the local sand retrieved from Pahang river as a bed material in FBC. The proposed work will study through experiments to deduce a new classification of local river sand in a fluidized bed based on the minimum fluidization velocity of inert particles according Geldart's classification operated at five different temperatures: 55, 82, 108, 124 and 148 °C respectively. River sand was chosen as a fluidizing medium due to its characteristics that can withstand high operating temperature of more than 1000 °C. Apart from that, its cheap cost and availability make it preferred choice for the operation. The results show that the minimum fluidizing velocity of the sand, U_{mf} was found increasing as the temperature increased for the sands operated at low temperature, but maintaining almost at the same value, for the sands operated higher operating temperature respectively. The findings agree well with the reported trend by previous authors, in which their results based on the bed operated at combustion temperature which is above 800 °C. Thus, it can be conclude that the local sands, will exhibits the same trend for the application of fluidized bed combustor.

Keywords Fluidized bed combustor · Geldart's classification · Sand particles

A. Shukrie (✉) · S. Anuar · A. Alias
Energy and Sustainability Focus Group, Faculty of Mechanical Engineering,
Universiti Malaysia Pahang, Kuantan, Pahang, Malaysia
e-mail: shukrie@ump.edu.my

S. Anuar
e-mail: shahrani@ump.edu.my

A. Alias
e-mail: azribalias@ump.edu.my

N.A. Yacob et al. (eds.), *Regional Conference on Science, Technology and Social Sciences (RCSTSS 2014)*, DOI 10.1007/978-981-10-0534-3_14

1 Introduction

Fluidized bed has been used widely in thermo chemical processes such as pyrolysis, gasification, liquefaction and combustion. The unique operational advantages such as fuel flexibility, intense solid mixing, and efficient heat transfer mechanism between solid fuel and inert particles, have nominated the fluidized bed combustor the most efficient reactor for the above processes (Fotovat et al. 2014). Many industrial processes utilizing the advantage of gas-solid mixing in fluidized bed that may operate in the following regimes: fixed bed, particulate fluidization, bubbling fluidization, slug flow regime, turbulent fluidization, fast fluidization, and pneumatic transport (Shukrie 2012; Basu Paudel 2011; Zhu et al. 2013) (Table 1).

It was the pioneered work of Professor Derek Geldart (Geldart 1972), who first categorized the behavior of particle fluidized by gases falls into four clearly recognizable groups A, B, C, D, characterized by density difference $\rho_s - \rho_f$ and mean particle size. The most easily recognizable features of the groups are: Geldart group A particles, demonstrate dense phase expansion after minimum fluidization and prior to the commenment of bubbling; and those in group B bubble at the minimum fluidization velocity. Group C was the smallest and most cohesive whilst Group D particle was the largest and demonstrated spoutable behavior. Particle mixing, reported by Geldart, was found to be much poorer in group C and D rather than the fluidized bed operated with powders of groups A or B. Thus, many commercial units operated by using particles classification of Geldart group A or B (Formisani et al. 2013; Oliveira et al. 2008; Rao et al. 2011; Zhu et al. 2013).

Combustion of biomass particles in fluidized bed combustor is a cumbersome process due to their irregular size, density and shape (Paudel and Feng 2013).

Table 1 Summary of fluidization regimes and its features (Shukrie 2012; Basu Paudel 2011; Zhu et al. 2013)

Fluidization regime	Velocity range	Fluidization features
Fixed bed	$0 \ll U < U_{mf}$	Particles are quiescent, gas flow through void spaces
Particulate fluidization	$U_{mf} \ll U < U_{mb}$	Bed expands smoothly, bed surface is well defined
Bubbling fluidization	$U_{mf} \ll U < U_{ms}$	Gas bubbles coalesce and grows as they rise, gas bubbles promote solids mixing
Slug flow regime	$U_{mf} \ll U < U_c$	Bubbles size spread across the bed cross-section, rises and falls with irregular frequency with corresponding pressure fluctuation
Turbulent fluidization	$U_k \ll U < U_{tr}$	Small gas voids and particles clusters, bed surface is diffused and difficult to distinguish
Fast fluidization	$U > U_{tr}$	Particles are carried out of the bed with gas
Pneumatic transport	$U > U_{tr}$	All particles are transported out

Generally, an inert material, which was used for start up of a bed, like silica sand, alumina and calcite, is added to assist and improve the heat transfer when the combustion process taking place (Arromdee and Kuprianov 2012; Fotovat et al. 2014; Han et al. 2008; Ninduangdee and Kuprianov 2013, 2014a, b). The measurements of physical properties of inert particle being used in the fluidized beds and the choice of its size directly influence the hydrodynamics of the combustor. In general, sand was chosen as an inert particle due to its characteristics that can withstand high operating temperature of more than 1000 °C despite of its availability and cheap cost make it preferred choice for the operation (Paudel and Feng 2013; Sanaei et al. 2010).

The minimum fluidization velocity of bed material is one of the most imperative parameters needed when designing a fluidized bed. It sets a lower limit on the flow rate needed by the blower and pressure drop required for fluidization. Extensive researches have been performed on the minimum fluidization velocity of inert particles of different Geldart's size classification which include sand, glass bead, and alumina (Chen et al. 2010; Goo et al. 2010; Jiliang et al. 2013; Lin et al. 2002; Paudel and Feng 2013; Wu and Baeyens 1991). To predict the minimum fluidization velocity, most of previous researches proposed the empirical correlations at room temperature condition, although many commercial fluidized bed combustors are operating at high temperature (Goo et al. 2010). Therefore, the reported correlations at higher temperatures have shown a marked disparity between the predicted and the measured minimum fluidization velocities with correction of gas physical properties at the given temperature (Chen et al. 2010; Goo et al. 2010; Wu and Baeyens 1991).

For that reason, the main focus of this study is to fully utilize the local sand as a bed material in FBC. The proposed work will study through experiments to deduce a new classification of local sand in a fluidized bed based on the minimum fluidization velocity of inert particles according Geldart's classification operated at both room temperature and higher temperature conditions. No such classification of local sand for the purpose of fluidization and fluidized bed combustor exists to date, though D. Geldart and other researchers have proposed classifications for conventional fluidized beds.

2 Experimental Procedure

Sand as Inert Particle

Sand particles used in the experiments were collected from the nearby river of Sungai Pahang, in the district of Pekan, Pahang, as shown in the Google Map™ of Fig. 1. The location of the collected sand is highlighted in box on the map. These particles were sieved using sieve shaker located at Civil Engineering Laboratory, University Malaysia Pahang. Tyler standard mesh sizes were used for the

Fig. 1 Google map of Pahang river

characterization of particles sizes. The Sauter mean diameter of particles, $\bar{d}_p$ is obtained from the sieve analysis of solid particles using the following equation (Shukrie et al. 2014; Paudel and Feng 2013)

$$\bar{d}_p = \frac{1}{\sum_i \left(\frac{w}{d_p}\right)_i} \tag{1}$$

The densities of sands were determined by water displacement where a known mass of sands was slowly added to a graduated cylinder containing a known quantity of water. The volume of absorbed water was then used when calculating the actual volume of displaced water. The volume of displaced water was used to determine the volume of the particle added (Paudel and Feng 2013). The phone of Scanning Electron Microscope (SEM) with 30× magnification and physical properties of the sand are tabulated in Table 2. Particle classifications were done in Ergun 6.2 Software™ and based on the work of Professor Derek Geldart (Geldart 1972) and tabulated in the Geldart's classification diagram as depicted in Fig. 2.

Fluidized Bed Set up

The experiments were conducted in the Energy and Sustainability Focus Group Laboratory in Faculty of Mechanical Engineering, University Malaysia Pahang by using a laboratory scale cylindrical glass column of 108 mm internal diameter and

Table 2 Properties of sand

Mean particle size $\bar{d}_p$= 341 µm

Particle density ρ_s= 2192 kg/m^3

Geldart Classification : B

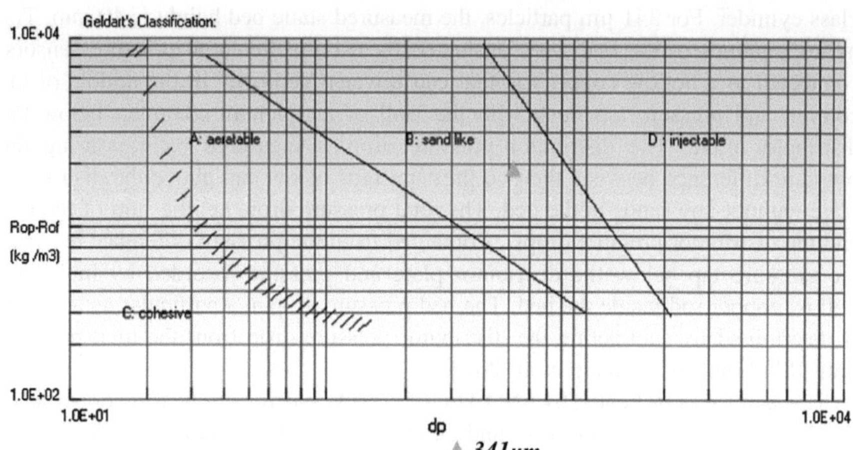

Fig. 2 Geldart classifications diagram of sand particle used in the experiment

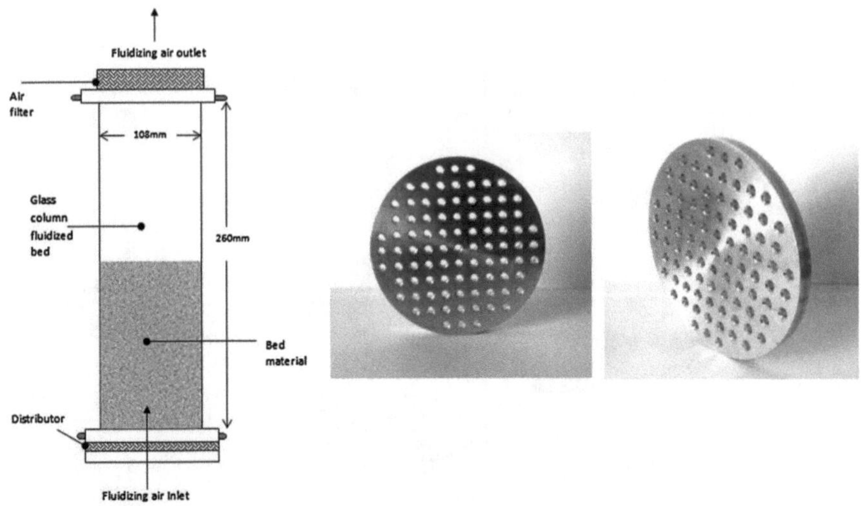

Fig. 3 Schematics of the fluidized bed and perforated distributor

260 mm long as shown in the Fig. 3. The details of the experimental procedures can be found elsewhere (Shukrie et al. 2014).

A centrifugal compressor supplies the fluidizing air at ambient conditions to the system and an orifice meter provides the measurement of air flow rate into the system. 115 mm diameter perforated-type air distributor plate of thickness 8 mm made of aluminium was used in the experiments. The perforated distributor had 89 holes with the diameter of 4 mm distributed over the plate with 10 mm pitch giving open area ratio of 3 %. 1200 g of sands were prepared accurately by a digital scale of resolution 0.1 g and was poured into the bed from the top. The static bed height in all experiments is measured by a strip of scale tape located on the outside of the glass cylinder. For 341 μm particles, the measured static bed height is 40 mm. The hydrodynamics of the bed were evaluated by using differential pressure sensors; connected to a hollow copper rod that can traverse vertically in the middle of the column and pressure tap flush with the wall of air plenum chamber, below the distributor plate. The distributor pressure drop is measured by measuring the pressure difference between the two pressure taps below and above the distributor plate without any sands in the bed. The total pressure drop, i.e. the sum of the bed and the distributor pressure drops, is obtained from the pressure difference between the pressure tap below the distributor plate and pressure recorded by immersed hollow copper rod inside the bed. The bed pressure drop at a particular air velocity is determined by subtracting the distributor pressure drop from the total pressure drop (Shukrie 2012; Jiliang et al. 2013).

The sands will be heated by SMATEC™, 750 W electrically heated copper tube, fitted with a T-type thermocouple and mounted vertically 25 mm above the distributor, capable of heating up the bed up to 400 °C. The ends of the tube are provided with Teflon support to reduce axial heat loss, which is estimated to be less than 1 %.

A DC power supply with a voltage regulation of 0.1 % is used to energize the heater. A voltmeter and an ammeter with an accuracy of 0.5 % are used to measure the electric power supplied to the heater. To measure bed and ambient temperature, two T-type thermocouples were located at the bulk of the bed and the outlet of the column respectively (Shukrie et al. 2014). The temperatures of the thermocouples were recorded simultaneously during the operation, via a computer controlled temperature measurement system, which incorporated a multiplexer that employed the electrical signals from the sensors, to feed the connected computer-plus-analog-digital (A/D) converter. Data were logged via 32-channel expansion board and analog-digital converter to a PC. The temperature and pressure data were recorded for a 15 min period with a 1 s sampling time. The minimum fluidizing velocity is established in the experiment by the intersection of the two linear plots describing the constant and decreasing pressure drop with decreasing fluidizing velocity.

3 Results and Discussions

Figure 4 shows the result of bed pressure drop for sand particle of 300 μm diameter operated at five different temperatures: 55, 82, 108, 124 and 148 °C respectively. It is shown that bed pressure drop increases with with air superficial velocity and bed temperature. The pressure drop with respect to air superficial velocity is lower for a sand particles operated at 55 °C temperature, and abruptly increased when the temperature is increased. The minimum fluidizing velocity is associated in the experiment by the intersection of the two linear plots describing the constant and decreasing pressure drop with decreasing air superficial fluidizing velocity. From Fig. 4, the minimum fluidizing velocity, U_{mf} was found increasing as the

Fig. 4 Bed pressure drop versus air superficial velocity of 300 μm sands operated at different temperature

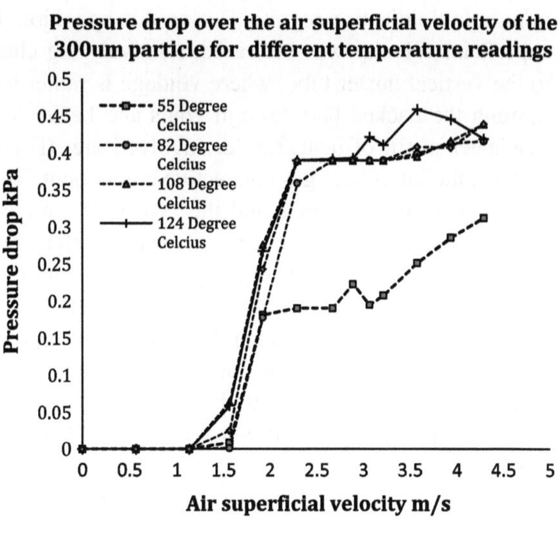

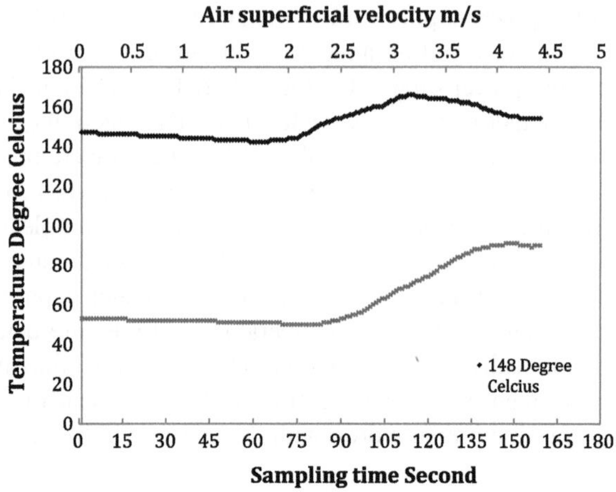

Fig. 5 Temperature profiles of the sand bed versus sampling time and air superficial velocity

temperature increased for the sands operated at 55, 82 °C, but maintaining almost at the same value, for the sands operated at 108, 124 and 148 °C respectively. The trend observed in this experiment agrees well with the reported trend by (Lin et al. 2002). The authors reported that the minimum fluidization velocity increased as the temperature increased for the bed operated at above 800 °C. However, the trend contradicted with (Lin et al. 2002) when the minimum velocity was found to be vice versa when the temperature of the bed plotted versus the air superficial velocity as plotted in the graph below.

The temperature profiles of bed operated at 55 °C (minimum) and 148 °C (maximum) recorded over time is depicted in Fig. 5. The increase of the air superficial velocity results in slightly drop of bed temperature at 2 m/s for sand bed operated at 148 °C and 2.5 m/s for sand bed operated at 55 °C respectively. The temperature of the sands started to gradually increased as the sands enter the minimum fluidization regime. During this operation, it is observed that when the air superficial velocity was increased, relatively big channels become visible adjacent to the vertical heater tube, where voidage is rather high. The hot air tended to pass through the packed bed through voids and heater located at the centre of the bed. The mechanism of heat transfer involved are: (i) particle convective components and, (ii) the interphase gas convective component. The first component involved the heat transfer by the sands and the heater. Meanwhile for the interphase gas convective component, the process of heat transfer are by convective mixing which enhances the heat transfer in the gas voidages and the heater surface and between neighboring sand particles. The temperature attains a maximum value at the excess of fluidizing velocity in which the same trend reported from the previous work of (Shukrie et al. 2014)—heat transfer coefficient increases with the increase of gas velocity. The temperature of the sands were recorded decreases with additional increase in air superficial velocity. This phenomena is due to the gas flow conditions are in the transitional or turbulent flow—could be expected to lead to some reduction in the overall heat transfer coefficient because of a possible increase in continuous phase void age and hence reduction in particle packing density at the heat transfer surface.

4 Conclusions

Series of experiments were performed to deduce a new classification of local sand in a fluidized bed based on the minimum fluidization velocity of inert particles according Geldart's classification operated at both room temperature and higher temperature conditions. Local sand was chosen as a fluidizing medium for fluidized bed combustor application due to its availability and its characteristics that can withstand high operating temperature. Result shows that the trend of minimum fluidization velocity operated at five different temperatures: 55, 82, 108, 124 and 148 °C observed in this experiment agrees well with the reported trend by previous authors, in which their results based on the bed operated at above 800 °C. Thus, it can be conclude that the local sands, will exhibits the same trend for the application of fluidized bed combustor.

Acknowledgements The authors wish to express their gratitude to the Faculty of Mechanical Engineering, Universiti Malaysia Pahang (UMP), Pekan, Pahang for providing the research facilities and supporting the research under University Research Grant RDU140528 and Post Graduate Research Scheme (PGRS) RDU140346.

References

Arromdee P, Kuprianov VI (2012) Combustion of peanut shells in a cone-shaped bubbling fluidized-bed combustor using alumina as the bed material. Appl Energy 97:470–482. doi:10.1016/j.apenergy.2012.03.048

Basu Paudel (2011) Experimental study on fluidization of biomass, inert particles, and biomass/sand mixtures. University of North Texas

Chen ZD, Chen XP, Wu Y, Chen RC (2010) Study on minimum fluidization velocity at elevated temperature. In: Zhongguo Dianji Gongcheng Xuebao/Proceedings of the Chinese Society of Electrical Engineering, 30(14), 21–25. http://www.scopus.com/inward/record.url?eid=2-s2.0–77953775691&partnerID=tZOtx3y1

Formisani B, Girimonte R, Vivacqua V, Rende IA (2013) Fluidization of mixtures of two solids: a unified model of the transition to the fluidized state. AIChE J 59(3):729–735. doi:10.1002/aic

Fotovat F, Chaouki J, Bergthorson J (2014) Distribution of large biomass particles in a sand-biomass fluidized bed: experiments and modeling. AIChE J 60(3):869–880. http://doi.wiley.com/10.1002/aic.14337

Geldart D (1972) The effect of particle size and size distribution on the behaviour of gas-fluidised beds. Powder Technol 6(4):201–215. doi:10.1016/0032-5910(72)83014-6

Goo JH, Seo MW, Kim SD, Son BH (2010) Effects of temperature and particle size on minimum fluidization and transport velocities in a dual fluidized bed experimental. In: Proceedings of the

20th international conference on fluidized bed combustion, pp 305–310. doi:10.1007/978-3-642-02682-9_43

Han J, Kim H, Minami W, Shimizu T, Wang G (2008) The effect of the particle size of alumina sand on the combustion and emission behavior of cedar pellets in a fluidized bed combustor. Bioresour Technol 99(9):3782–3786. doi:10.1016/j.biortech.2007.07.010

Jiliang M, Xiaoping C, Daoyin L (2013) Minimum fluidization velocity of particles with wide size distribution at high temperatures. Powder Technol 235:271–278. doi:10.1016/j.powtec.2012.10.016

Lin C, Wey M, You S (2002) The effect of particle size distribution on minimum fluidization velocity at high temperature. Powder Technol 126:297–301

Ninduangdee P, Kuprianov VI (2013) Study on burning oil palm kernel shell in a conical fluidized-bed combustor using alumina as the bed material. J Taiwan Inst Chem Eng 44 (6):1045–1053. doi:10.1016/j.jtice.2013.06.011

Ninduangdee P, Kuprianov VI (2014a) Combustion of oil palm shells in a fluidized-bed combustor using dolomite as the bed material to prevent bed agglomeration. Energy Procedia 52:399–409. doi:10.1016/j.egypro.2014.07.092

Ninduangdee P, Kuprianov VI (2014b) Combustion of palm kernel shell in a fluidized bed: optimization of biomass particle size and operating conditions. Energy Convers Manag 85:800–808. doi:10.1016/j.enconman.2014.01.054

Oliveira WP, Souza CRF, Lim CJ, Grace JR (2008) Evaluation of flow regimes in a semi-cylindrical spouted bed through statistical, mutual information, spectral and Hurst's analysis. Can J Chem Eng 86(3):582–597. http://doi.wiley.com/10.1002/cjce.20055

Paudel B, Feng Z-G (2013) Prediction of minimum fluidization velocity for binary mixtures of biomass and inert particles. Powder Technol 237:134–140. doi:10.1016/j.powtec.2013.01.031

Rao A, Curtis JS, Hancock BC, Wassgren C (2011) Classifying the fluidization and segregation behavior of binary mixtures using particle size and density ratios. AIChE J 57(6):1446–1458. http://doi.wiley.com/10.1002/aic.12371

Sanaei S, Mostoufi N, Radmanesh R, Sotudeh-Gharebagh R, Guy C, Chaouki J (2010) Hydrodynamic characteristics of gas-solid fluidization at high temperature. Can J Chem Eng 88 (1):1–11. doi:10.1002/cjce.20240

Shukrie A (2012) Studies on the residence time distribution of solids in a swirling fluidized bed. Universiti Teknologi Petronas

Shukrie A, Anuar S, Arshad A (2014) Experimental Investigation of heat transfer in a bubbling fluidized bed for Geldart A and B particles. In: 11th international conference on fluidized bed technology, Beijing

Wu SY, Baeyens J (1991) Effect of operating temperature on minimum fluidization velocity. Powder Technol 67(2):217–220. doi:10.1016/0032-5910(91)80158-F

Zhu J, Qi M, Barghi S (2013) Identification of the flow structures and regime transition in gas-solid fluidized beds through moment analysis. AIChE J 59(5):1479–1490. doi:10.1002/aic.13948

Chapter 15
High Aspect Ratio Micro-EDM Drilling with Nano Surface Finish

Mohammad Yeakub Ali, Mohamed Abd Rahman, Asfana Banu, Shakira Adnan and Fatin Nadia

Abstract This paper presents the issues related to high aspect ratio in micro drilling on metallic materials using micro electro discharge machining (Micro-EDM). Experimental investigation is carried out by drilling high aspect ratio micro holes on copper workpiece with 1 mm diameter tungsten carbide tool electrode using micro-EDM DT 110 with two varying parameters voltage and capacitance. For the measurement and observation of aspect ratio, roundness, and surface roughness were performed by scanning electron microscope, optical microscope, and WYKO NT1100 optical profiler respectively. The purpose of this paper is to achieve high aspect ratio, accurate roundness, and low surface roughness with the minimum wear of tool electrode within machining time. The data were analyzed by Design Expert software. Emperical models for aspect ratio, roundness, and surface roughness are developed. The analysis revealed that the most influential parameter is capacitance. Multiple optimization of responses showed that the optimum parameter values are 90 V gap voltage and 0.26 μF capacitance which provide 2.4 aspect ratio, 44 μm roundness, and 6.0 μm R_a surface roughness. The desirability achieved by optimization is 73.9%. The models are validated by experiments.

Keywords High aspect ratio · Micro-EDM · Micro-hole · Nano surface finish · Roundness

---

M.Y. Ali (✉) · M. Abd Rahman · A. Banu · S. Adnan · F. Nadia
Faculty of Engineering, Department of Manufacturing and Materials Engineering, International Islamic University Malaysia, Kuala Lumpur, Malaysia
e-mail: mmyali@iium.edu.my

M. Abd Rahman
e-mail: mabdrahman2@gmail.com

A. Banu
e-mail: asfanabanu.asharaf@yahoo.com

S. Adnan
e-mail: shakira_adnan@yahoo.com

F. Nadia
e-mail: rnadia_3102@yahoo.com

N.A. Yacob et al. (eds.), *Regional Conference on Science, Technology and Social Sciences (RCSTSS 2014)*, DOI 10.1007/978-981-10-0534-3_15

1 Introduction

In recent years, numerous developments in micro-EDM have focused on the productions of micro features,very small process forces and good repeatability of the process result have made micro-EDM the best means for achieving high aspect ratio micro features. In micro-EDM drilling, high aspect ratio, wear of electrode, low surface roughness, and roundness are critical issues.

Micro-EDM theory states that, when two electrodes are separated by a dielectric medium, come closer to each other, the dielectric medium that is initially non conductive breaks down and becomes conductive. Sparks are generated between the electrodes during this period. The released thermal energy will remove material by melting and evaporation. It is possible to machine micro features on any electrically conductive material by controlling the amount of energy released precisely (Murali and Yeo 2004; Billa et al. 2010). In micro-EDM, the dielectric fluid acts as a cutting medium to improve wear resistance, surface roughness, and corrosion resistance (Chow et al. 2008).

Micro-EDM drilling is a special type of micro die sinking EDM which uses the rotation pin electrodes in order to manufacture rotationally symmetrical bores and through holes. Mechanical drilling is widely used as a micro holes machining methods because the workpiece properties are independent, less thermal deformation, and minimum finishing work. Thus, micro drilling can generate deeper holes with better straightness and better roundness with smoother surfaces. Better finishing of drilled holes are due to many factors such as type of materials, capacitance, voltage, feed rate, and tool diameter (Azlan et al. 2009).

Micro-EDM drilling is limited to conductive and semi conductive materials. The process is ideal for handling both difficult-to-machine and easy-to-machine materials. The advantages of micro-EDM drilling over other micro drilling technologies are the walls of micro-EDM drilled holes have less or no burrs. Burr-free EDM drilling process is especially suited for machining difficult holes. When machining speed and discharge energy of EDM machines are controlled, small holes can be drilled in workpieces with high levels of accuracy. EDM drilling machines can be used to drill materials such as soft copper and aluminum that produce gummy chips when machined. The non-contact EDM drilling process is suited for drilling deep straight holes in workpieces, as opposed to conventional methods in which the drill bits tend to drift during deep hole drilling. However, micro-EDM drilling have few limitations which are low material removal rate (MRR) and high electrode wear ratio. The wear of electrode must be compensated; it is not encouraged to change the micro electrode during machining because it may reduce the accuracy (Lim et al. 2003).

In the micro-EDM process, to avoid clamping error, the tool electrode is fabricated on the machine. From an electrode thicker than the required diameter, a cylindrical electrode is fabricated by EDM process using a sacrificial electrode (Lim et al. 2003). Hence, the objectives of this paper is to determine the influence of capacitance and voltage on high aspect ratio, roundness, and surface roughness by micro-EDM drilling.

2 Experimental

Taguchi orthogonal array design was used as the experimental design to conduct 9 experiments. Tungsten carbide with 1 mm diameter was used as the electrode since it has the highest melting point and can withstand all forms of wear including sliding abrasion, erosion and corrosion. It exhibits a high degree of toughness, high compressive strength, and retains its hardness values at high temperatures. As for the workpiece, copper was used.

The experiments were conducted using micro-EDM drilling with multi-purpose micro machine tools, DT 110 (Mikrotools Inc., Singapore) which is shown in Fig. 1a. The constant time for drilling was four hours with feed rate of 5 μm/s. The time was constant in order to observe which parameters can obtain high aspect ratio, the accurate roundness, and the lowest surface roughness. The vary parameters

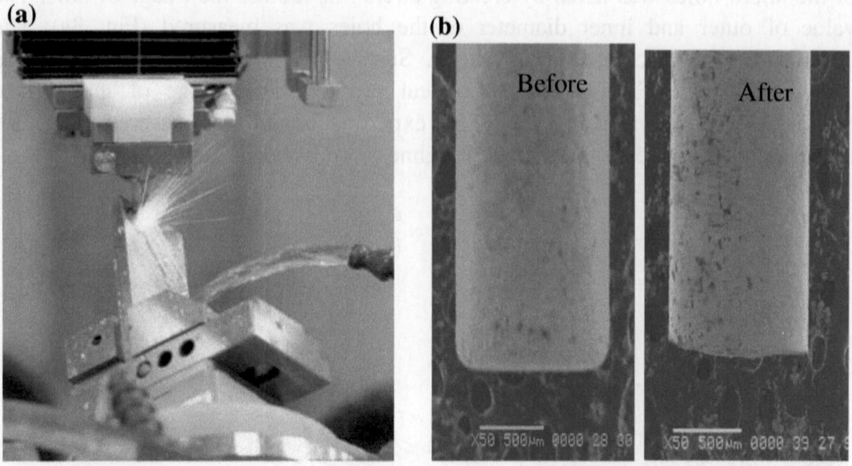

Fig. 1 **a** Micro-EDM drilling setup and **b** electrode before and after drilling

Table 1 Experimental parameters

Controlled parameters	Factors	Level		
		I	II	III
Capacitance (μF)	c	0.01	0.1	0.4
Voltage (V)	V	90	100	110
Fixed parameters				
Workpiece	Copper (45 mm × 15 mm × 3 mm)			
Electrode	Tungsten carbide (Ø1 mm × 40 mm)			
Drilling time (h)	4			
Feed rate (μm/s)	5			
Dielectric fluid	EDM-3 synthetic oil			

were capacitance and voltage. Figure 1b shows the scanning electron microscope (SEM) (JEOL, Japan) images of the electrode before and after the machining where electrode wear was found tom be insignificant. The experimental parameters are shown in Table 1.

3 Measurements

Cross sections of holes were done in order to measure the depth and diameter of the hole. The measurements were estimated by SEM. The image of a cross section hole is shown in Fig. 2a. The values of aspect ratio of the workpiece were calculated using Eq. 1. Roundness of the workpiece were observed and the measurement were estimated using optical microscope. The value was measured by comparing the enlargement of each hole that was drilled with the actual roundness. The dimension of the micro holes was taken by creating circle line around the image of hole. The value of outer and inner diameter of the holes was measured (Fig. 3a). The roundness were calculated using Eq. 2. Surface roughness of micro holes was measured using WYKO NT1100 optical profiler. The image of the surface roughness was shown in Fig. 4a. The 9 experiments and the measured values for aspect ratio, roundness, and surface roughness are listed in Table 2.

$$ar = \frac{d}{\emptyset} \tag{1}$$

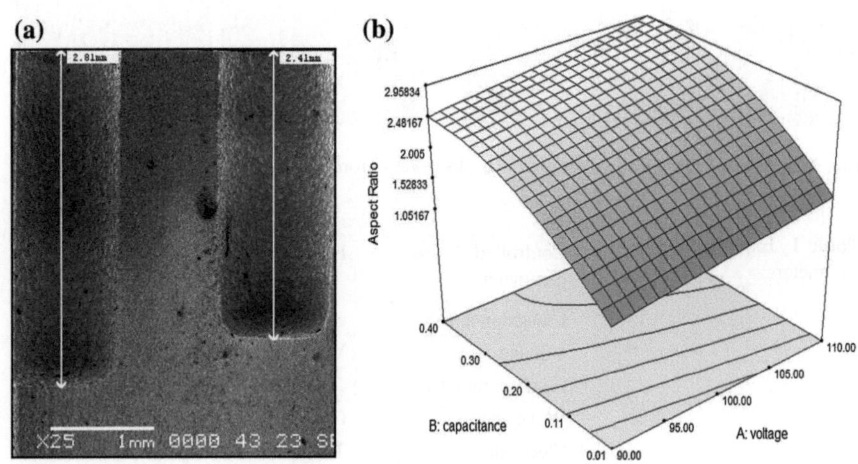

Fig. 2 **a** Cross-section of microhole, **b** 3D graph of aspect ratio versus voltage and capacitance

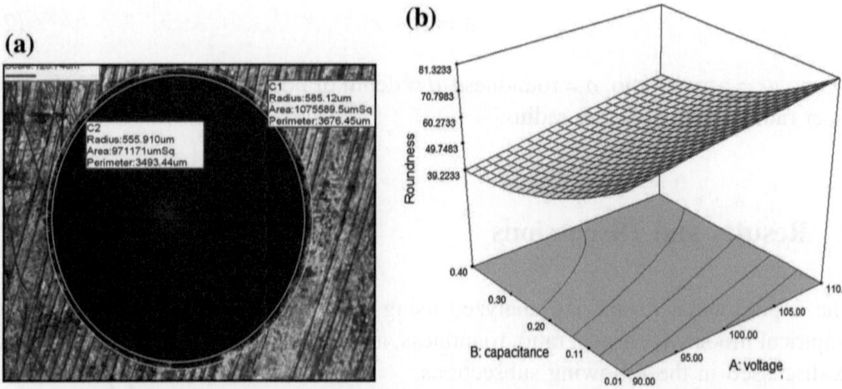

Fig. 3 **a** Roundness of micro hole and **b** 3D graph of roundness versus voltage and capacitance

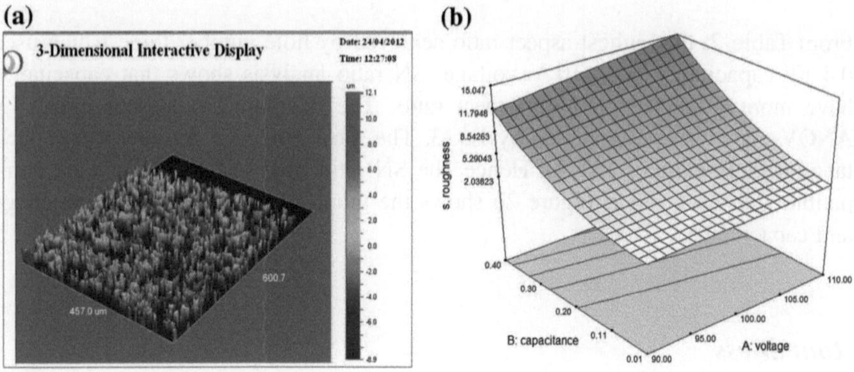

Fig. 4 **a** Surface roughness profile, and **b** surface roughness versus voltage and capacitance contour

Table 2 Experimental parameters and measured responses

Hole No.	Capacitance (μF)	Voltage (V)	Aspect ratio, *ar*	Roundness, δ (μm)	Surface roughness, Ra (μm)
1	0.4	90	2.41	36.86	11.80
2	0.4	100	2.81	46.70	13.38
3	0.4	110	2.95	53.12	14.96
4	0.1	90	1.77	58.42	2.52
5	0.1	100	1.93	64.74	2.78
6	0.1	110	1.96	67.68	4.59
7	0.01	90	0.98	70.36	2.34
8	0.01	100	1.22	71.74	2.37
9	0.01	110	1.66	82.86	2.47

$$\delta = R - r \tag{2}$$

where, ar = aspect ratio, δ = roundness, d = depth of hole, $\emptyset$ = diameter of hole, R = outer radius, and r = inner radius.

4 Results and Discussions

The experimental results are analyzed using analysis of variance (ANOVA) and empirical models for aspect ratio, roundness, and surface roughness were developed as discussed in the following subsections.

Aspect Ratio

From Table 2, the highest aspect ratio achieved by hole number three which used 0.4 μF capacitance and 110 V voltage. SN ratio analysis shows that capacitance have more influence on high aspect ratio. The developed empirical model by ANOVA analysis is expressed by Eq. 3. The most influence parameter is capacitance with F-value of 183.30. Hence, the SN ratio and ANOVA results are compatible with each other. Figure 2b shows the graph of aspect ratio versus voltage and capacitance.

Roundness

From Table 2, the minimum value of roundness show the level of accurate round achieved by micro-EDM drilling of holes. The minimum value of roundness achieve by holes number one which use 0.4 μF of capacitance and 90 V of voltage. From SN ratio analysis, capacitance gives more influence to achieve accurate roundness. The ANOVA analysis have developed the empirical model which is expressed by Eq. 4. Capacitance have high influence on achieving accurate roundness since the F-value is 203.35. Hence, the SN ratio and ANOVA results are well-matched with each other. Figure 3b shows the graph of roundness versus voltage and capacitance.

Surface Roughness

From Table 2, the lowest surface roughness achieved (2.34 μm) by hole number seven with 0.01 μF capacitance and 90 V voltage. The least value of surface roughness shows the fine surface roughness. Capacitance is more influential to achieve low surface roughness resulted by the SN ratio analysis. The developed empirical model expressed by Eq. 5 is based on ANOVA analysis. The most influential parameter on surface roughness would be the capacitance with the F-value of 833.25. Hence, the SN ratio and ANOVA results are suited with each other. Figure 4b shows the graph of surface roughness versus voltage and capacitance.

$$ar = 1.1399 + 0.0235V + 7.7604c - 9.9430c^2 \tag{3}$$

$$\partial = 13.0534 + 0.6336V - 145.0401c + 169.7246c^2 \tag{4}$$

$$Ra = 0.86128 + 0.032147V - 30.25115c + 60.44634c^2 + 0.33639Vc \tag{5}$$

where, ar = aspect ratio, ∂ = roundness, Ra = surface roughness, V = voltage (V), and c = capacitance (μF).

Optimization

Multiple optimization of responses showed that the optimum parameter values are 90 V gap voltage and 0.26 μF capacitance which provide 2.4 aspect ratio, 44 μm roundness and 6.0 μm R_a surface roughness with 74 % desirability. The models are validated by experiments.

5 Conclusion

This paper discusses the micro-EDM drilling of copper using tungsten carbide electrode on aspect ratio, roundness, and surface roughness. This experimental investigation showed:

1. The experimental result shows that 0.4 μF capacitance and 110 V voltages gives aspect ratio of 2.95; 0.4 μF capacitance and 90 V voltage gives roundness of 36.86 μm, and 0.01 μF capacitance and 90 V voltage gives lowest Ra surface roughness of 2.3 μm.
2. The image that was taken from SEM shows that the hole produced is almost vertical. High aspect ratio of hole has affected the electrode wear.
3. It was observed that capacitance gives high influence to achieve high aspect ratio, accurate roundness, and low surface roughness.
4. The optimum parameters are 90 V voltage and 0.26 μF capacitance for 2.3126 aspect ratio, 44.0264 μm roundness, and 6.02623 μm surface roughness.

5. The duration of the micro drilling process can be prolonged in order to achieve high aspect ratio and low surface roughness for microfluidic applications.

Acknowledgement This research was conducted under the Science Fund grant 03-01-08-SF0135 and Fundamental Research Grant Scheme FRGS12-074-0223 from Ministry of Science and Technology and Ministry of Education of Malaysia. The authors are grateful to the micromanufacturing laboratory where the experimental studies were conducted.

References

Azlan AR, Azuddin M, Wagiman A (2009) Effect of machining parameters on hole quality of micro drilling for brass. J Mod Appl Sci 3(5):27–30

Billa S, Sundaram MM, Rajukar KP (2010) A study on the high aspect ratio micro hole drilling using ultrasonic assisted micro electro discharge machining. In: Proceedings of the ASPE 2007 Spring Topical Meeting, vol 40, pp 32–36

Chow HM, Yang LD, Lin CT, Chen YF (2008) The use of SiC powder in water as dielectric for micro slit EDM machining. J Mater Process Technol 195(1–3):160–170

Lim HS, Wong YS, Rahman M, Edwin MKL (2003) A study on the machining of high aspect ratio microstructures using micro EDM. J Mater Process Technol 140(1–3):318–325

Murali M, Yeo SH (2004) Rapid biocompatible micro device fabrication by micro electro discharge machining. Biomed Microdevices 6(1):41–45

Chapter 16
Dumbbell-Shaped Inline Mach–Zehnder Interferometer for Glucose Detection

Asiah Lokman, Hamzah Arof and Sulaiman Wadi Harun

Abstract A simple inline Mach–Zehnder interferometer (MZI) with a dumbbell-shape structure is demonstrated for measurement of glucose concentration in distilled water. The MZI is fabricated using an arcing process of a fusion splicer to form two bulges, which are separated by a tapered waist. The MZI generates a good reflected interference spectrum where the dip wavelength is red-shifted with the increase of glucose concentration. This is due to the increase of the surrounding refractive index (RI), which reduces the phase difference between the core and cladding modes. As the glucose concentration increases from 0 to 12 %, the dip wavelength increases from 1554.419 to 1554.939 nm in a quadratic manner with the coefficient of determination of 0.9818. It is also found that the sensor has a sensitivity of 0.028 nm/% with a linearity of 99.5 % and limit of detection of 6.5 %. This preliminary result shows that the proposed probe can be used as a sensor to detect glucose concentration in distilled water.

Keywords Inline Mach–Zehnder interferometer (MZI) · Refractive index (RI) · Dumbbell-shape · Glucose detection

A. Lokman (✉) · H. Arof
Faculty of Engineering, Department of Electrical Engineering, University of Malaya, Kuala Lumpur, Malaysia
e-mail: lokman_asiah@yahoo.com

H. Arof
e-mail: ahamzah@um.edu.my

S.W. Harun
Photonics Research Center, University of Malaya, Kuala Lumpur, Malaysia
e-mail: wharun@um.edu.my

N.A. Yacob et al. (eds.), *Regional Conference on Science, Technology and Social Sciences (RCSTSS 2014)*, DOI 10.1007/978-981-10-0534-3_16

1 Introduction

Fiber-based sensors have been developed for many purposes including those used for the measurements of humidity, refractive index (RI), temperature, and strain (Zhao et al. 2006; Shu et al. 2001; Gwandu et al. 2002). Compared to conventional sensors based on mechanical and electrical properties, optical fiber sensors offer many advantages since they are compact, highly sensitive, immune to electromagnetic interference, resistant to corrosion, and placid to volatile surrounding. One of the most popular probes based on optical fiber is Mach–Zehnder interferometer (MZI) which is widely used for various sensing purposes (Tian et al. 2008; Wang et al. 2010; Jiang et al. 2011; Zhu et al. 2011). In the last 10 years, many MZI sensor designs have been proposed including those using tapered fibers (Villatoro et al. 2004; Kieu and Mansuripur 2006; Wu et al. 2011a), long-period gratings (LPGs) (Patrick et al. 1998; Chiang et al. 2000), two different single-mode fibers (SMFs) (Wu et al. 2011b; Cárdenas-Sevilla et al. 2013), and some other special configurations (Wu et al. 2012; Yang et al. 2011, 2012). For instance, Jiang et al. (2011) demonstrated a refractive index sensor using a photonic crystal fiber (PCF)-based interferometer which was realized by fusion splicing a short section of PCF (blaze photonics, LMA-10) between two standard single-mode fibers. The fully collapsed air holes of the PCF at the spice regions allow the coupling of PCF core and cladding modes that makes a Mach–Zehnder interferometer. But for this PCF, it required an expensive equipment for cavity fabrication which increases the cost of the fabrication.

In an MZI used for sensing refractive index (RI), a fraction of the fundamental core mode power leaks out as cladding modes that are sensitive to the RI of the surrounding medium. Therefore, based on this operating principle, the MZI sensor of any configuration can be employed to detect the refractive indices of liquids. Meanwhile, the difference in path lengths between the core and cladding modes (in some of the structures such as LPGs, core-offset, and other insertions) usually changes with the ambient temperature and strain. It alters the optical path difference (OPD) that makes the spectra red shift or blue shift. Therefore, a temperature and strain sensitive sensor can be realized.

In this paper, a simple inline MZI with a dumbbell-shape structure is proposed. The MZI is then used to detect various glucose concentrations in distilled water. The proposed MZI structure comprises two bulges connected by a tapered waist.

2 Fabrication and Characterization of Inline MZI

The first step in fabricating the MZI structure is to cleave a single-mode fiber (SMF) into two sections, each with a flat and smooth end surface, as illustrated in Fig. 1a. Then the end facets of the two sections are matched and fused using the manual "arc" function of DCM Fusion Splicer S Type 39. While fusing the two sections of the fiber, the "arc" function is first used to merge the two end surfaces and then used again (for a few times) to exert extra pressure from both the sides to form a bulge at the joint as shown in Fig. 1b. Once the first bulge is formed, the jointed fiber is then cleaved once again at 1 mm away from the center of the first bulge as depicted in Fig. 1c. Then, the second bulge is formed in a similar fashion. It should be noted that the fusion splicer softens the ends of the two fibers with heat before pulling them together to merge at the center. When the "arc" function is used repeatedly, more material is lumped at the center, thus forming a bulge, while the neck of the bulge gets thinner. When the second bulge is made, it is a bit smaller in size than the first one as less material is available from the waist area between the bulges since it has been used in forming the first bulge. The waist area also becomes thinner because it gets tapered even more. The completed dumbbell structure is shown in Fig. 2, where the diameters of the first and second bulges are 198 and 196 µm, respectively. The diameter of the waist section is about 97.5 µm and its length is approximately 1000 µm, while the length of the whole structure is around 2000 µm. The parameters of the fusion splicer program used to fabricate the dumbbell structure are given in Table 1.

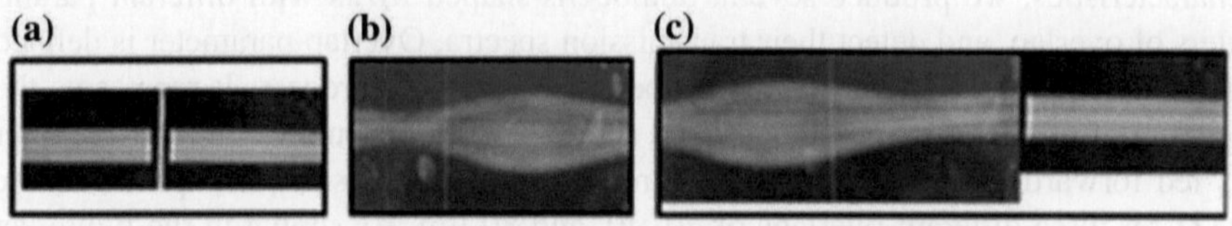

Fig. 1 Illustration of the fabrication procedure of the proposed inline MZI

Fig. 2 The experimental view of the proposed dumbbell shape MZI

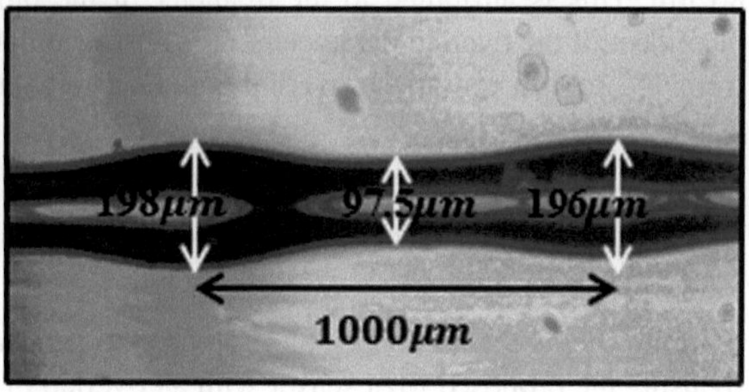

Table 1 Parameters for fabricating the bulges

Parameters	Units	Values
Fusion time	s	1.65
Prefusion	s	0.05
Arc gap	µm	20
Overlap	µm	80

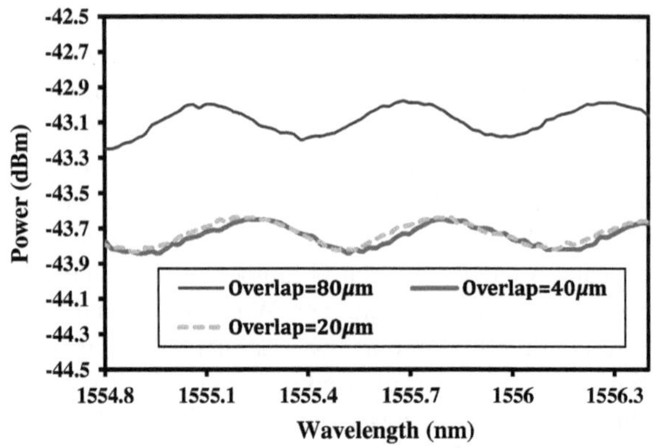

Fig. 3 The transmission spectra of the MZI for three different overlaps parameter of the fusion splicer used; 20, 40, and 80 μm

Due to optical path difference (OPD) between cladding mode and core mode, the interference pattern could be obtained in the output in case of a broadband spectrum input. Since the two bulges act as a beam splitter and combiner, changes in their diameters lead to the change in output transmission spectrum whereby the extinction ratios of the interference fringe can be controlled. To verify these characteristics, we produce several dumbbells-shaped MZIs with different parameters of overlap, and detect their transmission spectra. Overlap parameter is defined as an overlap distance between two fibers during arcing process. It represents the amount of overlap between the left and right fibers that occurs when the right fiber is fed forward during arc fusion. Figure 3 shows the transmission spectra of the MZI for three different overlaps of 20, 40, and 80 μm. As shown in the figure, an interference comb spectrum is obtained for all MZIs with a free spectral range of around 3.6 nm. It is observed that the extinction ratio of the interference spectrum increases from 0.18 to 0.22 dB as the overlap parameter is increased from 20 to 80 μm. This is attributed to the diameter of the bulge, which increases with the increment of the overlap parameters. It is observed that the bulge diameter increases from 177 to 196 μm as the overlap parameter increases from 20 to 80 μm.

3 Experimental Setup for the Glucose Sensor

Figure 4 shows the experimental setup used to measure changes in refractive index by using the proposed MZI sensor with an overlap of 80 μm. The input signal from an amplified spontaneous emission (ASE) laser source is launched into a sensor probe via a 3 dB coupler. The reflected signal from the sensor is then routed into an optical spectrum analyzer (OSA) through the same coupler. When the input single-mode light beam reaches the first bulge of the sensor probe, it is divided into two parts where the first part continues to propagate in the core, while the other part travels in the cladding of the SMF. Due to the OPD between cladding mode and core mode, an interference pattern is generated as shown in Fig. 5. In our experiment, the

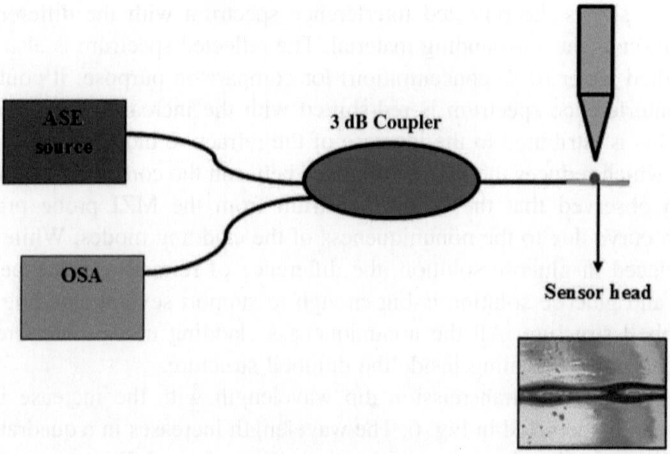

Fig. 4 The experimental setup of the proposed glucose sensor

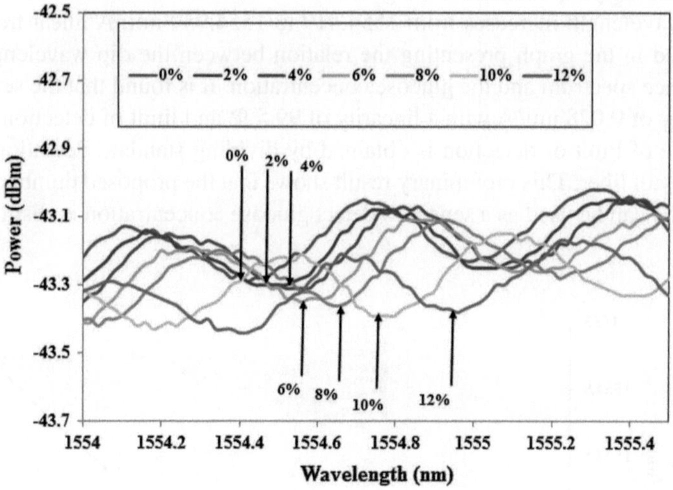

Fig. 5 The reflected interference spectrum with different glucose concentration as the surrounding material

dumbbell-shaped MZI probe is immersed into glucose solution of different concentrations ranging from 0 to 12 %. The solution was prepared by mixing glucose and distilled water with different proportions of 0, 2, 4, 8, 10, and 12 %, which corresponds to mixtures with refractive indices of 1.333, 1.336, 1.339, 1.3425, 1.3461, 1.3494, and 1.353, respectively. The solution refractive index is measured using ABBE refractometer. Due to the compact structure of the sensor, a single drop of the liquid solution is enough to surround the whole dumbbell-shape. The MZI was cleansed with deionized water and compressed air after each measurement.

Figure 5 shows the reflected interference spectrum with the different glucose concentration as the surrounding material. The reflected spectrum is also measured with distilled water (0 % concentration) for comparison purpose. It could be seen that the interference spectrum is red-shifted with the increase of glucose concentration. This is attributed to the increase of the refractive index of the surrounding medium, which reduces the phase difference between the core and cladding modes. It is also observed that the output spectrum from the MZI probe produces an unsmooth curve due to the nonuniqueness of the cladding modes. While the inline MZI is placed in glucose solution, the difference of refractive index between the cladding and glucose solution is big enough to support several cladding modes in the dumbbell structure. All the nonuniqueness cladding modes interfere with the core mode while oscillating inside the dumbell structure.

The change in the transmission dip wavelength with the increase in glucose concentration is depicted in Fig. 6. The wavelength increases in a quadratic manner with the increase in glucose concentration. The adjusted R-square value or the coefficient of determination is the measure of the goodness of fit which is 0.9818. The considerably high values of the adjusted R-square allow the prediction of unknown concentration by the model. As the glucose concentration increases from 0 to 12 %, the dip wavelength increases from 1554.419 to 1554.939 nm. A linear trend line is also added in the graph presenting the relation between the dip wavelength of the interference spectrum and the glucose concentration. It is found that the sensor has a sensitivity of 0.028 nm/% with a linearity of 99.5 % and limit of detection of 6.5 %. The value of limit of detection is obtained by dividing standard deviation with the sensitivity of fiber. This preliminary result shows that the proposed dumbbell-shaped MZI probe can be used as a sensor to detect glucose concentration in distilled water.

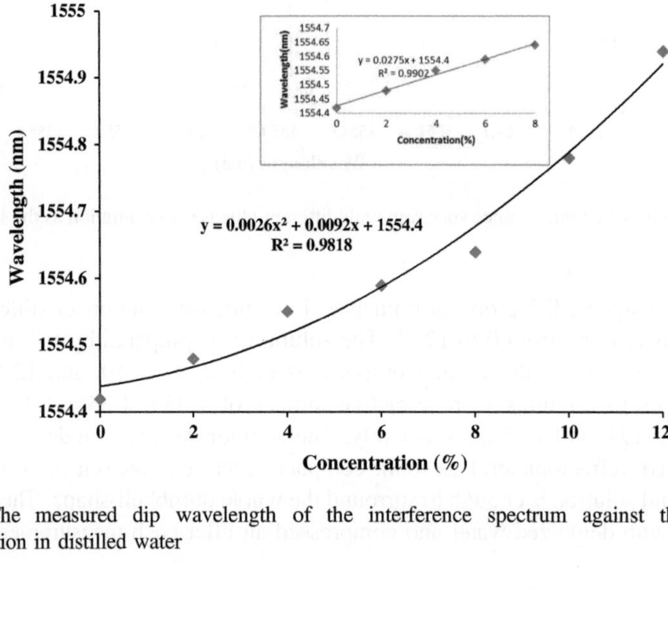

Fig. 6 The measured dip wavelength of the interference spectrum against the glucose concentration in distilled water

4 Conclusion

A new dumbbell-shaped inline MZI is developed using an arcing process of a fusion splicer for measurement of glucose concentration in distilled water. The sensor probe consists of two bulges separated by a tapered waist that generates a good reflected interference spectrum where the extinction ratio depends on the parameter of the overlap. The interference spectrum is red-shifted with the increase of glucose concentration due to the increase of the refractive index of the surrounding, which reduces the phase difference between the core and cladding modes. As the glucose concentration increases from 0 to 12 %, the dip wavelength increases from 1554.419 to 1554.939 nm in a quadratic manner with the coefficient of determination of 0.9818. It is also found that the sensor has a sensitivity of 0.028 nm/% with a linearity of 99.5 % and limit of detection of 6.5 %.

Acknowledgement The authors acknowledge the financial support from the University of Malaya and Ministry of Higher Education (MOHE) under the High Impact Research Grant Scheme (Grant No: D000009-16001).

References

Cárdenas-Sevilla GA, Fávero FC, Villatoro J (2013) High-visibility photonic crystal fiber interferometer as multifunctional sensor. Sensors 13(2):2349–2358

Chiang KS, Liu Y, Ng MN, Dong X (2000) Analysis of etched long-period fibre grating and its response to external refractive index. Electron Lett 36(11):966–967

Gwandu BA, Shu X, Allsop TD, Zhang W, Zhang L, Webb DJ, Bennion I (2002) Simultaneous refractive index and temperature measurement using cascaded long-period grating in double-cladding fibre. Electron Lett 38(14):695–696

Jiang L, Zhao L, Wang S, Yang J, Xiao H (2011) Femtosecond laser fabricated all-optical fiber sensors with ultrahigh refractive index sensitivity: modeling and experiment. Opt Express 19 (18):17591–17598

Kieu KQ, Mansuripur M (2006) Biconical fiber taper sensors. IEEE Photonics Technol Lett 18 (21):2239–2241

Patrick HJ, Kersey AD, Bucholtz F (1998) Analysis of the response of long period fiber gratings to external index of refraction. J Lightwave Technol 16(9):1606

Shu X, Gwandu BA, Liu Y, Zhang L, Bennion I (2001) Sampled fiber Bragg grating for simultaneous refractive-index and temperature measurement. Opt Lett 26(11):774–776

Tian Z, Yam SH, Loock HP (2008) Single-mode fiber refractive index sensor based on core-offset attenuators. IEEE Photonics Technol Lett 20(16):1387–1389

Villatoro J, Monzón-Hernández D, Talavera D (2004) High resolution refractive index sensing with cladded multimode tapered optical fibre. Electron Lett 40(2):106–107

Wang Y, Yang M, Wang DN, Liu S, Lu P (2010) Fiber in-line Mach-Zehnder interferometer fabricated by femtosecond laser micromachining for refractive index measurement with high sensitivity. J Opt Soc Am B 27(3):370–374

Wu D, Zhu T, Deng M, Duan DW, Shi LL, Yao J, Rao YJ (2011a) Refractive index sensing based on Mach-Zehnder interferometer formed by three cascaded single-mode fiber tapers. Appl Opt 50(11):1548–1553

Wu Q, Semenova Y, Wang P, Farrell G (2011b) High sensitivity SMS fiber structure based refractometer–analysis and experiment. Opt Express 19(9):7937–7944

Wu D, Zhu T, Chiang KS, Deng M (2012) All single-mode fiber Mach-Zehnder interferometer based on two peanut-shape structures. J Lightwave Technol 30(5):805–810

Yang R, Yu YS, Xue Y, Chen C, Chen QD, Sun HB (2011) Single S-tapered fiber Mach-Zehnder interferometers. Opt Lett 36(23):4482–4484

Yang R, Yu YS, Chen C, Xue Y, Zhang XL, Guo JC, Sun HB (2012) S-tapered fiber sensors for highly sensitive measurement of refractive index and axial strain. J Lightwave Technol 30 (19):3126–3132

Zhao CL, Yang X, Demokan MS, Jin W (2006) Simultaneous temperature and refractive index measurements using a 3 slanted multimode fiber Bragg grating. J Lightwave Technol 24(2):879

Zhu T, Wu D, Deng M, Duan D, Rao Y, Bao X (2011) Refractive index sensing based on Mach-Zehnder interferometer formed by three cascaded single-mode fiber tapers. In: 21st international conference on optical fibre sensors (OFS21) (pp 77532P–77532P), international society for optics and photonics

Part II
Robotics

Chapter 17
Novel Rehab Devices' Feature Extraction Analysis Using EMG Signal via Self-Organizing Maps (SOM)

Zul Hasrizal Bohari, Mohd Hafiz Jali, Tarmizi Ahmad Izzuddin and Mohamad Na'im Mohd Nasir

Abstract Rehabilitation device is designed to be an exoskeleton for people who had limb failure that proven beneficial toward rehab program. The device used to facilitate the tasks of the program is able to improve the electrical activity in the motor unit and minimize the mental effort of the user. Electromyography (EMG) is the technique to analyze the presence of electrical activity in musculoskeletal systems related to muscle movement. To prevent from the muscle paralyzed, it is becoming spasticity that the force of movements should minimize the mental efforts needed. To achieve this, the rehab device should analyze the surface EMG signal of normal people to be implemented to the rehab device. The EMG signal collected using noninvasive method is implemented to set the movements' pattern of the arm rehab device. The signal is filtered and extracted for three time-domain features of standard deviation (STD), mean absolute value (MAV), and root mean square (RMS). The features' combinations are important to produce the best classification result with less error. To determine the best combination features for any movements, several trials of movements are used by determining the right combination using self-organizing maps (SOM) for the classification process.

Keywords Electromyography · Rehabilitation device · Self-organizing maps · Time-domain features

---

Z.H. Bohari (✉) · M.H. Jali · T.A. Izzuddin · M.N. Mohd Nasir
Faculty of Electrical Engineering, Universiti Teknikal Malaysia Melaka, Durian Tunggal, Melaka, Malaysia
e-mail: zulhasrizal@utem.edu.my

M.H. Jali
e-mail: mohd.hafiz@utem.edu.my

T.A. Izzuddin
e-mail: tarmizi@utem.edu.my

M.N. Mohd Nasir
e-mail: mohamad.naim@utem.edu.my

N.A. Yacob et al. (eds.), *Regional Conference on Science, Technology and Social Sciences (RCSTSS 2014)*, DOI 10.1007/978-981-10-0534-3_17

1 Introduction

Human support system is endoskeleton. Endoskeleton plays a role as a framework of the body which is bone. Our daily movements fully depend on the functionality of our complex systems in the body. The disability one or more of the systems in our body will reduce our physical movements. The assistive device is a need for rehab as an exoskeleton. The functionality of the rehabilitation device has to smooth as the physical movement of normal human. The rehabilitation programs provide the suitable program for conducting the nerve and stimulating the muscles. People who have temporary physical disability have the chances to recover. Nowadays, rehabilitation program is using exoskeleton device in their tasks. The functionality of exoskeleton depends on muscle contraction. Electromyogram studies help to facilitate the effectiveness of the rehabilitation device by analyzing the signal transmitted from the muscle (Chan 2011; John 2005; Gauthaam 2011).

The technique of measuring electrical activity that is produced from the muscles during rest or contractions is known as electromyography (EMG) (Reaz et al. 2006; Jamal 2012; Cram 2011; Mahaphonchaikul 2010; Jali et al. 2014a, b). The electric signal generates from the brain and sends to the muscles via motor neuron. The EMG may detect the dysfunction of the muscles or failure in signal transmission from nerve to muscle. The failure of sending the electrical signal from the brain requires electrical stimulation from the external source to muscles. Electrodes are used for signal detection of electrical activity in muscles. The study of this electrical activity is important for combination of electromyogram and rehabilitation device.

The rehabilitation device is a tool that is used to help the movements for daily life activities of the patients who suffer from the failure of muscle contractions; due to the failure of the muscle's contractions the movements are limited. The ability of the patients to do the tasks in the rehabilitation programs needs to be measured. The rehabilitation programs have to assure whether the tasks will be effective or will it bring harm to the patients (Ada 2006).

Historically, the rehabilitation tasks have been avoided due to a belief that it would increase spasticity (Bobath 1990). In this research, the analysis of the data will be focusing on the upper limb muscles contraction consisting of biceps muscles only. The experiment is limited to certain upper limb movements that are used in training. EMG is a division of biosignal; the biosignal analysis is the most complex analysis. Thus, the signal analysis is a complicated process that has to be through many phases of analysis (Muthuswamy 2004; Jamal 2012). Therefore, the challenge of this study is to assure that the signal processing is conducted properly to overcome the environment noise during data collection.

A study has been conducted to do EMG recognition using artificial neural network (Ahsan 2011; Park 1998; Jali et al. 2014a, b). In this research work, unsupervised neural network is proposed. Unsupervised learning (learning by observation) refers to the method that learns by itself according to input attributes and also applies competitive learning that made the output nodes to compete to be activated. Only one of the node will be activated at any one time which we called winning neuron (Bohari et al. 2014; Jali et al. 2014a, b). Such a competition can be induced through negative feedbacks between neurons. The system is called self-organizing map (SOM) because all nodes are forced to be self-organized through the feedback path.

Based on neurological studies, all human sensory inputs are mapped onto certain areas at the cerebral cortex that form a map called topographic map (Bohari et al. 2014). It has two most essential assets:

1. At each stage of processing, every information is reserved in its proper environment.
2. Closely related information nodes will be close to each other to ensure short synaptic connections.

SOM's primary purpose is to transform incoming input patterns into a one- or two-dimensional discrete map. This process must be performing in orderly approach. In this research work, SOM is used to determine the feature combination for rehabilitation control.

2 Methodology

Figure 1 shows a chevron diagram of our research methodology that consists of four major phases. It started with the experimental setup sEMG data gathering. Second phase is EMG data processing. Third phase is the extraction time-domain features. Finally, SOM will do the clustering and the performances are observed.

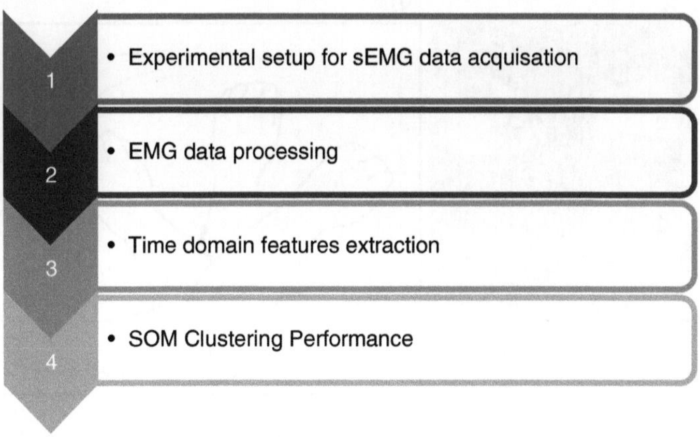

Fig. 1 Research strategy

Experimental Setup

Implementation of arm rehabilitation device based on movement is recorded from the EMG signal of healthy subjects. From the human anatomy studies, different angle movements of upper limb with elbow as the reference depends on relation of agonist and antagonist. This study is focusing on the behavior of the biceps muscle as agonist and the triceps as the antagonist, respectively. Muscle that involved in this movement is biceps and triceps; however, in this study to understand the electrical activity during muscle contraction, the biceps is the only muscle taken into account. The movements range between positions of arm flexion until the arm fully extends.

The environment is in a room with low lighting especially the fluorescent light, and any electromagnetic device away from the experiment equipment and the environment is in silent room. Then, the experiment is setup with the subject sitting on the chair while the hand is on the table. The subject has to complete the task of lifting up the dumbbell with 5 pounds of weight as shown in Fig. 2 for 5 times. Normally, the appearance of EMG signal is in chaos and noise depends on the type of electrodes and also the noise factor. To simplify the difference of amplitude response for the motion, the dumbbell is functioned to amplify the amplitude in analyzing the electrical activity during rest and contract. The rehabilitation devices (white in color in Fig. 2a) help to keep the position of the elbow joint and the wrist joint in line. Mostly, the EMG signal is obtained after several trials of the movements. These movements are specified from angle of 0° (arm in rest position), up to 120° (arm is fully flexion). These electrodes are connected to the combination of hardware Olimex EKG-EMG-PA and Arduino Mega for data collection.

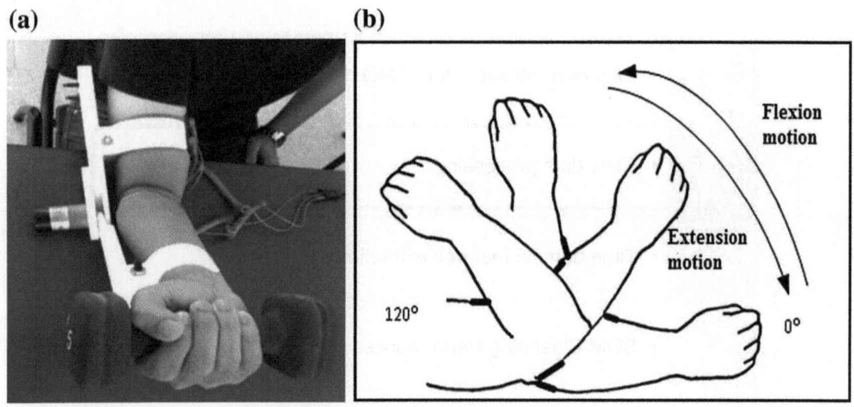

Fig. 2 Subject is setup with arm rehabilitation-assistive device for experiment (**a**). Simulation of subject is to lift up the dumbbell 5 pounds of weight (**b**)

EMG Data Processing

The filtering signal is accessed for features extraction. The raw sEMG signal is continued by filtering for removing the noise in purpose to obtain the clean features. Low-pass filter is chosen for removing the unwanted frequencies of a sEMG signal. The filter preserves the wave shape of any filtered signals across the group delay in the pass band. EMG signal is filtered in the range of above 10 Hz as a cut-off frequency for low-pass filter. The DC offset of the EMG signal is removed and is rectified to obtain its absolute value. Finally, the signal was smoothed and normalized passing it through a fifth-order butterworth-type low-pass filter with cut-off frequency 10 Hz.

Time-Domain Features Extraction

The goal of this study is to facilitate the future works for designing the rehabilitation devices system. The time-domain feature is the most commonly used by the researchers, among them are Phinyomark (2009) and Hamedi (2012). These researchers implement these features in their future studies in pattern recognition and classifier (Phinyomark 2009; Hamedi 2012). One of the advantages of using time-domain is that it can be implemented in real time. These features can be analyzed using the powerful tool MATLAB, as for this study, the analysis is conducted in MATLAB 2013a. There are many features related to time-domain, and below is described more on features in this study:

1. *Root mean square*. The RMS represents the square root of the average power of the EMG signal for a given period of time. It is known as a time-domain variable because the amplitude of the signal is measured as a function of time:

$$\mathrm{RMS} = \sqrt{\frac{1}{N}\sum_{i=1}^{N}\mathrm{EMG}(i)^2} \tag{1}$$

2. *Mean absolute value*. The MAV is the computer calculated equivalent of the average rectified value (ARV). The MAV is known as a time-domain variable because it is measured as a function of time. It represents the area under the EMG signal, once it has been **rectified**, it is defined by the **negative voltage values have been made positive**. The MAV is used as a measure of the amplitude of the EMG signal like the root mean square (RMS); the RMS is often preferred over the MAV because it provides a measure of the power of the EMG signal, while the MAV does not. The mean absolute value is calculated using a

moving window. It is calculated for each window of data according to the equation

$$\mathrm{MAV} = \frac{1}{S}\sum_{1}^{s}|f(s)| \tag{2}$$

3. *Standard Deviation.* The STD of a set of data is the square root of the variance, where $\bar{x}$ refers to the mean of the sample. The STD is often used as the fluctuation for a sample when data is being collected in an experiment. The results taken from data are frequently written as the mean ±STD.

$$\mathrm{STD}_{n-1} = \sqrt{\frac{1}{n-1}\sum_{i=1}^{n}(x_i - \bar{x})^2} \tag{3}$$

The combinations of selected features are extremely important to produce the best clustering result for SOM. This research work proposed four combinations that are shown in Table 1.

SOM Clustering Performances

For this SOM, clustering hexagonal lattice is selected and numbers of 200 neurons are selected for optimized SOM result. Normalization method is used in this research in 'range' normalization. Features extracted (RMS, MAV, and STD) from the sEMG signal are evaluated using SOM to determine the best combination. The performances of each combination of feature are evaluated according to their quantization error, topographic error, and training time.

Architecture of SOM Network. SOM structure built from two main layers: input and output layers that sort in two-dimensional preposition. SOM algorithms resemble learning vector quantization (LVQ). In LVQ all neurons are arranged on a grid together with selected neurons, whereas SOM has a feed-forward structure with a single computational layer arranged in rows and columns. Each neuron is fully connected to all the source nodes in the input layer. Neighboring nodes will be updated to perform neurons order. This indicates SOM as a multidimensional scaling method from input space to two-dimensional output space. Visual format of

Table 1 SOM clustering combination

No	Features combination
1	RMS and MAV
2	MAV and STD
3	STD and RMS
4	RMS, MAV, and STD

SOM helps researcher to define clusters, relations, and structures in complex input database (Bohari et al. 2014).

Matching Unit (BMU). Number of neurons may vary from a few dozen up to several thousands. Each neuron is represented by a *d*-dimensional weight vector (prototype vector, codebook vector) m = (m_1, …, m_d), where ***d*** is the dimension of input vectors. Neurons are connected to the adjacent neurons through neighborhood relation that dictates its topology.

SOM is then subjected to iteration for training the network. In the iteration process, one sample vector **s** from the input data will be selected randomly, and the distances between all the nodes are then calculated by particular distance measures. The neurons that have closest weight vector to the selected sample *s* is called best matching unit (BMU) and denoted by ***c***

$$\|s - m_c\| = \min_i \|s - m_i\| \tag{4}$$

where ***c*** is Euclidean distance measure (Bohari et al. 2014; Jali et al. 2014a, b).

Upon determining BMU, SOM weight vectors are updated so that the BMU will be closer to the input vector in input space. This adaptation process will stretch the BMU and its topological neighbors toward the sample vector.

3 Results and Discussion

The EMG signals are produced by the subjects during flexion and extension motion by lifting the 5 pounds of dumbbell with their right hand due to the rehabilitation device that designated for right hand. Figure 3 shows the raw EMG signal from one subject. Figure 3 shows that the amplitude signal is amplified when the arm is in flexion motion and shrinking during extension motion. Figure 4 presents a sEMG signal after removing the DC offset of the signal and rectification to obtain its absolute value. Five trials are done for every subject to assess the best combination of features.

From Table 2, all types of combinations produce near zero in both quantization and topographic errors except STD and RMS combination that got zero value for both errors. For STD and RMS, using the optimum value of neurons (200), this combination is proven to be the best when compared to other combinations (RMS and MAV, MAV and STD and RMS, MAV and STD) in terms of error produced and training time needed to produce a good result in clustering.

Next, for further analysis, the U-Matrix results for the four feature combinations are depicted in Fig. 5a, b, c, d.

In Fig. 5a, b, c, and d all the clustering results are depicted carefully. Using 200 neurons with hexagonal topology, the combination of STD and RMS is still the best contender for the U-Matrix mapping result. Theoretically, (RMS, MAV, and STD) combination is the best but according to this research work (STD and RMS)

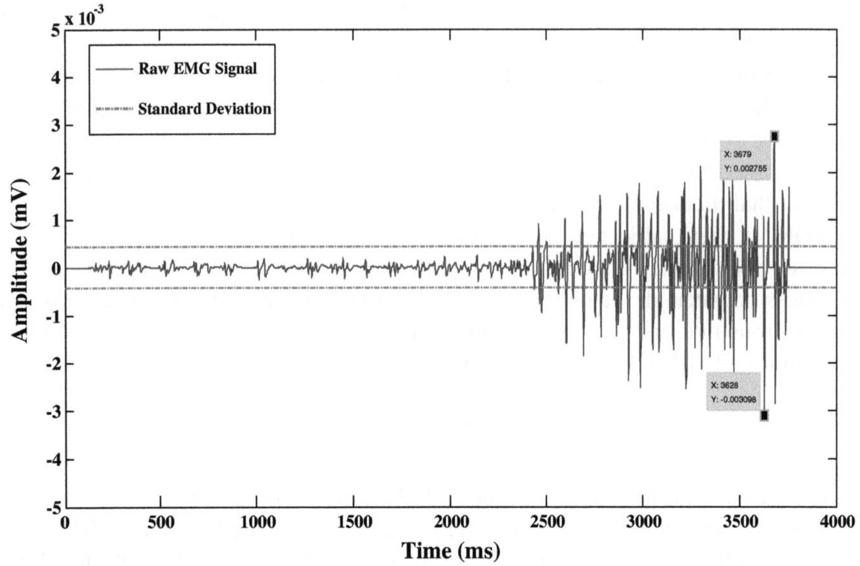

Fig. 3 Sample of raw sEMG signal

Fig. 4 Filtered and rectified signal

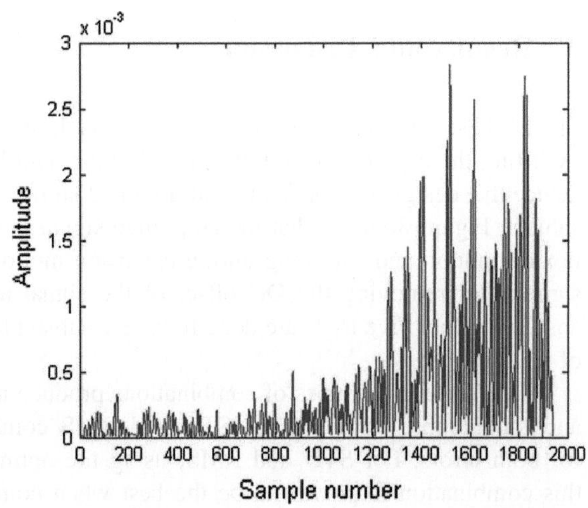

combination is the best. The clarity and quality of mapping and the produced boundary or separation between different trials are produced by SOM clustering. This combination will be further used in next research to finish the classification process.

Table 2 Summary of SOM Clustering Result

No	Results				
	Features combination	No of neuron	Quant. error	Topo. error	Train. time (s)
1	RMS and MAV	200	0.0000	0.0067	9
2	MAV and STD	200	0.0000	0.1330	10
3	STD and RMS	200	0.0000	0.0000	5
4	RMS, MAV and STD	200	0.0010	0.1330	5

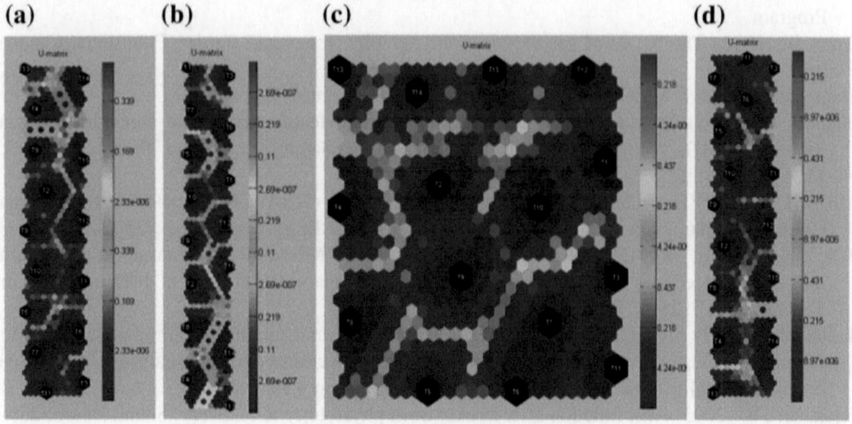

Fig. 5 U-Matrix for combination of RMS and MAV(**a**), MAV and STD (**b**), STD and RMS (**c**), and RMS, MAV and STD (**d**)

4 Conclusion

In conclusion, based on the experiment result a combination of RMS and STD produces the best result in term of error and training time. The combination of RMS, MAV, and STD with higher input neurons and longer training time failed to produce zero error for topographic compared to RMS and STD combination. Thus, it is proven that the right combination of sEMG signal features is crucially important in order to produce a good SOM clustering result and the inclusion of too many features will not guarantee producing a better clustering result especially when designing an appropriate rehab device.

Acknowledgments The authors would like to thank Universiti Teknikal Malaysia Melaka (UTeM) and Ministry of Education Malaysia for the financial support given through Research Grant.

References

Ada L (2006) Strengthening interventions increase strength and improve activity after stroke: a systematic review. Aust J Physiotherapy 52:241–248

Ahsan MR (2011) Electromygraphy (EMG) signal based hand gesture recognition using artificial neural network (ANN). In: International conference on mechatronics (ICOM), vol 4

Bobath B (1990) Adult hemiplegia: evaluation and treatment. Oxford, Butterworth-Heinemann

Bohari ZH, Ab Ghani S, Baharom MF, Md Thayoob YH (February 2014) Feature analysis (SFRA) traces using self organizing maps. Jurnal Teknologi Special Edition (Recent Research in Engineering) no. 66

Chan A (2011) An assessment platform for upper limb myoelectric prosthesis. Biomed Eng Program

Cram JR (2011) The basics of surface electromyography. Jones and Bartlett Publishers, Canada, pp 1–8

Gauthaam M (2011) EMG controlled bionic arm. B.E. Biomed Eng 111–114

Hamedi M (2012) Comparison of different time-domain feature extraction methods on facial gestures' EMGs, progress. In: Electromagnetics Research Symposium Proceedings, KL, Malaysia, p 1897

Jali MH, Bohari ZH, Sulaima MF, Nasir MNM, Jaafar HI (2014) Classification of EMG signal based on human percentile using SOM. Res J Appl Sci Eng Technol 8(2):235–242

Jali MH, Ibrahim IM, Bohari ZH, Sulaima MF, Nasir MNM (2014) Classification of arm movement based on upper limb muscle signal for rehabilitation device. J Theor Appl Inf Technol 68(1):125–137

Jamal MZ (2012) Signal acquisition using surface EMG and circuit design considerations for robotic prosthesis. In: Computational intelligence in electromyography analysis-a perspective on current applications and future challenges. InTech, pp 427–448

Krakauer JW (2005) Arm function after stroke: from physiology to recovery. Semin Neurol 25(4)

Mahaphonchaikul K (2010) EMG signal feature extraction based on wavelet transform

Muthuswamy J (2004) Biomedical signal analysis, in standard handbook of biomedical engineering and design. Mc-Graw Hill, pp 18.1–18.30

Park SP (1998) EMG pattern recognition based on artificial intelligence techniques. IEEE Trans Rehabil Eng 6(4)

Phinyomark A (2009) A novel feature extraction for robust EMG pattern recognition. J Comput 1(1):71–80

Reaz MBI (2006) Techniques of EMG signal analysis: detection, processing, classification and applications, cyberjaya, selangor. www.biologicalprocedures.com, pp 11–35

Part III
Architecture

Chapter 18
Town of Karai: The Only Coal Mining Site in the State of Perak and Its Contribution to the Urban Development

Mohd Hasrol Haffiz Aliasak, Mhd. Nor Osman, Siti Rahayu Zakaria, Mohd Farid Sa'ad and Nur Lesya Firsya Johaimi Ling

Abstract This study is to identify the history and traces the historical development of the town. The present generation has little knowledge on importance role of place as a commercial hub during the British colonialism. A rubber trader from Singapore discovered coal and this had transformed the isolated settlement into a popular business center due to coal mining activities. Realizing the importance of coal to the British, government constructed a network of modern transportation of the era, adding value to the development of Karai. Construction of the Enggor Bridge which is also known as the Victoria Bridge enabled this small town to connect with other modern colonial towns from all over Malaya and Singapore. However, when coal mining activities discontinued, the town had also loss its glory. The bridge ceased the function and was only used by the villages to connect with each other.

Keywords Coal mining industry · Economy development · Railway connection

1 Introduction

Town of Karai or Enggor was founded by the British circa 1890s following the discovery of coal nearby. Coal which was mined here was among the best mineral found in Malaya. Its quality was similar to the coal from Batu Arang, Rawang, Selangor. The British had built railway line and station in Enggor as the center for collecting coal. It generates coal-based engines used by train and steamship. Then, coal was supplied to the Federated Malay States Railway (FMSR) which was the operator for train services in Malaya and Singapore. At the end of the nineteenth

---

M.H.H. Aliasak (✉) · S.R. Zakaria · M.F. Sa'ad · N.L.F.J. Ling
Faculty of Architecture, Planning and Surveying, Universiti Teknologi MARA, Seri Iskandar, Perak, Malaysia
e-mail: haffiz677@perak.uitm.edu.my

Mhd.N. Osman
Faculty of Art and Design, Universiti Teknologi MARA, Seri Iskandar, Perak, Malaysia

N.A. Yacob et al. (eds.), *Regional Conference on Science, Technology and Social Sciences (RCSTSS 2014)*, DOI 10.1007/978-981-10-0534-3_18

century and the early twentieth century, FMSR actively expanded its train service network which connected the major colonial towns in Malaya and Singapore, the main hubs of trading during British colonialism (Fawzi Basri 1985; Kaur 1985).

However, Enggor golden years ended when the Enggor Coal Syndicate Ltd. had disposed company's interest on July 18, 1928. Liquidation of asset was affected by the drop in world's coal price. Since then, Enggor began to lose its glory until rubber was introduced and commercialized within the vicinity around Sg. Siput (Utara). Enggor was later named Karai in conjunction with the opening of rubber plantations by the European capitalists. Most of the rubber estates used Karai after the name of the Malay settlement (Khursiah and Fakhrul 2010; Loke 2013).

2 Objectives of Study

The main purpose of this study is to identify the historical elements on the development of Karai/Enggor town during British administration period. Therefore, the objectives of study are as follows:

(a) to identify the historical development of Karai/Enggor town, and
(b) to analyze the factors that were contributed to the development of Karai/Enggor town.

3 Research Methodology

This study applied the qualitative approach where the researcher collected the relevant secondary data related to this study. The data are mainly acquired from articles in newspapers, technical reports, and government publications in relation to the opening and development of Karai/Enggor town. Besides that, the researcher also acquired main data from unstructured interviews session and observation in fieldwork. This process also required the initial data which are derived from secondary data as recorded before.

At the end of study, the researcher will evaluate and analyze data using qualitative method through observation and narrative in order to explore the loss history of development of Karai/Enggor town as a colonial town during the British era.

4 Background of Karai/Enggor Town

"Karai" is a name taken from a tree which is also known as *mempisang*. As a type of jungle tree, this plant grows in many tropical jungles across southeast Asian countries such as Peninsular Malaysia, Borneo, Sumatra, Java, and the Philippines. It is the part of the species called *Annonaceae*. Another version on the origin of 'Karai' is derived from the discovery of an old grave belonging to a woman from Acheh known as Tok Karai. She migrated to Kampung Enggor from Acheh by easing her way along the Perak River. Her demise was not recorded but her grave can be identified in Kampung Enggor which is located approximately one kilometer from the town. "Enggor" was given an honor by the British colonials as the name of a street in Singapore, Enggor Street in 1898. The street is named after a town in Malaya, in Perak which is Enggor town, located at the bank of the Perak River in the District of Kuala Kangsar (Kolej Komuniti Sungai Siput 2013). The word 'Enggor' also refers to a type of freshwater fish which is pronounced in the local Malay dialect (Pejabat Penghulu Karai 2007).

Karai town is one of the towns located in the *Mukim* (Subdistrict) of Kota Lama Kanan, District of Kuala Kangsar, Perak. Located 10 km from the Royal town Kuala Kangsar, Karai town functions as the center for collection of local production such as rubber which is produced by the villagers from Kampung Karai, Kampung Enggor, Kampung Jawang, Kampung Perik, and Kampung Periang. Karai town is connected with complete roads. One can reach the town via the royal town of Kuala Kangsar through the Ipoh-Butterworth trunk roads and Jalan Kampung Enggor. There is a railway track closeby but at the time this research was conducted, the town did not have any train station or stops. The town is situated close to the Perak River basin which ranges from the north to the south (Tourism Kuala Kangsar 2012). Its topography is rugged and hilly (Fig. 1).

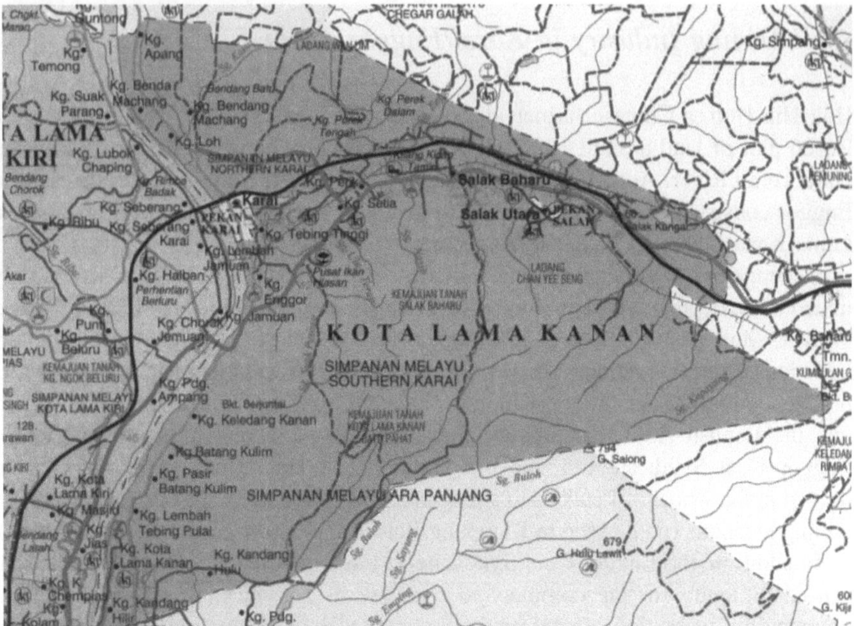

Fig. 1 Location of the Karai Town in the *Mukim* of Kota Lama Kanan

The population of Pekan Karai is 949 and it is made of Chinese, Malay, and Indian. According to the population survey in 2010, the number of people in Karai town was represented by 450 Chinese, 285 Indians, 209 Malays, five from other races, four non-Malaysian citizen, and one other native Malaysian (Jabatan Perangkaan Malaysia 2010). The main economic activity is the collection of all agricultural produce which is produced by various areas of the villages in Karai. Rubber is the dominant material for economic activity in Karai town. Small to moderate rubber plantation industry is run by the Malays from the villages which surround Karai town (Pejabat Penghulu Karai 2007).

With new source of commodity, Karai town continues to uphold its name till today. The uniqueness of the town is exemplified by various facilities which use both names Enggor and Karai at their respective premises. For instance, there are Karai Police Station, Kampung Baru Karai, Enggor Post Office, Enggor Bridge (Victoria Bridge), Enggor Train Stop, *Sekolah Kebangsaan Karai* (Karai Primary School), *Sekolah Jenis Kebangsaan Tamil Enggor* (Enggor Tamil Primary School), *Sekolah Agama Rakyat Karai* (Karai Religious Primary School), and many others. The town is equipped with various public facilities such as Karai Police Station, Enggor Post Office, *Penghulu* (Chieftain) for *Mukim* of Kota Lama Kanan Office, Telecommunication Station, public schools, and others (Jabatan Perancangan Bandar dan Desa 2005).

5 Data Analysis and Evaluation

Coal Mining Industry in Karai/Enggor Town

Hok Hin Hoh, a Chinese farmer, discovered coal while planting rubber trees on a leased plot of land around 1890s. It was located between upper Enggor River and Hulu Terap River. According to Scrivenor Report in 1917, the area of coal field in Enggor consisted of 21 acres with 40 m depth on Lot 1701, in the Mukim of Kota Lama Kanan, District of Kuala Kangsar (Yee 1990).

Nevertheless, when coal mining activities were carried out in Enggor, it was only mined at the depth of 16 m until the activities came stopped in July 18, 1928. Mr Boyd Walker, Managing Director of the Enggor Coal Syndicate Ltd, announced the termination of the mining activities as effects of a drop in coal demand even when its production reached 3000 ton per month. When the coal mining activities were carried out, the English Administrator had awarded concessions and mining rights to the Enggor Coal Syndicate Ltd, the only mining company which monopolized coal mining in the whole of Perak as to strengthen mining activities. Today, traces of these mining locations can be seen in Kampung Baru Karai (Perak Historical Crew 2010a).

In addition, the Enggor Coal Syndicate Ltd had incurred a deficit of $8,600 that caused difficulty for the company to continue its activities. It was further compounded by fluctuation of coal market price as well as innovation in machines run on petroleum energy in 1930s (Fig. 2).

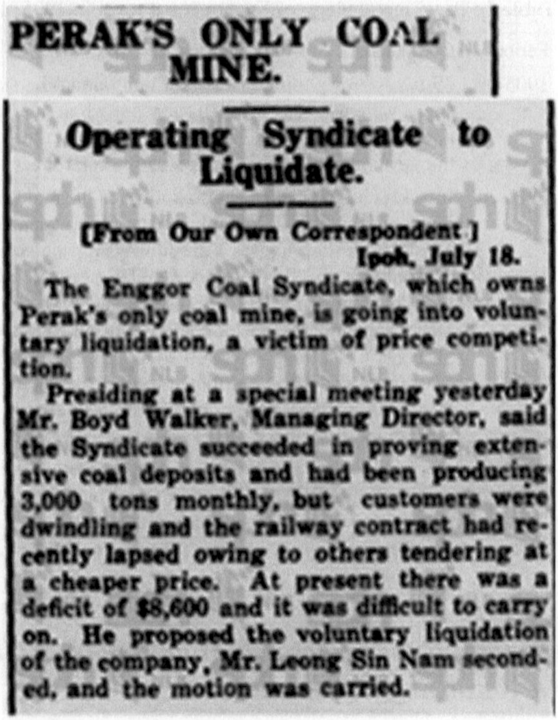

PERAK'S ONLY COAL MINE.

Operating Syndicate to Liquidate.

[From Our Own Correspondent.]

Ipoh, July 18.

The Enggor Coal Syndicate, which owns Perak's only coal mine, is going into voluntary liquidation, a victim of price competition.

Presiding at a special meeting yesterday Mr. Boyd Walker, Managing Director, said the Syndicate succeeded in proving extensive coal deposits and had been producing 3,000 tons monthly, but customers were dwindling and the railway contract had recently lapsed owing to others tendering at a cheaper price. At present there was a deficit of $8,600 and it was difficult to carry on. He proposed the voluntary liquidation of the company, Mr. Leong Sin Nam seconded, and the motion was carried.

Fig. 2 A newspaper cutting the straits times dated July 18, 1928

History of Development of Karai/Enggor Town

The town's development is characterized by shophouses during colonial era. Enggor was planned around 1900–1906 through Certified Plan dated on December 19, 1905 by the District Surveyor for the Federated Malay States (FMS). Kinta Valley railways network which connected Ipoh and Enggor in 1898 had major influence in the development of Karai. Upon completion of Victoria Bridge on March 21, 1900, the town linked to the other major towns in the State of Perak. The security and defence systems were provided to the people with the establishment of Enggor Police Station and British Army Barracks which protected the town from any subversive threats (Rokiah 2013; Arkib Negara Malaysia 2015; Khursiah and Fakhrul 2010).

Today, not all of these shop houses are used for business. Some of them have turned into residential buildings and others are just abandoned. All of the shop houses fail to contribute to any type of economic activities since there is no strong basic economy to revive the town to its former glory as a coal mining town. The distance which is far from main road between Ipoh and Butterworth also contributed. In all, Karai town has gone through three phases of development which took place in 1905, 1916, and 1928 (Table 1).

Table 1 Town morphology of Enggor/Karai town in 1905, 1916, and 1928

Period	Characteristics of the town
1905	39 two-storey shop houses were built along the station road and cross road. These shops were used as places to carry out wholesale rubber trade, sundry shop, eateries, pawn shop, liquor outlet, places to smoke opium, and gambling dents. At this phase, the development of this town was the forerunner to future progress
1916	It was a phase which was also known as the middle phase of the town's development due to its prosperous coal mining activities which were commercialized around 1910. Most of the shop houses were utilized for the same purposes from the first phase. 21 double-storey shop houses were built of which 11 were developed around Station road, in front of the Muslim reserve and next to the Enggor Train Complex. Development of this sector could be attributed to the progress in train services among all Federal States. Another 10 shop houses were erected on Cross Road. The Perak River jetty had also contributed to the construction of houses at the end of Cross Road. All of these shop houses were used exactly like the earlier ones which were built in the first phase
1928	During the last phase of the town's progress, 10 double-storey shop houses were constructed on the left side of Cross Road. They were built to serve the same functions as discussed in phase one and two

Source Modified from the Certified Plan of Enggor/Karai Town 2013

History of the Federated Malay States Railway (FMSR) in Karai/Enggor Town

Historically, the train services in Karai began with Kinta Valley train service projects which started as early as 1895 through the construction of railways that linked Ipoh and Telok Anson (Teluk Intan). By 1896, Kinta Valley line stretched all the way to Chemor and adding the length of the railway originally 51 miles to 63 miles. Owing to the revenue increased to $434,770, extension of the line was constructed from Chemor to Kuala Kangsar (Fig. 3).

By 1897, Kinta Valley railways had been extended to as far as Sg. Siput (Utara), making the current length of the track at 71 miles. It has contributed to an increase in revenue as much as $28,176, compared to the previous year. However, products which were transported using the Kinta River became a competition. Products such as rice, gasoline, and groceries in large amount were still ferried using river. The works on extending the railway track by 14 miles at Kampung Sungai Perak near Enggor was completed. Kinta Valley continued to record an increase in revenues in 1898 when the track was stretched to Enggor for another 8 miles. Revenue collected monthly from passengers traffic is increased to $221.80. Construction work of Enggor Bridge which went across the Perak River began in 1897 came to a halt in December due to the big flood (Fawzi Basri 1985).

The construction of Enggor Railway Station Complex was closely related to the development and planning of Karai. To cope with the town's role as the center of coal production, the British Colonial had built and developed a network of train services from Kinta Valley to the north. A piece of land which was situated close to

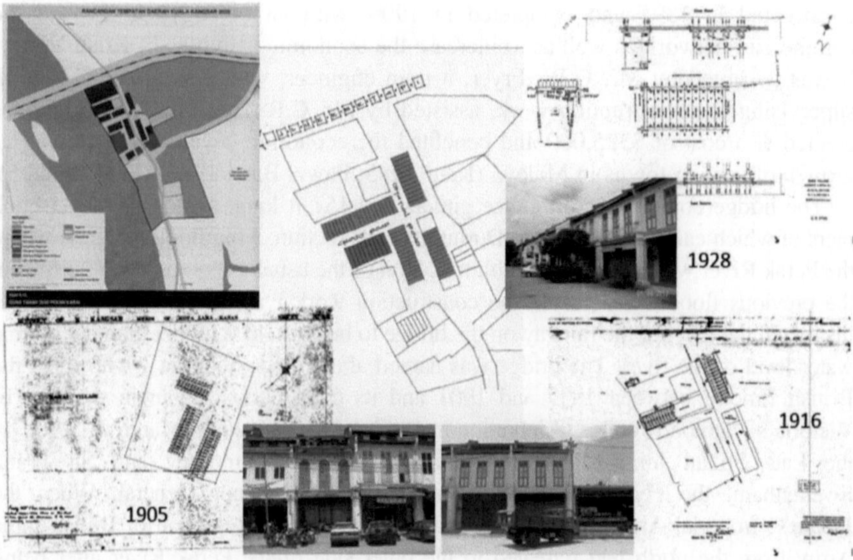

Fig. 3 Town Morphology of Karai/Enggor town in 1905, 1916, and 1928

the town was allocated and reserved for the construction of Enggor Train Station Complex which housed a train station equipped with communication facilities as well as ticketing counters and a sundry shop, a bay to unload coals, two railway tracks, three British army posts, a row of British army quarters, a house for the Enggor Railway Station Inspectorate, two rows of the train staff quarters (single-storey colonial era houses), and communication system facilities. The distance between Enggor Train Station Complex and the town was less than 1 km and the distance between the mining area and Enggor Train Station Complex was only 1.5 km (Kaur 1985; Fawzi Basri 1985).

Now, the areas of Enggor Railway Station Complex are abandoned and the remnants of its glory can still be seen such as rubbles of the train station, ruins of the neglected quarters, and the remains of the British posts. Ever since Karai ceased to be the hub for coal mining in 1928, things began to deteriorate (Fawzi Basri 1985).

The Construction of Victoria Bridge

Before the Victoria Bridge was built, the train services from the north and south would stop at Enggor Railway Station or at Beluru Railway Halt, an area in the Mukim of Kota Lama Kiri. Products transported by the trains would be uploaded and transferred by a pontoon bridge. However, due to big flood in Perak which occurred on December 28, 1897, a locomotive as well as the pontoon bridge was washed away. This drove the British to hasten building the bridge. The bridge was constructed in 1897 and completed in 1900, with an objective to expand the commercial network as well as to increase the economic activities of Kinta Valley. It was designed by Mr. G.W. Fryer, a train engineer, who was also involved in supervising the construction work, assisted by Mr. C.R. Hanson. The bridge was erected at a cost of $325,000 and benefited the economic development as well as growth of mining towns in Malaya (Kaur 1985; Fawzi Basri 1985) (Figs. 4 and 5).

The bridge comprised iron lattice girders at 1,157 ft long, resting on seven brick piers of which each is 150 ft long. During its construction, a big flood hit Karai where the Perak River water level rose as high as 50 ft of the usual as compared to 22 ft from the previous flood. As a result, the construction work was halted and the engineer instructed the level of the railway on the bridge to be built 40 ft higher than the normal water level of the river. The bridge was named after Queen Victoria 1 who ruled the British Empire between 1837 and 1901 and its construction began in the year of Victoria's Diamond Jubilee Celebration in 1897. It was opened on March 21, 1900 by the Late Sultan Idris Murshidul Adzam Shah at a ceremony and Sir Frank Swettenham, the Federated Malays States British Governor General. When the Japanese Imperial Army attacked Malaya in 1941, the 5th Regiment of British India Army from the north had retreated to the other side of the Perak River after being aggressively attacked by the Japanese Imperial Army in Kerian. To stall the Japanese advances to Singapore, Victoria Bridge was blown up by the British army engineers

Fig. 4 Victoria Bridge in 1914

Fig. 5 Victoria Bridge in 2014

on the dawn of December 22, 1941. After the Japanese Imperial Army occupied Karai, the Victoria Bridge was built using the local laborers with a payment of 55 cents per day. Materials used were trunks of durian, mangosteen, mangoes, coconut, and rubber trees which were all indigenous of the locality (Fawzi Basri 1985).

Victoria Bridge stands significance as it is situated close to the town and it is fast becoming a tourist attraction for both local and international. The bridge has been selected as site for filming shooting for local movie production. It is hoped that Victoria Bridge will remain strong for future generations to reflect on the nation's history (Hashim 2010).

Other Historical Structures in Karai/Enggor Town

Air Raid Shelter (Pillbox). Not too far from the Enggor Railway Station Complex, there are two British Malaya Air Raid Shelters known as pillboxes and the militari's barracks. Another bunker is located on the other side of Victoria Bridge in Kampung Seberang Karai. The bunkers also known as pillboxes are for defence. They were built due to several threats in the area between 1900 and 1989. At the onset of World War II, the British Army had courageously fought to protect Karai. The Japanese Imperial Army had used aircrafts to bomb on the British Malaya strongholds and barracks which could have housed 30–40 men. British post on the Victoria Bridge was crucial to stop the Japanese Imperial Army advances toward Singapore. After the end of World War II, the army's posts and barracks were used to control the Malayan Communist Party (MCP) threats as they were still active around Sg Siput (Utara). After Emergency was announced by the British in July 1948, the colonial had strengthened the security at Karai due to its close location where the three English planters who were murdered by MCP. The strongholds and barracks were once used by the Malaysian Army during the confrontation era from 1963 to 1965. However, the threats did not take place. Today, the posts and defence barracks can be seen from inside and outside even though they are in ruin (Fawzi Basri 1985).

***Waqf* land**. In the center of Karai, there is a *waqf* land of 0.362 ha which stands a surau (small mosque) built almost 100 year old. The land was donated by a religious man, a Syed descendant, who had wanted it to be put to good use for the people in the town and the surrounding villages. During the planning stages, the British had reserved the land for Muslim cemetery. However, as the Chinese began to dominate the area, the land was used for activities which benefited the Muslims. At present, the *waqf* land was erected with a double-storey surau and five units of shops. On its opening plaque, the building was named *Al-Madrasatul Karai* and was built in 1928. The locals contributed to the construction of surau (Pejabat Penghulu Karai 2007).

Tok Setia's Mansion. The mansion which was built before the World War II is located on the top of hill next to the Karai–Enggor road and at 200 m from the town. Originally, the house belonged to Tuan Syed Assin who was a brother of Tuan Syed Adam or Alim Besar Kuala Kangsar. Tuan Syed Assin passed away in 1940s and the house was inherited by his son, Syed Nordin. The mansion area was not affected when Kuala Kangsar hit by big flood at all due to its strategic location on the hill (Pejabat Penghulu Karai 2007).

6 Important Figures Behind the Establishment of Karai Town

The Late Sultan Idris Mursyidul Azzam Shah I, 28th Perak Sultanate (1887–1916)

Sultan Idris Murshidhul Adzam Shah was the Prince of Almarhum Raja Bendahara Alang Iskandar Ibni Almarhum Raja Kecil Tengah. He born on 1849 at Kuala Keboi, Kampar. He ascended the Perak Sultanate throne as the 28th Sultan on July 26, 1887 to replace the late Sultan Yusuf Sharifuddin Muzaffar Shah, upon his passing. During his reign, Perak had gone through modern developmental processes which were inclusive of progress in infrastructure, modernization of the state's administrative group, and many more. Sultan Idris passed away on January 14, 1916 at the age of 67 and was given royal burial ground at the Royal Mausoleum in Bukit Chandan, Kuala Kangsar, Perak (Perak Historical Crew 2010b).

Sir Frank Swettenham, the Federated Malay States British Resident General (1850–1946)

Sir Frank Athelstane Swettenham was born on the March 28, 1850 in Belper, Dershire, England. He was a British Senior Officer in Malaya who had influenced the policy making and British administrative structuring in Malaya. Swettenham was sent to Singapore as Public Service Trainee for the Straits Settlement in 1871. Having acquired *Bahasa Melayu*, he played an important role in being the moderator between British and Malays in every British intervention in the Malay states in 1870s. In 1882, he was appointed as the Resident to the Sultan of Selangor. He succeeded in bringing forth the establishment of coffee and tobacco plantations in Selangor as well as contributed to the increase of tin mining through the development of railway tracks from Kuala Lumpur to the Klang's port. In 1895, he served as the British Resident in Perak and materialized the Federal Agreement which combined Perak, Selangor, Negeri Sembilan, and Pahang as Federal States. He also led the Federation with a title as Resident General. In 1897, he was bestowed the title 'Sir' by Queen Victoria which saw him appointed as the British High commissioner in Malaya and the Federated Malay States British Resident General in 1901, which was 3 years before his retirement. Sir Frank Swettenham died on June 11, 1946 in London, England at the ripe age of 96 years old (Temin 1997; Allen 2010).

Ratu Victoria I, Ruler of the British Empire (1837–1901)

Queen Victoria 1 was one of the English Queens who held the British Empire throne from 1837 to 1901. She was born at the Kesington Palace, London on May 24, 1819 as a daughter to Edward Duke of Kent who was the son to King George III. Queen Victoria I became as the Queen of England in 1837 after three of her uncles who were King George IV, Frederick Duke of York, and King William IV passed away without any heir to take the throne. Her reign was closely related to the glory of the Great Britain as an industrialized empire which was fast expanding, strong economic development, and the golden era of the British Empire itself. Queen Victoria's era was known as the Victorian era. The Queen passed away on January 22, 1901 at Obsome Hoiuse, Isle of Wright after 64 years on the British throne which also made her the Queen with the longest term of ruling in the history of the British royals (Allen 2010).

Mr. G.W. Fryer, Resident Head of Engineer FMS Railways

Mr. G.W. Fryer was the Chief of Resident Engineer for the Federated Malay State Railway services. He was taken from Northern Indian Railways in 1891 to serve with the civil service in Perak for several years under the supervision of Mr. C.R Hanson. Mr G.W. Fryer was appointed the position in 1904 until his sudden demise on Thursday, March 2, 1909 in Seremban, Negeri Sembilan (Fawzi Basri 1985).

7 Karai Town—In Present

Today, not all of these shop houses are used for business purposes. Some of them have turned into residential buildings and others are just abandoned. All of the shop houses fail to contribute to any type of economic activity since there is no strong basic economy to revive the town to its former glory as a coal mining town. It is also located quite far from the main road between Ipoh and Butterworth. Tuesday Morning Market is known as a one-day market which operated as early as 7.00 am at the town. The concept of this market is similar to the night markets where one can find in many towns and cities in Peninsular Malaysia. On every Tuesday morning, the market offers a variety of items. They are sold by small traders. Among some of the items sold are chicken, beef, groceries, food for breakfast, and the list goes on. However, the antiques are what people hope to find there. Like many other night market across the country, the town's night market offers items such as chicken and meat, fish, vegetables, fruits, clothes, and many others (Pejabat Penghulu Karai 2007).

8 Conclusion

Research on and writing about the historical background of Karai town development is aimed at making it a valuable resource not only to the academicians. It can be read and appreciated by people from all walks of life. These historical elements must be preserved and relevant authorities must find a method or approach in keeping the nations' historical remnants available at the town to be cherished and understood by future generations. Nevertheless, preservation efforts seem to be neglected and it would be a total waste should all these elements continue to deteriorate and completely disappear. Looking at the surrounding area of the Victoria Bridge which is the monumental landmark of Karai town, the researchers found that there are some initiatives by some authorities to take care of this structure. It can become a factor that will draw a lot of tourists to the area and contributes to the economic activities of the locals. Other remaining structures and historical elements are not given any preservation efforts especially for the train transportation system which once put Karai town on the world map for its services.

References

Arkib Negara Malaysia (2015) Official Website. http://arkib.gov.my. Accessed 25 Aug 2015

Allen RC (2010) Why was the industrial revolution British?. Yale University, The Kuznets Lecture

Fawzi Basri MA (1985) Sejarah Keretapi di Malaysia. Keretapi Tanah Melayu, Kuala Lumpur

Hashim M (2010) Keistimewaan Pekan Karai. Utusan Malaysia. http://www.utusan.com.my. Accessed 21 Aug 2013

Jabatan Perancangan Bandar dan Desa (2005) Rancangan Tempatan Kuala Kangsar

Jabatan Perangkaan Malaysia (2010) Banci Penduduk Malaysia 2010

Kaur A (1985) Seabad Keretapi di Malaysia. Jabatan Muzium Malaysia, Kuala Lumpur

Khursiah Abd. Aziz and Fakhrul Zaman Abdullah (2010). Cultural Heritage Tourism Development in Kota Lama Kanan, Kuala Kangsar, Perak. Universiti Tun Abdul Razak E-Journal 7: 1–10

Kolej Komuniti Sungai Siput (2013) E-Komuniti—Mukim Kota Lama Kanan. http://www.ekomuniti.kkss.edu.my. Accessed 11 July 2013

Loke E (2013) Peaceful idylic living in sungai siput. The star online. http://www.thestar.com.my. Accessed 24 Aug 2013

Pejabat Penghulu Karai (2007) Anugerah Pengurusan Mukim Terbaik Peringkat Negeri Perak Darul Ridzuan Tahun 2007. Unpublished

Perak Historical Crew (2010a) Jambatan victoria and Karai. http://ieshamww2.webs.com. Accessed 21 Aug 2013

Perak Historical Crew (2010b) Kenali Kesultanan Perak-Sultan Idris Mursyidul Azzam Shah I. http://www.blogspot.com/2010/04/kenali-kesultanan-perak. Accessed 21 Aug 2013
Rokiah Abdullah (2013) Legasi jambatan kereta api tertua. Utusan Malaysia. http://www.utusan.com.my. Accessed 21 Aug 2013
Temin P (1997) Two views of the British industrial revolution. J Econ Hist 57:63–82
Tourism Kuala Kangsar (2012) Kuala Kangsar, the royal town. http://tourismkualakangsar.wordpress.com. Accessed 21 Aug 2013
Yee FK (1990) Geology and Mineral Resources of the Taiping-Kuala Kangsar area Perak Darul Ridzuan. Percetakan Zainon Kassim Sdn. Bhd, Ipoh

Chapter 19
Determination of New Bank Branch Location Using GIS Approach

Noorsazwan Ahmad Pugi, Halmi Zainol and Azlizan Adila Mohamad

Abstract In business, location of the premise plays an important role in ensuring the success of the business. It is important in terms of the business marketing and profit. The location of the competitors also should be taken into account. With the growing needs for computer-based system, geographic information system (GIS) can be utilized to determine the suitable location for business purpose including new bank branch location. This paper has summarized the method to carry out the research to determine the best location to place a new branch for Bank X, particularly in Shah Alam, Selangor. The process can be divided into three stages; planning, implementation, and decision stage. The planning stage was focused on the data methods and analysis to be used in this research. For the implementation stage, geospatial database was created and all required data for decision stage were prepared. In the decision stage, the location of new branch of Bank X was investigated based on the suggested criteria from the bank itself and location of other banks. Spatial modeling and Euclidean distance methods were used to obtain the desired decision. The results showed four suitable locations for new bank branch of Bank X; location near Stadium Shah Alam, location at Pusat Perniagaan Worldwide, location opposite Sek. Keb. Seksyen 7, and location behind SRJK Tamil Seksyen 7. Ranking technique was used in order to get the most suitable location from those four locations and the most recommended location was the location opposite to Sek. Keb. Seksyen 7. For the conclusion, GIS technology is a useful and powerful tool in planning and decision-making purpose.

Keywords Euclidean distance · GIS · Suitable location

N. Ahmad Pugi (✉) · H. Zainol · A.A. Mohamad
Faculty of Architecture, Planning and Surveying, Universiti Teknologi MARA, Seri Iskandar, Perak, Malaysia
e-mail: noors240@perak.uitm.edu.my

H. Zainol
e-mail: halmi461@perak.uitm.edu.my

A.A. Mohamad
e-mail: azliz1122@perak.uitm.edu.my

N.A. Yacob et al. (eds.), *Regional Conference on Science, Technology and Social Sciences (RCSTSS 2014)*, DOI 10.1007/978-981-10-0534-3_19

1 Introduction

Determining strategic new bank branch location is one of the important elements in marketing a business expansion. Finding the best new location for business expansion is really a challenging task. It requires substantial capital investment and with so much money on the line, management wants to make sure that they have selected the right expansion site (Jafrullah et al. 2003). With the growing needs for computer-based system, GIS can be utilized to determine the suitable location for business purpose including new bank branch location. GIS is masterful at analyzing hundreds of variables to quickly identify the top sites for consideration. GIS helped us to quickly obtain a snapshot in understanding of all markets from a remote location (Anthony 2007).

This paper is about to determine the best location for establishing new Bank X branch in Shah Alam City Center region which includes Seksyen 1 to Seksyen 14 by using the advancement of geographic information system (GIS) technology. This is because this area consists of two Bank X branches only and were predicted that it could not support the capacity of customers, mainly from Shah Alam city center residence due to the rapid development. Banks can use GIS-based solutions to ranks service area within market according to viability for the concept. Based on the bank's criteria, the analysis can consider competition, demographics and other concept factors to help prioritize options.

2 Data Layers and Sources

The base map used to fuse all available data sources is based on Shah Alam landuse digital map provided by Majlis Bandaraya Shah Alam (MBSA). This map has been registered via some ground control points based on Malaysia Rectified Skew Orthomorphic (MRSO) projection. This digital map provided all information about boundaries and all related information on blocks, sections, and villages.

The resultant of the above solution is the following list of the layers: Shah Alam Sub-districts, major and minor streets layers, banks, commercial areas including restaurants, hotels, gas stations, and police stations.

3 Bank Location Criteria

The suggested criteria by the bank to locate new bank branch location are as follows (Jabatan Perbankan Cawangan 2008):

1. Area with the population about 10,000 people
2. The location should be in trading area such as supermarket or malls that could develop a business matter in that area.

3. The maximum area coverage must be within 3 km from existing branch.
4. Must be located in the area that avails security, business establishments, and infrastructure support such as accessing road.

4 Implementation and Resultant Maps

For the methodology of this study, it was carried out as shown in Fig. 1. It was divided into three phases: planning stage, implementation stage, and decision stage (Figs. 2, 3 and 4).

Planning stage was focused on the creation and development of the GIS data model. It is also used to determine the criteria of suitable location for the new branch to be located. Besides that, it was also involved the decision to apply which method and analysis to be used in decision stages. For this research, the analysis method to be used is using Euclidean distance method. This method is supported in the ESRI's software that is ArcGIS.

Implementation stage discussed the building of geospatial data layers and the GIS data model implementation. This is the stage that GIS data model be built and

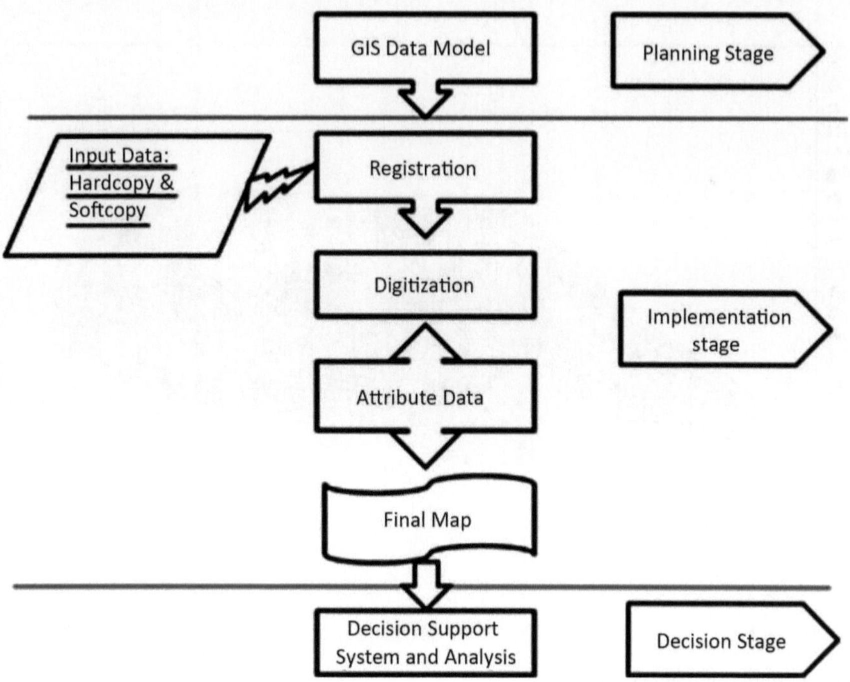

Fig. 1 Methodology chart

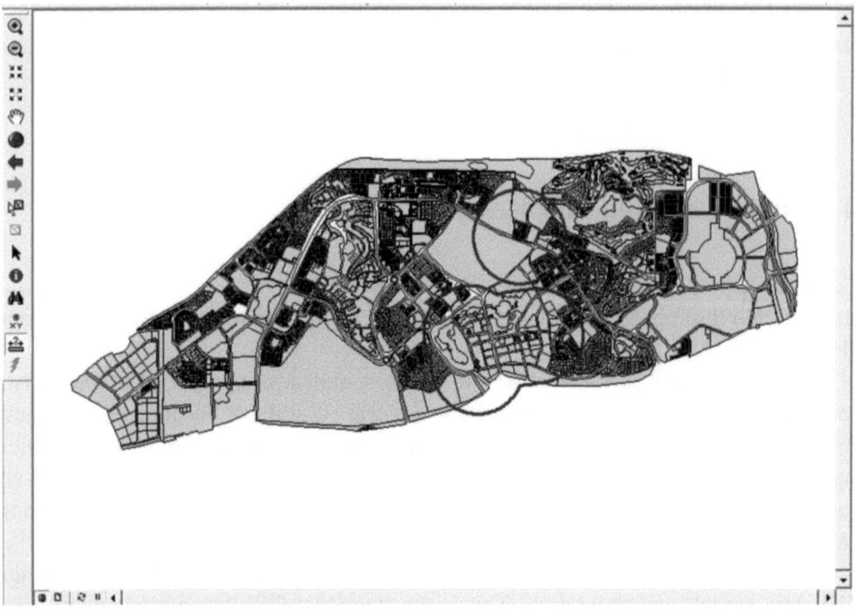

Fig. 2 Buffer of existing banks (competitors)

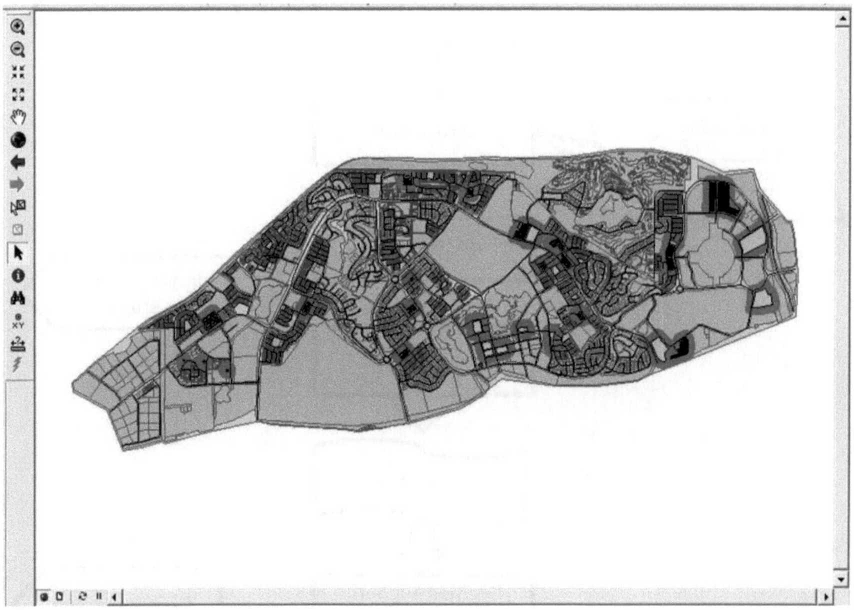

Fig. 3 Buffer of commercial area

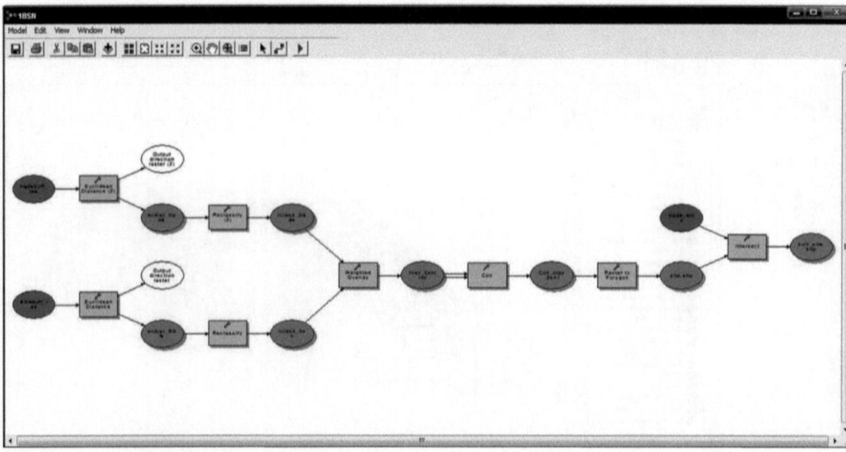

Fig. 4 Analysis workflow

implement. It was started from registering the input data (particularly the digital map) into Malaysia Rectified Skew Orthomorphic (MRSO) projection. Then, it followed by digitizing the road layers and other points of interest such as banks features. After that, the attribute of those features in the map were created on the particular features. Finally, it produced the final map that will be used for analysis process (decision stage).

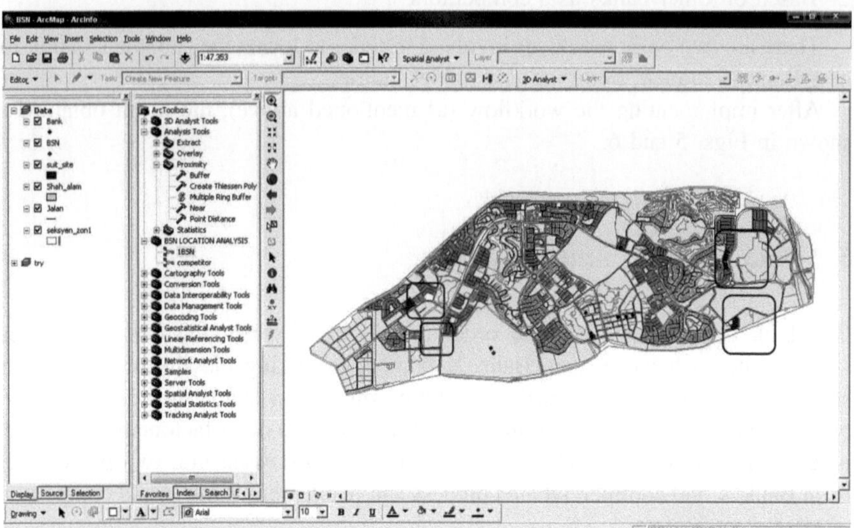

Fig. 5 Overall suitable location (Bank X and commercial area analysis process)

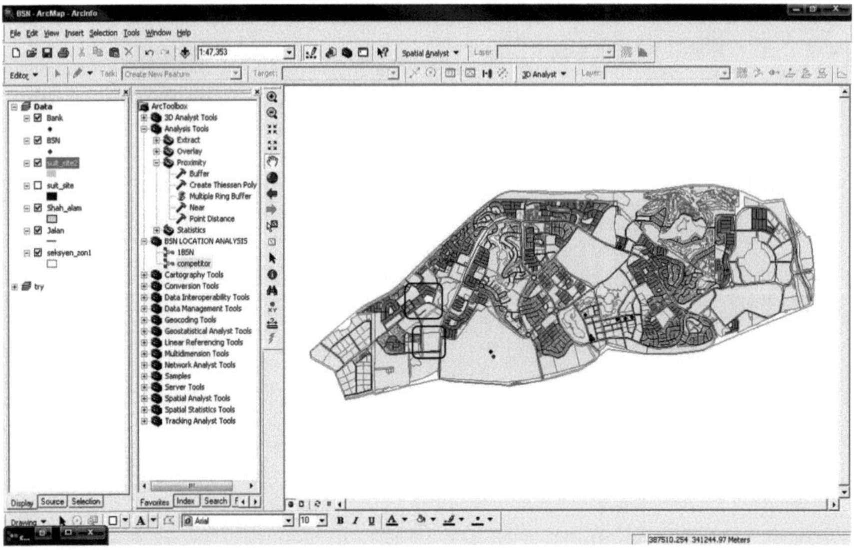

Fig. 6 Overall suitable location (Bank X and competitor's analysis process)

In the decision stage, it discussed the analysis that is being carried to determine the suitable location for new Bank X branch. In this stage, there are two types of investigation that have been carried out. They are

1. Based on bank X itself (no competitors were considered)
2. Based on other banks as a competitors

There are two basic processes involved in order to locate the suitable location for new Bank X branch; buffering and model builder.

After implementing the workflow (as mentioned above), the result obtained is shown in Figs. 5 and 6.

5 Data Analysis and Results

Based on the resultant map, in order to determine which location is the most suitable site to locate a new Bank X branch, a ranking classification was constructed. The analysis was carried out based on certain factor such as population, type of commercial area, accessibility, parking lots, and other factors including area potential and security of safeness. The result was discussed into two parts; result from Bank X and commercial area process and result from Bank X and competitors process.

From the resultant map, the location can be assigned as **A**: location near Stadium Shah Alam, **B**: location at Pusat Perniagaan Worldwide, **C**: location opposite to

Table 1 Total ranking locations' suitability

Result (location)	Total ranking suitability
A	5 + 3 + 2 + 2 + 2 + 5 = 19
B	5 + 3 + 2 + 2 + 2 + 5 = 19
C	4 + 4 + 2 + 2 + 1 + 6 = 19
D	4 + 1 + 1 + 1 + 1 + 5 = 13

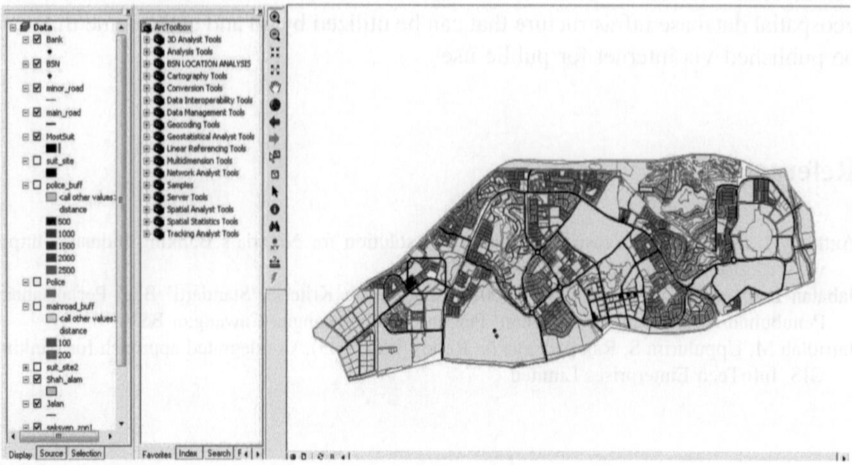

Fig. 7 The most suitable location based on ranking suitability (overall)

Sek. Keb. Seksyen 7 and **D**: location behind SRJK Tamil Seksyen 7. Since the result in Bank X and competitor analysis process was same as C and D, therefore it used the same category C and D.

Table 1 shows the total ranking suitability obtained from those locations. The most suitable location was **Location C** due to its maximum value in the total ranking suitability. Furthermore, this location is obtained after the result through the investigation in the decision stage (Fig. 7).

6 Conclusion and Recommendations

As a conclusion, the objective of this study was successfully achieved. Four recommended locations for new Bank X branch location was successfully determined by using GIS as a tool for decision-making. This was supported by the use of Euclidian distance analysis method in order to investigate the location based on the existing Bank X branch location and other banks location as a competitor. Finally, total ranking suitability was used in order to determine the most recommended suitable location for new Bank X branch location.

There were some recommendations that can be highlighted from this study. In order to maintain the accuracy and make the data more global, base map of Shah Alam needs to be updated. Additional information such as land costs, building availability and suitability, construction costs, local and state taxes, local and state development incentive, availability and cost of energy, transportation availability, availability of infrastructure such as telecommunications, sewer, and water can be included in order to get more information about the features. Banks and other commercial sectors in Shah Alam should organize their effort to build one geospatial database infrastructure that can be utilized by all and at the same time can be published via internet for public use.

References

Anthony L (2007) STL released integrated GIS solution for Nigeria's Banking Industry. http://www10.giscafe.com. Accessed 7 July 2008

Jabatan Perbankan Cawangan, BSN (2008) Penggunaan Kriteria 'Standard' Bagi Perlaksanaan Penubuhan/ Penutupan/ Pemindahan/ Percantuman Cawangan-Cawangan BSN

Jafrullah M, Uppulurim S, Rajopadhaye N, Reddy VS (2003) An integrated approach for banking GIS. InfoTech Einterprises Limited

Chapter 20
Application of Sustainable Site Planning and Management (SM) Criterion in Green Building Index (GBI) Assessment for Hill Land Development in Penang—A Case Study

Nadira Ahzahar, Intan Bayani Zakaria and Siti Ismahani Ismail

Abstract Demands for residential area in Penang Island have increased drastically from year to year. In contrast, the suitable lands for development sector become scarce. Those conditions lead to the encroachment to the few available development at the hill land development area. Regarding to this situation, various green rating tools namely Green Building Index (GBI), Leadership in Energy and Environmental Design (LEED), and BCA Green Mark are applied to evaluate sustainable site planning and management (SM) of the new development. However, application of these Green Rating Tools is still lacking at the hill land development area. As such, the aim of this case study is to apply GBI, LEED, and BCA Green Mark on SM criterion at the study area located at hill land development at Sungai Ara, Penang. There are several methods to obtain the points regarding the SM criterion such as site evaluation and inspection and also conduct interview sessions with planner and consultant. The result obtained from the evaluation shows that GBI achieved 79.7 % of the overall SM points while LEED and BCA Green Mark achieved 36.4 % and 20.7 %, respectively. From this result, it shows that GBI obtained the highest percentage for SM criterion and the most suitable green rating tool to rate SM criterion at the hill land development area in Penang.

Keywords Residential · Hill · Land · Development · SM criterion

N. Ahzahar (✉) · I.B. Zakaria
Department of Building Surveying, Faculty of Architecture, Planning & Surveying, Universiti Teknologi MARA, Seri Iskandar, Perak, Malaysia
e-mail: d_ra81@yahoo.com

I.B. Zakaria
e-mail: intan_bayani@yahoo.com

S.I. Ismail
Faculty of Civil Engineering, Universiti Teknologi MARA, Kepala Batas, Pulau Pinang, Malaysia
e-mail: sid_hany@yahoo.com

N.A. Yacob et al. (eds.), *Regional Conference on Science, Technology and Social Sciences (RCSTSS 2014)*, DOI 10.1007/978-981-10-0534-3_20

1 Introduction

Nowadays, land has become one of the limited source in Penang Island (PI) as regards to its hilly topography and restricted flat lands, which is about to exhaust (Anilarasu 2010). Since PI is one of the many rapidly industrializing states in Malaysia with a majority of urban population, many developers find alternatives to move to hilly areas for new development projects (Fig. 1). Such rapid development will put the environment at risk by natural disasters like flood, changes of climate; landslides etc., and become a safety threat to the life and property of local inhabitants. The Penang Structure Plan (PSP) 2020, forecasts a population growth from 1.6 million in 2010 to 2 million by 2020, with 40 % of this population anticipated to reside on the island itself by 2020 (Lim 2010).

As awareness and concern regarding environmental and sustainable issues from this hazardous circumstances, Malaysian Institute of Architect (PAM) and the Association of Consulting Engineers Malaysia (ACEM) have developed the rating tools known as Green Building Index (GBI) to encourage sustainability in built environment and raise the realization in the market including all three tiers of government, public and private sector developers, professional services providers, academia, product manufacturers, suppliers, and other industry stakeholder groups. Prior to GBI implementation in Malaysia, variety of green rating tools (GRTs) are accessible throughout the world such as Leadership in Energy & Environmental Design (LEED) by the United States, Building Research Environment Assessment Method Consultancy (BREEAM) by the United Kingdom, BCA Green Mark by Singapore, and etc. GRTs are mostly an independent, national, voluntary rating tool to drive the development of more sustainable, productive, and liveable cities. Unfortunately most of these GRTs are not designed to suits for Malaysia's local conditions (Shari et al. 2010) and since the establishment of GBI (2009), most of its application is for flat terrain development area. However, as we are aware that the landscape in PI is corrugated and hilly topography with very steep terrain rises up to

Fig. 1 Hilly area for new development project in Penang Island

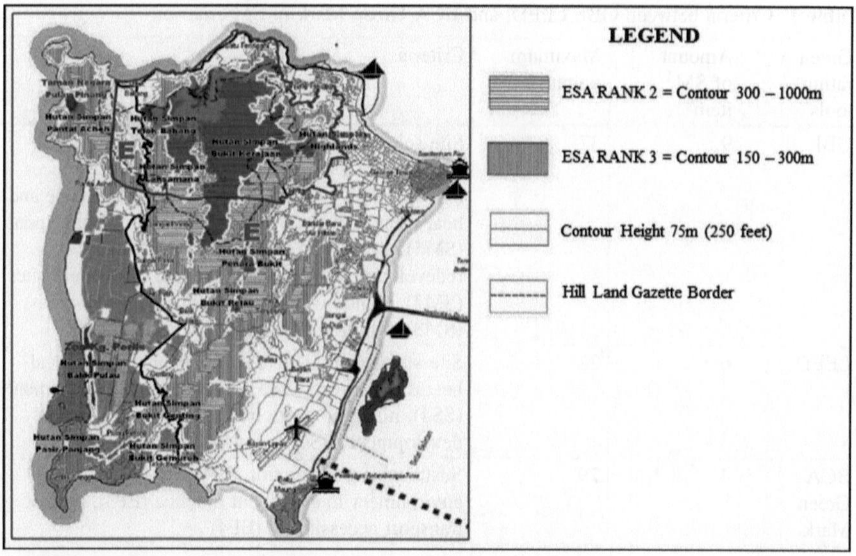

Fig. 2 Penang Island topographical information (PSP 2007)

830 m (Salma 2001) as shown in Fig. 2, which is obtained from Penang Structure Plan (PSP) 2020.

Hence, due to this current situation, this case study is conducted to check the suitability of applying GRTs namely GBI, LEED, and BCA Green Mark for sustainable site planning and management (SM) criterion at hill land development and conduct comparison studies between these GRTs in order to recommend the best GRTs to be implemented in Malaysia.

2 Literature Review

The buildings in which we live, work, and play protect us from nature's extremes, yet they also affect our health and environment in countless ways. As the environmental impact of buildings becomes more apparent, a new field called "green building" is gaining momentum (EPA 2013). The term "Green Building" is a loosely defined collection of land-use, building design, and construction strategies that reduce the environmental impact for buildings to their surroundings. Traditional building practices often overlook the inter-relationships between a building, its components, its surroundings, and its occupants. Typical buildings consume more of our resources than necessary and generate large amounts of waste.

Green buildings have many benefits, such as better use of building resources, significant operational savings, and increased workplace productivity (Leong 2009). Many countries have pushed forward countless best practices for sustainable

Table 1 Criteria between GBI, LEED, and BCA Green Mark in SM criterion

Green rating tools	Amount of SM item	Maximum points	Criteria
GBI	9	37	Site selection and planning (SM1), public transportation access (SM2), community services and connectivity (SM3), open space, landscape and heat island effect (SM4), construction management (SM5), storm water management (SM6), redevelopment of existing site and brownfield sites (SM7), avoiding environmentally sensitive areas (SM8), building user manual (SM9)
LEED	6	22	Site stewardship (SS1), landscaping (SS2), local heat island effects (SS3), surface water management (SS4), nontoxic pest control (SS5) and compact development (SS6)
BCA Green Mark	4	29	Sustainable construction (EP1), greenery (EP2), environment management practice (EP3), public transport accessibility (EP4)

development and environmental management while pursuing economic and social development (Chua and Oh 2011). One of the many best practices is by going "green" and applying sustainable site planning. In reviewing the existing research, this case study aims to apply SM criterion at a chosen study area located at hilly terrain by using GRTs namely GBI, LEED, and BCA Green Mark. GBI will emphasis on sustainable site planning and management (SM), LEED stress on sustainable sites (SS) and BCA Green Mark emphasis on environmental protection (EP) criterion. Even though the criterion for each of these differ in names, but the measures are in majority similar and fall under the same principle. Table 1 shows the distinction between these tools for SM criterion.

3 Methodology

The methodology for this case study begins with the development of background and desk studies focusing on SM between GBI, LEED, and BCA Green Mark followed by the determination of problem statement and main objectives. Hence, the evaluation is conducted via five main phases which are the site visit and measurements, site evaluation, data collection, discussion with planner and architect, and research of the background of study area. Panel discussion with the respective consultants is conducted in order to obtain the data in compliance with SM criterion. After the evaluation process, comparisons between the selected GRTs on SM criterion are conducted and the best GRTs are proposed to be implemented at the selected study area, which is located at Sungai Ara, Penang.

Site Evaluation, Data Collection, and Literature Review Research

Site evaluation consists of site measurement and inspection has been conducted in several SM criterions for this case study. The data obtained from the site evaluation process are then compared with the Google Maps to check the accuracy. Further discussion with planner and architect to obtain the detailed site information and materials for this project is used to evaluate all SM criterions. This session is conducted at BYG Architecture and Green Acre Studio and data such as location plan, key plan, local and structure plan, and many more layouts are then gathered. Literature review has been accomplished in acquiring more information on GBI, LEED, and BCA Green Mark in order to evaluate their similarities and differences. The implementations of methodology for GBI, LEED, and BCA Green Mark are illustrated in the flowchart as shown in Fig. 3.

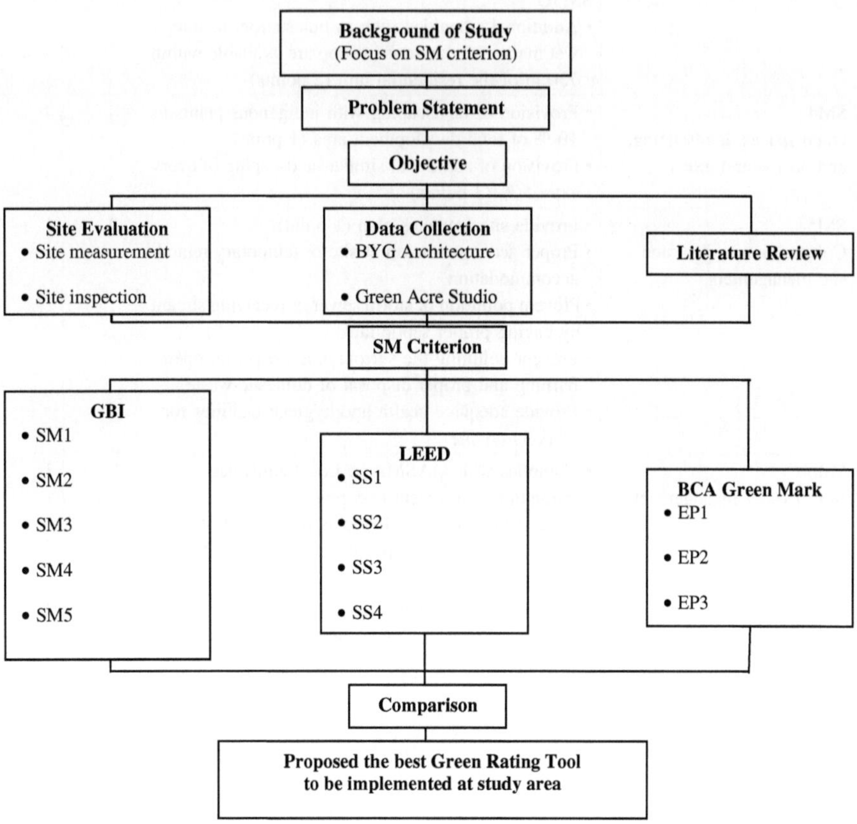

Fig. 3 Methodology for this case study

Table 2 GBI overall assessment on SM criterion

GBI			
Items	Criterion	Max. points	Score
SM1 Site selection and planning	• Structure plan or local plan for the area (1 point) • Infrastructure requirement is available	1	1
SM2 Public transportation access	SM2A • Within 251–500 m from the area (6 points) SM2B • Dedicated walkway with regular arrays of shaded trees covering at least 70 % of route (3 points)	12	9
SM3 Community services and connectivity	SM3A • Basic amenities such as grocery store, coffee house, surau, and playground are available within 750 m of residential unit (4 points) SM3B • Other amenities such as clinic, police station, and school are available within 750 m of residential unit (1.5 points) SM3C • Additional amenities such as bus station, laundry, wet market, and hardware shop are available within 750 m of the residential unit (2 points)	8	7.5
SM4 Open spaces, landscaping, and heat island effect	• Provision of landscaping with indigenous plants to 10 % of total development area (1 point) • Provision of additional similar landscaping of every extra 5 % (3 points)	4	4
SM5 Construction system and site management	• Provide site amenites plan (1 point) • Proper accommodation at site or temporary rented accommodation • Prevent pollution of storm sewer or receiving stream by having proper septic tank • Prevent polluting the surrounding area from open burning and proper disposal of domestic waste • Provide adequate health and hygiene facilities for workers on site	3	1
SM6 Storm water management	• Complies with MASMA or Local equivalent minimum requirement (1.5 point) • Exceed MASMA requirements by 30 %: Entitled to 2 additional points pro-rated for lower value (1.5 point)	3	3
SM7 Re-development of existing sites and brownfield sites	• Re-development, refurbishment or extension of exiting building or sites or rehabilation sites (0 point)	2	0

(continued)

Table 2 (continued)

GBI			
Items	Criterion	Max. points	Score
SM8 Avoiding environmentally sensitive areas	• Prime agriculture land as defined by the Town and Country Planning Act • Land that is specifically identified as habitat for any species threatened or endangered lists • Within 30 m of any wetlands as defined by the Structure Plan of the area	2	2
SM9 Building user manual	• Provide a building user manual which documents passive and active features that should not be downgraded (2 point)	2	2
Total score		37	29.5

Table 3 LEED overall assessment on SM criterion

LEED			
Items	Criterion	Max. points	Score
SS1 Site stewardship	• Leave undisturbed at least 40 % of the buildable lot area, not including area under roof. Only soft-scaped can be counted toward this credit (1 point)	1	1
SS2 Landscaping	• Any turf must be drought-tolerant (1 point) • Install drought-tolerant plant (2 points)	7	3
SS3 Local heat island effects	• Locate trees or other planting to provide shading for at least 50 % of sidewalks, patios, and drive-ways within 50 ft of the home (1 point)	1	1
SS4 Surface water management	• If portions of the lot are located on a steep slope, reduce long-term runoff effects through use of terracing and retaining walls (1 point)	7	1
SS5 Nontoxic pest control	• Keep all wood (structure) at least 12 in. above soil (0.5 point) • Include no wood-to-concrete connections or separate any exterior wood-to-concrete. Connections such as at posts, deck support, and stair stringers with metal or plastic fasteners or dividers (0.5 point) • Install a steel mesh barrier termite control system (0.5 point) • Install nontoxic termite bait system (0.5 point)	2	2
SS6 Compact development	• Build homes with an average housing density of 7 or more dwelling units per acre of buildable land. A single home on 1/7-acre buildable lot qualifies (0 points) • Build homes with an average housing density of 10 or more dwelling units per acre of buildable land. A single home on 1/10-acre buildable lot qualifies (0 points) • Build homes with an average housing density of 20 or more dwelling units per acre of buildable land. A single home on 1/20-acre buildable lot qualifies (0 points)	4	0
Total score		22	8

4 Results and Discussion

The results from the evaluation done are tabulated in the tables.

From the results shown in Tables 2, 3, and 4, respectively, GBI has contributed 29.5 points while LEED = 8 points, and BCA Green Mark = 6 points for SM criterion. The result in Table 5 shows that GBI has the highest score of 29.5 points out of 37 points follows by LEED which is only 8 points out of 22 points and BCA Green Mark, 6 points out of 29 points. Table 5 showsthe comparison activities for GBI, LEED, and BCA Green Mark on SM criterion obtained for this case study area with GBI.

Table 4 BCA Green Mark overall assessment on SM criterion

BCA Green Mark			
Items	Criterion	Max. points	Score
EP1 Sustainable construction	• More efficient concrete usage for building components (0 point) • Conservation of existing building structure. Applicable to existing structural elements or building envelope (0 point) • Use of sustainable materials and products in building construction (0 point)	12	0
EP2 Greener	• Restoration of trees on site, conserving, or relocating of existing trees on site (1 point)	6	1
EP3 Environmental management practice	• Developer, main builder, M and E consultant and architect who are ISO 14000 certified (1 point) • Provision of building users' guide including details of the environmental friendly facilities and features within the building and their uses in achieving the intended environmental performance during building operation (1 point) • Provision of facilities or recycling bins for collection and storage of different recyclable waste such as paper, glass, plastic etc. (1 point)	9	3
EP4 Public transport accessibility	• Good access to nearest MRT/LRT or bus stops (1 point) • Adequate bicycles parking lots (1 point)	2	2
Total score		29	6

Table 5 Comparison of total points and percentage of GBI, LEED, and BCA Green Mark on SM criterion

Green rating tools	Maximum points	Points scored	Percentage (%)
GBI	37	29.5	79.7
LEED	22	8	36.4
BCA Green Mark	6	6	20.7

5 Conclusion and Recommendation

Therefore, it can be concluded that GBI is the most likely and appropriate GRTs to be applied at the hill land development for SM criterion for this case study area. From this case study, also it is proven that GBI has been invented to suit the local conditions in our country.

In future, other researchers may explore all the criterion in GBI such as energy efficiency (EE), indoor environment quality (EQ), sustainable site planning and management (SM), material and resources (MR), water efficiency (WE), and also innovation (IN). The comparison can be made by adding more green rating tools such as Green Earth by Australia, HK-BEAM by Hong Kong, CASBEE by Japan, Green Star by New Zealand, BREEAM by the United Kingdom, Ecology, Energy Saving, Waste Reduction and Health (EEWH) by Taiwan, GBAS by China, and many more. This comparison will produce precise and accurate results. Moreover, this study can be enhanced by adding more sites with difference geographically conditions throughout Malaysia.

References

Anilarasu A (2010) Hill land development challenges in Penang Island. Faculty of Civil Engineering, Universiti Teknologi Malaysia

Chua SC, Oh TH (2011) Green progress and prospect in Malaysia. Renew Sustain Energy Rev 15 (2011):2850–2861

Leong CT (2009) Building structure for the future—the green way. Board of Engineers Malaysia (BEM). ISSN 0128 - 4347

Lim GE (2010) Fund raising dinner ‘the development and economic future of Penang on 10th May 2010 [Online]. http://dapmalaysia.org/english/2010/may10/lge/lge1096.htm. Accessed on June 2013

Salma NK (2001) The sustainable Penang initiative: creating state society partnership for sustainable development. IIED

Shari Z, Jaafar MFZ, Salleh E, Lim CH (2010) The potential of sustainable building rating system in the Malaysian Building Industry. Department of Architecture, Faculty of Design and Architecture, Universiti Putra Malaysia

United State Environmental Protection Agency (EPA) (2013) Green building. www.epa.gov/greenbuilding/. Accessed on 10/6/2013

Part IV
Mathematics and Statistics

Chapter 21
Secure Key Authentication Scheme Based on Discrete Logarithm and Factoring Problems

Azimah Suparlan, Asyura Abd Nassir, Nazihah Ismail, Fairuz Shohaimay and Eddie Shahril Ismail

Abstract Protecting public key from being forged or misused by enemies is very important in public key cryptosystems. Thus, key authentication process must be done to ensure that no intrusion occurred. Most researchers develop the key authentication scheme based on a single problem such as factoring, discrete logarithm, elliptic curve discrete logarithm or knapsack. Although some of these systems look secure, but due to technological advancement nowadays, it is possible that intruders can solve the single problem easily. Hence, in this paper we develop a key authentication scheme based on multiple problems; discrete logarithm and factoring problems. This research proposes a scheme with two phases; user registration phase and key authentication phase. The efficiency performance of this scheme requires $1442T_{\text{mul}} + T_h + T_{\text{qrt}}$ for user registration phase and $481T_{\text{mul}}$ for key authentication phase. Security of the scheme was proven mathematically in the security analysis and it is more secure compared to the scheme that is based on a single problem. This scheme is developed as an alternative to existing key authentication schemes and can contribute towards the development of the cryptography system based on multiple problems.

Keywords Key authentication · Discrete logarithm · Factoring

A. Suparlan (✉) · A. Abd Nassir · N. Ismail · F. Shohaimay
Faculty of Computer and Mathematical Sciences, Universiti Teknologi MARA, Jengka, Pahang, Malaysia
e-mail: azimahsuparlan@pahang.uitm.edu.my

A. Abd Nassir
e-mail: asyuraan@pahang.uitm.edu.my

N. Ismail
e-mail: nazihah@pahang.uitm.edu.my

F. Shohaimay
e-mail: fairuzshohaimay@pahang.uitm.edu.my

E.S. Ismail
School of Mathematical Sciences, Universiti Kebangsaan Malaysia, 43600 UKM, Bangi, Selangor, Malaysia
e-mail: esbi@ukm.my

N.A. Yacob et al. (eds.), *Regional Conference on Science, Technology and Social Sciences (RCSTSS 2014)*, DOI 10.1007/978-981-10-0534-3_21

1 Introduction

The most important thing in public key cryptosystem is to protect public keys from being attacked by enemies. In the past, many authentication schemes have been proposed on a single problem such as discrete logarithm. In 1996, Horng and Yang proposed a new scheme for public key cryptosystems based on discrete logarithms. The scheme is similar to the conventional certificate-based scheme that requires no authorities while most others require one or more authorities to authenticate keys. In 1999, Zhan et al. proved that Horng-Yang's scheme is not safe because an enemy can detect a user's password using the password guessing attack. For enhancing the security, many researchers such as Zhan et al. (1999), Lee et al. (2003) and Yoon and Yoo (2008) proposed an improved key authentication scheme based on the similar problem.

However, due to the technological advancement nowadays, it is possible that intruders can easily solve the authentication scheme based on single problem. Hence, this research is conducted to develop a key authentication scheme based on multiple problems; discrete logarithm and factoring problems. This scheme proposes two phases; user registration phase and key authentication phase that will be more secure because it is based on two hard problems.

The next section presents the notation and parameters used in the scheme. Then, Sect. 3 provides the explanation about the proposed scheme which consists of user registration phase and key authentication phase. In Sect. 4, the results regarding the security and efficiency of the scheme will be discussed in detail. Finally, discussion and conclusion are given in Sect. 5.

2 Notations and Parameters

The notations and parameters used to initialize in this scheme are shown in Table 1:

Table 1 Notations and parameters

Notation	Parameters and definition
P, p_1 and q_1	Three large strong primes (Gardon 1984), $n = p_1 \times q_1$
$h(\cdot)$	A cryptographic hash function (Scheiner 1996) the output is a t-bit length and for practical use, t = 128 (Ismail and Hijazi 2011)
g	A primitive element, in $\{1, 2, \ldots P-1\}$ and satisfying $g^n = 1 \bmod P$.
$f(\alpha) \equiv g^{\alpha} \bmod P$	One way function of f as a public function
K_{pub}	User's public key
K_{priv}	User's private key

3 Key Authentication Scheme

The proposed scheme has two phases: user registration phase and key authentication phase. Consider Abu as a manual system user and he will register into the system. Then, third authority, a server will store all public information of Abu. This scheme holds that the third authority, the server can be trusted. If Mimi wants to send a message to Abu, she must confirm that Abu's public key is not a forged key. Due to that, Mimi as a sender will perform calculations in key authentication phase to solve this problem. Key authentication scheme proposed is as follows:

User Registration Phase

The user registration phase of the scheme is illustrated in Fig. 1.

(i) Choose randomly a large prime number, P and two distinct prime numbers, p_1 and q_1.
(ii) Calculate

$$n = p_1 \times q_1 \quad \text{where } n|(P-1) \tag{1}$$

$$\phi(n) = (p_1 - 1)(q - 1) \tag{2}$$

(iii) Choose randomly any integer, e where $\gcd(\phi(n), e) = 1$

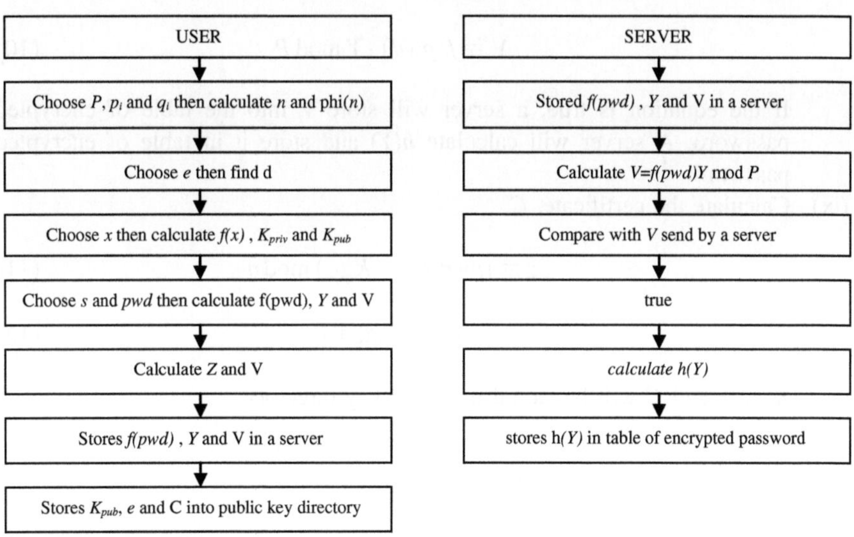

Fig. 1 User registration phase

(iv) Find d, where

$$ed \equiv 1 \bmod \phi(n) \tag{3}$$

(v) Randomly choose any integer, x where $X \in \{1, 2, \ldots, P-1\}$
(vi) Calculate

$$f(x) = g^x \bmod P \tag{4}$$

$$K_{\text{priv}} = xf(x) + d \bmod P \tag{5}$$

$$K_{\text{pub}} = g^{K^2_{\text{priv}}} \bmod P \tag{6}$$

(vii) Choose randomly any integer, s where $s \in Z^*p$ and a password, pwd where $pwd \in Z^*p$, then calculate:

$$f(pwd) \equiv g^{pwd} \bmod P \tag{7}$$

$$Y \equiv g^s \bmod P \tag{8}$$

$$\begin{aligned} V &\equiv f(pwd + s) \\ &\equiv g^{(pwd+s)} \bmod P \end{aligned} \tag{9}$$

$f(pwd)$ is the encrypted password. User will secretly send $f(pwd)$, Y and V to a server.

(viii) A server will identify either the user is a true user by verify following equation:

$$V \equiv f(pwd) \cdot Y \bmod P \tag{10}$$

If the equation is true, a server will store V into the table of encrypted password. A server will calculate $h(Y)$ and store it in table of encrypted password.

(ix) Calculate the certificate, C:

$$z \equiv (pwd + s + K^2_{\text{priv}}) \bmod n \tag{11}$$

$$C \equiv z^d \bmod n \tag{12}$$

K_{pub}, e and C will be stored in public key directory.

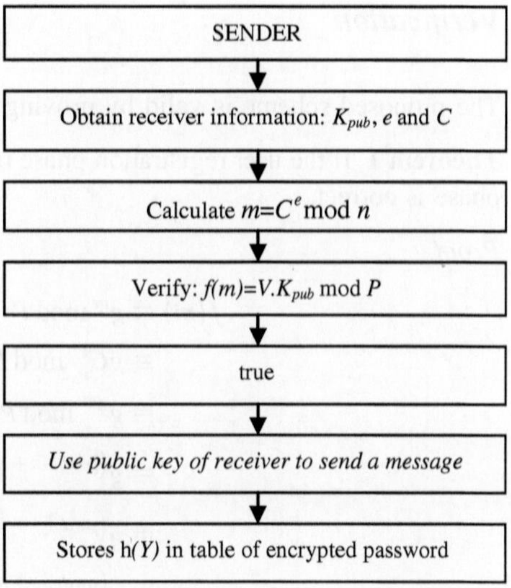

Fig. 2 Key authentication phase

Key Authentication Phase

The key authentication phase of the scheme is illustrated in Fig. 2

(i) Mimi will obtain Abu's information: K_{pub}, e and C. Then she will calculate

$$m \equiv C^e \bmod n \tag{13}$$

(ii) Mimi will verify that the public key of Abu is unaltered by validating whether the following equation is true or not:

$$f(m) \equiv V \cdot K_{\text{pub}} \bmod p \tag{14}$$

If it is valid, Mimi will use the public key to encrypt the message to be sent to Abu in cryptosystem.

4 Results

In this section, the result of the proposed key authentication scheme is discussed based on verification, security analysis and efficiency analysis.

Verification

The proposed scheme is valid by proving the following theorem:

Theorem 1 If the user registration phase runs smoothly then the key authentication phase is correct.

Proof

$$\begin{aligned} f(m) &\equiv g^{m} \bmod P \\ &\equiv gC^{c^{e}} \bmod P \\ &\equiv g^{Z^{de}} \bmod P \\ &\equiv g^{\left(pwd+s+K_{\text{priv}}^{2}\right)^{de}} \bmod P \\ &\equiv g^{\left(pwd+s+K_{\text{priv}}^{2}\right)^{\text{demod}\,\phi(n)}} \bmod P \\ &\equiv g^{\left(pwd+s+K_{\text{priv}}^{2}\right)} \bmod P \\ &\equiv g^{(pwd+s)} \cdot g^{\left(K_{\text{priv}}^{2}\right)} \bmod P \\ &\equiv V \cdot K_{\text{pub}} \bmod P \end{aligned} \quad (15)$$

Security Analysis

The proposed authentication scheme is shown to be heuristically secured by using ad hoc model. This model was also practically used in the previous literature including Zhan et al. (1999), Lee et al. (2003), Peinado (2004), Wu-Lin (2004), Sun et al. (2005), Yoon and Yoo (2005), Zhang and Kim (2005) and Shao (2005). The possible attacks by adversary (Adv) are as follows:

Factoring attack: If factoring problem can be solved by Adv, the value of d is known.

Based on Eq. (5)

$$K_{\text{priv}} = xf(x) + d \bmod P,$$

in order to obtain K_{priv}, Adv needs to know the value of x in advance. However, it is difficult to get the value of x because the discrete logarithm problem is computationally infeasible (Horng and Yang 1996).

Based on Eqs. (11) and (12)

$$\begin{aligned} z &\equiv \left(pwd+s+K_{\text{priv}}^{2}\right) \bmod n \\ C &\equiv z^{d} \bmod n, \end{aligned}$$

in order to obtain K_{priv} value, $(pwd + s)$ must be known in advance. Therefore, it is difficult to get $(pwd + s)$ because the discrete logarithm problem is hard to solve (Ismail and Hijazi 2011).

Discrete logarithm attack: If discrete logarithm can be solved by Adv, the value of x is known by referring to Eq. (4)

$$f(x) = f^x \bmod P$$

Therefore, the value of K_{priv} can be obtained if the value of d also known from Eq. (5):

$$K_{\text{priv}} = xf(x) + d \bmod P$$

However, it is difficult to get d because he needs to solve factoring problems in Eq. (3):

$$ed \equiv 1 \bmod \phi(n).$$

1. The value of K^2_{priv} can be obtained if the Eq. (6) $K_{\text{pub}} = g^{K^2_{\text{priv}}} \bmod P$ can be solved.
   However, to obtain the value of K_{priv}, he need to solve factoring problems. This is impossible since factoring problems is hard to solve (Ismail and Hijazi 2011).

Efficiency Analysis

Table 2 shows the definition of symbols that are being used while the efficiency of the scheme shown in Table 3 by calculating the computational complexity and communication cost.

From Table 2, during user registration phase and key authentication phase, time complexity needed are $1442T_{\text{mul}} + T_h + 2T_{\text{qrt}}$ and $481T_{\text{mul}}$, respectively, using the conversion $T_{\text{exp}} = 240T_{\text{mul}}$ (Koblitz et al. 2000). The communication cost of the scheme is $8|P| + 3|n|$.

Table 2 Symbol and definition

Symbol	Definition
T_{exp}	Time taken for a modular exponentiation
T_{mul}	Time taken for a modular multiplication
T_{qrt}	Time taken for a modular quadratic residue computation
T_{inv}	Time taken for a modular inverse computation
T_h	Time taken for performing a hash function

Table 3 Efficiency of the proposed key authentication scheme

	User registration phase	Key authentication phase
Computational complexity	$6T_{\exp} + 2T_{\text{mul}} + T_h + 2T_{\text{qrt}}$	$2T_{\exp} + T_{\text{mul}}$
Communication cost	$7\|P\| + 2\|n\|$	$\|P\| + 1\|n\|$

5 Conclusion

There are many key authentication schemes based on single problem. In this study, we are focusing on the key authentication scheme based on multiple problems which are discrete logarithms and factoring problems. Based on the security analysis, it shows that this scheme is protected since adversary needs to solve both problems in a single period of time. For the efficiency of this scheme, it requires $1442T_{\text{mul}} + T_h + 2T_{\text{qrt}}$ and $481T_{\text{mul}}$ for user registration phase and key authentication phase, respectively. The result shows that the communication cost of the scheme is $8|P| + 3|n|$. Although this scheme is said to be more secure, it is less efficient compared to the existing authentication scheme based on single problem. For future study, it is suggested to build a key authentication scheme based on multiple problems which provide the higher level of security and efficiency aspects.

References

Gardon J (1984) Strong RSA key. Electron Lett 20(12):514–516

Horng G, Yang CS (1996) Key authentication scheme for cryptosystems based on discrete logarithm. Comput Commun 19:848–850

Ismail ES, Hijazi MSN (2011) A new cryptosystem based on factoring and discrete logarithm problems. J Math Stat 7(3):165–168

Koblitz N, Menezes A, Vanstone S (2000) The state of elliptic curve cryptography. Des Codes Crypt 19:173–193

Lee CC, Hwang MS, Li LH (2003) A new key authentication scheme based on discrete logarithms. Appl Math Comput 139:343–349

Peinado A (2004) Cryptanalysis of LHL-key authentication scheme. Appl Math Comput 152:721–724

Scheiner B (1996) Applied cryptography: protocols, algorithms, and source code in C, 2nd edn. Wiley

Shao Z (2005) A new key authentication scheme for cryptosystems based on discrete logarithms. Appl Math Comput 167:143–152

Sun DZ, Cao ZF, Sun Y (2005) Remarks on a new key authentication scheme based on discrete logarithms. Appl Math Comput 167:572–575

Wu TS, Lin HY (2004) Robust key authentication scheme resistant to public key substitution attacks. Appl Math Comput 157:825–833

Yoon EJ, Yoo KY (2005) On the security of Wu-Lin's robust key authentication scheme. Appl Math Comput 169:1–7

Yoon EJ, Yoo KY (2008) Robust key authentication scheme. Int J Web Serv Pract 3(1–2):12–18
Zhan B, Li Z, Yang Y, Hu Z (1999) On the security of HY-key authentication scheme. Comput Commun 22:739–741
Zhang F, Kim K (2005) Cryptanalysis of Lee-Hwang-Li's key authentication scheme. Appl Math Comput 161:101–107

Chapter 22
Jaccard Ranking Index with Algebraic Product *t*-Norm Based on Second Function Principle in Handling Fuzzy Risk Analysis Problem

Nazirah Ramli, Norhuda Mohammed and Fairuz Shohaimay

Abstract Jaccard set theoretic index based on the first function principle approach has been proposed for ranking fuzzy numbers. However, the arithmetic operations of the first function principle is not straightforward which consists of determining both the lower and upper limit of the fuzzy numbers and also the minimum and maximum values of the fuzzy numbers' domain. In this paper, a simple point-wise arithmetic operation, namely the second function principle is applied in developing the Jaccard ranking index with algebraic product *t*-norm. Based on the proposed ranking index, a fuzzy risk analysis (FRA) is presented in dealing with FRA problem. The behavior of the proposed risk's ranking order is investigated and compared with some of the previous studies.

Keywords Algebraic product *t*-norm · Fuzzy risk · Jaccard index · Second function principle

1 Introduction

A number of studies related to fuzzy risk analysis (FRA) problem have been proposed such as by Chen and Chen (2007), Chen and Wang (2009), Chen and Sanguansat (2011) and Ramli et al. (2013). In order to deal with FRA, fuzzy numbers (FNs) are used to represent the evaluating values of the risk of each subcomponent. The procedure for solving FRA problem by the above-mentioned

N. Ramli (✉) · N. Mohammed · F. Shohaimay
Department of Mathematics, Faculty of Computer & Mathematical Sciences,
Universiti Teknologi MARA, Jengka, Pahang, Malaysia
e-mail: nazirahr@pahang.uitm.edu.my

N. Mohammed
e-mail: hudamohammed@pahang.uitm.edu.my

F. Shohaimay
e-mail: fairuzshohaimay@pahang.uitm.edu.my

N.A. Yacob et al. (eds.), *Regional Conference on Science, Technology and Social Sciences (RCSTSS 2014)*, DOI 10.1007/978-981-10-0534-3_22

studies consists of two main components which are calculation on probability of failure and determining the ranking order of the risk of manufacturer. Chen and Chen (2007) proposed a method for FRA based on ranking fuzzy numbers (RFNs) using the centroid point and standard deviation concepts. Chen and Wang (2009) used the α-cuts, belief features, and noise ratios concepts for RFNs in solving FRA. Further, Chen and Sanguansat (2011) presented an approach for RFNs based on area and height of the FNs. Recently, Ramli et al. (2013) proposed a ranking method based on Ochiai similarity measure index and Hurwicz criterion in dealing with FRA.

The concept of RFNs introduced by Jain (1976) plays an important role in decision-making. To date, research in RFNs is still very active and new research directions are continuing to emerge. Allahviranloo and Saneifard (2012) proposed a RFNs method based on centroid and maximum crisp of FNs concept. Their method improved some of the unreasonable ranking results by Cheng's (1998) and Chu and Tsao's (2002) methods. The concept of similarity measure index is also proposed for ranking fuzzy numbers where Setnes and Cross (1997) first presented a RFNs method based on Jaccard similarity measure index and extension principle concept (EP). However, the conventional arithmetic operation EP is only applicable to normal FNs and thus the RFNs methods by Setnes and Cross (1997) fails to rank the non-normal FNs. The EP will also change the type of membership functions and require complex and laborious operations. Further, Ramli and Mohamad (2010) proposed Jaccard ranking index with degree of optimism concept using the first function principle (FFP) arithmetic operation from Chen (1985). Their method can rank both normal and non-normal FNs. However, the arithmetic operations of the FFP is not straightforward which consists of determining both the lower and upper limit of the fuzzy numbers and also the minimum and maximum values of the fuzzy numbers' domain.

In this paper, a simple point-wise arithmetic operation, namely the second function principle (SFP) is applied in developing the Jaccard ranking index with algebraic product t-norm. The proposed RFNs method is then applied in dealing with FRA problems. This paper is organized as follows. In Sect. 2, basic concepts on FNs are presented. Section 3 presents the proposed Jaccard ranking index with algebraic product t-norm using SFP. In Sect. 4, the proposed ranking index is applied to the FRA problems. Lastly, the paper is concluded in Sect. 5.

2 Preliminaries

This section reviews some basic concepts of fuzzy numbers (Cheng 1998) and their arithmetic operation using SFP from Chen and Hsieh (1998).

Definition 1 (Cheng 1998)

A fuzzy number A is a fuzzy set in the universe of discourse $X = \Re$ that is both convex and normal, has bounded support and all α-cuts of A are closed intervals of X. The membership function of a fuzzy number A can be defined as

$$\mu_A(x) = \begin{cases} \mu_A^L(x), & a \le x \le b \\ 1, & b \le x \le c \\ \mu_A^R(x), & c \le x \le d \\ 0, & \text{otherwise} \end{cases},$$

where $\mu_A^L : [a, b] \to [0, 1]$, $\mu_A^R : [c, d] \to [0, 1]$, μ_A^L and μ_A^R denote the left and the right membership functions of the fuzzy number A respectively, and, a, b, c, and d denote the lower limit, lower modal, upper modal, and upper limit of the fuzzy number A, respectively.

A trapezoidal fuzzy number (TrFN) denoted as $A = (a, b, c, d)$ has a membership function defined as

$$\mu_A(x) = \begin{cases} \frac{x-a}{b-a}, & a \le x \le b \\ 1, & b \le x \le c \\ \frac{d-x}{d-c}, & c \le x \le d \\ 0, & \text{otherwise} \end{cases}.$$

For $b = c$, A becomes a triangular fuzzy number (TFN) denoted as $A = (a, b, d)$.

Definition 2 (Chen and Hsieh 1998)

Let g be an arithmetical mapping from n-dimensional real number $\Re^n$ into real line $\Re$ and f_g is a corresponding mapping from n-dimensional generalized trapezoidal fuzzy numbers into generalized trapezoidal fuzzy number. Suppose that $\tilde{A}_i = (a_i, b_i, c_i, d_i; h_i)$, $i = 1, 2, \ldots, n$ are n generalized trapezoidal fuzzy numbers. The generalized trapezoidal fuzzy number B on $\Re$ induced from these generalized trapezoidal fuzzy numbers $\tilde{A}_i$ through function f_g is $f_g(\tilde{A}_1, \tilde{A}_2, \ldots, \tilde{A}_n) = B = (a, b, c, d; h)$,

where

$$h = \min\{h_1, h_2, \ldots, h_n\}, T = \{g(x_1, x_2, \ldots, x_n) | x_i = a_i \quad \text{or} \quad d_i, i = 1, 2, \ldots, n\},$$
$$T_1 = \{g(x_1, x_2, \ldots, x_n) | x_i = b_i \quad \text{or} \quad c_i, i = 1, 2, \ldots, n\},$$
$$a = \min T, b = \min T_1, c = \max T_1,$$
$$d = \max T, \min T \le \min T_1 \text{ and } \max T_1 \le \max T.$$

3 Proposed Jaccard Ranking Index with Algebraic Product *t*-Norm Based on Second Function Principle

The procedure for fuzzy Jaccard ranking index with algebraic product *t*-norm using SFP is given as follows:

Step 1: Find the fuzzy maximum (MAX) and fuzzy minimum (MIN) of two TFNs $A_i = (a_i, b_i, c_i, d_i; h_i)$ and $A_j = \left(a_j, b_j, c_j, d_j; h_j\right)$ using the SFP concept as follows:

The fuzzy maximum of A_i and A_j is defined as

$$\mathrm{MAX}\left(A_i, A_j\right) = (a, b, c, d; h),$$

where

$$\begin{aligned} h &= \min\left\{h_i, h_j\right\},\ T = \left\{\max\left(a_i, a_j\right), \max\left(a_i, d_j\right), \max\left(d_i, a_j\right), \max\left(d_i, d_j\right)\right\}, \\ T_1 &= \left\{\max\left(b_i, b_j\right), \max\left(b_i, c_j\right), \max\left(c_i, b_j\right), \max\left(c_i, c_j\right)\right\}, \\ a &= \min T, b = \min T_1, c = \max T_1, d = \max T, \\ \min T &\leq \min T_1 \text{ and } \max T_1 \leq \max T. \end{aligned}$$

The fuzzy minimum of A_i and A_j is defined as

$$\mathrm{MIN}\left(A_i, A_j\right) = (a, b, c, d; h),$$

where

$$\begin{aligned} h &= \min\left\{h_i, h_j\right\},\ T = \left\{\min\left(a_i, a_j\right), \min\left(a_i, d_j\right), \min\left(d_i, a_j\right), \min\left(d_i, d_j\right)\right\}, \\ T_1 &= \left\{\min\left(b_i, b_j\right), \min\left(b_i, c_j\right), \min\left(c_i, b_j\right), \min\left(c_i, c_j\right)\right\}, \\ a &= \min T,\ b = \min T_1,\ c = \max T_1,\ d = \max T_1, \\ \min T &\leq \min T_1 \text{ and } \max T_1 \leq \max T. \end{aligned}$$

Step 2: Calculate the evidences of $E\left(A_i \succ A_j\right)$, $E\left(A_j \prec A_i\right)$, $E\left(A_j \succ A_i\right)$ and $E\left(A_i \prec A_j\right)$ which are defined based on fuzzy Jaccard index as

$$\begin{aligned} E\left(A_i \succ A_j\right) &= S_J\left(\mathrm{MAX}\left(A_i, A_j\right), A_i\right), \\ E\left(A_j \prec A_i\right) &= S_J\left(\mathrm{MIN}\left(A_i, A_j\right), A_j\right), \\ E\left(A_j \succ A_i\right) &= S_J\left(\mathrm{MAX}\left(A_i, A_j\right), A_j\right) \text{ and} \\ E\left(A_i \prec A_j\right) &= S_J\left(\mathrm{MIN}\left(A_i, A_j\right), A_i\right) \end{aligned}$$

with $S_J(A_i, A_j) = \frac{|A_i \cap A_j|}{|A_i \cup A_j|}$ is the fuzzy Jaccard index and $|A_i|$ denotes the scalar cardinality of fuzzy number A_i. To simplify, C_{ij} and c_{ji} are used to represent $E\left(A_i \succ A_j\right)$

and $E(A_j \prec A_i)$, respectively. Likewise, C_{ji} and c_{ij} are used to denote $E(A_j \succ A_i)$ and $E(A_i \prec A_j)$, respectively.

Step 3: Calculate the total evidences $E_{\text{total}}(A_i \succ A_j)$ and $E_{\text{total}}(A_j \succ A_i)$ which are defined based on the aggregation of evidences with algebraic product *t*-norm as $E_{\text{total}}(A_i \succ A_j) = C_{ij} \cdot c_{ji}$ and $E_{\text{total}}(A_j \succ A_i) = C_{ji} \cdot c_{ij}$. To simplify, $E_J(A_i, A_j)$ and $E_J(A_j, A_i)$ are used to represent $E_{\text{total}}(A_i \succ A_j)$ and $E_{\text{total}}(A_j \succ A_i)$, respectively.

Step 4: For each pair of the fuzzy numbers, compare the total evidences in Step 3 which will result the ranking of the two fuzzy numbers A_i and A_j as follows:

i. $A_i \succ A_j$ if and only if $E_J(A_i, A_j) > E_J(A_j, A_i)$.
ii. $A_i \prec A_j$ if and only if $E_J(A_i, A_j) < E_J(A_j, A_i)$.
iii. $A_i \approx A_j$ if and only if $E_J(A_i, A_j) = E_J(A_j, A_i)$.

4 Fuzzy Risk Analysis Using Jaccard Ranking Index with Algebraic Product *t*-Norm

In this section, the proposed ranking method is applied in dealing with FRA problem. Assume A_1, A_2, …, and A_n are n manufacturers with each manufacturer A_i produces component S_i. The component S_i consists of p subcomponents S_{i1}, S_{i2}, …, and S_{ip}, where $1 \le i \le n$. Two evaluating items $\tilde{R}_{ik}$ and $\tilde{W}_{ik}$ are used to evaluate each subcomponent S_{ik}, where $\tilde{R}_{ik}$ denotes the probability of failure of the subcomponent S_{ik} and $\tilde{W}_{ik}$ denotes the severity of loss of the subcomponent S_{ik} with $1 \le k \le p$ and $1 \le i \le n$.

The proposed FRA algorithm is presented as follows:

Phase 1: Calculate the probability of failure $\tilde{R}_i$ of each component S_i produced by manufacturer A_i by aggregating the evaluating values $\tilde{R}_{ik}$ and $\tilde{W}_{ik}$ as follows:

$$\tilde{R}_i = \frac{\sum_{k=1}^{p} \tilde{R}_{ik} \otimes \tilde{W}_{ik}}{\sum_{k=1}^{p} \tilde{W}_{ik}} \tag{1}$$

where $\tilde{R}_i = (r_{i1}, r_{i2}, r_{i3}, r_{i4}; w_{\tilde{R}_i})$ and $1 \le i \le n$.

Phase 2: Transform $\tilde{R}_i = (r_{i1}, r_{i2}, r_{i3}, r_{i4}; w_{\tilde{R}_i})$ into standard trapezoidal fuzzy numbers

$$\tilde{R}_i^* = \left(\frac{r_{i1}}{k}, \frac{r_{i2}}{k}, \frac{r_{i3}}{k}, \frac{r_{i4}}{k}; w_{\tilde{R}_i}\right) \tag{2}$$

where $k = \max\{||r_{ij}||\}$, $1 \le i \le n$, j = 1, 2, 3, 4 and $||r_{ij}||$ represents the ceiling function of $|r_{ij}|$.

Phase 3: By using the proposed fuzzy ranking index, determine the ranking of $\tilde{R}_i^*$ and $\tilde{R}_j^*$.

To illustrate the fuzzy risk analysis process of the proposed method, we use data from Chen and Chen (2007). Assume A_1, A_2 and A_3 are three manufacturers which produce components S_1, S_2 and S_3, respectively. Component S_i, produced by manufacturer A_i, consists of three subcomponents S_{i1}, S_{i2} and S_{i3}, where $1 \leq i \leq 3$. Two evaluating items $\tilde{R}_{ik}$ and $\tilde{W}_{ik}$ are used to evaluate each subcomponent S_{ik}, where $\tilde{R}_{ik}$ denotes the probability of failure of the subcomponent S_{ik} and $\tilde{W}_{ik}$ denotes the severity of loss of the subcomponent S_{ik} with $1 \leq k \leq 3$.

The structure of FRA is shown in Fig. 1.

Table 1 shows a nine-member linguistic term which is used for representing the linguistic terms and their corresponding generalized fuzzy numbers (GFNs).

Table 2 shows the linguistic values of evaluating items $\tilde{R}_{ik}$ and $\tilde{W}_{ik}$ of the subcomponent S_{ik} produced by manufacturer A_i. $\tilde{w}_{\tilde{R}_{ik}}$ denotes the degree of confidence of the decision-makers' perspective with respect to subcomponents S_{ik} produced by manufacturer A_i for $1 \leq i \leq 3$ and $1 \leq k \leq 3$.

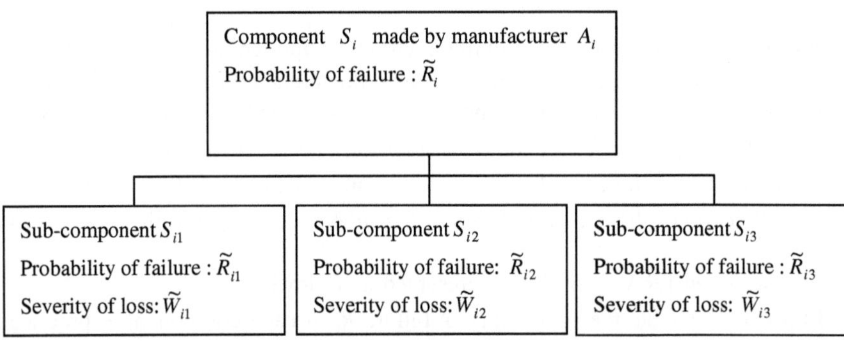

Fig. 1 Structure of fuzzy risk analysis (Chen and Chen 2007)

Table 1 Linguistic terms and their corresponding generalized fuzzy numbers (Chen and Chen 2009)

Linguistic terms	GFNs
Absolutely low	(0.0, 0.0, 0.0, 0.0; 1.0)
Very low	(0.0, 0.0, 0.02, 0.07; 1.0)
Low	(0.04, 0.1, 0.18, 0.23; 1.0)
Fairly low	(0.17, 0.22, 0.36, 0.42; 1.0)
Medium	(0.32, 0.41, 0.58, 0.65; 1.0)
Fairly high	(0.58, 0.63, 0.80, 0.86; 1.0)
High	(0.72, 0.78, 0.92, 0.97; 1.0)
Very high	(0.93, 0.98, 1.0, 1.0; 1.0)
Absolutely high	(1.0, 1.0, 1.0, 1.0; 1.0)

Table 2 Linguistic evaluating values of the subcomponents produced by manufacturers A_1, A_2 and A_3 (Chen and Chen 2009)

Manufacturer	Subcomponents	Linguistic values of the severity of loss $\tilde{W}_{ik}$	Linguistic values of the probability of failure $\tilde{R}_{ik}$
A_1	S_{11}	$\tilde{W}_{11}$ = Low	$\tilde{R}_{11}$ = Fairly Low $(w_{\tilde{R}_{11}} = 0.9)$
	S_{12}	$\tilde{W}_{12}$ = Fairly High	$\tilde{R}_{12}$ = Medium $(w_{\tilde{R}_{12}} = 0.7)$
	S_{13}	$\tilde{W}_{13}$ = Very Low	$\tilde{R}_{13}$ = Fairly High $(w_{\tilde{R}_{13}} = 0.8)$
A_2	S_{21}	$\tilde{W}_{21}$ = Low	$\tilde{R}_{21}$ = Very High $(w_{\tilde{R}_{21}} = 0.85)$
	S_{22}	$\tilde{W}_{22}$ = Fairly High	$\tilde{R}_{22}$ = Fairly High $(w_{\tilde{R}_{22}} = 0.9)$
	S_{23}	$\tilde{W}_{23}$ = Very Low	$\tilde{R}_{23}$ = Medium $(w_{\tilde{R}_{23}} = 0.9)$
A_3	S_{31}	$\tilde{W}_{31}$ = Low	$\tilde{R}_{31}$ = Fairly Low $(w_{\tilde{R}_{31}} = 0.95)$
	S_{32}	$\tilde{W}_{32}$ = Fairly High	$\tilde{R}_{32}$ = High $(w_{\tilde{R}_{32}} = 0.8)$
	S_{33}	$\tilde{W}_{33}$ = Very Low	$\tilde{R}_{33}$ = Fairly High $(w_{\tilde{R}_{33}} = 1.0)$

In the following, the proposed method is applied in dealing with the FRA problems.

Phase 1: Based on Eq. (1), the calculation for probability of failure $\tilde{R}_1$ of component S_1 produced by manufacturer A_1 is shown as follows:

$$\begin{aligned}\tilde{R}_1 &= \left[\tilde{R}_{11} \otimes \tilde{W}_{11} \oplus \tilde{R}_{12} \otimes \tilde{W}_{12} \oplus \tilde{R}_{13} \otimes \tilde{W}_{13}\right] ø \left[\tilde{W}_{11} \oplus \tilde{W}_{12} \oplus \tilde{W}_{13}\right],\\ &= (0.1659, 0.2803, 0.7463, 1.1545; 0.7).\end{aligned}$$

In similar manner, the probabilities of failures $\tilde{R}_2$ and $\tilde{R}_3$ of each component S_2 and S_3, produced by manufacturers A_2 and A_3, respectively, are obtained as follows:

$$\begin{aligned}\tilde{R}_2 &= (0.3221, 0.4949, 1.1392, 1.6373; 0.85), \text{ and}\\ \tilde{R}_3 &= (0.3659, 0.5134, 1.1189, 1.5984; 0.8).\end{aligned}$$

Phase 2: Based on Eq. (2), the standard trapezoidal fuzzy numbers $\tilde{R}_i^*$ are given as follows:

$$\begin{aligned}\tilde{R}_1^* &= (0.0830, 0.1402, 0.3732, 0.5773; 0.7),\\ \tilde{R}_2^* &= (0.1611, 0.2475, 0.5696, 0.8187; 0.85), \text{ and}\\ \tilde{R}_3^* &= (0.1830, 0.2567, 0.5595, 0.7992; 0.8).\end{aligned}$$

Phase 3: Based on the proposed fuzzy ranking index, the values of evidence are obtained as follows:

$$(C_{12}, c_{21}, C_{21}, c_{12}) = (0.4651, 0.4153, 0.8235, 1),$$
$$(C_{13}, c_{31}, C_{31}, c_{13}) = (0.4498, 0.4165, 0.875, 1),$$
$$(C_{23}, c_{32}, C_{32}, C_{23}) = (0.9113, 0.9673, 0.9688, 0.9127).$$

Thus, obtained the total evidences and pair wise ranking results as follows:

$$E_J(\tilde{R}_1^*, \tilde{R}_2^*) = 0.1932, \quad E_J(\tilde{R}_2^*, \tilde{R}_1^*) = 0.8235, \quad \therefore \tilde{R}_1^* \prec \tilde{R}_2^*,$$
$$E_J(\tilde{R}_1^*, \tilde{R}_3^*) = 0.1873, \quad E_J(\tilde{R}_3^*, \tilde{R}_1^*) = 0.8750, \quad \therefore \tilde{R}_1^* \prec \tilde{R}_3^*,$$
$$E_J(\tilde{R}_2^*, \tilde{R}_3^*) = 0.8815, \quad E_J(\tilde{R}_3^*, \tilde{R}_2^*) = 0.8842, \quad \therefore \tilde{R}_2^* \prec \tilde{R}_3^*,$$

The final ranking order for $\tilde{R}_1^*$, $\tilde{R}_2^*$ and $\tilde{R}_3^*$ is $\tilde{R}_1^* \prec \tilde{R}_2^* \prec \tilde{R}_3^*$.

Thus, the ranking of order of the risk for manufacturer A_1, A_2 and A_3 is "$A_1 \prec A_2 \prec A_3$".

5 Discussion and Conclusion

This paper proposed RFNs method based on Jaccard similarity measure index with algebraic product *t*-norm. The SFP is applied in determining the fuzzy maximum and minimum of the proposed method. The calculation using SFP is based on the point-wise operation and has straightforward calculation when compared with FFP. The proposed RFNs method is applied in FRA procedure for handling FRA problems. The ranking of risk of the manufacturer is "$A_1 \prec A_2 \prec A_3$" which is consistent with neutral and optimistic decision-makers by Ramli et al. (2013). This shows that the proposed method using Jaccard index has consistent result with Ochiai index which is also a similarity measure index. However, the result is not consistent with other methods such as Chen and Chen (2007), Chen and Wang (2009) and Chen and Sanguansat (2011) which produce "$A_1 \prec A_3 \prec A_2$".

References

Allahviranloo T, Saneifard R (2012) Defuzzification method for ranking fuzzy numbers based on center of gravity. Iran J Fuzzy Syst 9(6):57–67

Chen SH (1985) Operations on fuzzy numbers with function principle. Tamkang J Manage Sci 6(1):13–26

Chen SJ, Chen SM (2007) Fuzzy risk analysis based on the ranking of generalized trapezoidal fuzzy numbers. Appl Intell 26(1):1–11

Chen SM, Chen JH (2009) Fuzzy risk analysis based on ranking generalized fuzzy numbers with different heights and different spreads. Expert Syst Appl 36:6833–6842

Chen SH, Hsieh CH (1998) Graded mean representation of generalized fuzzy numbers. In: Proceedings of the sixth conference on fuzzy theory and its applications, pp 1–5. Taiwan, Republic of China

Chen SM, Sanguansat K (2011) Analyzing fuzzy risk analysis based on a new fuzzy ranking method between generalized fuzzy numbers. Expert Syst Appl 38:2163–2171

Chen SM, Wang CH (2009) Fuzzy risk analysis based on ranking fuzzy numbers using α-cuts, belief features and signal/ noise ratios. Expert Syst Appl 36:5576–5581

Cheng CH (1998) A new approach for ranking fuzzy numbers by distance method. Fuzzy Sets Syst 95:307–317

Chu TC, Tsao CT (2002) Ranking fuzzy numbers with an area between the centroid point and original point. Comput Math Appl 43:111–117

Jain R (1976) Decision making in the presence of fuzzy variables. IEEE Trans Syst Man Cybern 6:698–703

Ramli N, Mohamad D (2010) Fuzzy Jaccard with degree of optimism ranking index based on function principle approach. Majlesi J Electr Eng 4(4):9–15

Ramli N, Mohamad D, Ramli R (2013) Fuzzy risk analysis based on Ochiai ranking index with Hurwicz criterion for generalized trapezoidal fuzzy numbers. Appl Math Sci 7(134): 6669–6682

Setnes M, Cross V (1997) Compatibility based ranking of fuzzy numbers. In: Proceeding of fuzzy information processing society (NAFIPS 1997), pp 305–310. Syracuse, New York

Chapter 23
Analysis of Forensic Ballistic Specimens for Firearm Identification Using Supervised Naive Bayes and Decision Tree Classification Technique

Muhamad Hasbullah Mohd Razali and Balkiah Moktar

Abstract Every crime gun has a story to tell. Recently, the rise of crime cases involving firearms is quite alarming. In ballistics studies, the cartridge cases left by gunshots are important evidences since they act as "fingerprint" to the firearms used. In this study, the secondary data of Legendre Orthogonal Moment (LOM) for the firing pin impression (FPI) images captured from a total of 747 cartridge cases are used. The bullets were shot from five different pistols of the same model namely, type 9 mm Parabellum Vektor SP1, made in South Africa. The method of moment is used in this study since it is more accurate, robust, and less time consuming as compared to the traditional method via comparison microscope. Preliminary analysis using Pearson correlation shows that the features are highly correlated. Therefore, principal component analysis (PCA) was used to analyze the interrelationship among the features and combine them into eight significant component of features while maintaining the information of the original patterns. PCA has reduced the dimensionality of the features. The classification techniques used are naïve Bayes and decision tree analysis on J48 algorithm using the WEKA software due to the simplicity, ability for high dimensionality inputs, and outperformance of these methods than the other more sophisticated classification techniques. Classification results using the naïve Bayes and decision tree techniques show that about 85 and 78.4 % of the images were correctly classified respectively. Hence, the results demonstrate the potential of using numerical features generated by moment of FPI images for firearms identification.

Keywords Decision tree · Firearms identification · Firing pin impression · Geometric moment · Naïve Bayes · Statistical features

M.H. Mohd Razali (✉) · B. Moktar
Faculty of Computer and Mathematical Sciences, Universiti Teknologi MARA, Arau, Perlis, Malaysia
e-mail: hasbullah782@perlis.uitm.edu.my

B. Moktar
e-mail: balkiah@perlis.uitm.edu.my

N.A. Yacob et al. (eds.), *Regional Conference on Science, Technology and Social Sciences (RCSTSS 2014)*, DOI 10.1007/978-981-10-0534-3_23

1 Introduction

The number of crimes involving firearms are increasing and the identification of firearms used by criminals is a crucial evidence in the court (Chuan et al. 2013). When there are crimes involving gun, police will make inquiries to find any evidence at the crime scene. The criminals tend to flee off the crime scene along with the gun used. The only evidence left at the crime scene is the bullet case; if there are any. Bullet case is an important evidence for the identification of the gun used. It is found that cartridge cases can be used as evidence and they are taken to the laboratory for investigation and identification of particular firearms. The identification aims to identify the model and type of gun used by bullet case found. The characteristic indicated on the cartridge and projectile of the bullet fired from a gun can be recognized as a “fingerprint” for identification of the firearm (Smith 2001).

Basically, forensic team will face the problem of not having enough time to retrieve sufficient evidence and information on the particular firearm used. The traditional technique requires a long time for the classification of firearms. Recently, the crimes are becoming more common and frequent. So, the result of the classification of firearms is needed on an urgent basis. This method is also called as a comparison microscope technique that makes a difference through the naked eye. Using a specific microscope, comparisons of the cartridge case images are made. The comparison microscope has two different sides which projected the evidence and reference bullet case images side by side for comparison. It requires the experts who could classify it accordingly based on their experience. Meanwhile, it takes a longer time and it will lead to inaccurate results. Therefore, the significance of the classification rate is quite low as if the process is automated using the generated image moments. Using method of moment, there is a potential for the classification rate to be improved since the human intervention has been minimised by the automated procedure.

According to Thumwarin (2008), firearms identification method is based on rotation invariant feature of cartridge case image represented by absolute value. Ghani et al. (2012) used geometric moment and principal component analysis (PCA). The study also applied discriminant analysis (DA) to identify the types of pistol based on center of firing pin impression (FPI). These methods are also emphasized by Liong et al. (2012), who used PCA in their previous research with principal component analysis (PCA) to determine the factors that influence a set of measure and the strength of relationship between each factor. According to Khalid (2007), he proposed the Legendre Moment to compute the image reconstruction for gray level images. The method of moment is also applied by Anath and Bharathi (2012) based on their issues report. They responded to this situation using the moment method in face image retrieval.

This study focuses on the geometric moment as its basis in generating the numerical features of firing pin impression images. The moment will extract the numerical features that represents the pattern captured in the image. This paper intended to apply and compare the performance of popular classification technique in machine learning which are naïve Bayes and decision tree (J48) algorithm using

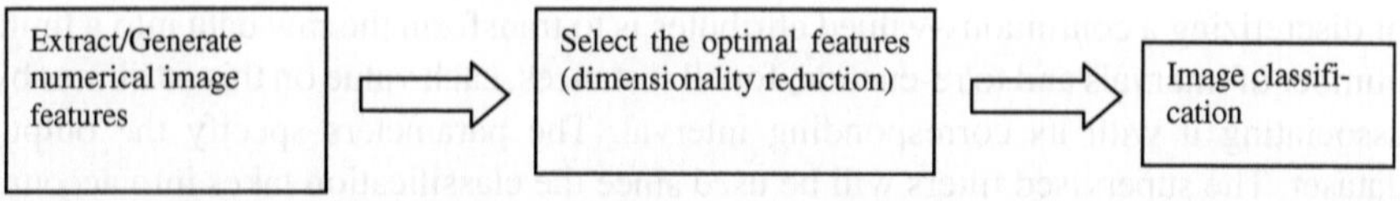

Fig. 1 Theoretical framework of the study

Weka software. The performance of each classifier is compared using cross validation to see their respective prediction accuracy. The theoretical framework of this study is illustrated in Fig. 1.

2 Methodology

Before identifying the five different types of pistols, initially, the Pearson Correlation Coefficients of the features will be examined first to investigate the relationship behavior. Unreliable findings will occur when multicollinearity and singularity exist in the data of study. When two or more features that are found to be highly correlated with each other ($r = 0.90$ or more), multicollinearity exist. The 47 features are then combined into meaningful components under the factor analysis using the principal component method as the way to overcome the multicollinearity problem.

In conducting the factor analysis, all the features are assumed to be normally distributed and gained small number of higher correlations between the variables. Initially, the Kaiser-Meyer-Olkin and the Bartlett's test will be performed to measure the appropriateness of using the factor analysis towards the data. Kaiser recommends that KMO values above 0.5 can be considered acceptable. As for the Bartlett's test, it is necessary to reject the null hypothesis in which the original correlation matrix is the identity matrix. Significant p-value, which is less than the significance level of 0.05, will be observed (Field 2005). In determining the number of component after features reduction, the use of scree plot will be employed. Factor rotation will be applied to uncover more meaningful pattern of item factor loadings or orthogonal transformation of factors loading. This paper used Varimax rotation method since PCA's components are assumed to be uncorrelated.

All the components (factors) obtained in the factor analysis will then be used for classification using naïve Bayes' and decision tree analysis on J48 algorithm using the Weka software. Whichever method occupies the highest classification rate in performance would be considered as the best method to classify the pistol accordingly. The value of Kappa statistic also will be observed. The closer the value of Kappa statistic to one indicates the higher accuracy of the classifier. Kappa statistic is used to assess the accuracy of measurement in firearm cases, and it is use to distinguish the reliability of the data collected and image of firing pin validity. Kappa result varies between 0–1 intervals. First, the data need to be discretized based on the attributes as a preliminary condition of machine learning process. The main purpose of discretizing a continuous-valued attributes is to transform the raw data into a finite number of intervals and to re-encode, for all instances, each value on this attributes by associating it with its corresponding interval. The parameters specify the output dataset. The supervised filters will be used since the classification takes into account of the instance labels of the class attributes; i.e., Pistol A, B, C, D, E.

Data Background

This study used the secondary data from previous studies by Ghani et al. (2012) initiatives. There are 47 features of firing pin impression (747 samples for each features) which consist of 20 basic statistical features and 27 Legendre Orthogonal Moment (LOM) features which are extracted from the cartridge case images. The 20 basic statistical numerical features were generated from each part of the images as illustrated in Fig. 2 and the 27 significantly different features of Legendre Orthogonal Moment features were obtained from extracted images of the three main part of firing pin impression.

All of the features are already been transformed into normalized values between 0–1. Normalizing the data attempts to give all features an equal weight in order to avoid dependency on the choice of measurement units. Therefore, regards to the objective of study, 47 features of basic statistical and LOM features will be the independent variables. These numerical features comprised of 150 samples from pistol A, B, and C, 149 samples of pistol D and 148 samples from pistol E, accounted a total of 747 samples for each of the features taken from FPI images of the Parabellum Vector SPI 9 mm model from Africa. Thus, the types of pistols used, A, B, C, D, and E, as the dependent variable. All of these 47 respective features of the basic statistical and LOM features are summarized in Tables 1 and 2. The analyses will be carried out using the Statistical Packages for Social Sciences (SPSS) and Weka software.

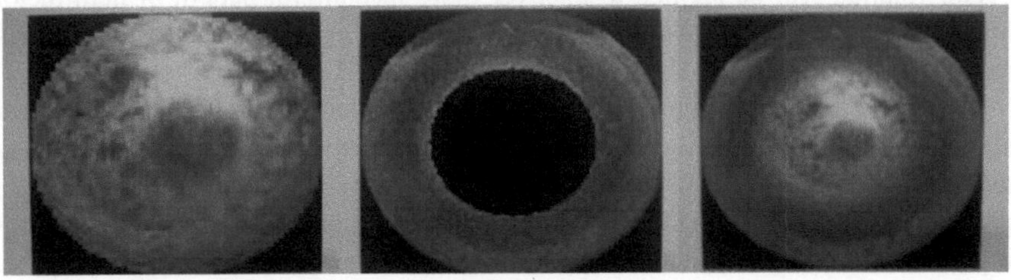

Fig. 2 Extracted region of the firing pin impression image (*whole*, *ring* and *center*)

Table 1 List of basic statistical features

No.	Basic statistical features	Code	No.	Basic statistical features	Code
1	Mean whole image	Mean_W	11	Third quartile center image	Q3_C
2	Median whole image	Med_W	12	Variance center image	Var_C
3	First quartile whole image	Q1_W	13	Skewness center image	Skew_C
4	Third quartile whole image	Q3_W	14	Kurtosis center image	Kur_C
5	Variance whole image	Var_W	15	Mean ring image	Mean_R
6	Skewness whole image	Skew_W	16	Median ring image	Med_R
7	Kurtosis whole image	Kur_W	17	Third quartile ring image	Q3_R
8	Mean center image	Mean_C	18	Variance ring image	Var_R
9	Median center image	Med_C	19	Skewness ring image	Skew_R
10	First quartile center image	Q1_C	20	Kurtosis ring image	Kur_R

Table 2 List of Legendre orthogonal moment features

No.	Legendre orthogonal moment features	Code	No.	Legendre orthogonal moment features	Code
1	Whole image 00	LW00	15	Center image 02	LC02
2	Whole image 01	LW01	16	Center image 12	LC12
3	Whole image 10	LW10	17	Center image 21	LC21
4	Whole image 11	LW11	18	Center image 22	LC22
5	Whole image 20	LW20	19	Ring image 00	LR00
6	Whole image 02	LW02	20	Ring image 01	LR01
7	Whole image 12	LW12	21	Ring image 10	LR10
8	Whole image 21	LW21	22	Ring image 11	LR11
9	Whole image 22	LW22	23	Ring image 20	LR20
10	Center image 00	LC00	24	Ring image 02	LR02
11	Center image 01	LC01	25	Ring image 12	LR12
12	Center image 10	LC10	26	Ring image 21	LR21
13	Center image 11	LC11	27	Ring image 22	LR22
14	Center image 20	LC20	–	–	–

3 Results and Discussion

Result of Pearson correlation coefficients between all pairs of variables were significant and highly correlated which give indication of multicollinearity problem. The features reduction process were then taking place to overcome this problem. The features were tested using the Kaiser-Meyer-Olkin measure of sampling adequacy and Bartlett's test of sphericity. Table 3 shows that, the KMO statistics is 0.671 which falls in the range of good items. Thus, it indicates that the correlations of the features were relatively densed and factor analysis further should yield distinct and reliable component factors. The Bartlett's test is also significant at 5 %

Table 3 The Kaiser-Meyer-Olkin measure of sampling adequacy and Bartlett's test of sphericity

KMO measure of sampling adequacy		0.671
Bartlett's test of sphericity	Approx. chi-square	1.359E5
	df	1081
	Sig.	0.000

level of significancy with p-value equals to 0.000, this demonstrate that the R-square matrix is not an identity matrix and there were some relationship between the items. Therefore, it can be said that factor analysis is appropriate to be performed to the respected features.

The number of components were determined using a scree plot and justification is made by observing the eigenvalues that are greater than one. The scree plot is shown in Fig. 3, the point of inflexion of the curve is known as the elbow of the curve that can be seen at component number eight. After this point, the remaining eigenvalues are relatively small and about the same sizes. Graphically, this study justifies that eight to nine factors that should be retained for further analysis. Checking on associated eigenvalues shows that their values are less than one after the eight component, conforming that only eight components will be used for further analysis. The eigenvalues associated with each factor represent the variance explained by the particular linear component and displays the eigenvalue in terms of the percentage of variance explained. Factor one explains 26.35 % of the total variance, while factor two and factor three explain 20.75 and 11.85 % of total variance, respectively. Factor four up to the eighth factor explain 9.16, 7.75, 6.9, 5.5, and 3.7 % of the total variance accordingly.

Varimax rotation were performed to identify which original variables load into which component. The factor loadings in the component matrix before rotation are

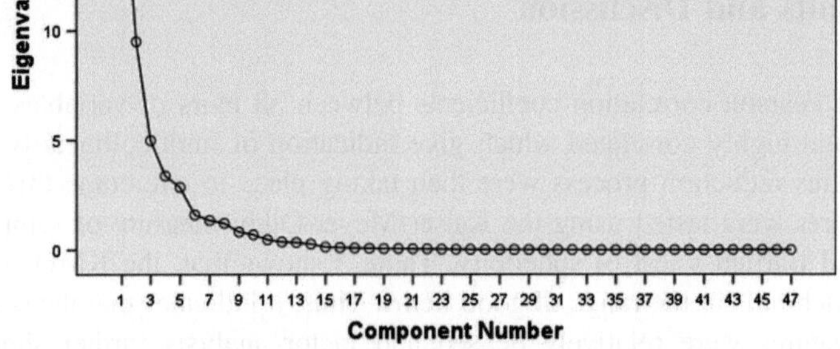

Fig. 3 Scree plot

Table 4 Summary of component factors

Factor 1		Factor 2	Factor 3	Factor 4	Factor 5	Factor 6	Factor 7	Factor 8
LW20	LC20	Mean_R	Skew_W	LR10	LR21	LC21	LR02	LW11
LR20	LC02	Var_R	Kur_W	LR12	LR01	LC10	Kur_R	LC11
LW02	LC22	Q3_R	Skew_C	LW10	LW21	LC01	–	LR11
LR22	LW22	Mean_W	Var_C	LW12	LW01	LC12	–	–
Q3_C	Q1_C	Q3_W	Kur_C	–	–	–	–	–
Mean_C	Skew_R	Med_W	–	–	–	–	–	–
Med_C	Med_R	–	–	–	–	–	–	–
LW00	LC00	–	–	–	–	–	–	–
Q1_W	LR00	–	–	–	–	–	–	–

difficult and complicated interpretation of the factors. After the rotation, the process of identifying which variables that belong to each eight components become easier. Then the eight components will be used to further analysis for classifying the type of pistols used. The summary of the eight factors with their respective component features is presented in Table 4.

Further analysis of classifying the five types of pistol were carried out via Weka software using naive Bayes and decision tree J48 classifier. The detailed of classification rate using cross validation for both classifier were presented in Tables 5 and 6. All the pistols were well classified as most of the pistol achieved at least 70 % of success rate.

Table 7 shows the summary output of classification rate and the Kappa statistic values using naive Bayes and decision tree J48. The classification rate using the Weka tools which are naïve Bayes and J48 has successfully presented the numerical features from the firing pin impression image in firearm identification. This study shows that the naïve Bayes performed an 85.01 % of classification rate while decision tree of J48 algorithm gave 78.45 % accuracy for classification of pistols.

Table 5 Classification rate using cross validation under naïve Bayes

Actual	Predicted				
Type of pistol	A	B	C	D	E
A	**139** **(92.7 %)**	0 (0 %)	5 (3.3 %)	2 (1.3 %)	4 (2.7 %)
B	1 (0.7 %)	**130** **(86.7 %)**	8 (5.3 %)	11 (7.3 %)	0 (0 %)
C	1 (0.7 %)	6 (4 %)	**123** **(82 %)**	13 (8.7 %)	7 (4.7 %)
D	1 (0.7 %)	9 (6 %)	16 (10.7 %)	**117** **(78.5 %)**	6 (4 %)
E	6 (4.1 %)	0 (0 %)	6 (4.1 %)	10 (6.8 %)	**126** **(85.1 %)**

Table 6 Classification rate using cross validation under decision tree J48

Actual	Predicted				
Type of pistol	A	B	C	D	E
A	**130 (86.7 %)**	1 (0.7 %)	5 (3.3 %)	3 (2 %)	11 (7.3 %)
B	5 (3.3 %)	**128 (85.3 %)**	9 (6 %)	8 (5.3 %)	0 (0 %)
C	7 (4.7 %)	2 (1.3 %)	**122 (81.3 %)**	14 (9.3 %)	5 (3.3 %)
D	7 (4.7 %)	9 (6 %)	22 (14.8 %)	**100 (67.1 %)**	11 (7.4 %)
E	18 (12.2 %)	1 (0.7 %)	7 (4.7 %)	16 (10.8 %)	**106 (71.6 %)**

Table 7 Summary of classification results

Classifier	Cross validation (tenfolds)		Kappa statistic
	Correctly classified (%)	Incorrectly classified (%)	
Naïve Bayes	85.01	14.9	0.8126
Decision tree	78.45	21.55	0.7306

The average of Kappa score from the selected algorithm is around 0.8. Based on the Kappa statistic criteria, the accuracy of the firearm classification purposes is greater than 0.6 which achieves the acceptable accuracy of classification. Since the Kappa values for using naïve Bayes and decision tree under J48 tools are 0.8126 and 0.7306, respectively, this indicates that a high level of accuracy has been achieved.

4 Conclusion

The overall conclusion for the classification method using the Weka tools via naïve Bayes and decision tree J48 has been successfully presented and the potential of using the numerical features from the firing pin impression image for firearm identification purposes. This paper is believed to contribute in terms of providing an insight on how statistical pattern recognition, computer vision and machine learning can improve the existing procedure of firearms identification. For future work, ensemble method such as random forest that consist of several decision trees is recommended in order to improve the classification rate.

References

Ananth JP, Bharathi VS (2012) Face image retrieval system using discrete orthogonal moments. In: 4th international conference on bioinformatics and biomedical technology, 29

Chuan ZL, Ghani NAM, Liong CY, Jemain AA (2013) Pendekatan pengesanan titik sauh secara automatic bagi kesan pin peletup senjata api. Sains Malays 42(9):1339–1344

Field A (2005) Factor analysis using SPSS. J Res Methods 639

Ghani NAM, Kamaruddin SA, Liong CY, Jemain AA (2012) Firearm identification using numerical features of center firing pin impression image. In: International symposium on computer applications and industrial electronics (ISCAIE 2012), Kota Kinabalu Malaysia, pp 293–296

Khalid MH (2007) Exact legendre moment computation for gray level images. Pattern Recognit 3597–3605

Liong CY, Ghani NAM, Kamaruddin SA, Jemain AA (2012) Firearms classification based on numerical features of the firing pin impression. Procedia Comput Sci 13:144–151

Smith CL (2001) Profile measurements as a technique for forensic ballistics identification. In: Proceeding 5th Australia security research, Perth, Australia, 153–162

Thumwarin (2008) Firearm identification based on rotation invariant feature of cartridge case. In: SICE annual conference 2008, The University Electro-Communications, Japan, pp 45–49

Chapter 24
Fuzzy Spatial Forecasting Model of Rainfall Distribution for Flood Early Warning

Mahmod Othman, Siti Nor Fathihah Azahari and Noor Atiqah Abu Massuut

Abstract Flood is commonly known as one of the most frequent type of natural disaster worldwide that occurred when the ground is not able to absorb and accommodate the heavy rainfall. Flood might be caused by increased levels of river water more than the river bank or the dams. Many methods have been proposed to forecast the rainfall distribution but mostly the forecasting accuracy of the existing method are questionable. In this paper, a fuzzy time series method was proposed to forecast the rainfall distribution. The objectives of this paper, first is to formulate fuzzy spatial forecasting model for rainfall distribution for each month in Perlis. Second, to predict the accurate rainfall values in future for early warning of flood in order to reduce flood issues. Using the fuzzy spatial forecasting method, the historical data of rainfall in Perlis were used to forecast. After that, several rules was applied to determine whether the rainfall forecasting trend value goes downward or upward movement Then, the mean square error (MSE) was calculated to compare the forecasting rainfall results of various forecasting method. The smaller the value of MSE, the better the forecasting model. The monthly historical rainfall distribution in Perlis for 4 years had been used to illustrate the forecasting algorithm of the new fuzzy time series method. The experimental results of this research exhibited higher forecasting accuracy for forecasting rainfall compare to existing methods.

Keywords Flood · Fuzzy forecasting · Rainfall distribution

1 Introduction

One of the most frequent types of natural disaster worldwide is flood. It is categorized as the occurrence of unexpected major disaster in the world (Flood List 2014). It occurs when the ground is not able to absorb and accommodate the heavy

---

M. Othman (✉) · S.N.F. Azahari · N.A. Abu Massuut
Faculty of Computer and Mathematical Sciences, Universiti Teknologi MARA, Arau, Perlis, Malaysia
e-mail: mahmod135085@perlis.uitm.edu.my

N.A. Yacob et al. (eds.), *Regional Conference on Science, Technology and Social Sciences (RCSTSS 2014)*, DOI 10.1007/978-981-10-0534-3_24

rainfall, which causes the remaining water to flow into the rivers and seas. Finally, when the river water level is higher than the river bank or the dams, the overflow water will meet a failure and flood may occur at any time (The Official Portal for Department of Irrigation and Drainage Malaysia 2013).

As in Malaysia, there has threatened by quite unprecedented flooding of abnormal magnitude and damage in the affected areas as well as Kangar, Perlis, Jasin Melaka and BatuPahat, Johor, and also Kuala Lumpur. Basically, one of the main causes of why these floods happened is that, the heavy and intense rainfall associated with high runoff (Ceballos et al. 2004; Tymvios et al. 2010; Ngongdo et al. 2011). However, the dam breaks, blockages in river channels, population pressure due to increasing number of people, failure in drainage systems and rapid development also contribute to the severe floods. To be fair, many previous flood control measures have received different degrees of success but generally have little effect in reducing the problem.

In this research, Perlis has been chosen as the research area since Perlis is among the states that frequently experience flash floods even though Perlis is known as the hottest state. The major floods that struck Perlis in recent years are due to a torrential rainfall that causes water levels in rivers to increase. A nonstop rain for 3–4 days had resulted in a severe flood in many areas of Perlis (Bernama 2011).

Several methods have been proposed to forecast the rainfall distribution but mostly the forecasting accuracy of the existing method are not good enough. The accuracy of the forecast data is immensely significant because it is one piece of the hydrologic data where if one mistake occurrs in the calculation, it will affect the entire system (El-Shafie et al. 2011; Chen and Liu 2012; Zaw and Naing 2009; Hung et al. 2008). In this paper, a new method of fuzzy time series is proposed to forecast the rainfall distribution. Therefore, there are two main objectives for this study; first is to formulate fuzzy spatial forecasting model for rainfall distribution. The fuzzy spatial forecasting allows efficient computation algorithms applied in managing the rainfall distribution (forecasting, locating). Furthermore, the rainfall distribution data that had been used in this study are vague and subjective (Zadeh 1965). Hence, it should be appropriate to apply fuzzy spatial forecasting model compare to the classical time series model (Song 2003). Second, the objective is to predict the rainfall reading of Perlis state for the year of 2013 using fuzzy spatial forecasting model. Based on the study that had been presented by Chen, he was using a method of fuzzy time series to forecast the enrollments of the University of Alabama. The implementation had reduced the calculation time and simplifying the calculation process (Chen 1996; Hwang et al. 1998). Then the same method had been improvized and was used to forecast the temperature (Chen and Hwang 2000).

2 Methodology

The proposed model consisted of embedding and integrating Chen and Hwang (2000) technique. There are several steps that had been improvized to make sure the method is suitable to be implemented for the rainfall distribution data. There were four steps proposed in this model starting with step 1: collect the rainfall distribution data, step 2: define the universe U and intervals, step 3: fuzzy logic relationship, and step 4: classification of rules, trends and forecasts.

Step 1: Collect the rainfall distribution data

The Perlis state's rainfall historical data starting from year 2009 to 2012 was collected from the Department of Irrigation and Drainage Malaysia. Based on the data, it was shown that the data only contain a single row of actual rainfall distribution reading, S_j. Hence, from the actual data, we need to find the average of three months rainfall distribution, S_i from the actual data. Equation (1) shows the formula to find the fuzzy set S_i.

$$S_i = \frac{n + (n-1) + (n-2)}{3} \tag{1}$$

where

S_i fuzzy set or average of three months,

n is first month, $n - 1$ is second month before, and $n - 2$ is third month before

S_i was considered as input variable while the historical data were considered as output variable, S_j

The equation above used is the enhancement of the Lazim (2011) average of Changes's equation.

Step 2: Define the universe U and intervals

Based on the Chen and Hsu study, the universe U, was defined from the rainfall data and divided it into several equidistant intervals $u_1, u_2, \ldots u_n$. Then, the intervals from the highest interval to the lowest interval were ranked. Next, the subintervals were obtained by dividing the interval according to the size of the interval, based on the frequency of numbers of intervals (Chen and Hsu 2004).

Table 1 shows the division of subinterval of universe U. Rank each of the intervals from the highest to the lowest. The interval was divided into the subintervals according to the size of the interval as Table 1.

Table 1 Division of subinterval

Level of frequency numbers of interval	Divide into
Highest	4 subinterval of equal length
Second highest	3 subintervals of equal length
Third highest	2 subintervals of equal length
Fourth highest	Length of the interval remains unchanged

Step 3: Fuzzy Logic relationship

The next step of this study was to create fuzzy logical relationship as follows:

$$S_j \rightarrow S_q,$$

where the fuzzy logical relationship "$S_j \rightarrow S_q$" represents "IF S_j is the rainfall of input variable THEN S_q is the rainfall of output variable." The assumption had been made that if the fuzzy logical relationship is

$$S_i \rightarrow S_j,$$

where S_i represents the average rainfall of three months before month n and S_j represent the rainfall of month n, then the following table depicted the rules that are applied in forecasting of the rainfall distribution.

Step 4: Classification of rules, trends, and forecasts

After creating the fuzzy logical relationship, the classification of fuzzy forecasting rule was implemented to the data. Then the trend of each data was marked with three type of trend; downward 0.25, or middle, or upward 0.75. Then, proceed with the forecasting process.

Based on the Table 2, the rules of forecasting rainfall had been determined by referring to the fuzzy relationship above. The three rules were used to find the trend of the forecasting. Each of the rules has their own requirement. Below are the requirements of each rule:

Rule 1: To forecast the rainfall for May 2009, we need the data three months before, hence we are not able to calculate the difference. Therefore, if | (the difference of March 2009 and April 2009) | /2 > half of the corresponding interval S_j, then the rainfall forecasting go upward at the 0.75-point of this interval; if | (the difference of March 2009 and April 2009) | /2 = half of the corresponding interval S_j, then the rainfall forecasting be the middle value of this interval; if | (the difference of March 2009 and April 2009) | /2 < half of the corresponding interval S_j, then the rainfall forecasting go downward at the 0.25-point of this interval.

Rule 2: If (| the difference of the difference of rainfall between $n - 1$ and $n - 2$ and between $n - 2$ and $n - 3$ | × 2 + the rainfall distribution of $n - 1$) falls in the corresponding interval S_j, then the rainfall forecasting go upward at the 0.75-point of this interval; If (| the difference of the difference of rainfall between $n - 1$ and $n - 2$ and between $n - 2$ and $n - 3$ | /2 + the rainfall distribution of $n - 1$) falls in the corresponding interval S_j, then the rainfall forecasting go downward at the 0.25-point of this interval; if neither is the case, the rainfall forecast will be the middle value of the corresponding interval S_j.

Rule 3: If (| the difference of the difference of rainfall between $n - 1$ and $n - 2$ and between $n - 2$ and $n - 3$ | /2 + the rainfall distribution of $n - 1$) falls

Table 2 Rule of fuzzy forecasting, (Chen and Hsu 2004)

Rule no.	If			Then
1	$j > i$	and	$(n_{S_j} - n_{S_i}) - \left((n-1)_{S_j} - (n-1)_{S_i}\right) - \left((n-2)_{S_j} - (n-2)_{S_i}\right) = +\text{ve}$	Rule 2 used for forecasting rainfall
2	$j > i$		$(n_{S_j} - n_{S_i}) - \left((n-1)_{S_j} - (n-1)_{S_i}\right) - \left((n-2)_{S_j} - (n-2)_{S_i}\right) = -\text{ve}$	Rule 3 used for forecasting rainfall
3	$j < i$		$(n_{S_j} - n_{S_i}) - \left((n-1)_{S_j} - (n-1)_{S_i}\right) - \left((n-2)_{S_j} - (n-2)_{S_i}\right) = +\text{ve}$	Rule 2 used for forecasting rainfall
4	$j < i$		$(n_{S_j} - n_{S_i}) - \left((n-1)_{S_j} - (n-1)_{S_i}\right) - \left((n-2)_{S_j} - (n-2)_{S_i}\right) = -\text{ve}$	Rule 3 used for forecasting rainfall
5	$j = i$		$(n_{S_j} - n_{S_i}) - \left((n-1)_{S_j} - (n-1)_{S_i}\right) - \left((n-2)_{S_j} - (n-2)_{S_i}\right) = +\text{ve}$	Rule 2 used for forecasting rainfall
6	$j = i$		$(n_{S_j} - n_{S_i}) - \left((n-1)_{S_j} - (n-1)_{S_i}\right) - \left((n-2)_{S_j} - (n-2)_{S_i}\right) = -\text{ve}$	Rule 3 used for forecasting rainfall

in the corresponding interval S_j, then the rainfall forecasting go downward at the 0.25-point of this interval; If (| the difference of the difference of rainfall between $n-1$ and $n-2$ and between $n-2$ and $n-3$ | × 2 + the rainfall distribution of $n-1$) falls in the corresponding interval S_j, then the rainfall forecasting go upward at the 0.75-point of this interval; if neither is the case, the rainfall forecast will be the middle value of the corresponding interval S_j,

where,

n rainfall of current month,
$n-1$ rainfall of one month before the current month,
$n-2$ rainfall of two months before the current month, and
$n-3$ rainfall of three months before the current month.

Assume that we want to forecast the rainfall of month n, S_n then the "difference of differences" of the rainfall distribution between month $n-1$ and $n-2$ and between month $n-2$ and $n-3$ = ($Sn-1-Sn-2$) − ($Sn-2-Sn-3$). There are several steps to forecast in fuzzy time series method. The trend value of 0.25-point and 0.75-point of the interval was identified, where the point was used to determine whether the trend of the forecasting goes up or down and forecast the rainfall distribution.

Step 5: Error measurement

Finally, the mean squared error (MSE) of forecasting rainfall was calculated. Mean squared error is an estimator measures the average of the squares of the "errors," that is, the difference between the estimator and what is estimated. MSE is a risk function, corresponding to the expected value of the squared error loss or quadratic loss. The difference occurs because of randomness or because the estimator does not account for information that could produce a more accurate estimate (Wackerly et al. 2007). In this research, MSE is used to compare the forecasting rainfall results of various forecasting method. Then high MSE means the average distance of the arrows from the bull's-eye is high, and low MSE means the average distance from the bull's-eye is low (Bol'shev 2001). The calculation of mean square error (MSE) as follows:

$$MSE = \frac{\sum_{i=1}^{n} (ActualRainfall_i - ForecastedRainfall_i)^2}{2}$$

3 Results and Discussion

Forecasting of a rainfall event was performed using fuzzy time series method that refers to the rainfall distribution data of the years 2009–2012 in Perlis. The fuzzy time series method was used to compare the actual data and the forecast data.

Table 3 shows the monthly rainfall distribution data in Perlis starting in 2009 until 2012. The universe U of data was defined as [20,500], which was the range of rainfall distribution from 2009 to 2012. The third column was the average of three

Table 3 Historical rainfall distribution data in Perlis

Year	Month	S_i (mm)	S_j (mm)
2009	1	–	21.6
	2	–	40.5
	3	111.1	271.2
	4	192.5	265.9
	5	239.8	182.3
	6	161.0	34.8
	7	148.6	228.7
	8	167.7	239.7
	9	229.0	218.5
	10	214.2	184.5
	11	268.2	401.5
	12	205.3	30.0
2010	1	174.0	90.4
	2	53.9	41.3
	3	83.0	117.3
	4	109.9	171.1
	5	120.9	74.2
	6	121.2	118.4
	7	130.9	200.2
	8	148.1	125.7
	9	179.8	213.4
	10	208.3	285.7
	11	294.4	384.2
	12	278.7	166.3
2011	1	204.4	62.8
	2	85.6	27.6
	3	155.7	376.7
	4	168.7	101.9
	5	200.2	121.9
	6	116.9	126.8
	7	130.4	142.4
	8	169.3	238.8
	9	247.8	362.2
	10	242.2	125.7
	11	219.5	170.7
	12	124.0	75.6

(continued)

Table 3 (continued)

Year	Month	S_i (mm)	S_j (mm)
2012	1	157.0	224.6
	2	145.4	136.1
	3	163.2	128.8
	4	151.0	188.0
	5	140.6	105.1
	6	120.1	67.1
	7	103.9	139.5
	8	108.9	120.2
	9	175.1	265.6
	10	182.4	161.5
	11	230.1	263.2
	12	198.0	169.4

S_i–Average 3 months rainfall distribution, S_j–Actual rainfall distribution

months rainfall distribution that calculated from the actual data of rainfall distribution. Meanwhile, the fourth column was the actual data of rainfall distribution.

Table 4 shows the equidistant interval attained based on the universe U = [20,500]. The universe U was divided into several equidistant. Song and Chissom (1994) divide the universe set based on scalar of Likert seven eyes. However, in this study, the data need to divide to eight divisions and produce eight intervals, with every interval having 60 lengths (Song and Chissom 1994).

Table 5 shows the distribution in different intervals and its division. The interval is then sorted based on the number of frequency in historical data for each interval from highest to lowest. The first, second, and third largest number of frequency need to be identified before dividing it into four, three, and two equidistant subinterval respectively. Meanwhile, for the fourth and above number of frequency, the data in this interval remain unchanged. However, if there are no data in the interval, then the interval must be removed. As shown in Table 5, the highlighted

Table 4 Equal length interval

Equal length intervals
$u1$ = [20,80]
$u2$ = [80,140]
$u3$ = [140,200]
$u4$ = [200,260]
$u5$ = [260,320]
$u6$ = [320,380]
$u7$ = [380,440]
$u8$ = [440,500]

Table 5 Distributions in different intervals and division

Equal length intervals	Frequency	Sort	Divide into
$u1 = [20,80]$	10	2	3
$u2 = [80,140]$	13	1	4
$u3 = [140,200]$	9	3	2
$u4 = [200,260]$	7	4	
$u5 = [260,320]$	5	5	
$u6 = [320,380]$	2	6	
$u7 = [380,440]$	2	6	
$u8 = [440,500]$	0	7	

interval $u8 = [440,500]$, did not have any data. So, the interval $u8$ was eliminated. After the division was determined, the universe $U = [20,500]$ was redivided as follows:

Table 6 shows the new subinterval. The interval has increased from 8 intervals to 13 subintervals. Each of the new subinterval, were defined as S1, S2, S3… S13. Next, the fuzzy logical relationship was obtained based on the rainfall distribution of Table 3 above. It was depicted in Table 7.

Based on the Table 7, the forecasting value was calculated starting at the fourth month or row number four. First and second columns show the input and output variable, while the third column shows the classification rule of fuzzy forecasting. The fourth column was the relationship of fuzzy logical, fifth column was the trend of each rainfall reading and last column shows the forecasting value. For the beginning of forecasting, it was considered as special case since there is no previous data and it should be categorized in rule 1. Then for the next column, each of them was calculated using the method that had been explained in the methodology above.

After the fuzzy forecasting model was applied to the historical data, the MSE between the actual result and the forecast result was calculated. The result of forecasting using fuzzy forecasting model were compared from other forecasting

Table 6 New subintervals

S1	$u1,1 = [20,40]$	$u1,2 = [40,60]$	*S2*
S3	$u1,3 = [60,80]$	$u2,1 = [80,95]$	*S4*
S5	$u2,2 = [95,110]$	$u2,3 = [110,125]$	*S6*
S7	$u2,4 = [125,140]$	$u3,1 = [140,170]$	*S8*
S9	$u3,2 = [170,200]$	$u4 = [200,260]$	*S10*
S11	$u5 = [260,320]$	$u6 = [320,380]$	*S12*
S13	$u7 = [380,440]$		

Table 7 Actual rainfall distribution, fuzzy logical relationship, and forecast reading rainfall distribution in Perlis

S_i (mm)	S_j (mm)	Rule	Si → Sj	Trend	Forecast
	21.6		–		
	40.5		–		
111.1	271.2		S6 → S11		
192.5	265.9	1	S9 → S11	middle	290
239.8	182.3	3	S10 → S9	middle	185
161.0	34.8	3	S8 → S1	middle	30
148.6	228.7	3	S8 → S10	middle	230
167.7	239.7	2	S8 → S10	middle	230
229.0	218.5	2	S10 → S10	middle	230
214.2	184.5	3	S10 → S9	middle	185
268.2	401.5	3	S11 → S13	middle	400
205.3	30	2	S10 → S1	middle	30
174.0	90.4	3	S9 → S4	middle	87.5
53.9	41.3	3	S2 → S2	middle	50
83.0	117.3	2	S4 → S6	middle	117.5
109.9	171.1	2	S5 → S9	middle	185
120.9	74.2	2	S6 → S3	middle	70
121.2	118.4	3	S6 → S6	middle	117.5
130.9	200.2	3	S7 → S10	middle	230
148.1	125.7	2	S8 → S7	middle	132.5
179.8	213.4	3	S9 → S10	down 0.25	215
208.3	285.7	3	S10 → S11	down 0.25	275
294.4	384.2	2	S11 → S13	down 0.25	380
278.7	166.3	3	S11 → S8	middle	155
204.4	62.8	3	S10 → S3	middle	70
85.6	27.6	3	S4 → S1	middle	30
155.7	376.7	2	S8 → S13	middle	400
168.7	101.9	2	S8 → S5	middle	102.5
200.2	121.9	3	S10 → S6	middle	117.5
116.9	126.8	3	S6 → S7	middle	132.5
130.4	142.4	2	S7 → S8	middle	155
169.3	238.8	2	S8 → S10	middle	230
247.8	362.2	2	S10 → S13	middle	380
242.2	125.7	2	S10 → S7	middle	132.5
219.5	170.7	3	S10 → S9	middle	185
124.0	75.6	3	S6 → S3	middle	70
157.0	224.6	2	S8 → S10	middle	230
145.4	136.1	2	S8 → S7	middle	132.5
163.2	128.8	3	S8 → S7	middle	132.5
151.0	188	3	S8 → S9	middle	185
140.6	105.1	2	S8 → S5	middle	102.5
120.1	67.1	3	S6 → S3	middle	70
103.9	139.5	3	S5 → S7	middle	132.5

(continued)

Table 7 (continued)

S_i (mm)	S_j (mm)	Rule	Si → Sj	Trend	Forecast
108.9	120.2	2	S5 → S6	middle	117.5
175.1	265.6	2	S9 → S11	down 0.25	275
182.4	161.5	2	S9 → S8	middle	155
230.1	263.2	3	S10 → S11	middle	290
198.0	169.4	3	S9 → S8	middle	155

Table 8 Comparison of the forecasting rainfall results of several forecasting methods

Forecast Model	MSE
Fuzzy forecasting	104.47
Average change model	3453.6
Single exponential smoothing (a = 0.1)	12036.7
Naive forecast	16995.7

methods such as the average change model, single exponential smoothing (a = 0.1), and naive forecast were shown in Table 8. Table 8 depicted that the fuzzy forecasting model provided the lowest value of MSE which was 104.47 compared to other methods. Therefore, the fuzzy forecasting model gave the nearest forecasting exactness for forecasting the rainfall distributions than the existing methods.

To forecast the rainfall in year 2013, the value of S_j needed to be obtained first before the forecasting value could be calculated. So, the average of the sum of 4 months was calculated to get the new prediction data for a few months of 2013. Then, the model of forecast as explained above was repeated. Table 9 showed the rainfall forecast for year 2013 in January, February, March, and April.

Table 9 displays the forecasting rainfall results using the proposed fuzzy forecasting model for January until April 2013. The forecasting values as compared to actual prediction data, S_j were quite accurate which were different less than 9.0 mm.

Table 9 Rainfall forecasting for the year 2013

Year	Month	S_i (mm)	S_j (mm)	Trend of the forecasting	Forecasting
2013	1	177.5	99.9	Middle	102.5
	2	110.2	61.4	Middle	70
	3	128.3	223.5	Down 0.25	215
	4	155.5	181.7	Middle	185

4 Conclusion

In this paper, the new model of fuzzy forecasting of rainfall distribution in Perlis has been introduced. By referring to Table 9, it shows that the MSE of the new fuzzy time series model provide a smallest MSE value compared to the other methods. The proposed model had exhibited a good model, where it had fulfilled three major properties: (1) formal consistency, (2) usefulness; (3) efficiency, in the desired function at a minimum effort, time and cost. Additionally, this model enabled to help the farmer, fish harvesters, or any job that is related to agriculture and maritime in planning to gain the optimum yield by knowing the information of rainfall distribution for throughout a year. Therefore, they can reduce the rate of loss in agriculture or maritime and avoid of losing anyone life while working. This study can be further extended in applying the proposed method to deal with other forecasting problem based on the fuzzy time series.

Acknowledgments The study was funded by "Long Term Research Grant (LRGS) (UUM/RIMPC/P-30)" and the authors also thank the Faculty of Computer and Mathematical Sciences, Universiti Teknologi MARA for providing the laboratory facilities for completing the study.

References

Bernama (2011) More people evacuated in flood-hit Perlis. Borneo Post http://www.theborneopost.com/2011/04/02/more-people-evacuated-in-flood-hit-perlis/

Bol'shev LN (2001) Encyclopedia of mathematic. Statistical estimator. http://www.encyclopediaofmath.org/index.php/Statistical_estimator

Ceballos A, Martinez-Fernandez J, Luengo-Ugidos MA (2004) Analysis of rainfall trends and dry periods on pluviometric gradient representative of Mediterranean climate in the Duero Basin, Spain. J Arid Env 58:215–233

Chen FW, Liu CW (2012) Estimation of the spatial rainfall distribution using inverse distance weighting (IDW) in the middle of Taiwan. Paddy Water Environ 10:209–222

Chen SM (1996) Forecasting enrollments based on fuzzy time series. Fuzzy Sets Syst 81:311–319

Chen SM, Hsu CC (2004) A new method to forecast enrollments using fuzzy time series. Int J Appl Sci Eng 2:234–244

Chen SM, Hwang JR (2000) Temperature prediction using fuzzy time series. IEEE Trans Syst Man Cybern Part B: Cybern 30:263–275

El-Shafie AH, El-Shafie A, El Mazoghi HG, Shehata A, Taha MR (2011) Artificial neural network techniques for rainfall forecasting applied to Alexandria, Egypt. Int J Phys Sci 6(6):1306–1316

Flood List (2014) About flood. http://floodlist.com/aboutfloods

Hung NQ, Babel MS, Weesakul S, Tripathi NK (2008) An artificial neural network model for rainfall forecasting in Bangkok, Thailand. Hydrol Earth Syst Sci Discuss 5:183–218

Hwang JR, Chen SM, Lee CH (1998) Handling forecasting problems using fuzzy time series. Fuzzy Sets Syst 100:217–228

Lazim M, Alias (2011) Introductory business forecasting a practical approach, 3rd edn. University Publication Center (UPENA), Shah Alam

Ngongondo C, Xu C, Gottschalk L (2011) Evaluation of spatial and temporal characteristics of rainfall in Malawi. A case of data scarce region. Theor Apply Climatol 106:79–93

Song Q (2003) A note on fuzzy time series model selection with sample autocorrelation functions. Cybern Syst Int J 34:93–107

Song Q, Chissom BS (1994) Forecasting enrolmentswith fuzzy time series—part 2. Fuzzy Sets Syst 62:1–8

The official portal for department of irrigation and drainage Malaysia(2013). Flood management—program activities. Definition of flood. http://www.water.gov.my/our-services-mainmenu-252/flood-mitigation-mainmenu-323/programme-aamp-activities-mainmenu-199?lang=en

Tymvios F, Savvidou K, Michaelides SC (2010) Association of geo potential height patterns with heavy rainfall events in Cyprus meteorological services, Nicosia, Cyprus. J Refereed Proc Spec Publ 23:73–78

Wackerly D, Mendenhall W, Acheaffier RL (2007) Mathematical statistics with application, 7th edn. Brooks/Cole Cengage Learning, USA

Zadeh LA (1965) Fuzzy sets. Inf Control 8:338–353

Zaw WT, Naing TT (2009) Modelling of rainfall prediction over Myanmar using polynomial regression. Int Conf Comput Eng Technol 1:316–320

Chapter 25
Applications of Travelling Salesman Problem in Optimizing Tourist Destinations Visit in Langkawi

Zakiah Hashim and Wan Rosmanira Ismail

Abstract This paper discusses about finding the best possible route for self-drive tourism in Langkawi which focused on land transportation and road network. The problem of finding the best tourism route was formulated based on the travelling salesman problem (TSP) model. This work was developed in order to determine the best route that includes all the 19 most attractive tourist destinations in Langkawi. This is because in this way, the time spent in travelling and the total travelling cost of the tourist can be minimized. It will also optimize the leisure moments as well as taking to know all the tourist destinations. The Langkawi tourism data were used to obtain the self-drive tourism route in Langkawi. This model has been solved by using LINGO12.0 software. Results from the study found that tourist need to spend 4 days to visit all 19 tourist destinations in Langkawi with the total travelling cost of RM 856.55.

Keywords Self-drive tourism · Tourism route · Travelling salesman problem (TSP)

1 Introduction

This paper deals with the real-world self-drive tourism route application in Langkawi Island, Kedah. According to World Tourism Organization (WTO), the terms of tourism refers to "Activities of an individual travelling to a place outside their original environment and living there for not more than one consecutive year

Z. Hashim (✉)
School of Quantitative Sciences, College of Art and Sciences,
Universiti Utara Malaysia, Sintok, Kedah, Malaysia
e-mail: zakiah@uum.edu.my

W.R. Ismail
School of Mathematical Sciences, Faculty of Science and Technology,
Universiti Kebangsaan Malaysia, Bangi, Selangor, Malaysia
e-mail: wrismail@ukm.my

N.A. Yacob et al. (eds.), *Regional Conference on Science, Technology and Social Sciences (RCSTSS 2014)*, DOI 10.1007/978-981-10-0534-3_25

for leisure, business and other purposes." Olsen (2002) defined self-drive tourism as travelling from home at least one night for the purpose of holiday, or visiting friends, or relatives by driving their own or rented vehicle as the primary mode of transport. Meanwhile, according to Dean et al. (2002), self-drive tourism is linked to the concept of special interest tourism, where the tourists have specific purposes (usually activity based) for visiting destinations.

The main purpose of travel is to visit tourist attractions around the location. Since there are many natural destinations, some tourists who drive their own vehicles can plan their own itinerary without being tied to a travel itinerary from a travel agency. Therefore, in this mode of travel, there are no needs to hire tourist guides because the tourists are free to choose their route itself is based on the available information, time, and cost. Normally, tourists will choose the best route that connects a destination to the next destination which gives the shortest distance. The best route selection is important for self-drive tourists because it will affect the amount of travel costs, time to travel and also the number of tourist attraction can be visited. Taplin and McGinley (2000), mentioned that, for most holiday makers travelling by car, the pursuit of satisfaction, and enjoyment are limited by the length of time available and by travel distance. Self-drive tourism has advantages to tourist in terms of: greater control over itinerary; often greater comfort and lower cost (Dean et al. 2002). According to Bruce (2000), increase in the distance travelled will generally lead to increase in transportation costs and it is an important factor for the calculation of the overall cost of the trip. The potential for interaction between any two places increases as the cost of movement between them either in terms of money or time decreases (Dziauddin et al. 2013).

A system of travel routes for a self-drive tour mode more efficient and competitive should be created as an effort to attract more tourists to visit Langkawi. Therefore, the study needs to be done to ensure effectiveness in implementing this mode of self-drive tour. In this paper we tried to adapt a mathematical model of travelling salesman problem (TSP) that can be applied to determine the best route for self-drive tourism in Langkawi Island, which focused on land transport and road network only. The ability of this model to find the best route in order to minimize the travel costs is an alternative approach to attract more tourists to visit Langkawi.

For any problems associated with the shortest chain, the TSP method is a method frequently used. TSP is the basis for optimization problems in transportation and aims to minimize the travel costs involved, such as distance, time, energy, and others. A privilege of this method is to make sure every place is visited only once and return back to where they started.

The paper is organized as follows: The next section is devoted to the travelling salesman problem (TSP) background and methods of resolution. In Sect. 3, we discuss the problem definition and Sect. 4 was discussed about mathematical formulation of the TSP. Section 5 shows the results and discussions with Sect. 6 giving the conclusion.

2 Travelling Salesman Problem

The travelling salesman problem (TSP) is a classical combinatorial optimization problem, which deals with finding the shortest distance or time for closed tour in an n-city situation, where each city is visited exactly once before returning back to the starting point. A common application of TSP is the movement of people, equipment, and vehicles around tours in aiming to minimize the total travelling cost. Mathematical problems related to TSP appeared in the 1800s by an Irish mathematician, Hamilton and an English mathematician, Kirkman. They studied the problem of minimum distance path in the Hamiltonion game. Hamiltonion game is a puzzle game based on the theory of Hamiltonion cycles. Based on this theory, Hamilton and his colleagues are trying to find the number of different paths that exist in a particular network (Lawler et al. 1985).

In 1950, Dantzig and his colleagues reviewed the TSP using the branch and cut algorithm to form the basis of modern integer programming computation systems which are widely used to solve optimization models in the supply chain, telecommunications, manufacturing, transportation and others. They also solved the TSP with 49 cities and prove that there is no other route is shorter than the optimum tour that they got (Dantzig et al. 1954). In the following decade, TSP was studied by many researchers from mathematics, computer sciences, chemistry, physics, and others. TSP is widely used in engineering, design of metal tools, electronic equipment, and communications network. TSP also was applied in the tourism industry in order to determine the optimum route to visit the tourist destinations such as the research that has been done by Neto et al. (2012).

The method used to solve TSP vary widely from the classical methodology based on linear programming (Diaby 2007; Jain and Bahnot 2012) and branch and bound (Balas and Crhistofides 1981) to artificial intelligent methods such as neural networks (Kedar and Tuzun 1995), tabu search (Alkallak and Sha'ban 2008), genetic algorithm (Sadiq 2012), evolutionary algorithm (Földesi and Botzheim 2008), and so on.

3 Problem Definition

This paper presents a study on a real road network in Langkawi Island, Kedah. Although there are more than 30 tourist destinations in Langkawi (refer Fig. 1), this study only covers 19 of the most attractive tourist destinations around Langkawi. The tourist destinations were selected based on recommendations acquired from the Ministry of Tourism Malaysia website (2012). In addition, we only consider the tourist destinations that can be reached by road networks, because of the scope of our study focused on self-drive tour. Table 1 shows the 19 selected tourist destinations around Langkawi and the estimated trip time in minutes which provided to

Fig. 1 Map of Langkawi island

Table 1 Estimated trip time (minutes) provided to visit 19 tourist destinations

Tourist destination	Abbreviation	Estimated trip times (min)
Eagle Bay	DL	30
Taman Lagenda	TL	90
Langkawi Crystal	KL	60
Mahsuri Mausoleum	MM	90
Kampung Buku Malaysia, Langkawi	KBL	90
Underwater World	UW	120
Pantai Chenang	PC	120
Paddy Rice Museum	LP	90
Field of Burnt Rice	BT	30
Oriental Village (Langkawi Cable Car)	OV	150
Seven Walls Waterfalls	TT	150
Crocodile Farm	TB	120
Temurun Waterfall	ATT	150
Langkawi Craft Complex	KKL	90
Air Hangat Village	TAH	120
Tanjung Rhu	TR	90
Galeria Perdana	GP	90
Mount Raya (Gua Cherita)	GR	90
Langkawi Wildlife Park	ZM	120

visit the tourist destinations. This study involved a ground transportation and road networks only.

In developing the mathematical model of TSP for this study, the total time for travelling and visiting time provided at least 10 h per day. The reason for only 10 h a day allocated to the holiday was because the operation hour in many of the tourist destinations in Langkawi was started from 8.00 am until 6.00 pm daily. Estimated travel time to visit places of interest that exist between origin and destination is obtained from the estimated time given by the Google Maps application for each selected path. The distance from one destination to another are also found using the Google Maps application.

As the current petrol prices are constantly changing according to global market price of petrol, the price of petrol to be considered in this study is the actual price of of petrol during the study. The price for RON95 petrol is RM1.95 per liter and RON97 petrol is RM2.25 per liter. So, RON95 petrol type was selected in this study because it is less expensive compared to RON97. The average estimate of travel for 1 L petrol is 10 km. This means that the fuel cost per kilometer is RM0.195. The cost of using petrol is calculated by multiplying the price of RON 95 petrol at RM0.195 per kilometer of distance data between tourist destinations involved. Estimated cost of car rental per day (24 h) is RM134.00. The accommodation that was located near to the tourist destinations in a radius of less than 5 km with the lowest price will be selected to stay for the day.

4 Mathematical Formulation of the TSP

The model in this study was developed to minimize total travel time to visit 19 tourist destinations that are included in this study. This model was constructed based on travelling salesman problem model and also involves the integer programming model (Schrage 2006). Due to major transportation often used by tourists to Langkawi is ferry ride, thus Kuah Jetty (JK), was chosen as a depot in developing the TSP model. This means that, the first and last destination for visiting all the 19 tourist destinations is the Kuah Jetty.

$$\text{Minimize} \quad \left(\sum_{i=1}^{n} \sum_{j=1}^{n} d_{ij} x_{ij} \right) \Big/ 600 \tag{1}$$

subject to:

$$\sum_{i=1}^{n} x_{ij} = 1; \quad j = 1, 2, \ldots, n, \quad i \neq j \tag{2}$$

$$\sum_{j=1}^{n} x_{ij} = 1; \quad i = 1, 2, \ldots, n, \quad j \neq i \tag{3}$$

$$u_i - u_j + n x_{ij} \leq n - 1; \quad i \neq j; i = 2, 3, \ldots, n; \quad j = 2, 3, \ldots, n \tag{4}$$

$$x_{ij} \in \{0, 1\}; \quad \forall_{i,j} = 1, 2, \ldots, n \tag{5}$$

$$\text{with} \quad x_{ij} = \begin{cases} 1; & \text{if there is a path from tourist destination } i \text{ to } j \\ 0; & \text{otherwise} \end{cases}$$

d_{ij} = travelling and visiting time (in minutes) from tourist destination i to tourist destination j.
u_i = the sequence number of tourist destination i on the trip.
u_j = the sequence number of tourist destination j on the trip.

Equation (1) is the objective which is to minimize the total travelling and visiting time of the journey in a day. It means that after 10 h (600 min) of journey a day, tourists will stop at the destination and choose the hotel that offers minimum price to stay for that day. Equations (2)–(5) are the constraints for this mathematical model. Constraint (2) means that tourists arrived once in each tourist destination. Constraint (3) ensures that tourists leave each tourist destination once. Meanwhile constraint (4) is to avoid sub-tours.

5 Results and Discussion

LINGO software version 12.0 was used in order to solve the mathematical formulation of TSP above. LINGO is a comprehensive computer software and an effective way to solve the problem of modeling linear or nonlinear faster, easier, and effectively (Schrage 2006). In addition, LINGO is also a mathematical modeling language that allows us to express the optimization problem in a form similar to standard mathematical notation. Travel routes that minimize total travel time to visit 19 tourist destinations are shown as follows:

JK → DL → TL → TAH → GP → TR → KKL → TB → ATT → OV → TT
→ BT → LP → PC → UW → MM → GR → KBL → ZM → KL → JK

Tourist needs to spend 4 days to visit all the 19 tourist destinations which is limited to travelling and visiting time of 10 h (600 min) per day. After reach the time provided for each trip a day, the tourist needs to find the accommodation that located near to the tourist destinations in a radius of less than 5 km with the lowest

Table 2 Route network solutions that minimize the total travelling costs

Day	Path	Travel cost (RM)	Accommo-dation cost (RM)	Car rental (RM)	Daily total travelling cost (RM)
		A	B	C	A + B + C
1	JK → DL → TL → TAH → GP → TR→ KKL	0.05 + 0.20 + 0.31 + 0.99 + 2.28 + 1.54 = 5.37	82	134	221.37
2	KKL → TB → ATT → OV	2.32 + 1.23 + 2.77 = 6.32	150	134	290.32
3	OV → TT → BT → LP → PC → UW	0.23 + 2.30 + 1.77 + 0.18 + 0.13 = 4.61	60	134	198.61
4	UW → MM → GR → KBL → ZM → KL → JK	2.50 + 4.25 + 3.16 + 1.74 + 0.23 + 0.37 = 12.25	–	134	146.25
Total travelling cost (RM)					856.55

price to stay for that day. For this case, three different accommodations will be selected for the 4 days holiday because we assume that the last day visit, tourist will leave the Langkawi Island without stay at that night. The total travelling cost for the whole journey of is RM856.55. Table 2 shows the route network solutions that minimize the total travelling cost for the whole journey.

We cannot apply the shortest-path problem in this study because the model only find the shortest path with the minimum length from one destination to other destination without visiting all destinations in the network and not return back to where they started. Even though in the minimum spanning tree problem, all nodes in the network is connected such that the sum of length of the arcs is minimized, but we still cannot used this model in this study because a group of arcs should contain no loop and not return back to where they started (Winston 2004).

6 Conclusion

Based on this study, the use of proposed TSP model is proven to give an idea to the tourists in choosing a suitable path when they want to visit tourist destinations in Langkawi. It was found that the proposed TSP model was expected to help tourists make their travel plan guidance in order to choose route that will minimize the travel costs. Based on the results obtained, it was found that the best route for a self-drive tour in Langkawi to visit all 19 tourist destinations to save travel costs are as given in Table 2. A total of 4 days are required by tourists to visit all tourist destinations by the overall cost of the trip which includes the costs to hire a car, petrol costs, and the accommodation costs are RM856.55. If tourists choose to use regular path without using the TSP model, the total travelling costs is RM1191.10. Thus, this TSP model has been proven to help tourists in choosing the best route that can minimize the total travelling costs. However, this TSP model cannot obtain the accommodation to stay each day directly through the LINGO solution, but only counted manually. As the suggestion for improvement this model in the future work, it should be considered to add the appropriate constraints such as constraint that can provide directly an accommodation to stay each day after the tourists reached a certain travel period. It can be solved using the model of travelling salesman problem with hotel selection.

Acknowledgments The authors would like to thank to Department of Tourism Malaysia for providing the tourist destination data in Langkawi Island for this study. Special thanks also to Universiti Utara Malaysia, Universiti Kebangsaan Malaysia, and Ministry of Higher Education, Malaysia for their sponsorship.

References

Alkallak IN, Sha'ban RZ (2008) Tabu search method for solving traveling salesman problem. Al-Rafiden J Comput Sci Math 5(2):141–153

Balas E, Christofides N (1981) A restricted lagrangean approach to the traveling salesman problem. Math Program 21:19–46

Bruce P (2000) The role of the transport system in destination development. Tour Manag 21:53–63

Dantzig G, Fulkerson R, Johnson S (1954) Solution of a large-scale traveling-salesman problem. J Oper Res Soc Am 2(4):393–410

Dean C, Iain W, Noel S (2002) Drive tourism: up the wall and around the bend. Common Ground Publishing Pty Ltd, Australia

Diaby M (2007) The traveling salesman problem: a linear programming formulation. WSEAS Trans Math 6(6):745–753

Dziauddin MF, Alvanides S, Powe N (2013) Estimating the effects of light rail transit (LRT) system on the property values in the Klang Valley, Malaysia: a hedonic house price approach. J Teknologi (Sci Eng) 61(1):35–47

Földesi P, Botzheim J (2008) Solution for modified traveling salesman problem with variable cost matrix using bacterial evalutionary algorithm. Acta Tecnica Jaurinensis Series Logistica 1 (2):159–171

Jain L, Bahnot A (2012) Traveling salesman problem: a case study. Int J Comput Technol 3 (1):167–169

Kedar SN, Tuzun D (1995) Initializing the hopfield-tank network for the tsp using a convex hull: a computational study. In: Proceedings of the artificial neural networks in engineering (ANNIE'95) conference, pp 399–404

Lawler EL, Lenstra JK, Rinnooy AHGK, Shmoys DB (1985) The traveling salesman problem. A Wiley-Interscience Publication

Ministry of Tourism website. http://www.tourism.gov.my/destinations. Accessed 30 January 2012

Neto JC, Almeida CAD, Roveroto G, Tavares JDH (2012) Applications of operational research techniques in optimization to visit tourist points of Viña del Mar. Informatica Economică 16 (2):5–13

Olsen M (2002) Keeping track of the self drive market. In: Dean C, Iain W, Noel S (eds) Drive tourism: up the wall and around the bend, pp 10–24. Common Ground Publishing Pty Ltd, Australia

Sadiq S (2012) The traveling salesman problem: optimizing delivery routes using genetic algorithms. SAS Global Forum, Chicago, IL, USA, pp 1–6

Schrage L (2006) Optimization modelling with lingo, 6th edn. Lindo System Inc, USA

Taplin JHE, McGinley C (2000) A linear program to model daily car touring choices. Ann Tourism Res 27(2):451–467

Winston WL (2004) Operations research: application and algorithm, 4th edn. Brooks/Cole-Thomson Learning, Canada

Chapter 26
Digraph Representations of Machine Vision Workspace for Monitoring Worker's Behavior

Ahmad Yusairi Bani Hashim, Nur Sufiah Akmala Ramdan, Seri Rahayu Kamat and Siti Azirah Asmai

Abstract If motions are repeated normally for every few seconds, and for prolonged periods about an 8-h shift, fatigue and muscle strain may occur. Suppose that we have a worker who is operating a machine. We need to monitor the worker's behavior through real-time video streaming. Our challenge is to approximate the locations to place the tripods. We propose workspace geometry and workspace configuration so that, there is a proper methodology for designating the operator's initial position, the machine position as the origin of the workspace. We deduce that, for a typical local manufacturing setting, there is a generalized workspace configuration usable for a wide-ranging machine operators' height measurement. Implementing the proposed procedures, the camera placement is therefore defined by digraphs.

Keywords Machine vision · Workspace · Digraph

1 Introduction

Manual coping is the primary cause of musculoskeletal disorders, such as sprains and strains and joint disorders, which account for more than half of all injuries proclaims. If motions are repeated normally, such as for every few seconds, and for prolonged periods for an 8-h shift, fatigue and muscle strain may occur. Working hours are wastedevery year by ways of injury and personal injury cases, and compensation packets are pricey. Employers could skyrocket productivity by

A.Y. Bani Hashim (✉) · N.S.A. Ramdan · S.R. Kamat
Faculty of Manufacturing Engineering, Universiti Teknikal Malaysia, Melaka, Malaysia
e-mail: yusairi@utem.edu.my

S.A. Asmai
Faculty of Information and Communication Technology, Universiti Teknikal Malaysia, Melaka, Malaysia
e-mail: azirah@utem.edu.my

N.A. Yacob et al. (eds.), *Regional Conference on Science, Technology and Social Sciences (RCSTSS 2014)*, DOI 10.1007/978-981-10-0534-3_26

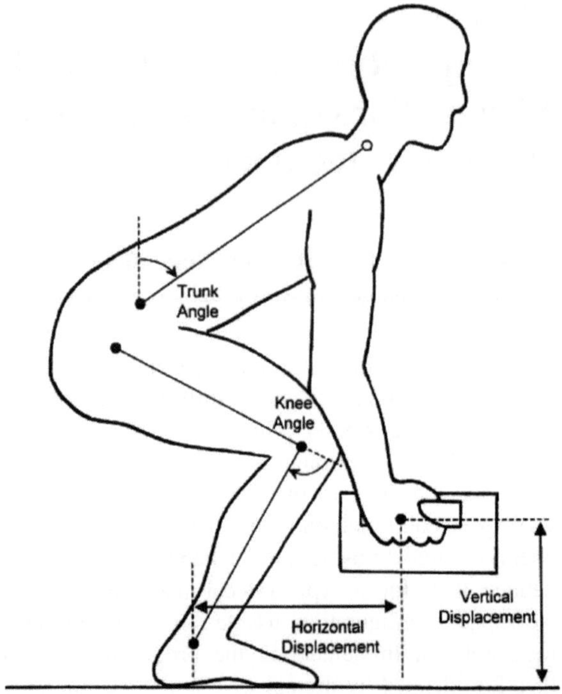

Fig. 1 The posture movement

training their employees to handle loads in a proper way, or by providing material handling equipment to aid them with the job (Ramdan et al. 2014). A worker's typical posture movements (PM) and the associated anthropometric parameters are shown in Fig. 1 (McKean and Potvin 2001).

In Fig. 1, suppose that a worker needs to lift a work piece. There is a resultant due to horizontal and vertical displacements. The knee and the trunk rotational movements change gradually and this depends on the speed of lifting. These movements, however, are determined by the conditions that the worker needs adaptation such as adjusting the posture movements to the constrained or freestyle's conditions. In this paper, we assume that the PM, while the operator is working, dictates the worker's behavior.

2 Background

Suppose that a worker operates a machine. We need to monitor the worker's behavior through a real-time video streaming. Say, there are three cameras on tripods that are to be installed on the site as shown in Fig. 2. Our challenge is to determine the approximate locations to place the tripods that in some ways, the configured workspace relate to the worker's anthropometric measurements, or

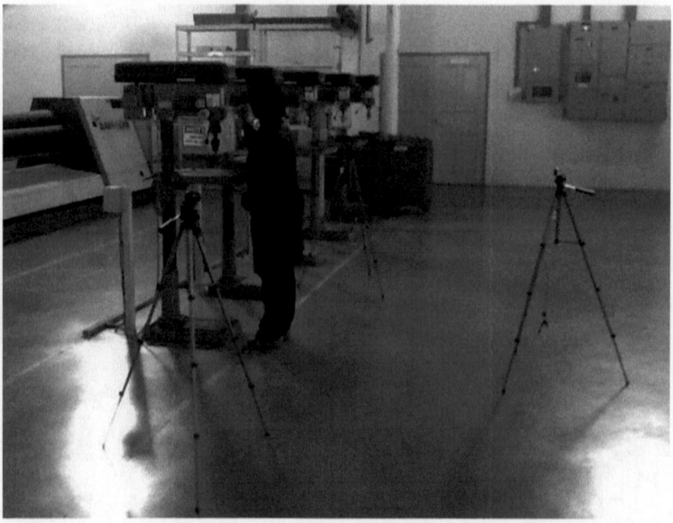

Fig. 2 A worker is operating a drilling machine. The three cameras on tripods are located within a specified workspace

atleast the height. Consider the operator and the machine function within a prescribed workspace. A workspace function as a planned space within which forms a boundary where the activities take place. As an example, obstacles that barricade a robot manipulator themselves are located in their workspace confined within circles (Capisani et al. 2012). In addition, a functional workspace is designed to suit the constraints caused by robot mechanical limits, collision avoidance, and industrial security (Gracia et al. 2012).

Method

We propose workspace geometry (WG) for a machine and an operator as shown in Fig. 3. It depicts a mannequin with height measurement, eight distinct (x, y) coordinates marked on the floor, three unique length measurements, two angle measurements, x-axis, and y-axis. The origin is at P_0. The initial position for the operator is at P_7.

Definition 1

Let H be the operator's height and he is a production worker of a production line in a manufacturing setting. We have $0.5H$ that defines the height measured from the

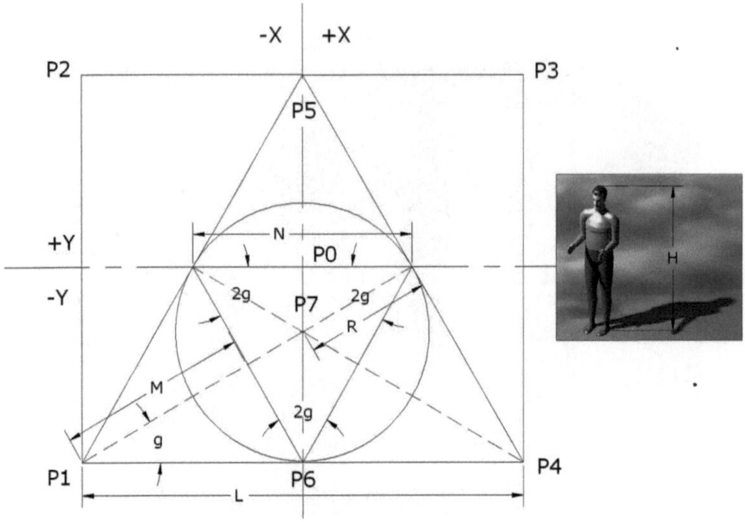

Fig. 3 The geometrical representation of the fundamental dimensions and the points in relation to human anthropometric measurement: height

floor to the body's centroid. Therefore, the radius $R = 0.5H$ and is based on the da Vinci's Vitruvian Man drawing that also inspired some of the works in (Cesare 1521). We have Eq. (1) that defines the radius of the circle in Fig. 3.

$$R = \frac{1}{2}H \tag{1}$$

Definition 2

Once the circle has been defined, let N be the length of an equilateral triangle drawn within the circle that coincides with the arc. The angle, by definition is $2\gamma = 60°$. Thus, Eq. (2) defines N. Therefore, Eqs. (3) and (4) define L and M, respectively. So that Eq. (5) defines the workspace's area.

$$N = 2R\cos\gamma \tag{2}$$

$$L = 2N \tag{3}$$

$$M = \frac{L}{R}\cos\gamma \tag{4}$$

$$A = 4H^2\cos^2\gamma \tag{5}$$

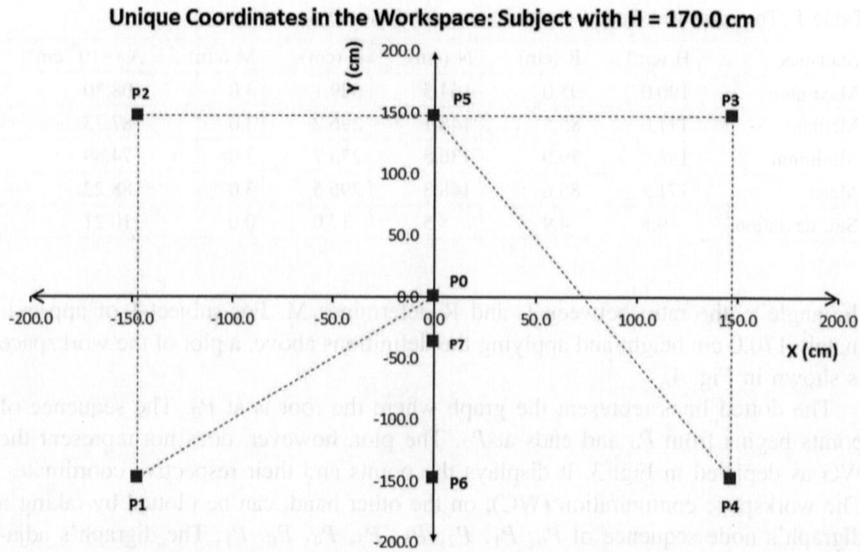

Fig. 4 The plotted coordinates of the workspace that relates to subject-1

Definition 3

There are atleast eight points that belong to the workspace which would be marked on the floor. Based on the geometry in Fig. 4, the points are selected in such a conservative way that they are located at common edges and intersections. They are set as P_0, P_1, P_2, P_3, P_4, P_5, P_6, P_7 where P_0 is the origin, whereas P_7 is the initial position for the operator. Therefore, Eq. (6) defines the points.

$$\left.\begin{array}{l} P_0(x_0, y_0); P_1(x_1, y_1) = \left(-\frac{L}{2}, -\frac{L}{2}\right); P_2(x_2, y_2) = \left(-\frac{L}{2}, \frac{L}{2}\right); \\ P_3(x_3, y_3) = \left(\frac{L}{2}, \frac{L}{2}\right); P_4(x_4, y_4) = \left(\frac{L}{2}, -\frac{L}{2}\right); P_5(x_5, y_5) = \left(x_0, \frac{L}{2}\right); \\ P_6(x_6, y_6) = \left(x_0, -\frac{L}{2}\right); P_7(x_7, y_7) = \left(x_0, -\frac{N}{2}\sin\gamma\right) \end{array}\right\} \quad (6)$$

3 Results

We need to replace the mannequin with a human so that the actual implementation of the proposed WG is validated. Ten male subjects participated in this study. The age ranges from 20 to 27 years old, all of whom have had no trauma. Table 1 shows the statistics for the subjects' height and their corresponding WG's measurements. For mean height of approximately 171.2 cm, we have a desired workspace area of 88.22×10^3 cm^2 and length measurements of 85.6, 148.3, 296.5, and 3.0 cm for R, N, L, and M, respectively. Notice that the M maintains 3.0 cm throughout. Given

Table 1 The computed data of the workspace fundamental dimensions

Statistics	H (cm)	R (cm)	N (cm)	L (cm)	M (cm)	A ($\times 10^3$ cm^2)
Maximum	190.0	95.0	164.5	329.1	3.0	108.30
Median	171.0	85.5	148.1	296.2	3.0	87.73
Minimum	158.0	79.0	136.8	273.7	3.0	74.89
Mean	171.2	85.6	148.3	296.5	3.0	88.22
Std. deviation	9.8	4.9	8.5	17.0	0.0	10.21

the angle γ, the ratio between L and R determines M. For subject-1 of approximately 170.0 cm height and applying the definitions above, a plot of the workspace is shown in Fig. 4.

The dotted lines represent the graph where the root is at P_0. The sequence of points begins from P_0 and ends at P_7. The plot, however, does not represent the WG as depicted in Fig. 3. It displays the points and their respective coordinates. The workspace configuration (WC), on the other hand, can be plotted by taking a digraph's node sequence of P_0, P_1, P_2, P_5, P_3, P_4, P_6, P_1. The digraph's adjacency matrices for the WG and WC can be derived as (7) and (8). The complete matrix solutions for (7) and (8) are shown in Eqs. (A1) and (A2), respectively. In addition, the plots for the nodal sequence for the geometry and configuration are shown in Fig. 5a and b. Node-1 (circle 1) represents P_0 and the other nodes represent the rest of the points, consecutively.

There is a significant difference between the two digraphs. The digraph of Fig. 5b depicts the nodes that make up WC. It represents the operational area. Notice that node-8 stands by itself. Conversely, Fig. 5a characterizes WG. It represents a graph version of the WG. Node-8, for example, represents P_7, which happens to be the operator's initial point. Following the same procedures, the plot

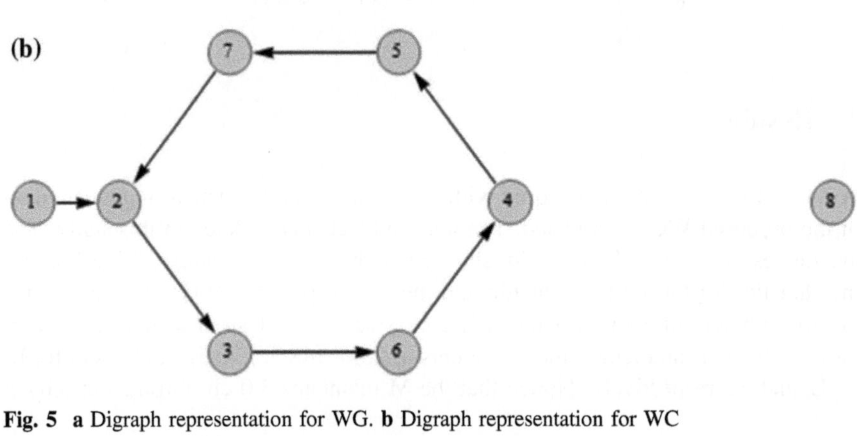

Fig. 5 **a** Digraph representation for WG. **b** Digraph representation for WC

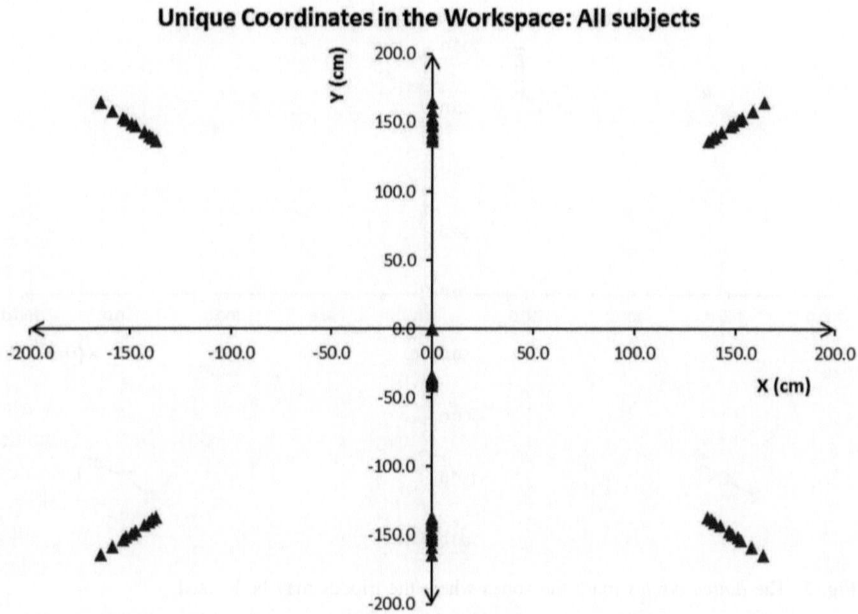

Fig. 6 The plotted chart of the workspaces for all subjects

for all subjects is shown in Fig. 6. Regardless of the subjects' degrees of tallness, the points somehow converge into groups of localized coordinates. By looking at the patterns, it is obvious that the data is naturally statistical.

$$\mathrm{A_{gometry}} = \begin{cases} 1 \text{ if } \mathbf{V} \\ 0 \text{ otherwise} \end{cases} \Bigg| \; V_{ij} = \begin{pmatrix} P_0 \to P_1 \to P_2 \to P_3 \\ P_3 \to P_4 \to P_5 \to P_6 \\ P_6 \to P_7 \end{pmatrix} \tag{7}$$

$$\mathrm{A_{config}} = \begin{cases} 1 \text{ if } \mathbf{V} \\ 0 \text{ otherwise} \end{cases} \Bigg| \; V_{ij} = \begin{matrix} P_0 \to P_1 \to P_2 \to P_5 \\ P_5 \to P_3 \to P_4 \to P_6 \\ P_6 \to P_1 \end{matrix} \tag{8}$$

We have Eq. (9) that defines the maximum point between the elements of the group of points. It follows that there is also the minimum point, hence Eq. (10). Assuming that these definitions mark the max–minimum of an arc, Eq. (11) thus defines the central tendency of the group. In other words, Eq. (11) is the center point of a circle where the value of n dictates the sample size. Applying the center point, the r_{XY} of Eq. (12) defines the radius of a circle. Similar approach has been performed when searching for the missing airplane MH370 deduced from the Inmarsat data (Zweck 2014).

As a result, Fig. 7 is a plot that renders: the machine or equipment's Center Point; the zones of the dotted circle that is the space allocated for tripod's positioning. The

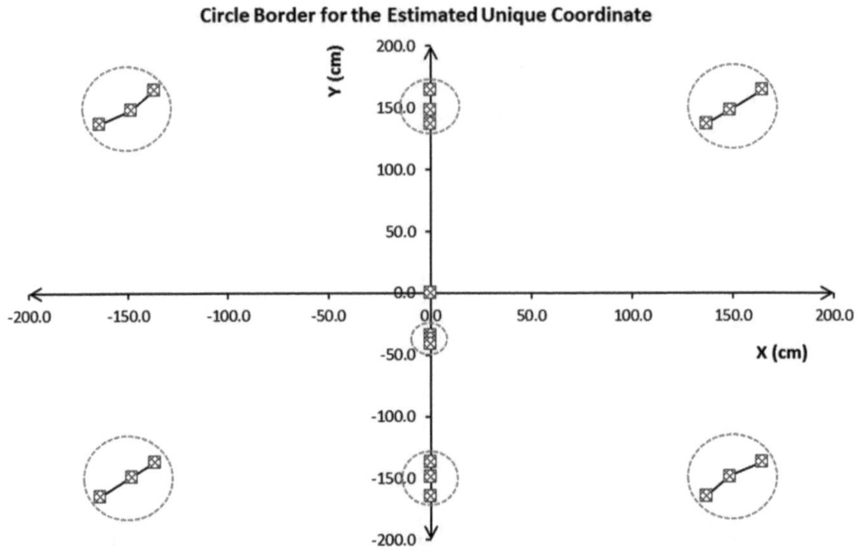

Fig. 7 The *dotted circles* mark the zones where the tripods may be located

zone could become the reference for a tripod's new location within the WC. A tripod location, however, is dependent upon the machine orientation and work volume, the machine standard operating procedure (SOP), and the operator's PM in relation to the SOP.

Let there be atmost four views in a configuration workspace within which the operator's PM is observable. We name the views as the side view, the rear view, the front view, and the top view. There are three digital cameras on tripods placed in workspace. We arrange them in such a way that the cameras record the subject's PM from some of the named views mentioned in the preceding. There are atmost 2^6 possible combinations for a three-camera installation based on the number of nodes in WC. Solving Eq. (8) results in the adjacency matrix for any one of the combinations.

$$P_{\max}(X,Y) = \{((\max(x), \max(y)))|\, x, y \in X, Y\} \tag{9}$$

$$P_{\min}(X,Y) = \{((\min(x), \min(y)))|x, y \in X, Y\} \tag{10}$$

$$P_{\text{center}}(X,Y) = \left\{\left(\frac{\sum x}{n}, \frac{\sum y}{n}\right)|x, y \in X, Y\right\} \tag{11}$$

$$r_X = \sqrt{\frac{\sum (x-\bar{x})^2}{n-1}};\ r_Y = \sqrt{\frac{\sum (y-\bar{y})^2}{n-1}};\ r_{XY} = \begin{cases} r_x = r_y \\ r_x|(r_y < r_x) \\ r_y|(r_x < r_y) \end{cases} \tag{12}$$

4 Discussion

Suppose we have decided to install the tripods at P_1, P_4, P_5 following the geometry depicted in Fig. 3. The locations to place the tripods can be somewhere nearby the dotted circles of Fig. 8. The cameras, nonetheless, must stay within the circles. The initial position for the operator maintains. We have Fig. 8 that represents the relationship between the points in WG and the nodes of WC. Three nodes display slings. These are purposely created to illustrate the desired tripod locations. This is a configuration where the planned tripods' installation is described as CONFIG-1, in which Eq. (A3) defines and characterizes with slings. The matrix form is solved in Eq. (A4). The specified zones may be obtained given the subject's height measurement.

From D5, it is straightforward that among the possible camera combinations mentioned in D6, one must consider D4 such that the video streaming runs as the tripod placement follows the expected locations. We implement a three-camera streaming so that the view coverage is extensive. We claim that a multiple camera system surpasses a one-camera system, in terms of, image coverage. On the contrary, it has been proven that an optical system composed of an LCD projector, a digital camera, and a microcomputer has the capability to measure three-dimensional human foot model (Del Vecchio et al. 2012). The idea of applying more cameras is to capture an image from different angles and views. This is to allow us to come up with multiple deductions from the information recorded. Our system is static where the algorithm will have to identify the best views and the next best views. This process continues throughout the operation. Similar approach has been tested where a camera was attached onto the robot's end effector and the manipulator with a laser range scanner. The system would model the in situ object in a three-dimensional view (Torabi and Gupta 2011).

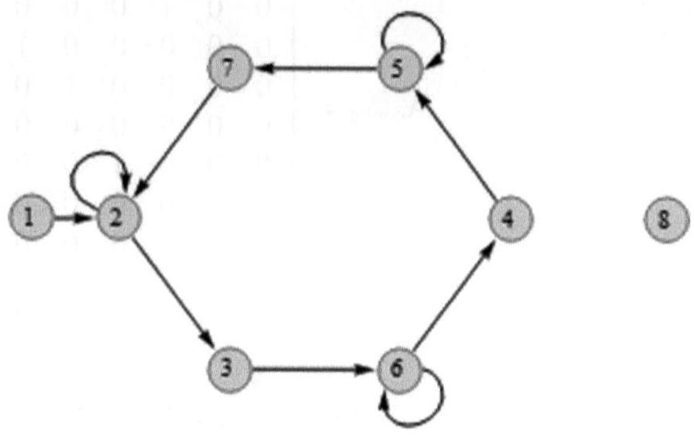

Fig. 8 The digraph representation of WC-1, where self-looping occur on nodes 2, 5, and 6

5 Conclusion

We propose the procedures to approximate the locations to place the tripods, in some way, to provide a clear relationship between the anthropometric measurement and the workspace geometry. It has been observed that regardless of the subjects' degrees of tallness, the proposed mathematical models would produce results of some patterns. The most noticeable pattern is the convergence of the coordinates. We deduce that, for a typical local manufacturing setting, there is a generalized workspace configuration usable for a wide-ranging machine operators' height measurement. In fact, it is obvious that they are Asians. Implementing the proposed procedures, the camera placement is therefore defined by digraphs.

Acknowledgment This work was supported in part by the Universiti Teknikal Malaysia Melaka under Grant PJP/2013/FKP(5D)/S01173. The authors thank Mohamad Farhan Mohamad Razif for conducting the survey of subjects' anthropometric data.

Appendix

$$\mathrm{A}_{\text{geometry}} = \begin{pmatrix} 0 & 1 & 0 & 0 & 0 & 0 & 0 & 0 \\ 0 & 0 & 1 & 0 & 0 & 0 & 0 & 0 \\ 0 & 0 & 0 & 1 & 0 & 0 & 0 & 0 \\ 0 & 0 & 0 & 0 & 1 & 0 & 0 & 0 \\ 0 & 0 & 0 & 0 & 0 & 1 & 0 & 0 \\ 0 & 0 & 0 & 0 & 0 & 0 & 1 & 0 \\ 0 & 0 & 0 & 0 & 0 & 0 & 0 & 1 \\ 0 & 0 & 0 & 0 & 0 & 0 & 0 & 0 \end{pmatrix} \tag{A1}$$

$$\mathrm{A}_{\text{config}} = \begin{pmatrix} 0 & 1 & 0 & 0 & 0 & 0 & 0 & 0 \\ 0 & 0 & 1 & 0 & 0 & 0 & 0 & 0 \\ 0 & 0 & 0 & 0 & 0 & 1 & 0 & 0 \\ 0 & 0 & 0 & 0 & 1 & 0 & 0 & 0 \\ 0 & 0 & 0 & 0 & 0 & 0 & 1 & 0 \\ 0 & 0 & 0 & 1 & 0 & 0 & 0 & 0 \\ 0 & 1 & 0 & 0 & 0 & 0 & 0 & 0 \\ 0 & 0 & 0 & 0 & 0 & 0 & 0 & 0 \end{pmatrix} \tag{A2}$$

$$\mathrm{A}_{\text{config}-1} = \begin{cases} 1 & \text{if } \mathbf{V} \\ 0 & \text{otherwise} \end{cases} \Bigg| \; V_{ij} = \begin{pmatrix} P_0 \to P_1 \to P_1 \\ P_1 \to P_2 \to P_5 \\ P_5 \to P_3 \to P_4 \\ P_4 \to P_4 \to P_6 \\ P_6 \to P_1 \end{pmatrix} \tag{A3}$$

$$A_{\text{config}-1} = \begin{pmatrix} 0 & 1 & 0 & 0 & 0 & 0 & 0 & 0 \\ 0 & 1 & 1 & 0 & 0 & 0 & 0 & 0 \\ 0 & 0 & 0 & 0 & 0 & 1 & 0 & 0 \\ 0 & 0 & 0 & 0 & 1 & 0 & 0 & 0 \\ 0 & 0 & 0 & 0 & 1 & 0 & 1 & 0 \\ 0 & 0 & 0 & 1 & 0 & 1 & 0 & 0 \\ 0 & 1 & 0 & 0 & 0 & 0 & 0 & 0 \\ 0 & 0 & 0 & 0 & 0 & 0 & 0 & 0 \end{pmatrix} \quad \text{(A4)}$$

References

Capisani LM et al (2012) Obstacle modelling oriented to safe motion planning and control for planar rigid robot manipulators. J Intell Rob Syst 71(2):159–178

Cesare C (1521) Architectura - Les livres d'Architecture, Como, G. da Ponte. http://architectura.cesr.univ-tours.fr/traite/Notice/BPNME276.asp?param=en. Accessed 16 June 2014

Del Vecchio S et al (2012) 3D measurement of human plantar foot by projection moiré technique. Int J Mechatron Manuf Syst 5(1):3–16

Gracia L, Sala A, Garelli F (2012) A supervisory loop approach to fulfill workspace constraints in redundant robots. Robot Auton Syst 60(1):1–15

McKean CM, Potvin JR (2001) Effects of a simulated industrial bin on lifting and lowering posture and trunk extensor muscle activity. Int J Ind Ergon 28(1):1–15

Ramdan NSA et al (2014) On lower-back pain and its consequence to productivity. J Industr Intell Inf 2(2):83–87

Torabi L, Gupta K (2011) An autonomous six-DOF eye-in-hand system for in situ 3D object modeling. Int J Robot Res 31(1):82–100

Zweck J (2014) How did inmarsat deduce possible flight paths for MH370? J Soc Ind Appl Math 47(4)

Chapter 27
Mango Size Classification Using RGB Color Sensor and Fuzzy Logic Technique

Ab Razak Mansor, Mahmod Othman, Mohd Nazari Abu Bakar, Khairul Adilah Ahmad and Tajul Rosli Razak

Abstract Fruit size is one of the most important criteria for classification and grading of mangoes. Currently, the size of a mango is determined by weight. In this project, a new model for classifying mango size using RGB color sensor and fuzzy logic are designed as an alternative automated grading of fruit size based on weight. RGB color sensor was used to measure the dimensions of mango such as major length, width and height in terms of reference color intensity value. Then, the fuzzy logic was used to classify mango size into small, medium, and large size. The proposed model is able to distinguish the three different classes of mango size automatically with more than 85 % accuracy.

Keywords Fuzzy logic · Mango size · RGB color sensor

1 Introduction

Mango is one of the famous fruits and one of the major export commodities of Malaysia. The planted area of mango in 2011 approximately 7,175.6 ha with a production value of 2,1021.5 tonnes or 111 million ringgit (FAO 2013). This situation explains that the demand of mango for local market and export are

A.R. Mansor (✉) · K.A. Ahmad
Faculty of Computer and Mathematical Sciences,
Universiti Teknologi MARA, Merbok, Kedah, Malaysia
e-mail: arazman@kedah.uitm.edu.my

M. Othman · T.R. Razak
Faculty of Computer and Mathematical Sciences,
Universiti Teknologi MARA, Arau, Perlis, Malaysia
e-mail: mahmod135085@perlis.uitm.edu.my

M.N. Abu Bakar
Faculty of Applied Sciences, Universiti Teknologi MARA,
Arau, Perlis, Malaysia
e-mail: mohdnazari@perlis.uitm.edu.my

N.A. Yacob et al. (eds.), *Regional Conference on Science, Technology and Social Sciences (RCSTSS 2014)*, DOI 10.1007/978-981-10-0534-3_27

increasing from year to year. In Malaysia, the Federal Agricultural Marketing Authority (FAMA) is responsible to ensure only a quality mango goes to the market. FAMA has launched Malaysia's best label as a benchmarking for quality fruits (FAMA 2006). Under this label, all fruits are required to be graded and inspected according to Malaysia's best quality specifications such as size, appearance, and maturity index.

Fruit size is one of the most important physical quality parameters for grading. Traditionally the method of determining the size of a mango fruit by weight was done by a human grader (Spreer and Müller 2011). The factor they consider in determining the size of mango is by looking at the shape of that fruit. This is the easiest and fastest way to determine the size of fruits but the level of accuracy in deciding whether that fruit is small, medium, or large is low. Furthermore, the problem with the traditional method is difficulty for humans to maintain their focus when dealing with a large quantity of mangoes. So, the appropriate method to determine the size of a mango is needed to solve those problems above.

In recent years, computer visions are most widely used for fruits sorting and size grading. Computer vision system has been used to develop classification algorithm for papaya size grading using shape characteristic analysis (Riyadi et al. 2007). Image captured on a mango was analyzed based on counted number of pixels on the mango region and the relationship between mango pixels and mango weights is used to estimate weight of Chokanan mangoes (Teoh and Syaifudin 2007). Tajul Rosli et al. (2012) developed a methodology and algorithm to grade a mango using computer vision and classifiy them using fuzzy logic technique. Even though computer vision has become a success in fruits grading, illuminating is one of a drawbacks of that system (Jyoti and Balaji 2012).

To classify fruits, there are lots of methods to identify and classify the fruits such as statistical algorithms, neural network, and fuzzy logic classifiers (Iraji et al. 2011). Fuzzy logic allows the use of linguistic values of variables and imprecise relationships for modelling system manner. In recent years, more and more applications of fuzzy theory to fruits have been reported such as watermelon, banana, pineapple. They used Mamdani fuzzy inference system to classify automatically the fruits and found good agreement between the results from fuzzy prediction and human experts (Naganur and Sannakki 2012).

This research used color sensor to detect the intensity of red color on reference color for determining the size of mango. This research is a first attempt to determine the fruit size by using RGB color sensor. Currently, RGB color sensor has been used to determine the mango ripeness and has given a good classification result (Ab Razak et al. 2014). The objectives of this paper were thus to define the mango size based on intensity of red color on reference color and evaluate the calibration efficiency by classifying the mango size index using fuzzy logic. This paper is organized as follows: Sect. 2 gives description about methodology we are using here. Section 3 gives results of fuzzy size classification. Finally, some conclusions are represented in Sect. 4.

2 Methodology

RGB Color Sensor Data Acquisition

To obtain the size of mango, data acquisition system using RGB color sensor based on printed reference color is designed. The reference color is created from the hue saturation lightness (HSL) color scheme and red color with hue 0° is chosen. The changes in color intensity will be measured by RGB color SUPER I detection mode. This mode is able to detect changes in color intensity more accurately. Next, we calibrate the intensity value of the increased RGB sensor with the lightness of printed reference color as shown in Fig. 1. This technique was assigned to give higher intensity value for large dimension of mangoes.

This measurement technique is shown in Fig. 2. The sensor is set up vertically and positioned with a constant distance of 70 mm so as to get the required data.

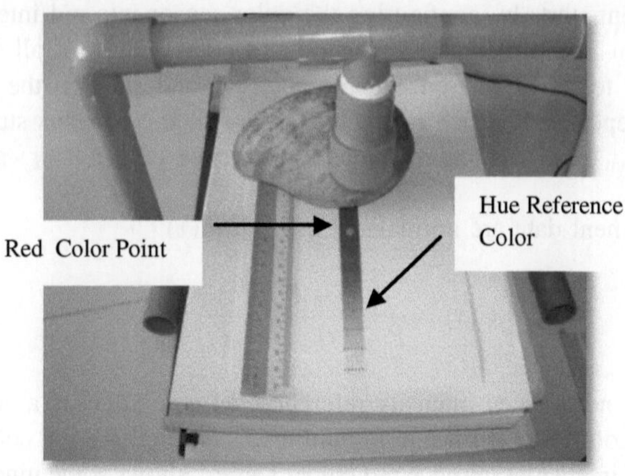

Fig. 1 RGB color sensor calibration

Fig. 2 RGB color sensor data acquisition

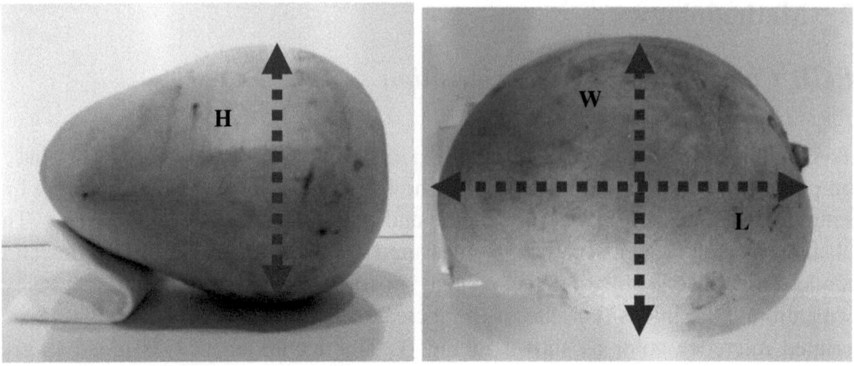

Fig. 3 Mango dimension

All the fruits were measured with respect to their length (L), maximum width (W), and maximum thickness (H) as shown in Fig. 3.

Measurement starts by configuring the color sensor into red intensity based on the steps given in the manual (Keyence 2012). The values of all dimensions are determined in terms of reference color intensity and store in the amplifier. The process was repeated for each mango for each category. In this study, we set the minimum (R_{min}) and maximum intensity reference color (R_{max}) are 0 and 600, respectively.

The experiment data are normalized using Eq. (1)

$$C(x) = \frac{I(x) - R_{min}(x)}{R_{max}(x) - R_{min}(x)} \times 100 \quad (1)$$

where $C(x)$ is normalized intensity reference red color. $I(x)$ is a current intensity reference red color. $R_{min}(x)$ is a minimum intensity reference red color and $R_{max}(x)$ is a maximum intensity reference red color. Table 1 shows some nine mangoes data collected using RGB color sensor and normalized data using Eq. (1).

Table 1 Measured range value of width, length, height

Mango	Width	Length	Height
1	34	85	40
2	44	90	45
3	45	95	46
4	51	95	64
5	62	96	65
6	64	97	66
7	64	97	70
8	65	97	72
9	73	98	81

Table 2 Chokanan size grade (FAMA 2006)

Size	Code	Weight (grams/pieces)
Large	L	301–400
Medium	M	201–300
Small	S	150–200

Sample

The size of mango is counted and classified according to weight classes as recommended by the FAMA. Size classification is essential for determining mango grading and also for pricing. The following grading scheme set by the FAMA is shown in Table 2.

The mangoes are collected from a plantation and graded manually by human grader based on weight in Table 1. About 60 mangoes contains 20 mangoes from each category are collected and weighed. The type of mango being used in this research is *Chokanan* because of their high demand in local and international market.

3 Fuzzy Size Classification

To classify the mango fruit into small, medium, and large categories, we need to obtain the range of width, length, and height for each fruit in each category. These range values are used as a reference and a range input of fuzzy classification system. A total of 60 mangoes are used in determining the range value of width, length, and height of each category as shown in Table 3.

The input variables are then divided into linguistic variables low, medium, and high. Membership functions are then formed assigning the proper range to respective linguistic variables as shown in Table 4. In this paper we have used the

Table 3 Measured min max value of width, length, height

	Width		Length		Height	
Category	Min	Max	Min	Max	Min	Max
Small	34	45	85	95	40	46
Medium	51	64	95	97	64	66
Large	64	73	97	98	70	81

Table 4 The range value of INPUT linguistic variable

Linguistic term	Range		
	Width	Length	Height
Low	0–30	0–30	0–30
Medium	20–60	20–70	20–50
High	50–100	60–100	40–100

trapezoidal membership function for converting the crisp set into fuzzy set and membership function of the input variables length (L) as shown in Fig. 4.

The rules determine input and output membership function that will be used in inference process. These rules are linguistics and are entitled "IF-THEN" rules. The if-then rules generated based on Chokanan size grade in Table 2 and the rules are shown in Table 5.

The output variable is the mango size classification, which has three linguistic variables. The degree of membership functions is given by Eq. (2).

$$\mu_s(x) = \underset{k}{\text{Max}}[\min[\mu_A(L_1), \mu_B(W_1), \ldots.]], \quad k = 1, 2, 3\ldots r \tag{2}$$

This expression determines an output membership function value for each active rule. When one rule is active, an AND operation is applied between inputs. The fuzzy linguistic variables of output variable are shown in Table 6 and the membership function of the output variable of mango size is shown in Fig. 5.

Fig. 4 The membership function of length input

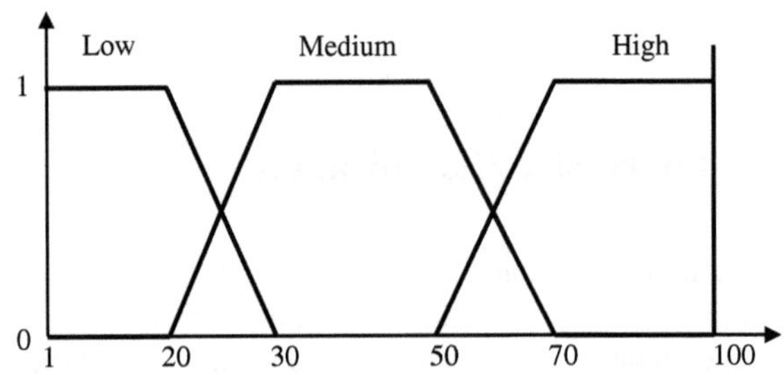

Table 5 Rules for the fuzzy system

No of rules	Input variables			Output variable
	Length	Width	Height	Size
1	Low	Low	Low	Small
2	Low	Low	Medium	Small
3	Low	Low	High	Small
27	High	High	High	Large

Table 6 The range value of output linguistic variable

Linguistic term	Range
Small	<30
Medium	$20 \leq S \leq 60$
Large	$50 \leq S \leq 100$

Fig. 5 The membership function of output variable

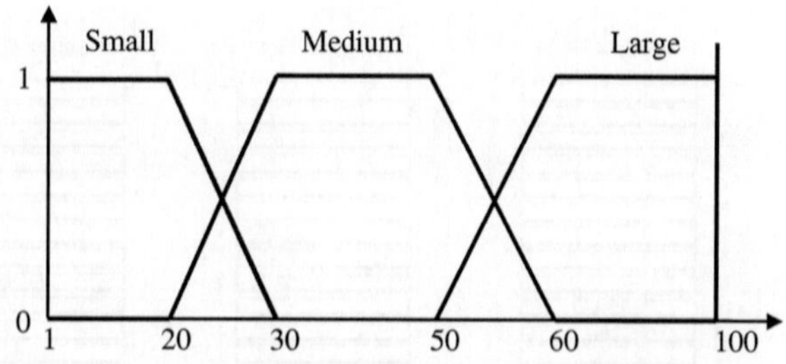

We implement our algorithm using MATLAB's Fuzzy Logic Toolbox. Figure 6 shows the fuzzy inference system (FIS) consists of the three inputs (width, length, and height) and one which is output (mango size) from MATLAB.

All these membership functions are developed by using FIS editor and membership function editor from MATLAB. A total of 27 rule statements have been used to classify the mango size. Figure 7 shows the rule viewer which consists of the system's input and output. It also consists of the defuzzification results. The first to third column represent the width, length, and height values while the last column is the category of mango size column. The last row of the last column which is the category column shows the defuzzification results where the category of mango fruit is obtained.

Based on the deffuzzification results from the Fig. 7, the mango fruit fulfilled Rule 13 in Table 5, where the width is medium, the length is medium, and the height is low. The value of category is calculated based on centroid method, and then the classification of mango is determined based on the crisp logic given in Table 7.

For evaluation of the developed FIS, 15 mangoes are selected randomly from the testing sample. The dimension of 15 mangoes were measured in order to validate the mango fruit into small, medium, and large categories. Developed input data were transferred to the rule viewer for classification process. Results of the comparison between human grader and proposed method are shown in Table 8.

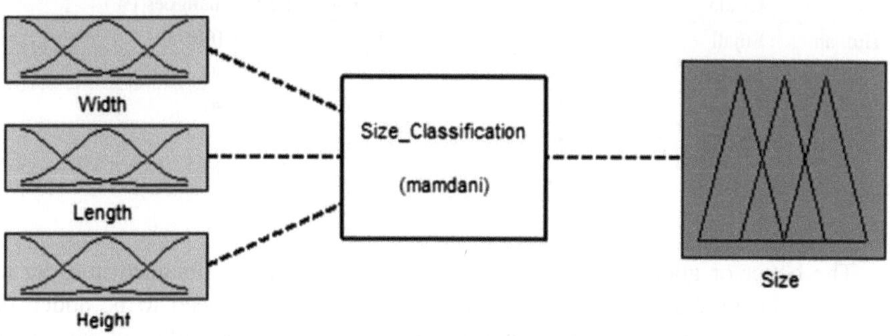

Fig. 6 The fuzzy inference system consists of *three* inputs and *one* output

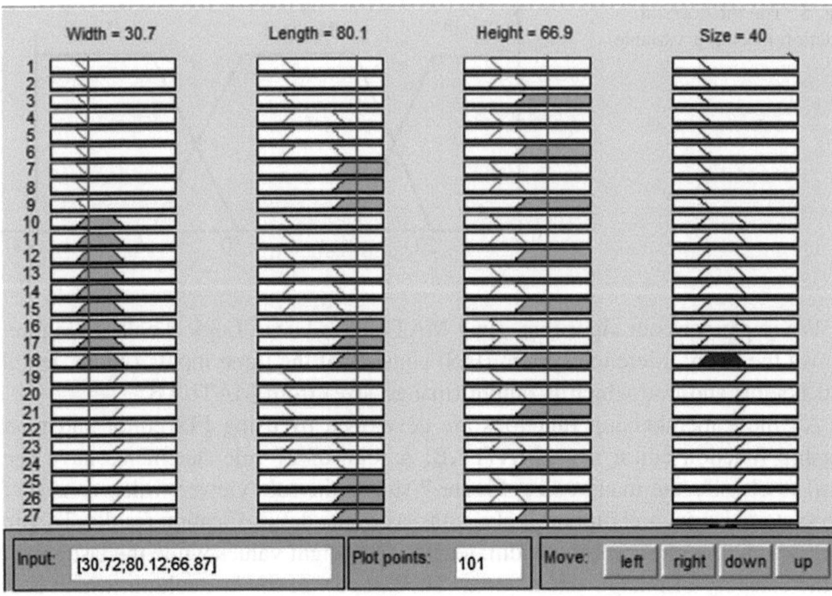

Fig. 7 Deffuzzification result from the rule viewer

Table 7 Defuzzification range

Deffuzzification output (%)	Mango fruit category
(category <= 30)	Small
(category >= 31) && (category <= 50)	Medium
(category >= 51)	Large

Table 8 Comparison of proposed fuzzy system and human expert in grading of Chokanan mangoes

Fuzzy logic prediction						
	Mango Grade	Small	Medium	Large	No of mangoes used	Correctly classified mangoes (%)
Human expert	Small	5	0	0	5	100
	Medium	0	4	1	5	80
	Large	0	1	4	5	80
Total observed (%)		5	5	5	15	–
		100 %	80 %	80 %	–	87

The better or good range for each dimension needs to be obtained in order to improve the grading system performance. The fuzzy rules need to be added or improved especially for the fruit that lies closely between two different categories and the additional feature that incorporate differences between the categories such as color need to be added. The results of the fuzzy logic system evaluated are against the human graders method to measure the accuracy for mango size. Based on the results in Table 8, our proposed model has increased the accuracy of grading by as much as 87 %. This shows that the grading system using fuzzy logic has a high potential of accurateness in grading the mango fruit.

4 Conclusion

The research has proven that by using RGB color sensor and fuzzy logic as classification algorithm, the accuracy of mango grading is more than 85 %. Fuzzy logic is successfully applied to serve as a decision support technique in mango size classification. Grading results obtained from fuzzy logic shows a very good agreement with the results from the human expert. Performance of the system can be improved by creating better membership function. Furthermore, this research is a first attempt to determine mango size using RGB color sensor. The result shows that RGB color sensor can be used accurately as data acquisition and can be applied to other classification fruits.

Acknowledgments The authors would like to thank Ministry of Education Malaysia on the financial support and MARA University of Technology, Malaysia on the research infrastructure. The authors would like to express appreciation to Mdm Noor Rasidah Ali and Miss Suraiya for helping us with data collection and analysis.

References

Ab Razak M, Mahmud O, Mohd Nazari AB, Khairul Adilah A, Tajul Rosli R (2014) Fuzzy Ripening mango index using RGB colour sensor model. Researchers World J Art Sci Commer v(2):1–9

FAMA (2006) Menuju kearah Kualiti Malaysia's Best, Siri Panduan Kualiti Manga

FAO (2013) Statistical division, food and agricultural organization of united nations (FAO). http://www.fao.org/statistics/en

Iraji MS, Tosinia A, Azad I, Branch A (2011) Classification tomatoes on machine vision with fuzzy the mamdani inference, adaptive neuro fuzzy inference system based (Anfis-Sugeno). Aust J Basic Appl Sci 5(11):846–853

Jyoti AK, Balaji S (2012) Computer vision and image analysis based techniques for automatic characterization of fruits—a review. Int J Comput Appl IJCA 50(6)

Keyence (2012) Digital fiber sensor FS-N10 series user's manual

Naganur H, Sannakki S (2012) Fruits sorting and grading using fuzzy logic. Int J Adv Res Comput Eng Technol 1(6):117–122

Riyadi S, Rahni AAA, Mustafa MM, Hussain A (2007) Shape characteristics analysis for papaya size classification. In: Paper presented at the 5th student conference on research and development, 11 Dec 2007 (SCOReD 2007)

Spreer W, Müller J (2011) Estimating the mass of mango fruit (Mangifera Indica, cv. Chok Anan) from its geometric dimensions by optical measurement. Comput Electron Agric 75:125–131

Teoh C, Syaifudin ARM (2007) Image processing and analysis techniques for estimating weight of Chokanan mangoes. J Trop Agric Food Sci 35(1):183

Tajul Rosli R, Mahmud O, Mohd Nazari AB, Khairul Adilah A, Ab Razak M (2012) Mango grading by using fuzzy image analysis. In: Paper presented at the international conference on agricultural, environment and biological sciences (ICAEBS'2012) Phuket, May 26–27, 2012

Chapter 28
Performance Analysis of 2-Point Explicit Group (2-EG) Method for Solving Second-Order Composite Closed Newton-Cotes Quadrature System

Mohana Sundaram Muthuvalu, Elayaraja Aruchunan and Jumat Sulaiman

Abstract In this paper, the effectiveness of 2-Point Explicit Group (2-EG) iterative method with second-order composite closed Newton-Cotes (2-CCNC) quadrature scheme for solving second kind linear Fredholm integral equations is investigated. The formulation and implementation of the proposed method are presented. In addition, computational complexity analysis and numerical results of test examples are also included to verify the performance of the proposed method.

Keywords Computational complexity · Dense system · Fredholm equation · Iterative solver · Quadrature scheme

1 Introduction

In the present paper, numerical solutions for linear Fredholm integral equations of the second kind, given by

$$\varphi(x) - \int_{\alpha}^{\beta} K(x,t)\varphi(t)\mathrm{d}t = f(x), \quad x \in [\alpha, \beta] \tag{1}$$

M.S. Muthuvalu (✉)
Faculty of Science and Information Technology, Universiti Teknologi PETRONAS, Perak, Malaysia
e-mail: msmuthuvalu@gmail.com

E. Aruchunan
Department of Mathematics and Statistics, Curtin University, Perth, Australia
e-mail: earuchunan@yahoo.com

J. Sulaiman
Faculty of Science and Natural Resources, Universiti Malaysia Sabah, Sabah, Malaysia
e-mail: jumat@ums.edu.my

N.A. Yacob et al. (eds.), *Regional Conference on Science, Technology and Social Sciences (RCSTSS 2014)*, DOI 10.1007/978-981-10-0534-3_28

are considered. The function $f(x)$ is given, $K(x,t)$ is the kernel of the integral equation and $\varphi(x)$ is the solution to be determined. It is assumed that the $f(x)$ and $K(x,t)$ are continuous, and problem (1) has a unique solution. Integral Eq. (1) are encountered in numerous applications in many fields such as continuum mechanics, potential theory, geophysics, and heat transfer problems (Wang 2006).

The basic concept of the numerical methods is to reduce the problem (1) in the form of a finite-dimensional linear system which is mostly dense. In recent years, some efficient numerical methods for solving problem (1) have been proposed. Some recent references are Allouch and Sablonnière (2014), Chen and Wong (2013), Mastroianni and Milovanović (2009), Muthuvalu and Sulaiman (2011, 2013) and Zhong (2013). In one of the recent works, the performance of the 2-Point Explicit Group (2-EG) iterative methods for solving first order composite closed Newton–Cotes (1-CCNC) algebraic equations was investigated (Muthuvalu et al. 2014). Consequently, in this paper, an application of the standard 2-EG method for solving second-order composite closed Newton–Cotes (2-CCNC) algebraic equations associated with the numerical solutions of the problem (1) will be investigated.

The rest of this paper is organised as follows. In Sect. 2, method of solution based on 2-CCNC scheme and 2-EG method for solving problem (1) are presented. The computational complexity analysis and numerical results from the simulations are given in Sects. 3 and 4 respectively. Finally, discussions and concluding remarks are summarized in Sect. 5.

2 Method of Solution

In the next subsections, numerical approaches based on 2-CCNC scheme and 2-EG method to solve problem (1) will be elaborated.

2-CCNC Algebraic Equation

As aforementioned, 2-CCNC scheme is utilized in order to construct 2-CCNC algebraic equations for problem (1). Let the interval $[\alpha, \beta]$ be divided uniformly into N even subintervals and the discrete set of points of x and t, respectively, be given by $x_i = \alpha + ih \quad (i = 0, 1, 2, \ldots, N-2, N-1, N)$ and $t_j = \alpha + jh (j = 0, 1, 2, \ldots, N-2, N-1, N)$. Meanwhile, h is the constant step size defined as

$$h = \frac{\beta - \alpha}{N}. \tag{2}$$

For simplicity, the following notations, i.e., $K_{i,j} \equiv K(x_i, t_j)$, $\hat{\varphi}_i \equiv \hat{\varphi}(x_i)$, $\hat{\varphi}_j \equiv \hat{\varphi}(t_j)$ and $f_i \equiv f(x_i)$ will be applied subsequently.

An application of the 2-CCNC scheme reduces problem (1) to

$$\hat{\varphi}_i - \sum_{j=0}^{N} w_j K_{i,j} \hat{\varphi}_j = f_i \tag{3}$$

for $i = 0, 1, 2, \ldots, N-2, N-1, N$. The solution $\overset{\wedge}{\varphi}$ is an approximation of the exact solution φ to (1) and w_j is the weights of 2-CCNC scheme that satisfies the following condition

$$w_j = \begin{cases} \frac{h}{3}, & j = 0, N \\ \frac{4h}{3}, & j = 1, 3, 5, \ldots, N-1 \\ \frac{2h}{3}, & otherwise \end{cases}. \tag{4}$$

Moreover, algebraic Eq. (3) can be represented in matrix form as

$$A \overset{\wedge}{\varphi} = f \tag{5}$$

where $A = \left[a_{i,j}\right] \in \Re^{(N+1)\times(N+1)}$ is a real coefficient matrix with elements

$$a_{i,j} = \begin{cases} 1 - w_j K_{i,j}, & i = j \\ -w_j K_{i,j}, & i \neq j \end{cases}. \tag{6}$$

2-Point Explicit Group Iterative Method

In this subsection, the formulation and implementation of the 2-EG method to solve corresponding 2-CCNC algebraic equations are discussed. Based on algebraic Eq. (3), let us consider any group of two points, i.e., x_i and x_{i+1} that are used simultaneously to calculate the values of $\overset{\wedge}{\varphi}$. Thus, at point x_i, the solution is approximated by

$$\hat{\varphi}_i - \sum_{j=0}^{N} w_j K_{i,j} \hat{\varphi}_j = f_i \text{ (i.e. Eq. (3))}. \tag{7}$$

Whereas, at point x_{i+1} the solution is given by

$$\hat{\varphi}_{i+1} - \sum_{j=0}^{N} w_j K_{i+1,j} \hat{\varphi}_j = f_{i+1}. \tag{8}$$

Now, the Eqs. (7) and (8) can be written simultaneously in the matrix form as follows:

$$\begin{bmatrix} 1+a_{i,i} & a_{i,i+1} \\ a_{i+1,i} & 1+a_{i+1,i+1} \end{bmatrix} \begin{bmatrix} \hat{\varphi}_i \\ \hat{\varphi}_{i+1} \end{bmatrix} = \begin{bmatrix} f_i \\ f_{i+1} \end{bmatrix} - \begin{bmatrix} \sum_{j=0}^{i-1} a_{i,j}\hat{\varphi}_j \\ \sum_{j=0}^{i-1} a_{i+1,j}\hat{\varphi}_j \end{bmatrix} - \begin{bmatrix} \sum_{j=i+2}^{N} a_{i,j}\hat{\varphi}_j \\ \sum_{j=i+2}^{N} a_{i+1,j}\hat{\varphi}_j \end{bmatrix}, \tag{9}$$

where coefficient matrix with size (2×2) can be easily inverted. Thus, the Eq. (9) can be written in explicit form as

$$\begin{bmatrix} \hat{\varphi}_i \\ \hat{\varphi}_{i+1} \end{bmatrix} = \frac{1}{|B|} \begin{bmatrix} 1+a_{i+1,i+1} & -a_{i,i+1} \\ -a_{i+1,i} & 1+a_{i,i} \end{bmatrix} \left(\begin{bmatrix} f_i \\ f_{i+1} \end{bmatrix} - \begin{bmatrix} \sum_{j=0}^{i-1} a_{i,j}\hat{\varphi}_j \\ \sum_{j=0}^{i-1} a_{i+1,j}\hat{\varphi}_j \end{bmatrix} - \begin{bmatrix} \sum_{j=i+2}^{N} a_{i,j}\hat{\varphi}_j \\ \sum_{j=i+2}^{N} a_{i+1,j}\hat{\varphi}_j \end{bmatrix} \right), \tag{10}$$

where $|B| = \det B = (1+a_{i,i})(1+a_{i+1,i+1}) - (a_{i+1,i})(a_{i,i+1})$. This simplifies Eq. (10) to the formulae

$$\begin{bmatrix} \hat{\varphi}_i \\ \hat{\varphi}_{i+1} \end{bmatrix} = \frac{1}{|B|} \begin{bmatrix} (1+a_{i+1,i+1})(f_i - C_{i,j} - D_{i,j}) - a_{i,i+1}(f_{i+1} - C_{i+1,j} - D_{i+1,j}) \\ -a_{i+1,i}(f_i - C_{i,j} - D_{i,j}) + (1+a_{i,i})(f_{i+1} - C_{i+1,j} - D_{i+1,j}) \end{bmatrix} \tag{11}$$

where

$$C_{i,j} = \sum_{j=0}^{i-1} a_{i,j}\hat{\varphi}_j,$$

$$C_{i+1,j} = \sum_{j=0}^{i-1} a_{i+1,j}\hat{\varphi}_j,$$

$$D_{i,j} = \sum_{j=i+2}^{N} a_{i,j}\hat{\varphi}_j,$$

and

$$D_{i+1,j} = \sum_{j=i+2}^{N} a_{i+1,j}\hat{\varphi}_j.$$

For an even subintervals, N, the number of discrete points is odd (i.e., $N+1$), which results in one ungrouped point. Therefore, for an ungrouped point, i.e., x_N, will be computed based on the following point iteration (12)

$$\hat{\varphi}_N = \frac{1}{1+a_{N,N}}\left[f_i - \sum_{j=0}^{N-1} a_{N,j}\hat{\varphi}_j\right]. \quad (12)$$

By considering formulations (11) and (12), algorithm of 2-EG method associated with 2-CCNC algebraic equations for both cases, i.e., complete grouped (Case 1) and incomplete grouped (with one single point ungrouped) (Case 2) are described in Algorithm 1 and 2, respectively.

Algorithm 1. 2-EG method for Case 1

Step i. Set all the parameters

Step ii. Iteration cycle

for $i = 0,2,4,\cdots,N-3,N-1$

$$\begin{bmatrix}\hat{\varphi}_i \\ \hat{\varphi}_{i+1}\end{bmatrix}^{(k+1)} \leftarrow \frac{1}{|B|}\begin{bmatrix}(1+a_{i+1,i+1})\left(f_i - C_{i,j}^{(k+1)} - D_{i,j}^{(k)}\right) - a_{i,i+1}\left(f_{i+1} - C_{i+1,j}^{(k+1)} - D_{i+1,j}^{(k)}\right) \\ -a_{i+1,i}\left(f_i - C_{i,j}^{(k+1)} - D_{i,j}^{(k)}\right) + (1+a_{i,i})\left(f_{i+1} - C_{i+1,j}^{(k+1)} - D_{i+1,j}^{(k)}\right)\end{bmatrix}$$

Step iii. Convergence test. If the converge criterion i.e., the maximum norm $\left\|\hat{\varphi}^{(k+1)} - \hat{\varphi}^{(k)}\right\| \le \varepsilon$ (where ε is the convergence criterion) is satisfied, go to Step iv. Otherwise, repeat the iteration cycle (i.e., go to Step ii).

Step iv. Stop.

Algorithm 2. 2-EG method for Case 2

Step i. Set all the parameters

Step ii. Iteration cycle

for $i = 0,2,4,\cdots,N-4,N-2$

$$\begin{bmatrix}\hat{\varphi}_i \\ \hat{\varphi}_{i+1}\end{bmatrix}^{(k+1)} \leftarrow \frac{1}{|B|}\begin{bmatrix}(1+a_{i+1,i+1})\left(f_i - C_{i,j}^{(k+1)} - D_{i,j}^{(k)}\right) - a_{i,i+1}\left(f_{i+1} - C_{i+1,j}^{(k+1)} - D_{i+1,j}^{(k)}\right) \\ -a_{i+1,i}\left(f_i - C_{i,j}^{(k+1)} - D_{i,j}^{(k)}\right) + (1+a_{i,i})\left(f_{i+1} - C_{i+1,j}^{(k+1)} - D_{i+1,j}^{(k)}\right)\end{bmatrix}$$

for $i = N$

$$\hat{\varphi}_N^{(k+1)} \leftarrow \frac{1}{1+a_{N,N}}\left[f_i - \sum_{j=0}^{N-1} a_{N,j}\hat{\varphi}_j^{(k+1)}\right]$$

Step iii. Convergence test. If the converge criterion i.e., the maximum norm $\left\|\hat{\varphi}^{(k+1)} - \hat{\varphi}^{(k)}\right\| \le \varepsilon$ (where ε is the convergence criterion) is satisfied, go to Step iv. Otherwise, repeat the iteration cycle (i.e., go to Step ii).

Step iv. Stop.

3 Computational Complexity Analysis

An estimation amount of the computational work has been conducted in order to evaluate the computational complexity of 2-EG method for solving generated 2-CCNC algebraic Eq. (3). The computational work is estimated by considering the arithmetic operations performed per iteration. In estimating the computational work, the values of elements in A and $|B|$ are stored in advance. The number of arithmetic operations involved for 2-EG method is (excluding the convergence test)

$$\left(2N^2+2N\right)ADD/SUB+\left(2N^2+4N+2\right)MUL/DIV \tag{13}$$

per iteration if N is odd. Meanwhile, for the case of N is even, the number of arithmetic operations required is

$$\left(2N^2+N+1\right)ADD/SUB+\left(2N^2+3N+2\right)MUL/DIV \tag{14}$$

per iteration. The ADD/SUB and MUL/DIV represent additions/subtractions and multiplications/divisions operations respectively.

4 Numerical Results

The following three test problems are used to verify the performance of the 2-EG method.

Test Problem 1 (Wang 2006)
Consider the Fredholm integral equation of the second kind

$$\varphi(x)-\int_0^1\left(4xt-x^2\right)\varphi(t)\,\mathrm{d}t=x,\quad x\in[0,1], \tag{15}$$

and the exact solution is given by

$$\varphi(x)=24x-9x^2.$$

Test Problem 2 (Muthuvalu and Sulaiman 2011)
Consider the Fredholm integral equation of the second kind

$$\varphi(x)-\int_0^1\left(x^2+t^2\right)\varphi(t)\,\mathrm{d}t=x^6-5x^3+x+10,\quad x\in[0,1], \tag{16}$$

with the exact solution

$$\varphi(x) = x^6 - 5x^3 + \frac{1045}{28}x^2 + x + \frac{2141}{84}.$$

Test Problem 3 (Kanwal and Liu 1989)
Consider the Fredholm integral equation of the second kind

$$\varphi(x) - \int_{-1}^{1} \left(xt + x^2t^2\right)\varphi(t)\mathrm{d}t = \left(x^2+1\right)^2, \quad x \in [-1, 1], \tag{17}$$

with the exact solution

$$\varphi(x) = \frac{125}{9}x^2 + 6x + 1.$$

For the numerical simulations, parameters such as number of iterations (*k*), computational time (*CPU*), and root mean square error (*RMSE*) are measured. The value of initial datum, $\hat{\varphi}^{(0)}$, is set to zero for all the test problems. The computations are performed on a personal computer with Intel(R) Core(TM) i3-2120 (3.30 GHz, 3.30 GHz) and 4.00 GB RAM, and the programs are compiled by using C language. Throughout the simulations, the convergence test considered is $\varepsilon = 10^{-10}$ and carried out on several different values of *N*. The numerical results of number of iterations and computational time for test problems 1–3 are tabulated in Table 1. Meanwhile, Figs. 1, 2 and 3 illustrated the logarithm of *RMSE* of the test problems 1–3, respectively. Also, results of logarithm of *RMSE* based on 1-CCNC scheme are included for the comparative analysis.

Table 1 Numerical results for test problems 1 to 3

N	Test problem 1		Test problem 2		Test problem 3	
	k	*CPU*	*k*	*CPU*	*k*	*CPU*
60	183	0.13	54	0.03	32	0.03
120	189	0.25	55	0.08	33	0.08
240	192	1.02	55	0.29	34	0.25
480	193	4.09	56	1.20	34	1.03
960	194	16.32	56	4.64	34	4.09
1920	194	64.59	56	18.39	34	16.28
3840	195	258.85	56	73.25	35	66.97
7680	195	1037.11	56	292.75	35	267.73

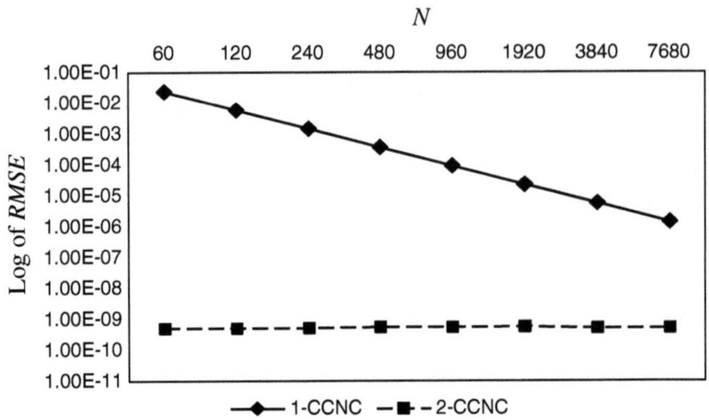

Fig. 1 Log of *RMSE* for test problem 1

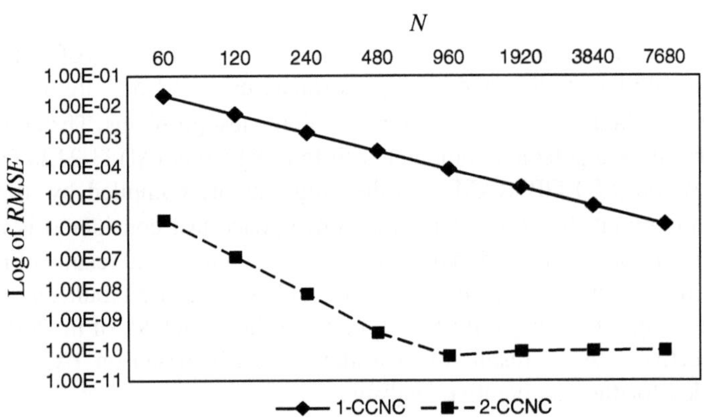

Fig. 2 Log of *RMSE* for test problem 2

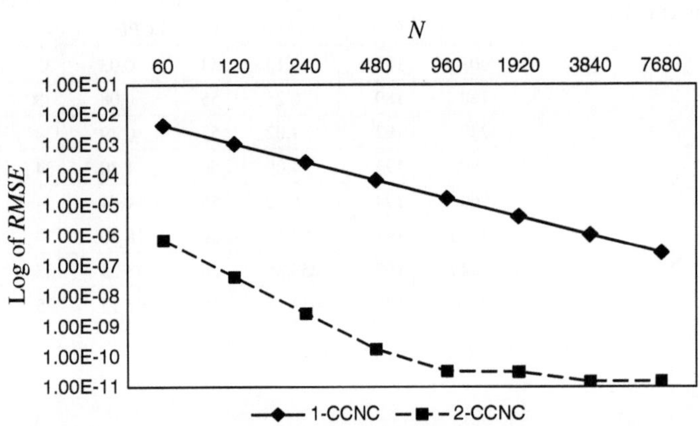

Fig. 3 Log of *RMSE* for test problem 3

5 Discussions and Conclusion

In this paper, the standard 2-EG iterative method has been successfully implemented in solving 2-CNCC algebraic equations associated with the numerical solution of problem (1). Some numerical results have been presented to illustrate the performance of the 2-EG method. Through the observation in Figs. 1, 2 and 3, increment of N improved the accuracy of numerical solutions. Also, implementations of the 2-CCNC scheme increased the accuracy of the numerical solutions compared to the 1-CCNC scheme.

Acknowledgment The first author gratefully acknowledges the financial support from the Universiti Teknologi PETRONAS for this research work.

References

Allouch C, Sablonnière P (2014) Iteration methods for Fredholm integral equations of the second kind based on spline quasi-interpolants. Math Comput Simul 99:19–27

Chen F, Wong PJY (2013) Solutions of Fredholm integral equations via discrete biquintic splines. Math Comput Model 57(3–4):551–563

Kanwal RP, Liu KC (1989) A Taylor expansion approach for solving integral equations. Int J Math Educ Sci Technol 20(3):411–414

Mastroianni G, Milovanović GV (2009) Well-conditioned matrices for numerical treatment of Fredholm integral equations of the second kind. Numer Linear Algebra Appl 16(11–12): 995–1011

Muthuvalu MS, Sulaiman J (2011) Half-Sweep Arithmetic Mean method with composite trapezoidal scheme for solving linear Fredholm integral equations. Appl Math Comput 217(12):5442–5448

Muthuvalu MS, Sulaiman J (2013) The Quarter-Sweep Geometric Mean method for solving second kind linear Fredholm integral equations. Bull Malays Math Sci Soc 36(4):1009–1026

Muthuvalu MS, Dass SC, Guan BH, Ching DLC, Sulaiman J (2014) Numerical solutions for linear Fredholm integral equations of the second kind using 2-Point Half-Sweep Explicit Group method. In: Zin WZW et al. (eds) Proceedings of the 3rd international conference on mathematical sciences. AIP Publishing LLC, New York, pp 186–191

Wang W (2006) A new mechanical algorithm for solving the second kind of Fredholm integral equation. Appl Math Comput 172(2):946–962

Zhong X-C (2013) A new Nyström-type method for Fredholm integral equations of the second kind. Appl Math Comput 219(17):8842–8847

Chapter 29
Effect of Inclination on Natural Convection Porous Cavity

Mat Salim Selamat and Anis Rosninawati Idayu Abd Rahim

Abstract A numerical investigation of natural convection in cavity filled with a fluid-saturated porous medium has been performed. A finite difference method is used to discretize the nondimensional governing equations. The resulting algebraic equation was solved by successive over-relaxation method. The results show that the inclined angle of the cavity has significant effect to heat transfer process. The maximum average Nusselt number is attained at about $\phi = 30^{\circ}$.

Keywords Natural convection · Porous media · Inclined cavity · Heat transfer

1 Introduction

Convective heat transfer has attracted significant attention over the last few decades. This interest is due to its wide range of applications, for example, in high performance insulation for building, chemical catalytic reactors, packed sphere beds, grain storage, and solar collector. Many studies with application to the previous research areas may be found in the books by Nield and Bejan (2006), Ingham and Pop (2005) and Vafai (2005).

Most of the studies on natural convection are devoted to the classical Rayleigh-Benard model (hot bottom wall and cold top wall) or to the case of rectangular or square cavity with one vertical wall heated and the opposite one cooled (Bejan 1979; Goyeau et al. 1996). However, in some engineering applications, enclosures are inclined to the direction of gravity. Hence, the flow structure

M.S. Selamat (✉)
Faculty of Computer and Mathematical Sciences, Universiti Teknologi MARA, Kuala Pilah, Negeri Sembilan, Malaysia
e-mail: matsalimselamat@ns.uitm.edu.my

A.R.I. Abd Rahim
Faculty of Computer and Mathematical Sciences, Universiti Teknologi MARA, Jengka, Pahang, Malaysia
e-mail: anis.nina@yahoo.com

N.A. Yacob et al. (eds.), *Regional Conference on Science, Technology and Social Sciences (RCSTSS 2014)*, DOI 10.1007/978-981-10-0534-3_29

and the heat transfer within the enclosure are modified by the components of buoyancy forced. The effects of inclination on natural convection in an enclosure have been discussed by several investigators, for example, Hart (1971), Holst and Aziz (1972), Ozoe et al. (1974), Rasoul and Prinos (1997) and Chamkha and Al-Mudhaf (2008). A good review on the study of the inclination of natural convection can be seen in Yang (1998).

The aim of this work is to study the effect of inclination angle of enclosure on natural convection in fluid-saturated porous media.

2 Mathematical Formulation

The heat transfer by natural convection across porous media is considered as shown in Fig. 1. The right wall is kept at low temperature (T_c) and the left wall at high temperature (T_h) where $T_h > T_c$. The bottom and top walls are adiabatic.

Applying Darcy's flow model and the Boussinesq approximation, the governing equations after nondimensionalization can be written as follows (Selamat et al. 2012):

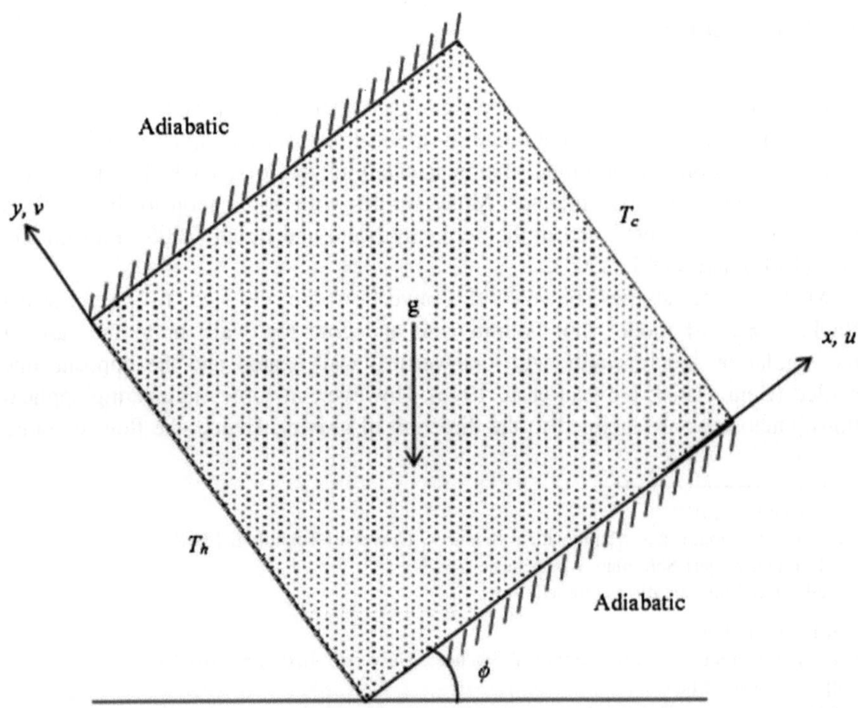

Fig. 1 Schematic diagram of the physical model and coordinate system

$$\frac{\partial^2 \Psi}{\partial X^2} + \frac{\partial^2 \Psi}{\partial Y^2} = Ra\left(\frac{\partial \theta}{\partial Y}\sin\phi - \frac{\partial \theta}{\partial X}\cos\phi\right) \tag{1}$$

$$\frac{\partial \Psi}{\partial Y}\frac{\partial \theta}{\partial X} - \frac{\partial \Psi}{\partial X}\frac{\partial \theta}{\partial Y} = \frac{\partial^2 \theta}{\partial X^2} + \frac{\partial^2 \theta}{\partial Y^2} \tag{2}$$

where Ψ is nondimensional stream functions, θ is nondimensional temperature and Ra is the Rayleigh number and defined as:

$$Ra = \frac{g\beta K(T_h - T_c)L}{\nu\alpha} \tag{3}$$

where g is magnitude of gravitational acceleration, β is coefficient of thermal expansion, K is permeability of the porous medium, ν is kinematic viscosity of the fluid and α is effective thermal diffusivity.

The boundary conditions can be written as:

$$\Psi(0, Y) = 0, \quad \theta(0, Y) = 0.5, \tag{4}$$

$$\Psi(1, Y) = 0, \quad \theta(1, Y) = -0.5, \tag{5}$$

$$\Psi(X, 0) = 0, \quad \frac{\partial \theta(X, 0)}{\partial Y} = 0, \tag{6}$$

$$\Psi(X, 1) = 0, \quad \frac{\partial \theta(X, 1)}{\partial Y} = 0. \tag{7}$$

The local Nusselt number along the hot wall is given, by

$$\mathrm{Nu}_{\mathrm{loc}} = \left(\frac{\partial \theta}{\partial X}\right)_{X=0} \tag{8}$$

The average Nusselt number is defined as

$$\mathrm{Nu}_h = \int_0^1 \mathrm{Nu}_{\mathrm{loc}}\, dY \tag{9}$$

3 Numerical Scheme

The couple system of Eqs. (1) and (2) subject to boundary conditions (4)–(7) was solved numerically using a finite difference method. The central difference method was applied for discretizing the equations. The resulting algebraic equations were

Table 1 Comparison average Nusselt number, Nu_h, against some previous results for isothermal vertical walls in a square cavity in the case $\phi = 0°$

References	Mesh size	Ra	
		100	1000
Bejan (1979)		3.110	15.800
Goyeau et al. (1996)		3.081	13.470
Saeid and Pop (2004)		3.002	13.726
Saeid and Mohamad (2005)		3.108	13.531
Kumari and Nath (2009)		3.114	13.675
Present results	20 × 20	3.005	9.865
	40 × 40	3.080	12.157
	80 × 80	3.100	13.180
	100 × 100	3.103	13.333
	160 × 160	3.106	13.498

solved by using the successive over-relaxation. The unknowns Ψ and θ were calculated until the following convergence criterium is fulfilled.

$$\frac{\sum_{i,j}\left|\xi_{i,j}^{n+1} - \xi_{i,j}^{n}\right|}{\sum_{i,j}\left|\xi_{i,j}^{n+1}\right|} \leq \varepsilon \tag{10}$$

where ξ is either Ψ or θ, n represents the iteration number and ε is the convergence criterium. In this study, the convergence criterium was set at $\varepsilon = 10^{-6}$.

In order to validate the computation coding, accuracy test had been performed for $Ra = 1000$ for several grid sizes as shown in Table 1. It can be seen that the results in Table 1, show a good agreement between our results and the existing results for a porous square enclosure for $\phi = 0°$. In the way to compromise the accuracy and cost in calculations, the grid size 100 × 100 was used in all calculations.

4 Results and Discussion

Figure 2 shows the streamlines and isotherms for different inclination angle of cavity for $Ra = 1000$. As can be seen in Fig. 2, when $\phi = 0$, a large recirculating cell or vortex elongated, parallel to the horizontal axis, exists. The flow is stably stratified and the isotherms are increased with height of cavity. Fluid near the hot wall is heated and rises up due to the buoyancy force. Then, the fluid moved along the top wall and reached the cold wall. As the fluid cooled by the cold wall, it moves downward and along the bottom wall to complete the flow cycle.

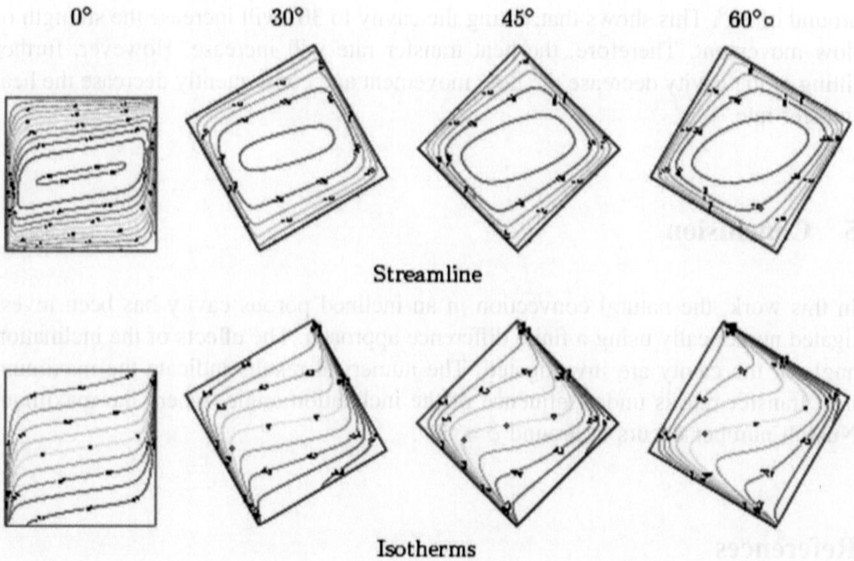

Fig. 2 Variations of streamlines (*above*) and isotherms (*below*) for *Ra* = 1000

As the cavity inclined, hot wall approached the bottom position and the gravity components started to assist and accelerate the flow motion along the top and bottom wall. Hence, the cell is stretched around walls and dominates the center region of the cavity. The isotherms stratification start to breakdown and no longer completely orthogonal to the gravitational field.

Figure 3 shows the effect of the tilt angle on the heat transfer rate given by the average Nusselt number for different Rayleigh numbers. As expected, the value of Nusselt number will increase with the increasing in Raleigh number. When the cavity is tilted, the heat transfer rate increases until it reaches the optimal tilt angle

Fig. 3 Variations of average Nusselt number against inclination angle for different *Ra*

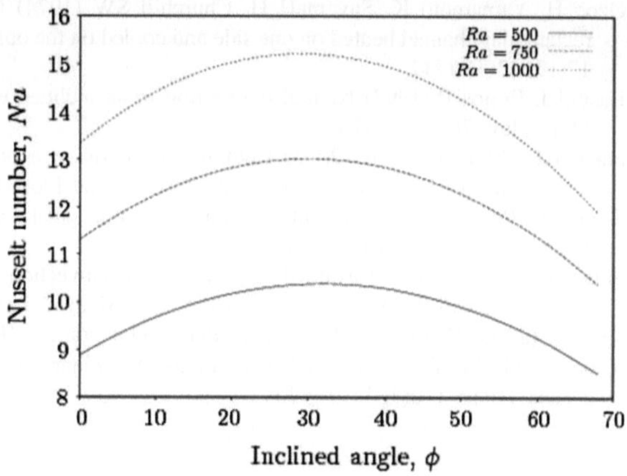

around of 30°. This shows that, tilting the cavity to 30° will increase the strength of flow movement. Therefore, the heat transfer rate will increase. However, further tilting of the cavity decrease the flow movement and consequently decrease the heat transfer rate.

5 Conclusion

In this work, the natural convection in an inclined porous cavity has been investigated numerically using a finite difference approach. The effects of the inclination angle of the cavity are investigated. The numerical results indicate the maximum heat transfer rate is under influence of the inclination angle, where the maximum Nusselt number occurs at around $\phi = 30^{\text{o}}$.

References

Bejan A (1979) On the boundary layer regime vertical enclosure filled with a porous medium. Lett Heat Mass Transf 6(2):93–102

Chamkha AJ, Al-Mudhaf A (2008) Double-difusive natural convection in inclined porous cavities with various aspect ratios and temperature-dependent heat source or sink. Heat Mass Transf 44:679–693

Goyeau B, Songbe JP, Gobin D (1996) Numerical study of double-diffusive natural convection in a porous cavity using Darcy-Brinkman formulation. Int J Heat Mass Transf 39(7):1363–1378

Hart JE (1971) Stability of the flow in a differentially heated inclined box. J Fluid Mech 47 (3):547–576

Holst PH, Aziz K (1972) Transient three-dimensional natural convection in confined porous media. Int J Heat Mass Transf 15(1):73–90

Ingham DB, Pop I (2005) Transport phenomena in porous media. Elsevier Science, Oxford

Kumari M, Nath G (2009) Unsteady natural convection flow in a square cavity filled with a porous medium due to impulsive change in wall temperature. Transp Porous Media 77:463–474

Nield DA, Bejan A (2006) Convective in porous media. Springer, New York

Ozoe H, Yamamoto K, SayamaH H, Churchill SW (1974) Natural convection in an inclined rectangular channel heated on one side and cooled on the opposing side. Int J Heat Mass Transf 17(12):1209–1217

Rasoul J, Prinos P (1997) Natural convection in an inclined enclosure. Int J Numer Meth Heat Fluid Flow 7(5):438–478

Saeid NH, Mohamad AA (2005) Natural convection in a porous cavity with spatial sidewall temperature variation. Int J Numer Meth Heat Fluid Flow 15(6):555–566

Saeid NH, Pop I (2004) Transient free convection in a square cavity filled with a porous medium. Int J Heat Mass Transf 47:1917–1924

Selamat MS, Roslan R, Hashim I (2012) Natural convection in an inclined porous cavity with spatial sidewall temperature variations. J Appl Math. doi:10.1155/2012/939620

Vafai K (2005) Handbook of porous media. Taylor-Francis, Boca Raton

Yang KT (1998) Transitions and bifurcations in laminar bouyant flows in confined enclosures. J Heat Transf 110(4):1191–1204

Chapter 30
Dice Index with Algebraic Product and Minimum t-Norm for Ranking Fuzzy Numbers

Nazirah Ramli, Fairuz Shohaimay and Nurhalijah Bachik

Abstract In fuzzy environments, the ranking of fuzzy numbers (RFNs) is important for solving decision-making problems. Many ranking methods have been developed based on various techniques but no method can provide satisfactory solution to every situation and case. Some methods lack in certain aspects such as inconsistency with human intuition, non-discriminating results and difficulty of interpretation. In this paper, fuzzy preference relation ranking methods based on Dice index with algebraic product and minimum t-norm are proposed. The procedure of the ranking methods involves six steps which are determining fuzzy maximum and fuzzy minimum, intersection and union of fuzzy numbers (FNs), scalar cardinality of FNs, fuzzy evidences and total fuzzy evidences. The findings show that the type of t-norm used affects the ranking results of some FNs.

Keywords Decision making · Dice index · Fuzzy numbers (FNs) · Ranking fuzzy numbers (RFNs)

1 Introduction

Fuzzy numbers (FNs) have been widely used in fuzzy control, decision making, approximate reasoning, optimization and statistics with imprecise possibilities. The process of ranking fuzzy numbers (RFNs) is not an easy task due to the representation of FNs as probability distributions that can overlap with other FNs. Since Jain (1976) first presented the concept of RFNs, various methods have been proposed

---

N. Ramli · F. Shohaimay (✉) · N. Bachik
Faculty of Computer and Mathematical Sciences,
Universiti Teknologi MARA, Jengka, Pahang, Malaysia
e-mail: fairuzshohaimay@pahang.uitm.edu.my

N. Ramli
e-mail: nazirahr@pahang.uitm.edu.my

N. Bachik
e-mail: nurhalijah@gmail.com

N.A. Yacob et al. (eds.), *Regional Conference on Science, Technology and Social Sciences (RCSTSS 2014)*, DOI 10.1007/978-981-10-0534-3_30

but no method can provide satisfactory solution for all situations and cases. Some methods lack in certain aspects such as inconsistency with human intuition, non-discriminating results and difficulty of interpretation.

Chen and Hwang (1992) have categorised RFNs into four major classes namely fuzzy scoring, fuzzy preference relation, fuzzy mean and spread and linguistic expression. According to Akyar et al. (2012), centroid index that is a particular fuzzy scoring class is the most widely used approach for RFNs. A number of researchers have proposed various techniques based on centroid approach such as Cheng (1998), Chu and Tsao (2002), Chen and Chen (2007) and Wang and Lee (2008). Cheng (1998) has proposed a distance and coefficient of variation (CV) indices based on centroid approach for RFNs. However, the distance index contradicts with the result of CV index. Hence, Chu and Tsao (2002) proposed an area between the centroid and original points as the ranking method. Furthermore, Chen and Chen (2007) found that Chu and Tsao's (2002) index could not discriminate the ranking of two FNs having the same mode and symmetric spread. Wang and Lee (2008) then also found that Chu and Tsao (2002) index produced counterintuitive results. Thus, Wang and Lee (2008) proposed a revised method to improve Chu and Tsao's (2002) method.

Although the fuzzy scoring class is relatively simple and easy to implement, it loses some information during the defuzzification process, and thus unable to grasp the sense of uncertainty. The second major class of RFNs namely as fuzzy preference relation (Chen and Hwang 1992) can provide the linguistic meanings from the membership function with the uncertainties of FNs are kept during the ranking process. A number of researchers have proposed various ranking methods based on fuzzy preference relations such as Setnes and Cross (1997), Gao and Wang (2009) and Ramli et al. (2013). Setnes and Cross (1997) have proposed Jaccard with mean aggregation to rank FNs. Gao and Wang (2009) proposed a ranking method based on the endpoint and possibility degree concepts. Furthermore, Ramli et al. (2013) has extended Jaccard index by Setnes and Cross (1997) method using Yager class t-norms. The ranking results for Ramli et al. (2013) are affected by the parameter w of Yager class t-norm. However, the method also cannot discriminate the ranking of two FNs having the same mode and symmetric spread. Besides, the method by Ramli et al. (2013) also cannot discriminate non-symmetric embedded fuzzy numbers.

In this paper, fuzzy preference relation ranking methods based on Dice index with algebraic product and minimum t-norm is proposed. Dice is a set theoretic similarity measure index that is commonly used in image segmentation. t-norm is a binary algebraic operation on the unit interval that is associative, commutative, non-decreasing and satisfying the boundary condition. Minimum is the largest t-norm and algebraic product is smaller than the minimum operator (Wang 1997). Keresztfalvi (1991) stated that minimum t-norm is the most pessimistic approach, while algebraic product is not too much benevolent or too much aggressive (Meza et al. 2009). The ranking behaviour of the proposed ranking methods is investigated. These two methods can discriminate the ranking of non-symmetric embedded fuzzy numbers.

2 Preliminaries

In this section, some basic definitions on FNs are reviewed.

Definition 2.1 (Shohaimay et al. 2012). A fuzzy number is a fuzzy subset that is both convex and normal. The membership function of a fuzzy number A is defined as

$$f_A = \begin{cases} f_A^L(x), & a \leq x \leq b \\ w, & b \leq x \leq c \\ \mu_A^R(x), & c \leq x \leq d \\ 0, & \text{otherwise} \end{cases} \tag{1}$$

where $f_A^L : [a,b] \rightarrow [0,1], \quad f_A^R : [c,d] \rightarrow [0,1]$, f_A^L and f_A^R are the left and right membership functions of the fuzzy number $\tilde{A}$, respectively. If f_A^L and f_A^R are linear and continuous, then $\tilde{A}$ is a trapezoidal fuzzy number denoted by (a,b,c,d). Triangular fuzzy numbers, which are special cases of trapezoidal fuzzy numbers with, $b = c$ are denoted as (a,b,d).

Definition 2.2 (Kaufmann et al. 1985). For two fuzzy numbers, A_i and A_j with $A_{i\alpha} = \left[a_\alpha^-, \alpha_\alpha^+\right]$ and $A_{j\alpha} = \left[b_\alpha^-, b_\alpha^+\right]$ be their α-cuts with $\alpha \in [0,1]$.

The fuzzy maximum and fuzzy minimum of A_i and A_j are defined as

$$\left[\mathrm{MAX}(A_i, A_j)\right]_\alpha = \left(\max\left(a_\alpha^-, b_\alpha^-\right), \max\left(a_\alpha^+, b_\alpha^+\right)\right) \tag{2}$$

$$\left[\mathrm{MIN}(A_i, A_j)\right]_\alpha = \left(\min\left(a_\alpha^-, b_\alpha^-\right), \min\left(a_\alpha^+, b_\alpha^+\right)\right) \tag{3}$$

Definition 2.3 (Wang 1997). The intersection of fuzzy numbers A_i and A_j with membership functions μ_{A_i} and μ_{A_j}, respectively, is defined as

$$\mu_{A_i \cap A_j} = \min\left[\mu_{A_i}(x), \mu_{A_j}(x)\right], \quad \forall x \in X. \tag{4}$$

The union of fuzzy numbers A_i and A_j with membership function μ_{A_i} and μ_{A_j}, respectively, is defined as

$$\mu_{A_i \cup A_j} = \max\left[\mu_{A_i}(x), \mu_{A_j}(x)\right], \quad \forall x \in X. \tag{5}$$

Definition 2.4 (Dubois and Prade 1980). The scalar cardinality of a fuzzy number A in the universe of discourse X with $\text{supp}(A)$ is not finite, is defined as

$$|A| = \int_{x} \mu_A(x)\ \text{d}x \tag{6}$$

or the area under the membership function of A.

3 Proposed Fuzzy Dice Ranking Index with Algebraic Product and Minimum t-Norms

The procedure for fuzzy Dice ranking index with algebraic product and minimum t-norms consists of six steps as follows:

Step 1. Find the fuzzy maximum and fuzzy minimum for each pair of FNs A_i and A_j.

Step 2. Find $A_i \cap \text{MAX}(A_i, A_j)$, $A_i \cap \text{MIN}(A_i, A_j)$, $A_j \cap \text{MAX}(A_i, A_j)$, and $A_j \cap \text{MIN}(A_i, A_j)$.

Step 3. Calculate $|A_i|$, $|A_j|$, $|\text{MAX}(A_i, A_j)|$, $|\text{MIN}(A_i, A_j)|$, $|A_i \cap \text{MAX}(A_i, A_j)|$, $|A_i \cap \text{MIN}(A_i, A_j)|$, $|A_j \cap \text{MAX}(A_i, A_j)|$ and $|A_j \cap \text{MIN}(A_i, A_j)|$.

Step 4. Calculate the evidences of $E(A_i \succ A_j)$, $E(A_j \prec A_i)$, $E(A_j \succ A_i)$ and $E(A_i \prec A_j)$, which are defined based on fuzzy Dice index as $E(A_i \succ A_j) = S_D(\text{MAX}(A_i, A_j), A_i)$, $E(A_j \prec A_i) = S_D(\text{MIN}(A_i, A_j), A_j)$, $E(A_j \succ A_i) = S_D(\text{MAX}(A_i, A_j), A_j)$ and $E(A_i \prec A_j) = S_D(\text{MIN}(A_i, A_j), A_i)$ where $S_D(A_i, A_j) = \frac{2|A_i \cap A_j|}{|A_i| + |A_j|}$ is the fuzzy Dice index and $|A_i|$ denotes the scalar cardinality of fuzzy number A_i.

For simplicity, C_{ij}, c_{ji}, C_{ji} and c_{ij} are used to denote $E(A_i \succ A_j)$, $E(A_j \prec A_i)$, $E(A_j \succ A_i)$ and $E(A_i \prec A_j)$, respectively.

Step 5. Calculate the total evidences by aggregating the evidence based on algebraic product t-norm and minimum t-norm as in Eqs. 7 and 8, respectively.

$$E_{\text{total}}(A_i \succ A_j) = C_{ij} \cdot c_{ji} \text{ and } E_{\text{total}}(A_j \succ A_i) = C_{ji} \cdot c_{ij} \tag{7}$$

$$E_{\text{total}}(A_i \succ A_j) = \min(C_{ij}, c_{ji}) \text{ and } E_{\text{total}}(A_j \succ A_i) = \min(C_{ji}, c_{ij}) \tag{8}$$

To simplify, $E_D(A_i, A_j)$ and $E_D(A_j, A_i)$ are used to represent $E_{\text{total}}(A_i \succ A_j)$ and $E_{\text{total}}(A_j \succ A_i)$, respectively.

Step 6. Rank each pair of fuzzy numbers A_i and A_j by comparing the total evidences as follows:

$$A_i \succ A_j \text{ if and only if } E_D(A_i, A_j) > E_D(A_j, A_i).$$
$$A_i \prec A_j \text{ if and only if } E_D(A_i, A_j) < E_D(A_j, A_i).$$
$$A_i \approx A_j \text{ if and only if } E_D(A_i, A_j) = E_D(A_j, A_i).$$

4 Implementation

Six sets of numerical examples taken from the literature are presented to illustrate the validity of fuzzy Dice ranking index with algebraic product and minimum t-norms. Table 1 shows the ranking results.

For all the fuzzy numbers in Table 1 except Set 2, Dice with algebraic product t-norms (DAPT) and Dice with minimum t-norms (DMT) produce consistent ranking results. For Set 2, the ranking order for DAPT is $A_1 \succ A_2$, while DMT is $A_1 \approx A_2$. The values of evidence for DAPT indicate that A_1 is slightly better than A_2. Both ranking results are not consistent with Setnes and Cross (1997) method which ranked as $A_1 \prec A_2$.

For Set 1, intuitively the ranking order is $A_1 \prec A_2$. However, Sun and Wu (2006) produced the ranking as $A_1 \succ A_2$, which is unreasonable. Both DAPT and DMT produced $A_1 \prec A_2$ with A_1 dominates A_2. The proposed results are consistent with Cheng's (1998), Chu and Tsao's (2002) indices and human intuition.

For Set 3, both DAPT and DMT produced $A_1 \succ A_2$ which is consistent with Chu and Tsao's (2002) index. The values of evidence indicate that A_1 is slightly better than A_2 for both DAPT and DMT. However, Cheng's (1998) index produced $A_1 \prec A_2$ and Ramli et al. (2013) produced $A_1 \approx A_2$ for $w \in (0, 0.318]$ and $A_1 \succ A_2$ for $w \in [0.319, 1000]$. This shows that DPAT and DMT can discriminate the ranking of non-symmetric embedded fuzzy numbers. For Set 4, both DAPT and DMT produced $A_1 \succ A_2$ which is consistent with Lu and Wang (2005) index.

The fuzzy numbers in Set 5 have the same mode and symmetric spread. Both DAPT and DMT produced $A_1 \approx A_2$ which is consistent with Chu and Tsao's (2002) index and Allahviranloo and Firozja's (2010) method. However, Xu and Wei (2009) ranked the FNs as $A_1 \prec A_2$. For Set 6, both DAPT and DMT produce $A_1 \succ A_2$ but Chen and Lu (2002) index ranked as $A_1 \prec A_2$.

Table 1 Comparison of ranking results of fuzzy Dice index with algebraic product and minimum t-norms

Fuzzy numbers	Proposed ranking index				Ranking result of previous studies
	Evidences	t-norm	Total evidences	Ranking result	
Set 1: $A_1 = \left(0, \frac{3}{8}, 3\right)$ $A_2 = \left(\frac{1}{4}, \frac{3}{2}, \frac{5}{3}, 9\right)$	$C_{12} = 0.326$ $c_{21} = 0.326$ $C_{21} = 1.000$ $c_{12} = 1.000$	Algebraic product	$E_D(A_1, A_2) = 0.106$ $E_D(A_2, A_1) = 1.000$	$A_1 \prec A_2$	Cheng (1998), Chu and Tsao (2002): $A_1 \prec A_2$ Sun and Wu (2006): $A_1 \succ A_2$
		Minimum	$E_D(A_1, A_2) = 0.326$ $E_D(A_2, A_1) = 1.000$	$A_1 \prec A_2$	
Set 2: $A_1 = (0.3, 0.5, 0.9)$ $A_2 = (0.155, 0.645, 0.8)$	$C_{12} = 0.908$ $c_{21} = 0.913$ $C_{21} = 0.910$ $c_{12} = 0.908$	Algebraic product	$E_D(A_1, A_2) = 0.829$ $E_D(A_2, A_1) = 0.826$	$A_1 \succ A_2$	Setnes and Cross (1997): $A_1 \prec A_2$
		Minimum	$E_D(A_1, A_2) = 0.908$ $E_D(A_2, A_1) = 0.908$	$A_1 \approx A_2$	
Set 3: $A_1 = (0, 1, 2)$ $A_2 = \left(\frac{1}{5}, 1, \frac{7}{4}\right)$	$C_{12} = 0.947$ $c_{21} = 0.939$ $C_{21} = 0.925$ $c_{12} = 0.933$	Algebraic product	$E_D(A_1, A_2) = 0.889$ $E_D(A_2, A_1) = 0.863$	$A_1 \succ A_2$	Cheng (1998): $A_1 \prec A_2$ Chu and Tsao (2002): $A_1 \succ A_2$ Ramli et al. (2013): $A_1 \approx A_2$, $w \in (0, 0.318]$, $A_1 \succ A_2$, $w \in [0.319, 1000]$
		Minimum	$E_D(A_1, A_2) = 0.939$ $E_D(A_2, A_1) = 0.925$	$A_1 \succ A_2$	
Set 4: $A_1 = (0.4, 0.5, 0.6, 0.7)$ $A_2 = (0.3, 0.4, 0.9)$	$C_{12} = 0.888$ $c_{21} = 0.910$ $C_{21} = 0.758$ $c_{12} = 0.704$	Algebraic product	$E_D(A_1, A_2) = 0.808$ $E_D(A_2, A_1) = 0.534$	$A_1 \succ A_2$	Lu and Wang (2005): $A_1 \succ A_2$
		Minimum	$E_D(A_1, A_2) = 0.888$ $E_D(A_2, A_1) = 0.704$	$A_1 \succ A_2$	
Set 5: $A_1 = (0.2, 0.5, 0.8)$ $A_2 = (0.4, 0.5, 0.6)$	$C_{12} = 0.800$ $c_{21} = 0.667$ $C_{21} = 0.667$ $c_{12} = 0.800$	Algebraic product	$E_D(A_1, A_2) = 0.533$ $E_D(A_2, A_1) = 0.533$	$A_1 \approx A_2$	Xu and Wei (2009): $A_1 \prec A_2$ Chu and Tsao (2002), Allahviranloo and Firozja (2010): $A_1 \approx A_2$
		Minimum	$E_D(A_1, A_2) = 0.667$ $E_D(A_2, A_1) = 0.667$	$A_1 \approx A_2$	

(continued)

Table 1 (continued)

Fuzzy numbers	Proposed ranking index				Ranking result of previous studies
	Evidences	t-norm	Total evidences	Ranking result	
Set 6: $A_1 = (0.35, 0.5, 1)$ $A_2 = (0.15, 0.7, 0.8)$	$C_{12} = 0.905$ $c_{21} = 0.905$	Algebraic product	$E_D(A_1, A_2) = 0.819$ $E_D(A_2, A_1) = 0.716$	$A_1 \succ A_2$	Chen and Lu (2002): $A_1 \prec A_2$
	$C_{21} = 0.846$ $c_{12} = 0.846$	Minimum	$E_D(A_1, A_2) = 0.905$ $E_D(A_2, A_1) = 0.846$	$A_1 \succ A_2$	

5 Conclusion

This paper presents new ranking methods based on Dice index with algebraic product and minimum t-norms. DAPT and DMT cannot discriminate the ranking of FNs having the same mode and symmetric spread but their ranking results are consistent with human intuition. The DAPT and DMT not only provide the conclusion of preferred or not preferred of the alternatives, but can also represent the imprecise relation between alternatives such as A_1 dominates A_2, A_1 is slightly better than A_2 and A_1 is more or less the same as A_2, and others. The ranking results of some FNs are affected by the type of t-norms used that contradict with results by Pap et al. (2000) which stated that the priority of crafting strategy is similar for algebraic product and minimum t-norm.

DAPT has better performance for RFNs compared to DMT since it can discriminate the ranking of some FNs while DMT cannot discriminate the FNs. Besides, the membership of algebraic product is neither the largest nor the smallest and also not too much benevolent or too much aggressive (Meza et al. 2009).

Acknowledgments The presentation of this paper is supported by the Department of Research & Industrial Linkages (PJI), Universiti Teknologi MARA Pahang under *Tabung Penyelidikan Am.*

References

Akyar E, Akyar H, Duzce SA (2012) A new method for ranking triangular fuzzy numbers. Int J Uncertainty, Fuzziness Knowl Based Syst. doi:10.1142/S021848851250033X

Allahviranloo T, Firozja MA (2010) Ranking fuzzy numbers by a new metric. Soft Comput 14:773–782

Chen L-H, Lu H-W (2002) The preference order of fuzzy numbers. Comput Math Appl 44(10–11):1455–1465. doi:10.1016/S0898-1221(02)00270-5

Chen SJ, Chen SM (2007) Fuzzy risk analysis based on the ranking of generalized trapezoidal fuzzy numbers. Appl Intell 26:1–11. doi:10.1007/s10489-006-0003-5

Chen S-J, Hwang C-L (1992) Fuzzy multiple attribute decision making: methods and applications. In: Lecture notes in economics and mathematical systems, vol 375, pp 289–486. doi:10.1007/978-3-642-46768-4

Cheng C-H (1998) A new approach for ranking fuzzy numbers by distance method. Fuzzy Sets Syst. doi:10.1016/S0165-0114(96)00272-2

Chu TC, Tsao CT (2002) Ranking fuzzy numbers with an area between the centroid point and original point. Comput Math Appl 43:111–117. doi:10.1016/S0898-1221(01)00277-2

Dubois D, Prade H (1980) Fuzzy sets and systems: theory and applications. Academic Press, New York

Gao X, Wang G (2009) A new method for ranking fuzzy numbers—Endpoint method. In: 6th International conference on fuzzy systems and knowledge discovery, FSKD 2009, vol 5, pp 145–149. doi:10.1109/FSKD.2009.696

Jain R (1976) Decisionmaking in the presence of fuzzy variables. IEEE Trans Syst Man Cybernet 6(10):698–703. doi:10.1109/TSMC.1976.4309421

Kaufmann A, Gupta MM, Kaufmann A (1985) Introduction to fuzzy arithmetic: theory and application. Van Nostrand Reinhold Company, New York

Keresztfalvi T (1991) t-norm based product of LR fuzzy numbers. http://www.polytech.univ-savoie.fr/fileadmin/polytech_autres_sites/sites/listic/busefal/Papers/49.zip/49_03.pdf

Lu H-W, Wang CB (2005) An index for ranking fuzzy numbers by belief feature. Int J Inf Manage Sci 16(3):57–70. http://cat.inist.fr/?aModele=afficheN&cpsidt=17153629

Meza LA, Gomes EG, Soares de Mello JCCB, Neto LB (2009) Different T-norms for DEA-fuzzy efficiency computations. In: XV international conference on industrial engineering and operations management, pp 1–12. http://www.alice.cnptia.embrapa.br/bitstream/doc/659436/1/DifferentTnorms.pdf

Pap E, Bošnjak Z, Bošnjak S (2000) Application of fuzzy sets with different t-norms in the interpretation of portfolio matrices in strategic management. Fuzzy Sets Syst 114(1):123–131. doi:10.1016/S0165-0114(98)00196-1

Ramli N, Mohamad NH, Mohamad M, Shohaimay F (2013) Jaccard index with yager class t-norm for ranking fuzzy numbers. Int J Undergraduates Stud 2(3):9–14

Setnes M, Cross V (1997) Compatibility-based ranking of fuzzy numbers. In: 1997 annual meeting of the north american fuzzy information processing society—NAFIPS (Cat. No.97TH8297). doi:10.1109/NAFIPS.1997.624057

Shohaimay F, Ramli N, Mohamed SR (2012) Fuzzy evaluation in IT supplier selection. In: IEEE Symposium on Computer Applications and Industrial Electronics (ISCAIE), pp 94–98. http://ieeexplore.ieee.org/xpls/abs_all.jsp?arnumber=6482076

Sun H, Wu J (2006) A new approach for ranking fuzzy numbers based on fuzzy simulation analysis method. Appl Math Comput 174:755–767. doi:10.1016/j.amc.2005.04.113

Wang LX (1997) A course in fuzzy systems and control. Prentice-Hall, Englewood Cliffs

Wang YJ, Lee HS (2008) The revised method of ranking fuzzy numbers with an area between the centroid and original points. Comput Math Appl 55:2033–2042. doi:10.1016/j.camwa.2007.07.015

Xu C-L, Wei L-L (2009) A new method for ranking fuzzy numbers based on area of a triangle. In: 2009 International Conference on Machine Learning and Cybernetics, vol 1. IEEE, pp 532–536. doi:10.1109/ICMLC.2009.5212525

Part V
Computer Science/Information Technology

Chapter 31
Investigating the Optimise *k*-Dimensions and Threshold Values of Latent Semantic Indexing Retrieval Performance for Small Malay Language Corpus

Roslan Sadjirin, Noli Maishara Nordin, Mohd Ikhsan Md Raus and Zulazeze Sahri

Abstract Presenting users with relevant feedback is the main aim and core in information retrieval (IR). Due to the poor relevance feedback returned by simple exact term-matching technique, a latent semantic indexing (LSI) based IR has come into place to overcome the retrieval drawback, and improve the effectiveness of retrieval performance. In other words, LSI-based IR aims in satisfying users rather than satisfying a given query. However, in developing an LSI-based information retrieval application, there are parameters that need to be considered in order to produce relevant feedback which optimise the precision and recall in retrieval process. Therefore, this paper investigates two important parameters that characterised the retrieval performance, which are the optimise *k*-dimension to represent terms and documents in corpus, and the optimise threshold values for the documents to be accepted, judged and returned as relevant for a given term query. A small Malay corpus which comprises of 1395 Malay language documents and terms were used as the test collection. The analyses suggest that the effective performance of the retrieval which satisfied as well as balanced the precision and recall, is obtained for *k*-dimension is $k = 4$ and threshold value is $\varepsilon = 0.8$ The study helps the software developers particularly the IR application developers in

R. Sadjirin (✉) · M.I. Md Raus · Z. Sahri
Faculty of Computer and Mathematical Sciences, Universiti Teknologi MARA, Jengka, Pahang, Malaysia
e-mail: roslancs@pahang.uitm.edu.my

M.I. Md Raus
e-mail: mohdikhsan@pahang.uitm.edu.my

Z. Sahri
e-mail: azeze@pahang.uitm.edu.my

N.M. Nordin
Academy of Language Studies, Universiti Teknologi MARA, Jengka, Pahang, Malaysia
e-mail: nolinordin@pahang.uitm.edu.my

N.A. Yacob et al. (eds.), *Regional Conference on Science, Technology and Social Sciences (RCSTSS 2014)*, DOI 10.1007/978-981-10-0534-3_31

designing and choosing the optimise value of the *k*-dimension and the threshold in the search engine.

Keywords Information retrieval · Latent semantic analysis · Singular value decomposition · Optimisation

1 Introduction

Information retrieval (IR) is a part of computer science studies which finding for documents from a collection of stored documents that are relevant to a user's need for information (Russell and Norvig 2010). The retrieved document aims at satisfying user's information need usually expressed in natural language (Baeza-Yates 2004; Ricardo and Berthier 2011). The best known examples of information retrieval system are search engines on the World Wide Web. A Web user can type a query in natural language into a search engine and see a list of relevant pages (Russell and Norvig 2010). Therefore, the representation and organisation of the information items should provide users with easy access to the information through characterisation of the user information need, which is not a simple problem (Ricardo and Berthier 2011).

Thus, in order to provide users with easy access to the information in which the user is interested and to characterise as well as to summarise the description of the information need, a timely and accurately text retrieval mechanism is required (Mckinley 2000). One of the mechanisms is indexing which is the process of transformation from received item to the data structured which helps to speed up the retrieval task (Kowalski 1997).

Indexing is divided into two types which are manual and automated indexing. Manual indexing refers to cataloguing which is common in libraries and learning institutions such as name of author, subject heading or title of a book are selected to represent the description of items. In contrast, automated indexing is the capability for a computer system to automatically determine the index to be assigned for item representation. In automated indexing, sometimes all words in a document were used as possible index term (Kowalski 1997).

There are many techniques in automated indexing such as signature files, suffix arrays, inverted files and latent semantic. The first three techniques—signature files, suffix arrays, inverted files—are categorised as indexing by term whilst latent semantic indexing is a type of indexing by concept (Kowalski 1997). Suffix array faster for phrase search with less common queries but harder to build and maintain. While signature files are no longer efficient in searching big data. Inverted files are the most common and outperform all techniques of indexing by term (Baeza-Yates 2004; Trotman 2003).

Although inverted files outperformed the other technique in indexing by term (Baeza-Yates and Ribeiro-Neto 1999), it still fails in satisfying the user because of the retrieval method used in the technique. The retrieval technique used in inverted files indexing technique is an exact term-matching or also known as hashing algorithm. This will be an issue if for instance, a user wants to retrieve documents collection X in which collection X consist of documents X_1, X_2, X_3, X_4. Moreover, each document $\{X_1, X_2, X_3, X_4\}$ might or might not have common terms or each document is described using different term. Example of sets in Eq. (1) depicts the respective circumstances.

$$X = \{X_1 = \{x_{00}, x_{01}, x_{03}\}, X_2 = \{x_{04}, x_{05}, x_{06,}\}, X_3 = \{x_{01}, x_{05}, x_{08}\}, X_4 = \{x_{06}, x_{08}, x_{09}\}\} \quad (1)$$

In Eq. (1) X is collection of document, whereas X_1, X_2, X_3 and X_4 are the documents in collection X. On the other hand, x_{00}, x_{01}, x_{03}, x_{04}, x_{05}, x_{06}, x_{07}, x_{08} and x_{09} are the terms that described each document. Let us say user performs query x_{01}, the document retrieved are X_1 and X_3 because x_{01} is the term in documents X_1 and X_3 respectively. However, X_2 and X_4 are unable to be retrieved because the absence of term x_{01} in these documents. Thus, this situation leads to poor and loss retrieval of documents because the searching process only matches exact terms or words.

The term-matching retrieval technique which is used in inverted files indexing technique has drawback due to the ambiguous meaning of single word, the lack of precision, the personal style, the knowledge background as well as the individual preferences in the word usage (Deerwester et al. 1990; Hofmann 1999; Hutchison and Mitchell 2009; Kokiopoulou and Saad 2004; Quan et al. 2008; Syu et al. 1996). To overcome these problems, Latent semantic indexing (LSI) approach can be used for automatic indexing and retrieving information by mapping documents and terms that associated with concepts in LSI space (Foltz and Dumais 1992; Foltz 1990; Gee 2003; Kolda and Leary 2008; Tang et al. 2004), despite the indexing process taken is longer than the inverted files indexing techniques (Sadjirin and Rahman 2010).

Latent semantic indexing (LSI) is an algebraic model of document retrieval based on a singular value decomposition (SVD) of a vector space (k-dimension) of index term (Baeza-Yates 2004; Hofmann 1999). LSI, however, does not require explicit translation capabilities between languages or terms (Kim and Khudanpur 2004) and does not use any external dictionaries because it employs a mathematical representation (Hutchison and Mitchell 2009; Kowalski 1997) and statistical technique that derives a statistical correlation between all terms and documents in corpus (Gee 2003). For example, if the pairing of documents is derived from the basis of contents similarity, the term neigbourhood of the other terms, are semantically related words although there is no word in common (Kim and Khudanpur 2004) because the system would identify the documents with the highest cosine similarity to the query and return these documents as the best matches to the query (Dasgupta et al. 2005; Hofmann 1999).

The performance of retrieval method for LSI depends on the number of k-dimensions (Bartell et al. 1992; Buckeridge and Sutcliffe 2002; Gee 2003;

Hutchison and Mitchell 2009; Yu et al. 2005) and the threshold value (Marcus and Maletic 2003) over recall and precision of the documents in response to the user's query. However, a review of literatures suggested that there were no definite number of *k*-dimension that would give efficient retrieval. The best possible approximation of *k*-dimension of any matrix using SVD in which each document can be viewed as a vector in a low-dimensional space of a few hundred *k*-dimensions (Dasgupta et al. 2005).

Futhermore, to avoid noise effect due to excessive variability in vocabulary usage, the derived matrix of SVD is estimated and reduced to an arbitarary *k*-dimension space (Gee 2003; Kolda and Leary 2008). On the contrary, the term-document matrix that decomposed into a set of *k*-dimension, typically 200–300 orthogonal factors can increase the retrieval performance in terms of the precision of retrieved documents over query. If the number of *k*-dimensions is too many, the data will be overfitting and incurring the vocabulary mismatch problem, but if the number of *k*-dimensions is reduced too far, the important information will be loosed (Furnas et al. 1988).

To achieve the best precision in retrieval, the optimal *k*-dimensions is currently determined by exhaustive evaluation (Ding 1999; Furnas et al. 1988) and there is no single choice of *k*-dimensions (Efron 2007), but to calculate it from the term-document matrix remains an open question (Ding 1999). Nevertheless, for the small corpora or corpus, LSI usually uses 50–350 *k*-dimensions space (Tang et al. 2004) in which the number of *k*-dimensions cannot be higher than $l = O(\log n)$, where n is the number of terms multiply by the number of documents in LSI space.

The retrieval performance of the LSI does not only depend on the number of *k*-dimensions of the SVD but also relies on the computation of the ranking of the similarity of the query over documents and terms in *k*-dimensions space in LSI. The computation typically uses cosine similarity (Gee 2003; Marcus and Maletic 2003) as shown in Eq. (2).

$$\begin{aligned} \text{Similarity}\,(A,B) &= \text{Cosine}\,\theta = A \cdot B/|A||B| \\ &= [(x_1 * x_2) + (y_1 * y_2)]/[\sqrt{}(x_{12} + y_{12})][\sqrt{}(x_{22} + y_{22})] \end{aligned} \quad (2)$$

Equation (2) depicts that A is the vector of query or term or document while B is another vector of term or documents in LSI space. Using cosine similarity computation, SVD will produce the relevant documents based on the predetermined threshold (ε) value. A good and widely used heuristic is $\varepsilon = 0.7$. This measure corresponds to a 45° angle between the corresponding vectors (Marcus and Maletic 2003).

Therefore, this paper investigates the optimise number of *k*-dimensions and the values of threshold which produced the efficient retrieval of latent semantic indexing technique.

2 Methods

The study employs Malay corpus which consists of 1395 collection of documents and terms and 12 samples of queries. Sample of queries are shown in Table 1. The evaluation for recall (*R*), precision (*P*) and the effectiveness (*E*) of the retrieval are shown in Eqs. (3)–(5), respectively.

$$\text{Recall}(R) = \sum\left(\text{Retrieved} \bigcap \text{Relevent}\right) / \sum(\text{Relevant}) \quad (3)$$

$$\text{Precision}(P) = \sum\left(\text{Retrieved} \bigcap \text{Relevent}\right) / \sum(\text{Retrieved}) \quad (4)$$

$$\text{Effectiveness}(E) = 1 - \left[\left((1+\beta^2)\cdot P \cdot R\right) / \left(\beta^2 \cdot (P+R)\right)\right] \quad (5)$$

In Eq. (5), β is a variable value to indicate the importance between recall and precision. In this study, the value of $\beta = 2.0$ was used to indicate that recall is twice as important as precision. To develop LSI, the following preprocessing was accomplished (Sadjirin and Rahman 2010) as follows:

1. Extracting terms from documents. During the process of term extraction, the following steps were performed:
   (a) Remove Malay stopword
   (b) Stem Malay vocabulary into its root word by removing its prefixes and suffixes using Ahmad's Stemmer

Table 1 List of queries used in this study for our small Malay corpus

Query #	Query words
1	Kewajipan menuntut ilmu di dalam Islam
2	Bersuci dari hadas sebelum mengerjakan solat
3	Dosa-dosa meninggalkan solat wajib
4	Jenis-jenis zakat dan kepentingannya
5	Rukun mengerjakan haji
6	Kelebihan berpuasa
7	Adab-adab makan mengikut sunnah
8	Perhiasan dan pakaian yang dibenarkan di dalam Islam
9	Kisah-kisah Rasullullah
10	Tanda-tanda kiamat
11	Ciri-ciri orang yang beriman
12	Minuman yang diharamkan di dalam Islam

2. Creating term-document matrix. Term-document matrix is created by comparing a term that has been extracted in (i) against the document from documents in the Malay corpus. During the process of creating term-document, the following steps were performed:
   (a) Remove Malay stopword
   (b) Stem Malay vocabulary into its root word by removing its prefixes and suffixes using Ahmad's Stemmer
   (c) Count the occurrences of each term in the Malay corpus documents.

Step (i) and (ii) above will create a matrix, say matrix A. The following is the algorithm of SVD computation (Sadjirin and Rahman 2010) for matrix A:

1. Compute the transpose of $\boldsymbol{A}$ and $\boldsymbol{A^T A}$ ($\boldsymbol{A^T}$ is the transpose of matrix $\boldsymbol{A}$).
2. Determine the eigenvalues of $\boldsymbol{A^T A}$ and sort these in descending order, in the absolute sense. Square root these eigenvalues to obtain the singular values of $\boldsymbol{A}$.
3. Construct diagonal matrix $\boldsymbol{S}$ by placing singular values in descending order along its diagonal. Compute the inverse of $\boldsymbol{S}$, which is $\boldsymbol{S^{-1}}$.
4. Use the ordered eigenvalues obtained in step 2 and compute the eigenvectors of $\boldsymbol{A^T A}$. Place these eigenvectors along the column of $\boldsymbol{V}$ and compute its transpose, $\boldsymbol{V^{-1}}$.
5. Compute $\boldsymbol{U}$ as $\boldsymbol{U = AVS^{-1}}$ ($\boldsymbol{A}$ is the original matrix, $\boldsymbol{V}$ is the eigenvectors, and $\boldsymbol{S^{-1}}$ is the inverse diagonal matrix $\boldsymbol{S}$).
6. To complete the proof of $\boldsymbol{U = AVS^{-1}}$, compute the full SVD using $\boldsymbol{A = USV^T}$. ($\boldsymbol{A}$ is the original matrix, $\boldsymbol{S}$ is the singular values and $\boldsymbol{V^T}$ is the transpose of eigenvectors).
7. To complete the proof of $\boldsymbol{U = AVS^{-1}}$, compute the full SVD using $\boldsymbol{A = USV^T}$. ($\boldsymbol{A}$ is the original matrix, $\boldsymbol{S}$ is the singular values and $\boldsymbol{V^T}$ is the transpose of eigenvectors).

Experimental Phase

In experimental phase, five (5) experiments have been conducted which involve various number of k-dimension which are 2, 3, 4, 5, 6 as well as various values of threshold. Experiment on dimensions space for $k = 1$ is omitted because in matrix studies, the least dimensional space is 2. In the literature reviews, other researchers stated that there were no indefinite numbers of k-dimensions space that determined the retrieval performance. Nevertheless, because of the size of Malay corpus in this study was small, the experiments of the k-dimensions space was started with $k = 2$ and up to $l = O(\log n) = 5$. Moreover, the values of threshold selected were 0.5, 0.6 and 0.7, because literature stated that the best value of threshold was 0.7 which was equal to 45°. However, to control or leverage the outcome of the experiments, the system was also executed for $k = 6$ and the value of the threshold was extended to 0.8 which was equivalent to 37° (degree) angle.

Experiment 1 for k = 2 and ε = {0.5, 0.6, 0.7, 0.8}. In Experiment 1, the LSI-based IR system was executed where the *k*-dimensions was set to $k = 2$ and the threshold values used were, ε = {0.5, 0.6, 0.7, 0.8}. The result in Table 2 suggested that for $k = 2$, the optimal threshold value that produced better retrieval performance was ε = 0.8.

Table 2 Percentage of recall, precision and effectiveness for k = 2, ε = {0.5, 0.6, 0.7, 0.8}

Query #	ε = 0.5			ε = 0.6			ε = 0.7			ε = 0.8		
	R	P	E	E	P	E	R	P	E	R	P	E
1	95.0	10.2	64.3	95.0	11.1	62.2	90.0	11.5	61.9	70.0	10.7	66.8
2	100.0	9.5	65.6	100.0	9.6	65.3	95.0	9.4	66.3	85.0	9.1	68.1
3	100.0	9.5	65.6	100.0	9.5	65.6	95.0	9.6	65.8	80.0	9.1	68.7
4	95.0	9.5	66.1	90.0	9.5	66.6	90.0	10.4	64.4	85.0	11.0	63.8
5	100.0	9.9	64.5	100.0	10.1	64.0	95.0	10.1	64.6	90.0	10.7	63.7
6	95.0	9.2	65.8	95.0	9.2	66.8	90.0	8.9	68.1	80.0	8.4	70.4
7	100.0	4.8	79.9	100.0	4.8	79.9	100.0	4.9	79.5	90.0	4.9	79.9
8	90.0	4.7	80.6	90.0	4.7	80.6	90.0	4.8	80.2	60.0	3.4	86.1
9	95.0	9.2	66.8	95.0	9.3	66.6	95.0	9.5	66.1	80.0	8.6	69.9
10	95.0	10.0	64.8	90.0	10.2	64.9	80.0	9.8	67.1	60.0	8.4	73.1
11	100.0	10.0	64.3	100.0	10.6	62.8	100.0	11.6	60.4	100.0	13.0	57.2
12	100.0	4.9	79.5	100.0	5.6	77.1	100.0	5.4	77.8	100.0	5.6	77.1
Average	97.1	8.5	69.0	96.3	8.7	68.5	93.3	8.8	68.5	81.7	8.6	70.4

Table 3 Percentage of recall, precision and effectiveness for k = 3, ε = {0.5, 0.6, 0.7, 0.8}

Query #	ε = 0.5			ε = 0.6			ε = 0.7			ε = 0.8		
	R	P	E	R	P	E	R	P	E	R	P	E
1	60.0	12.6	65.8	40.0	11.6	73.1	30.0	14.0	75.6	10.0	6.7	90.9
2	95.0	9.6	65.8	95.0	9.8	65.3	95.0	10.3	64.1	95.0	11.0	62.4
3	95.0	9.6	65.8	90.0	9.2	67.4	85.0	8.2	70.4	85.0	9.7	66.7
4	90.0	9.2	67.4	85.0	9.7	66.7	85.0	11.5	62.7	55.0	10.3	70.6
5	95.0	10.1	64.6	90.0	10.8	63.5	80.0	12.2	62.1	45.0	10.1	73.4
6	95.0	9.4	66.3	19.5	10.2	83.5	75.0	9.6	68.3	75.0	4.6	81.5
7	90.0	4.6	80.9	90.0	4.8	80.2	90.0	4.9	79.9	80.0	5.0	80.0
8	90.0	4.7	80.6	80.0	4.7	81.0	50.0	3.3	86.9	10.0	0.9	96.7
9	95.0	9.3	66.6	85.0	8.7	69.1	85.0	10.3	65.3	50.0	8.3	75.1
10	45.0	33.3	58.0	25.0	31.3	74.0	15.0	27.3	83.5	10.0	50.0	88.1
11	35.0	20.0	69.6	20.0	17.4	80.6	20.0	28.6	78.7	15.0	37.5	83.0
12	90.0	4.5	81.3	90.0	4.7	80.6	80.0	4.7	81.0	60.0	4.2	83.6
Average	81.3	11.41	69.4	67.5	11.08	73.7	65.8	12.08	73.2	49.2	13.2	79.3

Experiment 2 for $k = 3$ and $\varepsilon = \{0.5, 0.6, 0.7, 0.8\}$. Whereas in Experiment 2, the k-dimensions was set to $k = 3$ and the threshold values used were same as the values used in Experiment 1. The results shown in Table 3 suggested that for $k = 3$, an effective retrieval performance was achieved when threshold value was $\varepsilon = 0.8$. The threshold value was identical to the value in Experiment 1, where $\varepsilon = 0.8$ for $k = 2$.

Experiment 3 for $k = 4$ and $\varepsilon = \{0.5, 0.6, 0.7, 0.8\}$. In Experiment 3, the k-dimensions was set to $k = 4$ and various values of threshold were also used (refer to the values in Experiment 2). The results shown in Table 4 suggested that for $k = 4$, the optimal threshold was, $\varepsilon = 0.8$, which is also identical to the values that were conducted in Experiment 1 and Experiment 2, but with different k-dimensions which were $k = 2$ and $k = 3$, respectively.

Experiment 4 for $k = 5$ and $\varepsilon = \{0.5, 0.6, 0.7, 0.8\}$. In Experiment 4, the k-dimensions was set to 5 and various values of threshold were used (refers to the values in Experiment 3). The result shows in Table 5 suggested that for $k = 5$, the optimal threshold value that produced effective retrieval performance was $\varepsilon = 0.7$ which was slightly lower than the threshold values that was produced in the experiments conducted in Experiment 1, Experiment 2 and Experiment 3.

Experiment 5 for $k = 6$ and $\varepsilon = \{0.5, 0.6, 0.7, 0.8\}$. In experiment 5, the k-dimensions was set to $k = 6$ with various values of threshold. The result shown in Table 6 suggested that for $k = 6$, the optimal threshold value that produced effective retrieval performance was $\varepsilon = 0.7$.

Table 4 Percentage of recall, precision and effectiveness for $k = 4$, $\varepsilon = \{0.5, 0.6, 0.7, 0.8\}$

Query #	ε = 0.5			ε = 0.6			ε = 0.7			ε = 0.8		
	R	P	E	R	P	E	R	P	E	R	P	E
1	45.0	18.0	65.4	20.0	16.0	81.0	5.0	12.5	94.3	5.0	12.5	94.3
2	95.0	11.0	62.4	90.0	11.4	62.2	90.0	12.2	60.4	80.0	12.5	61.5
3	90.0	9.4	66.8	80.0	9.7	67.3	50.0	6.5	78.6	20.0	3.7	89.4
4	90.0	19.8	47.3	75.0	18.8	53.1	65.0	18.8	56.4	55.0	84.6	40.9
5	90.0	10.1	65.1	75.0	11.9	63.6	40.0	10.7	74.2	20.0	11.8	82.4
6	90.0	11.9	61.1	65.0	11.4	66.5	30.0	7.2	81.6	20.0	8.2	84.5
7	90.0	5.7	77.3	90.0	6.1	76.0	60.0	4.5	82.7	60.0	4.8	81.8
8	90.0	5.6	77.6	90.0	6.7	74.2	80.0	7.6	72.5	20.0	3.7	89.4
9	70.0	8.0	72.5	40.0	5.3	82.7	35.0	5.6	82.9	15.0	4.1	90.2
10	25.0	4.8	86.4	25.0	4.9	86.3	10.0	2.1	94.3	10.0	2.1	94.3
11	15.0	16.7	84.7	15.0	20.0	84.2	15.0	50.0	82.6	15.0	60.0	82.4
12	80.0	5.1	79.7	70.0	4.8	81.2	70.0	6.0	77.7	70.0	8.5	71.4
Average	72.5	10.5	70.5	61.3	10.6	73.2	45.8	12.0	78.2	32.5	18.0	80.2

Table 5 Percentage of recall, precision and effectiveness for $k = 5$, $\varepsilon = \{0.5, 0.6, 0.7, 0.8\}$

Query #	ε = 0.5			ε = 0.6			ε = 0.7			ε = 0.8		
	R	P	E	R	P	E	R	P	E	R	P	E
1	20.0	18.2	80.4	10.0	9.1	90.2	5.0	25.0	94.0	5.0	25.0	94.0
2	95.0	11.4	61.5	90.0	11.9	61.1	85.0	13.0	59.7	70.0	16.1	58.1
3	75.0	8.8	70.1	65.0	9.3	70.4	20.0	4.9	87.6	5.0	2.6	95.8
4	75.0	18.1	54.0	70	17.7	56.0	55.0	16.9	62.1	55.0	20.8	58.6
5	90.0	11.2	62.6	50.0	10.2	71.9	20.0	9.1	83.9	15.0	16.7	84.7
6	85.0	13.0	59.7	25.0	5.9	84.8	20.0	6.7	85.7	5.0	4.2	95.2
7	60.0	3.9	84.5	60.0	4.4	83.0	50.0	4.6	83.2	40.0	6.3	80.7
8	80.0	7.1	73.8	50.0	6.5	78.6	30.0	5.8	83.6	0.0	0.0	0.0
9	45.0	6.5	79.4	45.0	7.3	77.9	30.0	6.5	82.6	4.2	6.4	95.5
10	25.0	41.7	72.8	0.0	0.0	0.9	0.0	0.0	0.0	0.0	0.0	0.0
11	15.0	23.0	83.9	15.0	37.5	83.0	15.0	60.0	82.4	15.0	75.0	82.1
12	80.0	5.8	77.5	80.0	7.0	74.1	70.0	7.7	74.1	70.0	10.6	67.0
Average	62.1	14.1	71.7	46.7	10.6	69.2	34.2	13.3	73.2	23.7	15.3	67.6

Table 6 Percentage of recall, precision and effectiveness for $k = 6$, $\varepsilon = \{0.5, 0.6, 0.7, 0.8\}$

Query #	ε = 0.5			ε = 0.6			ε = 0.7			ε = 0.8		
	R	P	E	R	P	E	R	P	E	R	P	E
1	15.0	15.8	84.8	10.0	20.0	88.9	5.0	25.0	94.0	0.0	0.0	0.0
2	95.0	12.3	59.5	85.0	12.5	60.6	80.0	14.5	58.0	50.0	14.9	66.0
3	70.0	8.8	70.7	55.0	8.8	73.2	15.0	4.1	90.2	5.0	2.9	95.6
4	70.0	17.9	55.8	0.7	18.7	99.1	50.0	16.9	64.1	30.0	14.0	75.6
5	90.0	12.0	60.9	45.0	10.0	73.5	20.0	9.5	83.6	15.0	21.4	84.0
6	75.0	11.5	64.4	25.0	5.6	85.2	15.0	5.1	89.2	5.0	3.5	95.4
7	70.0	46.7	36.4	70.0	53.8	34.0	70.0	70.0	30.0	40.0	66.7	56.5
8	80.0	9.0	69.0	50.0	7.9	75.8	50.0	7.9	75.8	30.0	8.8	79.8
9	50.0	7.3	77.0	30.0	7.6	81.1	30.0	6.8	82.2	20.0	6.8	85.6
10	25.0	45.5	72.5	15.0	42.9	82.8	5.0	25.0	94.0	5.0	33.3	94.0
11	15.0	37.5	83.0	15.0	42.9	82.8	15.0	60.0	82.4	15.0	75.0	82.1
12	80.0	6.1	76.6	80.0	7.5	72.7	80.0	9.5	67.8	60.0	10.3	69.5
Average	61.3	19.2	67.5	40.1	19.9	75.8	36.3	21.2	75.9	22.9	21.5	73.7

3 Results and Discussion

Table 7 shows that the performance of the retrieval using latent semantic indexing was achieved if *k*-dimension was $k = 4$ and threshold value, $\varepsilon = 0.8$ which was approximately 37° angle. The result of the calculation signifies that given 1395 total numbers of document collections in Malay corpus which scattered in 4-dimensional

Table 7 Summary for average of effectiveness for the experiments

Average of effectiveness, E							
Threshold value, ε	*k*-dimensional (factor)					Max	Min
	2	3	4	5	6		
0.5	69.0	69.4	70.5	71.7	67.5	71.7	67.5
0.6	68.5	73.7	73.2	69.2	75.8	75.8	69.2
0.7	68.5	73.2	78.2	73.2	75.9	78.2	73.2
0.8	70.4	79.3	**80.2**	67.6	73.7	**80.2**	67.6
Max	70.4	79.3	**80.2**	73.2	75.9	–	–
Min	68.5	69.4	70.5	67.6	67.5	–	–

space of LSI, the terms and the documents are considered conceptually and semantically related, although there is no word in common to reveal the relationship between term and term, term and document and document and document, provided the angles between those terms and documents neighbourhood distance is 37° angles to each others. Otherwise, term–term, term–document and document–document relationships do not exist. In other words, they are not related.

Nevertheless, the results obtained from these experiments tend to score high result, even though the document is not a relevant document (based on the relevant judgment) for the document which contains many words and term occurrences. The analyses show that long documents are favoured because it tends to have more words and terms occurrences. Moreover, some terms are repeated within different contexts or topics in a document which does not ensure whether there is an existing relationship between term and documents. Therefore, another study needs to be undertaken to investigate the flaw in LSI techniques.

4 Conclusion

Experiments that were conducted in this study aim to investigate the optimise number of *k*-dimension and the value of threshold in order for the LSI-based information retrieval system to perform effectively hence produce efficient output, thus satisfying the user's information need rather than satisfying a given query. The study shows that to balance the percentage of recall and precision hence produce efficient retrieval, the *k*-dimension and threshold value of the LSI retrieval system which use singular value decomposition technique needs to be tuned to cater the number of collection in the corpus so that the retrieving for the information satisfied user's information need. In this study, we found out that, for a small Malay corpus which consist of 1395 total test collections, the best and efficient retrieval can be achieved if the *k*-dimension is $k = 4$ and the threshold value is $\varepsilon = 0.8$. This study helps the software developers, particularly the information retrieval application developer in designing and choosing the optimise value of the *k*-dimension as well as the value of threshold for the search engine.

References

Baeza-Yates R (2004) Challenges in the interaction of information retrieval and natural language processing. In: Computational linguistics and intelligent text processing, vol 2945, pp 445–456. http://link.springer.com/chapter/10.1007/978-3-540-24630-5_55

Baeza-Yates R, Ribeiro-Neto B (1999) Modern information retrieval, vol 9. ACM press, New York. doi:10.1080/14735789709366603

Bartell BT, Cottrell GW, Belew RK (1992) Latent semantic indexing is an optimal special case of multidimensional scaling. In: Proceedings of the 15th annual international ACM SIGIR conference on research and development in information retrieval—SIGIR'92. ACM Press, New York, USA, pp 161–167. doi:10.1145/133160.133191

Buckeridge AM, Sutcliffe RFE (2002) Disambiguating noun compounds with latent semantic indexing. In: COLING-02 on COMPUTERM 2002 second international workshop on computational terminology, vol 14. Association for Computational Linguistics, Morristown, NJ, USA, pp 1–7. doi:10.3115/1118771.1118772

Dasgupta A, Kumar R, Raghavan P, Tomkins A (2005) Variable latent semantic indexing. In: Proceeding of the eleventh ACM SIGKDD international conference on knowledge discovery in data mining—KDD'05. ACM Press, New York, USA, p. 13. doi:10.1145/1081870.1081876

Deerwester S, Dumais S, Landauer T (1990) Indexing by latent semantic analysis. JASIS. http://www.cob.unt.edu/itds/faculty/evangelopoulos/dsci5910/LSA_Deerwester1990.pdf

Ding CHQ (1999) A similarity-based probability model for latent semantic indexing. In: Proceedings of the 22nd annual international ACM SIGIR conference on research and development in information retrieval—SIGIR'99. ACM Press, New York, USA, pp. 58–65. doi:10.1145/312624.312652

Efron M (2007) Model-averaged latent semantic indexing. In: Proceedings of the 30th annual international ACM SIGIR conference on research and development in information retrieval—SIGIR'07. ACM Press, New York, USA, p 755. doi:10.1145/1277741.1277893

Foltz PW (1990) Using latent semantic indexing for information filtering. ACM SIGOIS Bull 11 (2–3):40–47. doi:10.1145/91474.91486

Foltz P, Dumais S (1992) Personalized information delivery: an analysis of information filtering methods. Commun ACM 12. http://dl.acm.org/citation.cfm?id=138866

Furnas GW, Deerwester S, Dumais ST, Landauer TK, Harshman RA, Streeter LA, Lochbaum KE (1988) Information retrieval using a singular value decomposition model of latent semantic structure. In: Proceedings of the 11th annual international ACM SIGIR conference on Research and development in information retrieval—SIGIR'88. New York, USA, pp 465–480. doi:10.1145/62437.62487

Gee KR (2003) Using latent semantic indexing to filter spam. In: Proceedings of the 2003 ACM symposium on applied computing—SAC'03. ACM Press, New York, USA, p 460. doi:10.1145/952532.952623

Hofmann T (1999) Probabilistic latent semantic indexing. In: Proceedings of the 22nd annual international ACM SIGIR conference on research and development in information retrieval—SIGIR'99, pp 50–57. doi:10.1145/312624.312649

Hutchison D, Mitchell JC (2009) Lecture notes in computer science. String processing and information retrieval, vol 5721. Springer, Berlin. doi:10.1007/978-3-642-03784-9

Kim W, Khudanpur S (2004) Lexical triggers and latent semantic analysis for cross-lingual language model adaptation. ACM Trans Asian Lang Inf Process 3(2):94–112. doi:10.1145/1034780.1034782

Kokiopoulou E, Saad Y (2004) Polynomial filtering in latent semantic indexing for information retrieval. In: Proceedings of the 27th annual international conference on research and development in information retrieval—SIGIR'04, vol 104. doi:10.1145/1008992.1009013

Kolda TG, Leary DPO (2008) A semidiscrete matrix decomposition for latent semantic indexing in information retrieval 16(4):322–346

Kowalski G (1997) Information retrieval systems: theory and implementation. Springer, New York

Marcus A, Maletic JI (2003) Recovering documentation-to-source-code traceability links using latent semantic indexing. In: 25th international conference on software engineering, 2003. IEEE Proceedings. pp 125–135. doi:10.1109/ICSE.2003.1201194

Mckinley KS (2000) The effect of collection organization and query locality on information retrieval system performance. Adv Inf Retrieval 7:173–202. doi:10.1007/0-306-47019-5_7

Quan X, Chen E, Luo Q, Xiong H (2008) Adaptive label-driven scaling for latent semantic indexing. In: Proceedings of the 31st annual international ACM SIGIR conference on research and development in information retrieval—SIGIR'08, pp 827–828. doi:10.1145/1390334.1390525

Ricardo B, Berthier R (2011) Modern information retrieval: the concepts and technology behind search second edition, vol 82. Addision Wesley, p 944. http://www.amazon.com/Modern-Information-Retrieval-Concepts-Technology/dp/0321416910

Russell S, Norvig P (2010) Natural language processing. In: Artificial intelligence: a modern approach, 3rd edn. Pearson, Prentice Hall, pp 861–887

Sadjirin R, Rahman NA (2010) Efficient retrieval of Malay language documents using latent semantic indexing. In: 2010 international symposium on information technology, vol 3. IEEE, pp. 1410–1415. doi:10.1109/ITSIM.2010.5561613

Syu I, Lang SD, Deo N (1996) Incorporating latent semantic indexing into a neural network model for information retrieval. In: Proceedings of the fifth international conference on information and knowledge management—CIKM'96, pp 145–153. doi:10.1145/238355.238475

Tang C, Dwarkadas S, Xu Z (2004) On scaling latent semantic indexing for large peer-to-peer systems. In: Proceedings of the 27th annual international conference on research and development in information retrieval—SIGIR'04. ACM Press, New York, USA, p 112. doi:10.1145/1008992.1009014

Trotman A (2003) Compressing inverted files. Inf Retrieval 6(1):5–19. doi:10.1023/A:1022949613039

Yu K, Yu S, Tresp V (2005) Multi-label informed latent semantic indexing. In: Proceedings of the 28th annual international ACM SIGIR conference on research and development in information retrieval—SIGIR'05. ACM Press, New York, USA, p 258. doi:10.1145/1076034.1076080

Chapter 32
Critical Review of Measurement for Multipartite Entanglement: Detection and Quantification

Siti Munirah Mohd, Bahari Idrus, Muriati Mukhtar and Hishamuddin Zainuddin

Abstract Quantum entanglement is one of the fields in quantum mechanics. Recently, quantum entanglement becomes the heart of many tasks in quantum information such as quantum cryptography, quantum teleportation and quantum computing due to the capability it to compute data more efficient compare classical computer. Therefore, the measurement of entanglement become important to determine either the state is entangled or separable. The aim of this paper is to view all possible methods of measurement in term of detection and quantification for multipartite entanglement cases. The outcome of this paper is to classify the method of detection and quantification including summarize the criteria and advantage or disadvantage of each method.

Keywords Entanglement · Detection · Measurement · Multipartite · Quantification

S.M. Mohd (✉) · B. Idrus · M. Mukhtar
Industrial Computing Research Group, Faculty of Information Science and Technology, Universiti Kebangsaan Malaysia, Bangi, Selangor, Malaysia
e-mail: smunirahm@gmail.com

B. Idrus
e-mail: bahari@ftsm.ukm.my

M. Mukhtar
e-mail: mm@ftsm.ukm.my

H. Zainuddin
Laboratory of Computational Sciences and Mathematical Physics Institute for Mathematical Research, Universiti Putra Malaysia, Serdang, Selangor, Malaysia
e-mail: hisham@fsas.upm.edu.my

N.A. Yacob et al. (eds.), *Regional Conference on Science, Technology and Social Sciences (RCSTSS 2014)*, DOI 10.1007/978-981-10-0534-3_32

1 Introduction

The Oxford dictionary defines entanglement as a 'state of being involved in complicated circumstances'. It is also denotes an affair between two people. Hence, in quantum mechanics, quantum entanglement can be explained as a 'complicated affair' between two or more particles (Krammer 2005). Quantum entanglement becomes an important component in quantum information and it is at the heart of quantum mechanics. It is known and often represented after Einstein, Podolsky and Rosen (EPR) and Schrödinger (Guhne and Toth 2009; Jingshui and Wenbo 2011; Kinsella 2006; Ryszard et al. 2009). Preceding to that, John Bell discovered that entanglement could be testable in experiments of quantum mechanics from classical (Guhne and Toth 2009). Entanglement is often applied in many fields such as quantum cryptography, quantum teleportation and super dense coding (Guhne and Toth 2009; Ma et al. 2011; Ryszard et al. 2009). Entanglement provides an alternative way to process information in quantum information field better than classical information theory. The use of this process has become more and more efficient with less physical material needed to transfer the information (Krammer 2005).

The basic problems in the theory of quantum entanglement involves the measurement of the entanglement in the state involved. Most measurements deals with bipartite cases compared to multipartite cases due to the fact that operational criterion towards it is not similar to simple bipartite cases (Amico et al. 2008). There are no general criteria to detect entanglement especially in multipartite cases. Even up till now, there are still many open questions regarding the properties and types of multipartite entanglement. Usually, measurements for multipartite is considered for a pure state and very limited to a mixed state since mixed states are more complicated than pure states.

In contrast to the bipartite cases, multipartite cases involves many types of situation. As an example in the simplest case involves three qubits, entanglement can exist either in a state of fully separable, biseparable or genuinely entangled. The dimensions represent that multipartite cases are more complicated, hence it quite difficult to detect and quantifying the entanglement includes distinguishing them from each other completely.

In this paper, we give meaning of multipartite entanglement in mathematical form in different situation, both in pure states and mixed states. Next, we explained the crucial properties needed, when measuring the entanglement. After that, we discuss the measurement of multipartite entanglement in term of detections and quantifications. Lastly, we have summarize the criteria, advantages and disadvantages of the measurement.

2 Multipartite Entanglement

Entanglement is a correlation between particles that cannot be separated regardless the distance among them. Multipartite entanglement involves more than two qubits that are mutually correlated. In both cases either bipartite or multipartite, the states that represent the entanglement exist in pure states and mixed states. Here, we define the multipartite entanglement in a mathematical form in both states and discuss the requirement needed for the states to become entangled.

Definition of Entanglement

Pure States: Pure states is the ground state representing quantum system. It is represented by a ket vector, $|\rangle$. Superposition of the pure states also produces the pure states (Chuang and Neilsen 2000). For pure n-partite entanglement, the system is said to be entangled if there exist entanglement between n-particle and separable if one is removed from the system.

Definition 1 (Abaffy 2011; Eltschka and Siewert 2014a) Assume that $|\psi\rangle \in H = H_{A_1} \otimes \cdots \otimes H_{A_n}$ is a pure state for quantum system $A_1 \otimes \cdots \otimes A_n$ is said to be fully separable to the system $A_1 \otimes \cdots \otimes A_n$ if and only if $|\psi\rangle$ can be expressed in terms of tensor product $|\psi\rangle = |\psi_{A_1}\rangle \otimes \cdots \otimes |\psi_{A_n}\rangle$ with $|\psi_{A_i}\rangle$ is a pure state of H_{A_i}. Otherwise, $|\psi\rangle$ entangled.

Example 1 $|\psi\rangle = \frac{1}{\sqrt{2}}(|000\rangle + |111\rangle)$ is an entangled state of three qubits in $H_1 \otimes H_2 \otimes H_3$. It can be proved by assuming it is not entangled. Therefore, it can be written in term of tensor product of three qubits:

$$\begin{aligned}(a \quad b)^{\mathrm{T}} \otimes (c \quad d)^{\mathrm{T}} \otimes (e \quad f)^{\mathrm{T}} &= (ace \quad acf \quad ade \quad adf \quad bce \quad bcf \quad bde \quad bdf)^{\mathrm{T}} \\ &= \left(\tfrac{1}{\sqrt{2}} \quad 0 \quad 0 \quad 0 \quad 0 \quad 0 \quad 0 \quad \tfrac{1}{\sqrt{2}}\right)^{\mathrm{T}}\end{aligned}$$

ace and bdf is non-zero. Hence, a, b, c, d, e, f must be all non-zero but the value of acf, ade, adf, bce, bcf, bde are zero and it is a contradiction.

Definition 2 (Abaffy 2011) A pure state is said to be k-separable if and only if $|\psi\rangle$ can be expressed as the tensor product of k substates: $|\psi\rangle = |\psi_1\rangle \otimes \cdots \otimes |\psi_k\rangle$.

Mixed States: Mixed states is represented by density matrix, ρ. It cannot be expressed by ket vector. Superposition in mixed states does not necessarily produce mixed states. Mixed states can be expressed by $\rho = \sum_i p_i |\psi_i\rangle\langle\psi_i|$, $\sum_i p_i = 1$, $p_i \geq 0$ with p_i denote as probability of each state.

Definition 3 (Abaffy 2011) Assume that $H = H_{A_1} \otimes \cdots \otimes H_{A_n}$ Hilbert space. Mixed state,ρ is said fully separable against the system $A_1 \otimes \cdots \otimes A_n$ if and only if ρ can be expressed in the form of $\rho = \sum_{i=1}^{k} p_i \rho_{A_1}^{i} \otimes \cdots \otimes \rho_{A_n}^{i}$ with $\sum_{i=1}^{k} p_i = 1$. Otherwise, ρ entangled.

Sheikholeslam and Gulliver (2012) reported that the states are genuine multipartite entanglement if the n-partite is a set of all m-partite, $m<n$ entangled. If the system is not fully separable, then exist any classical correlation in the system of biseparable. This condition is defined in Definition 4 till Definition 6.

Definition 4 (Abaffy 2011; Eltschka and Siewer 2014a) Mixed state ρ in a Hilbert space $\mathrm{H} = \mathrm{H}_{A_1} \otimes \cdots \otimes \mathrm{H}_{A_n}$ is said to be separable with respect to partition $\{B_1, B_2, \ldots, B_k\}$ if and only if it can be written as $\rho = \sum_{i=1}^{l} p_i \rho_{B_1}^{i} \otimes \cdots \otimes \rho_{B_k}^{i}$ with $\sum_{i=1}^{l} p_i = 1$ for each $l \in \{1, \ldots, k\}$ is a non-empty set that contains some A_n of the original system and for each $j \in \{1, \ldots, m\}$, A_j is a set that is in the set B_i. For each $\rho_{B_j}^{i}$ defined as tensor product in Hilbert space H_{A_n} for A_n in B_j. If the mixed state ρ not separable for any partition $\{B_1, B_2\}$, it called as m-partite entangled.

Definition 5 (Abaffy 2011) Mixed state ρ in a Hilbert space $\mathrm{H} = \mathrm{H}_{A_1} \otimes \cdots \otimes \mathrm{H}_{A_n}$ is said semi-separable if and only if ρ separable refers to all partition $\{B_1, B_2\}$ with $B_1 = \{A_j\}$ and $B_2 = \{A_1, \ldots, A_{j-1}, A_{j+1}, \ldots, A_m\}$, $j \in \{1, \ldots, m\}$.

Properties of Entanglement Measures

Good entanglement measures, $E(\rho)$ must fulfill certain conditions to make sure it can measure entanglement efficiently. These conditions are listed below. However, not all of the following condition satisfy by each measurement because is not always easy to adhere all these properties. Some of the condition are taken as an option (Abaffy 2011; Bruß 2002; Guhne and Toth 2009; Hiesmayr et al. 2009):

- Monotone under local operations and classical communication (LOCC). Entanglement cannot exceed LOCC.
- If ρ separable then $E(\rho) = 0$.
- Normalization: $E(|\psi_d^{+}\rangle) = \log_2 d$, d is dimension of the system.
- Convexity: $E(\sum_i p_i \rho_i) \leq \sum_i p_i E(\rho_i)$.
- Additivity: $E(\rho^{\otimes n}) = nE(\rho)$.
- Strong additivity: $E(\rho_1 \otimes \rho_2) = E(\rho_1) + E(\rho_2)$.
- Asymptotic continuity: $\|\rho_n - \sigma_n\| \to 0$ implies $\frac{|E(\rho_n) - E(\sigma_n)|}{\log_2 d_n} \to 0$ with ρ_n, σ_n in Hilbert space H, d_n is dimension for H and $\|X\|$ trace norm.

3 Detection of Multipartite Entanglement

Entanglement measurement is performed to determine the existence of quantum correlations between particles. There are two forms of measurement: detection and quantification. Entanglement detection (Abaffy 2011; Guhne and Toth 2009; Ryszard et al. 2009) is an instrument used to determine whether the states entangled or separable. Detection just only determines the condition of the state either satisfy properties of entangled or separable. This measure is useful to measure entanglement experimentally. In multipartite cases, six detections always implemented to measure the states. There are positive partial transpose (PPT) criterion, completely positive maps (CPM), conditional entropy, negativity, entanglement witness and range criteria.

Positive partial transpose (PPT) criterion is also known as the Peres-Horodecki Criterion (Guhne and Toth 2009). These criteria depend on the product of basis density matrix of mixed states. Density matrix that represents a subsystem represented as $\rho = \sum_{i,j}^{N} \sum_{k,l}^{M} p_{ij,kl} |i\rangle\langle j| \otimes |k\rangle\langle l|$ where $p_{ij,kl}$ is probability of qubits. PPT detects the value of separability in the state. If the eigenvalue of density matrix is positive mean the state separable. Otherwise it is entangled. Completely positive maps (CPM) (Abaffy 2011) involve with value of spectrum. Not all positive maps can detect the states either entangled or separable. It can detect separabality if and only if any positive mapping $\Lambda : B \rightarrow A$ for operator $(I \otimes \Lambda)(\rho)$ is positive and $A(B)$ represents the set of operator $II_{A(B)}$.

Conditional entropy (Idrus et al. 2010; Zulkarnain et al. 2006) defined as $E(i|j) = S(\rho_{ij}) - S(\rho_j)$ with $S(\rho) = -\mathrm{Tr}\rho \log(\rho)$ is von Neumann entropy. The state entangled when the value of entropy is negative but it cannot detect either the state is entangled or separable when the value is positive. Next detection is negativity. Negativity is a measurement that failed towards PPT criterion. It is a simple calculation compared with other measurements (Guhne and Toth 2009). It also satisfies entanglement monotone that not increase under local operation and classical communication (LOCC) (Abaffy 2011) concept. Negativity (Eltschka and Siewert 2014b; Guhne and Toth 2009) can be written as $N(\rho) = \frac{\|\rho^{T_B}\|_1 - 1}{2}$ where ρ^{T_B} is partial transpose. If negativity is equal to zero, the state is separable. Otherwise it is entangled.

Other usual detection is the entanglement witness. The function of this detection is to distinguish a certain entangled state from separable ones. When the value of this measure is positive, it determines the state separable and otherwise the state become entangled (Eltschka and Siewert 2014a; Horodecki et al. 1996; Lewenstein et al. 2000; Terhal 2000). Similarly to Schmidt measure (Eisert and Briegel 2001), entanglement witness is also one the tools that is useful to classify classes in multipartite entanglement. It is also identical with concept of geometric because it can separate bound operator into two space namely, positive and negative. Besides, entanglement witness is mostly applied in experiments that involve generating of entanglement (Eltschka and Siewert 2014a; Lanyon et al. 2013). Lastly, range criteria defined as $Ran(\rho) := \{|\phi\rangle \in \mathrm{H}|\rho|\psi\rangle = |\phi\rangle \text{ and } |\psi\rangle \in \mathrm{H}\}$ is one of the necessary conditions to detect separability states. Table 1 summarized the list of detection for multipartite entanglement in terms of criteria and advantages and disadvantages of each of the detection (Abaffy 2011; Bruß 2002; Eltschka and Siewert 2014a; Guhne and Toth 2009; Kinsella 2006; Ryszard et al. 2009).

4 Quantification of Multipartite Entanglement

In contrast, the quantification measure (Abaffy 2011; Eltschka and Siewert 2014b; Guhne and Toth 2009; Ryszard et al. 2009) refers to the degree of entanglement in terms of the strength of correlation between the particles. Quantification is a measure that gives certain value of entanglement for the state either the states is more entangled or less. These measurements are also used to distinguish classes in multipartite entanglement. There are a several methods of quantification that are usually used to measure entanglement. One of the latest quantification is concurrence (Jurkowski and Chruściński 2010). It very useful in pure states compared to mixed states (Zhu and Fei 2013). General mathematical notation of concurrence is $C_A(\psi) = \sqrt{2(1 - \mathrm{tr}\rho_A^2)}$ for party A (Eltschka and Siewert 2014b). It can measure as two qubits in multipartite by consider bipartition of two party. This concept applied here is monogamy relation (Jiao-Jiao 2010). In (Zhu and Fei 2013) concurrence denoted as $C(|\psi\rangle) := \min_{\gamma_i \in \gamma} \sqrt{2(1 - \mathrm{Tr}(\rho_{A_{\gamma_i}}^2))}$ with $\gamma = \{\gamma_i\}$ represent all possible set bipartition $\{A_i|B_i\}$ for $\{1, 2, \ldots, n\}$. The state separable when concurrence is equal to zero and become entangled when it is positive. It is a measurement on the contrary to negativity. At certain cases, concurrence is equal to negativity especially in pure states. Three tangle measurements have relation with concurrence. Three tangle is a measurement generalized from residual measure (Zhu et al. 2012). It is not suitable to measure genuine tripartite because of its properties. Formula for three tangle denoted as $\tau(|\psi\rangle) = 4\lambda_0^2\lambda_4^2$ (Ryszard et al. 2009) with λ represent eigenvalues. It also can be written as $\tau_{ABC} = C_{A(BC)}^2 - C_{AB}^2 - C_{AC}^2$ with right hand side is square of concurrence (Eltschka and Siewert 2014a; Yu and Song 2005). Advantages of this measure are invariant permutation and measure the state of three particle simultaneously.

Convex-roof extended negativity (CREN) has been extended in monogamy relation by considering convex roof extension negativity. Jiao-Jiao (2010) showed

Table 1 Detection of multipartite entanglement

Detection	Advantages and criteria	Disadvantages
Positive partial transpose (PPT) criterion	• Necessary for separability criteria • Measure in bipartition of qubit • Apply in mixed states • Separable when eigenvalue of density matrix positive	• Inconclusive result in higher dimension • Need other advanced test to compute
Completely positive maps (CPM)	• Separability criteria • Preserve dimension in any dimension	• Complex calculation
Conditional entropy	• Used von Neumann entropy in calculation • Entangled when value is negative • Apply in pure states and mixed states	• Inconclusive result when positive value
Negativity	• Simple computable measure • Convex and monotone • Can be used to quantify the degree of entanglement	• Fail to measure entanglement in PPT states
Entanglement witnesses	• Experimental implementations • Distinguish classes in multipartite entanglement • Can use both in pure and mixed states	
Range criteria	• Separability criterion • Apply concept linear span	• Difficult to find the product states in the range of a matrix

that CREN is stronger than concurrence involving quantum system on multipartite higher dimension. In mixed states, CREN formula defined as $N_c(\rho_{AB}) \equiv \min \sum_k p_k N(|\phi_k\rangle)$ considering all minimum pure states.

Geometric measure (Abaffy 2011; Barnum and Linden 2001; Guhne and Toth 2009; Wei and Goldbart 2003) is based on the measurement of the distance between the set separable states. It is defined as $G(|\phi\rangle) := -\log_2\left\{\sup_{|\alpha\rangle\in \mathrm{SEP}}|\langle\phi|\alpha\rangle|^2\right\}$ where SEP is set for all separable states. Other quantifier is average linear entropy (ALE) (Kinsella 2006; Yu and Song 2005) defined as $S_L(\rho_A) = \frac{D}{D-1}\left(1 - tr(\rho_A)^2\right)$ where D is dimension of the particle. The higher the non-negative value, the states become more entangled.

Distance measure is quite different from other quantifier because it is based on the distance between the systems. Distance measure can be measured using two types of measurement namely relative entropy and robustness measure. Relative entropy (Jingshui and Wenbo 2011; Kinsella 2006) is defined as $E_R(\sigma) := \min_{\rho \in D} S(\sigma || \rho)$ with $S(\sigma||\rho) = \mathrm{tr}\sigma[\ln \sigma - \ln \rho]$ and D is set of separable states while robustness measures that introduced by Vidal and Tarrach (Ryszard et al. 2009) defined as $R(\rho) = \inf_{\sigma_{\text{sep}}} R(\rho|\sigma_{\text{sep}})$ where ρ is separable states of σ_{sep} and $R(\rho|\sigma_{\text{sep}})$ is minimum t such that $\frac{(\rho + t\sigma_{\text{sep}})}{1+t}$ separable.

Schmidt measure is one of the famous quantifier that always use to determine entanglement and classify the classes of multipartite entanglement in pure states. It is an extension of convex roof in pure states. Schmidt measure (Eisert and Briegel 2001) for pure states is defined as $P(|\psi\rangle\langle\psi|) = \log_2 r$ where r is a minimum number of the product R in decomposition of $|\psi\rangle = \sum_{i=1}^{R} \alpha_i \left|\psi_{A_1}^{(i)}\right\rangle \otimes \cdots \otimes \left|\psi_{A_N}^{(i)}\right\rangle$. In contrast, Schmidt number (Guo and Fan 2013) involve rank of the density matrix of the pure states and mixed states. Same as Schmidt measure, Schmidt number also can classify classes in multipartite entanglement based on its rank.

Finally, these two quantification, Ingarden−Urbanik Entropy (IUE) (Sheikholeslam and Gulliver 2012) and Manhattan distance of averaged partial entropies (MAPE) (Li et al. 2013) applied existing quantifier. IUE measure applied concept Schmidt number (Sperling and Vogel 2011) and coefficient $C_{i_1 \cdots i_n}$(Carteret et al. 2000) and defined as

$$\begin{aligned} M(\rho) &= S_{IU}\left(\left|\psi_{\text{decompose}}\right\rangle\right) = S_{IU}\left(U_1 \otimes \cdots \otimes U_n\left|\psi_{\text{pure}}\right\rangle\right) \\ &= -\sum_{i_1 \cdots i_n} |C_{i_1 \cdots i_n}|^2 \log|C_{i_1 \cdots i_n}|^2. \end{aligned} \tag{1}$$

where $\left|\psi_{\text{pure}}\right\rangle = \sum_i \sqrt{\lambda_i} |\psi_i\rangle |0^1 \cdots 0^{n-1} i^n\rangle$ is generic purification. The higher the positive value of Schmidt number, the higher the value IUE. Meanwhile, MAPE is a modification of average partial entropy (APE) and implement von Neumann entropy. It is defined as

$$\mathrm{M} = |S_1| + |S_2| + \cdots + \left|S_{[n/2]}\right| = S_1 + S_2 + \cdots + S_{[n/2]} \tag{2}$$

where $S_l = \left[\prod_{q_1,\ldots,q_l=1}^{n} E_{q_1,\ldots,q_l}\right]^{\frac{1}{C_n^l}}$, $1 \leq l \leq [n/2], E_{q_1,\ldots,q_l} = -\mathrm{Tr}\left(\rho_{q_1,\ldots,q_l} \log_2{}_{q_1,\ldots,q_l}\right)$ is reduced von Neumann entropy, $C_n^l = \frac{n!}{(n-1)!l!}$ and $S_l \geq 0$ (Liu et al. 2010). Last of all, Table 2 summarized the quantification of multipartite entanglement.

Table 2 Quantification of multipartite entanglement

Quantification	Advantages and criteria	Disadvantages
Concurrence	• Noiseless detection • Currently most powerful to measure entanglement	• Hard to measure in mixed states
Three tangle	• Noiseless detection • Permutationally invariant	• Measure residual
Convex-roof extended negativity (CREN)	• Stronger entanglement measure than concurrence • Can apply in higher-dimensional system	• Hard to calculate in high dimension
Geometric measure (GM)	• Generalization from Shimony measure • Use in separable states • Easy to compute than the entanglement formation	• Difficult to evaluate all possible separable state • Harder in convex roof extension of mixed states
Distance measure: relative entropy	• Use in separable states • Important function in quantum information theory • Powerful upper bound of entanglement distillation • Strongly nonadditive	
Schmidt measure	• Its algorithm is strongly nonadditive • Can apply concept of upper bound in its complexity	• Cannot distinguish truly multipartite entanglement from bipartite entanglement
Ingarden-Urbanik entropy (IUE)	• Use concept generalized Schmidt decomposition (GSD) • Apply Schmidt number	• Mixed states only
Manhattan distance of averaged partial entropies (MAPE)	• Modification from averaged partial entropies (APE) • Use von Neumann entropy • Use coefficient matrices	

5 Conclusions

In conclusion, entanglement in any state can be determined by measuring it either in pure states or mixed states. Measurement for entanglement is divided into two, namely the detection and quantification. Entanglement detection only determine whether the situation is entangled or separable while quantification more to discover strength of entanglement of each states. Normally, detection is easier to compute compare quantification since calculation involve in quantification details. Some measurement can used to classify multipartite entanglement to some kind of concept such as local unitary (LU), local operations and classical communication (LOCC) and stochastic local operations and classical communication (SLOCC).

Till now, the specific method to classify multipartite entanglement is not fully discovered since each measurement listed has its own advantages and disadvantages. Some of them are complement to each other if it failed to determine entanglement of any states. In many papers, most of the states used is pure states rather than mixed state due to computation more difficult.

Acknowledgements This work is supported by the research grant provided by Malaysian Ministry of Higher Education code: FGRS/1/2011/SG/UKM/03/5.

References

Abaffy M (2011) Multipartite entanglement. Bachelor. Masaryk university, Brno

Amico L, Fazio R, Osterloh A, Vedral V (2008) Entanglement in many-body systems. Rev Mod Phys 80(2):517–576

Barnum H, Linden N (2001) Monotones and invariants for multi-particle quantum states. J Phys A: Math Gen 34(35):6787

Bruß D (2002) Characterizing entanglement. J Math Phys 43(9):4237–4251. doi:10.1063/1.1494474

Carteret HA, Higuchi A, Sudbery A (2000) Multipartite generalization of the Schmidt decomposition. J Math Phys 41(12):7932–7939. doi:10.1063/1.1319516

Chuang IL, Neilsen MA (2000) Quantum computation and quantum information. Cambridge University Press, Cambridge

Eisert J, Briegel HJ (2001) Schmidt measure as a tool for quantifying multiparticle entanglement. Phys Rev A 64(2):022306

Eltschka C, Siewert J (2014a) Practical method to obtain a lower bound to the three-tangle. Phys Rev A 89(2):022312

Eltschka C, Siewert J (2014b) Quantifying entanglement resources, arXiv:quant-ph/1402.6710v

Guhne O, Toth G (2009) Entanglement detection. J Phys Rep 1–75. doi:10.1016/j.physrep.2009.02.2004

Guo Y, Fan H (2013) Generalized Schmidt number for multipartite states, arXiv:quant-ph/1304.1950v2

Hiesmayr BC, Huber M, Krammer P (2009) Two computable sets of multipartite entanglement measures. Phys Rev A 79(6):062308

Horodecki M, Horodecki P, Horodecki R (1996) Separability of mixed states: necessary and sufficient conditions. Phys Lett A 223(1–2):1–8

Idrus B, Konstadopoulou A, Vourdas A (2010) Correlations in a chain of three oscillators with nearest neighbour coupling. J Mod Opt 57(7):1–6. doi:10.1080/09500341003789926

Jiao-Jiao L, Zhi-Xi W (2010) Monogamy relations in tripartite quantum system. Chin Phys B 19(10):100310–100310. doi:10.1088/1674-1056/19/10/100310

Jingshui Y, Wenbo X (2011) Calculation of quantum entanglement. In: Paper presented at the tenth international symposium on distributed computing and applications to business, engineering and science (DCABES), 2011

Jurkowski J, Chruściński D (2010) Estimating concurrence via entanglement witnesses. Phys Rev A 81(5):052308

Kinsella S (2006) Online measurement of entanglement of a quantum state (D. o. I. Technology, Trans). National University of Ireland, Galway, pp 1–50

Krammer P (2005) Quantum entanglement: detection, classification, and quantification Master University of Vienna. Austria, Wien

Lanyon BP, Jurcevic P, Zwerger M, Hempel C, Martinez EA, Dür W, Roos CF (2013) Measurement-based quantum computation with trapped ions. Phys Rev Lett 111(21):210501
Lewenstein M, Kraus B, Cirac JI, Horodecki P (2000) Optimization of entanglement witnesses. Phys Rev A 62(5):052310
Li H, Wang S, Cui J, Long G (2013) Quantifying entanglement of arbitrary-dimensional multipartite pure states in terms of the singular values of coefficient matrices. Phys Rev A 87 (4):042335
Liu D, Zhao X, Long G-L (2010) Multiple entropy measures for multi-particle pure quantum state. Commun Theor Phys 54(5):825
Ma Z-H, Chen Z-H, Chen J-L, Spengler C, Gabriel A, Huber M (2011) Measure of genuine multipartite entanglement with computable lower bounds. Phys Rev A 83(6):062325
Ryszard H, Paweł H, Michał H, Karol H (2009) Quantum entanglement. Rev Mod Phys 81 (2):865–942
Sheikholeslam SA, Gulliver TA (2012) Classification and measurement of multipartite quantum entanglements. arXiv:quant-ph/1205.2339v1
Sperling J, Vogel W (2011) The Schmidt number as a universal entanglement measure. Phys Scr 83(4):045002
Terhal BM (2000) Bell inequalities and the separability criterion. Phys Lett A 271(5):319–326
Wei T-C, Goldbart PM (2003) Geometric measure of entanglement and applications to bipartite and multipartite quantum states. Phys Rev A 68(4):042307
Yu C-S, Song H-S (2005) Multipartite entanglement measure. Phys Rev A 71(4):042331
Zhu X-N, Fei S-M (2013) Lower bound of concurrence for qubit systems. In: Quantum information processing, pp 1–9. doi:10.1007/s11128-013-0693-7
Zhu X-N, Zhao M-J, Fei S-M (2012) Lower bound of multipartite concurrence based on subquantum state decomposition. Phys Rev A 86(2):022307
Zulkarnain ZA, Konstadopoulou A, Vourdas A (2006) Measurements and correlations in tri-partite systems. Int J Mod Phys B 20:1551–1563

Chapter 33
Toward Developing an Enhanced Hough Transform Technique for Circle and Semicircle Detection

Ismariani Ismail, Adeline Engkamat and Abang Feizal Abang Ibrahim

Abstract Image segmentation is one of the techniques to localize shapes in an image. There are three components of image segmentation which are; image thresholding, edge-based segmentation, and region-based segmentation. In this research, we focused only on one of edge-based segmentation technique which is Hough transform, that can detect circular shapes in an image. However, it was found that there are problems with the existing Hough transform where it is unable to detect a circular shape with unknown radius and it is also unable to detect semicircular shapes. Thus, in this research, Hough transform is enhanced to solve these problems by comparing the pattern of graphs and by looking at the Hough peaks in the Hough transform matrix. In the testing part, results for both semicircle and circle images underwent the process of measuring the quantity of accepted and rejected images using confusion matrix. The results revealed that the accuracy of circle and semicircle detection in modified Hough transform has better performance compared to the existing Hough transform. Thus, this new enhanced technique could be used in the development of a methodology that will be of value in future studies of circle detection in image segmentation; such as in medical area to locate tumors and other pathologies, to locate objects in satellite images, iris recognition, and face recognition.

Keywords Circle detection · Hough transform · Image segmentation · Semicircle detection · Shape detection

---

I. Ismail (✉) · A. Engkamat
Faculty of Computer and Mathematical Sciences,
Universiti Teknologi MARA, Kota Samarahan, Malaysia
e-mail: ismariani@sarawak.uitm.edu.my

A. Engkamat
e-mail: adeline@sarawak.uitm.edu.my

A.F. Abang Ibrahim
Faculty of Business Management, Universiti Teknologi MARA,
Kota Samarahan, Malaysia
e-mail: abangfeizal@sarawak.uitm.edu.my

N.A. Yacob et al. (eds.), *Regional Conference on Science, Technology and Social Sciences (RCSTSS 2014)*, DOI 10.1007/978-981-10-0534-3_33

1 Introduction

The Hough transform is a feature extraction technique used in image analysis, computer vision, and digital image processing (Shapiro and Stockman 2001). This method is using voting procedure to find the not perfect instances in the objects within many types of shapes. Hough transform is also a technique that can extract features of a specific shape in images. Hough transform is widely used in detecting shapes of circles, lines, ellipse, square, and rectangle.

In this research, Hough transform is used for two types of shape detections, which are circle detection and semicircle detection. Circle detection is an important task in image processing. Circular patterns may suggest the existence of interesting objects, e.g., the wheels of a vehicle, or the eyes of an individual to be identified (Guo et al. 2006).

The existing Hough transform that was studied by Rabindra and Meena (2009), shows that the result of their research on circle detection is not efficient where there is compulsory to give the information of the radius. This Hough transform needed to be provided with minimum and maximum values of radius in detecting a circle. Circle with no information of radius value cannot be detected using this method. Another problem occurs is when it detects semicircle as circle as well. This method also cannot detect the line in the image and it concluded that the shape of the image is a fully circle.

2 Hough Transform

There are some weaknesses that have been identified in existing Hough transform methods regarding circles detection and semicircles detection, obtained from the literature reviews, as listed below:

1. large memory/space is used
2. the quality of input data is dependent on the efficiency of the Hough transform
3. slow-speed processing when voting is executed in every pixel
4. many similar lines are always recognized instead of the correct line
5. high computational and storage requirements
6. the detection is easy to be missed

Referring to Seul et al. (2000), they identified that Hough space is two-dimensional but circle detection requires examination of three-dimensional parameter spaces. This extra dimension is the cause of additional computational and memory expense, and it is for this reason that the Hough transform is far less popular for geometrical objects other than lines. Ellipse detection with Hough transform in one-dimensional parametric space was investigated by Chia et al. (2007). They found that a novel ellipse detection algorithm can retains the original advantages of the Hough transform and minimizing the storage and computation complexity. Furthermore, this algorithm also can use one-dimensional accumulator and can detect complete and incomplete ellipses in synthetic and real images when the end points of the major axes are available.

An improvement of the standard Hough transforms to detect line segment by Thuy et al. (2008), was as described the proposal for a new Hough transform method that contains three extensions: the technique of accumulation, the application of the local maxima rule in SHT (Standard Hough Transform), and detection of line segments. Hence, the authors got the results highly accurate position in detecting line rather than Progressive Probabilistic Hough Transform (PPHT). Moreover, the proposed method is better in terms of high computational and storage requirements.

Based on the image segmentation using Hough transform by Rabindra and Meena (2009), it described on how Hough transform is used to detect different shapes such as rectangle, circle, rhombus, and triangle in an image. Hence, the authors can achieve their objectives but the results shows that the method cannot be used for detecting semicircles and circles with unknown radius. The authors need to specify the coordinates of the circle centers, and the radius.

According to Shuai et al. (2010), they proposed an advanced Hough transform based on gradient for circle detection and this method has the characteristics of high-real time and high precision. Besides that, this proposed method can perform well in environmental changes and anti-interference where the results is desirable. The general test results of this algorithm is desirable except for several edge points, the amount of occupied memory space is low and the detection is easy to be missed. Sirisak et al. (2011) showed that the Hough method for circle detection can detect the defect of circle pattern that occurs in hard disk drive. Furthermore, the proposed method also will increase speed and accuracy in failure analysis.

Based on Yue et al. (2012), it was founded that tradional Hough transform (HT) does not give good result for concentric circle detection based on parameters decomposition. The reasons are because of its large amount of calculations, high demand of space storage, and low efficiency in real-time application. The detection method was further improved through adding an edge detection procedure based on global threshold. The experimental results showed that this algorithm meets the online detection of high accuracy requirement, with superior real-time performance and stronger anti-interference ability.

Last but not least, there are lots of rooms to improve the weaknesses that were found from the previous researchers even though Hough transform is widely used nowadays. This research was conducted to solve the two main problems that addressed by Rabindra and Meena (2009), which is unable to detect a semicircle correctly and it also unable to detect a circle with unknown radius. The outcome of this research can enhance the performance of Hough transform where the proposed Hough transform can be able to detect a circle with unknown radius and able to detect semicircle correctly. The results obtained are also compared with Hough transform by Rabindra and Meena (2009).

3 Quantitative Measurement

The method for quantitative measurement that has been used in this research is a confusion matrix that typically called as matching matrix. Each row of the matrix represents the instances in a predicted class, while each column represents the instances in an actual class (Zapata et al. 2011).

A confusion matrix shown in Fig. 1, contains information about actual and predicted classifications done by a classification system. Performance of such systems is commonly evaluated using the data in the matrix. The entries in the confusion matrix have the following meaning:

- True positive (TP) is the number of correct predictions that an instance is negative
- False negative (FN) is the number of incorrect predictions that an instance is positive
- False positive (FP) is the number of incorrect predictions that an instance is negative
- True negative (TN) is the number of correct predictions that an instance is positive

There are several standard terms that have been defined for the two class matrix such as the accuracy (AC), the true positive rate (TP), the false positive rate (FP), the true negative rate (TN), and the false negative rate (FN). The AC is the proportion of the total number of predictions that were correct.

It is determined using the equation:

$$AC = \frac{TP + TN}{TP + FN + FP + TN} \tag{1}$$

The accuracy that were determined by the Eq. (1) may not be an adequate performance measure when the number of negative cases is much greater than the number of positive cases (Kubat et al. 1998). For example, if the accuracy is 0.8 meaning that the accuracy of predictions is 80 % correct.

Fig. 1 Confusion matrix (Kohavi and Provost 1998)

		Predicted class	
		Yes	No
Actual class	Yes	TP	FN
	No	FP	TN

4 The Proposed Method

The proposed methodology consists of three main parts: data collection, preprocessing, and proposed Hough transform. In the Hough transform part, it is further divided into two subparts, the first subpart is about the enhanced Hough function to detect semicircle and the second subpart is to set parameter R as variable to detect circle with unknown radius.

The complete steps involved in the proposed methodology are shown as follows:

Step 1: In the data collection part, the input used would be images that consist of different shapes like circle and semicircle taken from the Internet.

Step 2: Every input of the circle image will be converted to binary image, in order to easily read the value 0's and 1's in the image. The total number of 1's will be used as totalpix in the graph for y-axis and semicircle images to be converted into grayscale image to reduce the computational complexity to easily find a single line of semicircle in the image.

Step 3: After that, in the edge detection process, Canny edge detector is used to detect the edge and to adjust the parameter of the processed image

Step 4: In the first part of the proposed Hough transform method, the information of the minimum and maximum radiuses is given to detect semicircle. Then, Hough function is applied to the image. If Hough transform (HT) matrix shows the right shape for circle, then the shape of image has round shape. If no, terminate the process. After that, if HT matrix shows the peaks appear at the same x and y coordinate, then it is approve that the shape of the image is a semicircle. If no, the shape of the image is a circle.

Step 5: For detecting the circle with unknown radius, the parameter R is set as variable for radius of the image. Then, reserve range 1–50 for HM matrix. If numbers of lines that cross the points until n points in which n is the diameter of the round object (approximately) increase in exponently, then the graph will show the pattern for circle image. If not, the shape of image is not a circle. Besides, HT matrix shows the peaks appear at different x and y coordinate.

Step 6: Finally, the result for both of circle and semicircle images will undergo the process to measure the quantity of accepted images and rejected images. A confusion matrix that contains information about actual and predicted classifications have standard terms that have been defined for the two class matrix such as the accuracy (AC), the true positive rate (TP), the false positive rate (FP), the true negative rate (TN), and the false negative rate (FN). The formula of AC at Eq. (1) is used to find the total number of predictions that were correct for each circle and semicircle images.

With this proposed methodology, the problems in semicircle and circle detection would be solved by comparing the pattern of graph and by looking at the Hough peaks in the Hough transform matrix, to check whether the pattern is the same with the data in Hough transform matrix by Tcl (2005). In the testing part, results for both circle and circle images would undergo the process of measuring the quantity of accepted and rejected images using confusion matrix.

In addition, this research only give focus on images that contained basic circle and semicircles. This experiment used 100 images for circle and 100 images for semicircle to test whether the programs and the proposed method works.

5 Experimental Results

In this research, the proposed Hough transform method is compared with the previous Hough transform by Rabindra and Meena (2009), as this method also focus on images that contained basic circle and semicircles. The comparison of techniques used in the experiments for both methods are shown in Table 1.

The comparison of the results for both methods using confusion matrix are shown in Table 2.

Based on Table 2, by using the formula of accuracy (AC), the accuracy of circle and semicircle detection performance for proposed Hough transform is better than Rabindra and Meena (2009), as it shows 0.89 which is equal to 89 % of prediction is correct for circle images and 0.93 is equal to 93 % of prediction is correct for semicircle images. It also shows that Rabindra and Meena (2009) could not detect semicircle images, thus the accuracy value is 0.00.

The findings shows that the proposed methods did detect semicircle images correctly and detect circle with unknown radius. Meanwhile, Rabindra and

Table 1 Hough transform by Rabindra and Meena (2009) versus proposed Hough transform

Hough transform by Rabindra and Meena (2009)	Proposed Hough transform
Less number of images used	100 images of circles and semicircle used
Use clear circles that do not have other shape of objects inside the image	General circles and semicircles are used which mean there are some other shape of objects inside the image
Do not have quantitative measurement to know the percentage of accepted images	Use confusion matrix as quantitative measurement to know the percentage of accepted and rejected images

Table 2 Method performance of circle and semicircle images by Rabindra and Meena (2009) and proposed method using confusion matrix

	Accuracy (AC)	
	Rabindra and Meena (2009)	Proposed Hough transform
Circle	0.79	0.89
Semicircle	0.00	0.93

Table 3 Comparison of Hough transform by Rabindra and Meena (2009) and proposed Hough transform

Units	Hough transform by Rabindra and Meena (2009)	Proposed method
Information of radius	Given	Given—semi circle detection Not given—circle detection
Change image to grayscale	Yes	Yes
Change image to binary	Yes	No
Edge detection	Yes	Yes
Detect circle with unknown radius	No	Yes
Detect semicircle correctly	No	Yes

Meena (2009) can only detect circles with radius and it cannot detect semicircles. The overall comparison of detection performance between Hough transform by Rabindra and Meena (2009) and proposed Hough transform is shown in Table 3.

6 Conclusion

The previous research studied by Rabindra and Meena (2009) shows that the Hough transform can only detect circles with radius and unable to detect circles with unknown radius. Furthermore, the previous Hough transform also unable to detect semicircles correctly. In order to improve the previous method, we proposed new Hough transform thas was able to detect a circle with unknown radius and able to detect semicircle correctly. It can be determined by the accuracy of circle and semicircle detection performance for proposed Hough transform is better than Rabindra and Meena (2009). The results obtained can be used to verify the performance of enhanced Hough transform in order to solve the problems of circle detection and semicircle detection. This new modified approach could be used in the development of new methodology that will be of value in future studies of circle detection and semicircle detection in image segmentation.

References

Chia AYS, Leung MKH, Eng H-L, Rahardja S (2007) Ellipse detection with Hough transform in one dimensional parametric space. In: Conference: IEEE international conference on image processing, 2007 (ICIP 2007), 333–336

Guo S, Zhang X, Zhang F (2006) Adaptive randomized Hough transform for circle detection using moving window. In: Proceedings of the fifth international conference on machine learning and cybernetics, 13–16 Aug, IEEE, Dalian, pp 3880–3885

Kubat M, Holte R, Matwin S (1998) Machine learning for the detection of oil spills in satellite radar images. Mach Learn 30, 195–215

Kohavi R, Provost F (1998) On applied research in machine learning. In: Editorial for the Special Issue on Applications of Machine Learning and the Knowledge Discovery Process, vol 30. Columbia University, New York, pp 1–6

Rabindra KM, Meena J (2009) Image segmentation using Hough transform. Bachelor, National Institute of Technology, Rourkela

Seul M, O'Gorman L, Sammon MJ (2000) Practical algorithm for image analysis description, examples and code. Cambridge University Press, Cambridge, UK

Shapiro LG, Stockman GC (2001) Computer vision. Prentice-Hall, Inc, USA

Shuai Z, Ze Z, Hai-Tao W (2010) Research of robot color logo orientation based on Hough transform. In: 2010 second international conference on intelligent human-machine systems and cybernetics. IEEE, pp 56–60

Sirisak L, Boonruang M, Ratchadaporn O, Anant O (2011) Extracted circle Hough transform and circle defect detection algorithm. World Academy of Science, Engineering and Technology, International Science Index 12

Tcl (2005) Object detection using Hough transform. Retrieved from http://basiceng.blogspot.com/2005/12/object-detection-using-hough-transform.html

Thuy TN, Xuan DP, Jae WJ (2008) An improvement of the standard Hough transform to detect line segments. In: Proceedings IEEE

Yue GH, Lu CH, Sheng LQ, Liu YN (2012) A combined method for concentric circles detection in image of o-shape rubber ring. Adv Mat Res 488–489:1619–1623 (Trans Tech Publications, Switzerland)

Zapata J, Vilar R, Ruiz R (2011) Automatic inspection system of welding radiographic images based on ANN under a regularisation process. Springer Science + Business Media, LLC 2011

Part VI
Forestry

Chapter 34
Properties of Particleboard from Oil Palm Trunk (*Elaeis guineensis*) and Resam (*Dicranopteris linearis*)

Nurrohana Ahmad, Jamaludin Kasim, Siti Noorbaini Sarmin, Zaimatul Aqmar Abdullah, Mazlin Kusin and Norhafizah Rosman

Abstract The increasing number of timber producers in Malaysia causes the increments in wood waste. The sources of wood wastes in Malaysia come from wood residues from logging, wood processing, and agriculture. Wood waste has the potential to be utilized. This can minimize the wood waste and maximize the wood waste into a value–added product to help the industries. The purpose of this study is to investigate the physical and mechanical properties of particleboard from oil palm trunk (OPT) and resam. Oil palm trunk mixed with resam can be utilized to produce various types of value–added products which are the resources of the substitute's material on wood-based industry. Single-layered and hybrid particleboard from OPT and resam were fabricated with 600 kg/m^3 and 10 % resin content. Phenol formaldehyde (PF) was used as a binder. The properties of bending strength (MOR & MOE), internal bonding strength (IB), thickness swelling (TS), and water absorption (WA) were evaluated based on Japanese Industrial Standard; JIS A 5908:2003. Results showed that both resam and OPT particles can be used in co-mixtures to produce particleboards. Particleboard from single-layered boards made from OPT met the MOR and MOE requirement of the JIS standard. Utilization of OPT and resam is one innovative way to reduce the usage of wood-based material and at the same time reduced the abundance of agricultural waste.

N. Ahmad (✉) · J. Kasim · S.N. Sarmin · Z.A. Abdullah · M. Kusin · N. Rosman
Faculty of Applied Sciences, Universiti Teknologi MARA,
Jengka, Pahang, Malaysia
e-mail: nurrohana@gmail.com

J. Kasim
e-mail: djamal@pahang.uitm.edu.my

S.N. Sarmin
e-mail: baini1212@yahoo.com

Z.A. Abdullah
e-mail: zaimatul@pahang.uitm.edu.my

M. Kusin
e-mail: mazlin@pahang.uitm.edu.my

N. Rosman
e-mail: hafizah29@pahang.uitm.edu.my

N.A. Yacob et al. (eds.), *Regional Conference on Science, Technology and Social Sciences (RCSTSS 2014)*, DOI 10.1007/978-981-10-0534-3_34

Keywords Oil palm trunk · Particleboard · Phenol · Resam

1 Introduction

For years, the demands for particleboard show that 57 % of the total volume of wood-based panels has risen intensely in the world, mostly for furniture manufacturing and housing construction (Maloney 1996). Recently, wood-based industries are facing problems in finding wood raw material because of deforestation, forest degradation, and increasing wood demand for a long time (Nadir et al. 2012). As a result, the use of renewable resources such as agricultural residue and wood wastes gains the interest in the production of wood-based composites.

Many researchers in composite manufacturing investigate nonwood lignocellulosic biomass utilization because of the increasing demands for wood resources in wood composite industry including particleboard. The new alternatives for the production of particle are agricultural residues and annual plant fiber. Research on the various nonwood fibers has been done for the following examples: kiwi pruning, wheat straw, corn pith, durian peel, coconut coir, almond shells, flax shiv, cotton stalks, baggase, rice straw, sunflower stalks, eggplant stalks, cotton stalks, kenaf stalks, hazelnut husk, bamboo chips, kenaf core, and waste grass clippings (Mohammad et al. 2011). The general use of composite particleboard was associated to the economic benefits such as small charge of wood natural resource, reasonable means, and simple handling (Gürü et al. 2006).

Oil palm trunk (OPT) is one of the most important agriculture crops and is a waste product of the agricultural industry. Oil palm is harvested and replanted every 25 years rotation because of the decrease in fruit production and not economical for harvesting as it reaches above 30 foot height. The OPT is chopped and left for natural degradation which attends the purpose biomass as fertilizer, as fuels, burning, landfill, and left unused. These agricultural residues are able to be value–added as sources for composite, particleboard, pulp and paper, lumber, plywood, and laminated veneer lumber because of the lignocellulosic material from OPT (Yeoh et al. 2013).

Dicranopteris linearis (family Gleicheniaceae) is a common fern distributed through humid subtropical and tropical regions of the world (Chong and Ismail 2006; Maheswaran and Gunatilleke 1990). These species are considered as early-stage colonizers in primary succession of acidic and oligotrophic soils. By rapid clonal growth form, *Dicranopteris* often forms a mat-like dense understory layer, covering a large area of open canopy. *Dicranopteris* is well adapted to highlight and nutrient-depleted habitats but is eventually shaded out by overstory trees (Kochummen 1977). Ratios of lignin to nitrogen of *Dicranopteris* surpassed those of the other tropical plant types (Amatangelo and Vitousek 2009; Russell and Vitousek 1997).

The objective of this study is to determine the effect of single layer and hybrid board on mechanical and physical properties of OPT-resam composite.

2 Materials and Methods

The oil palm trunk was collected from FELDA Jengka 21. The particles were then screened using vibrator screening machine and only particle with 0.5–2.0 mm were used in this study. The particles were dried in a special oven until they achieved 5 % moisture content.

Resam was collected from the Reserved Forest of UiTM Jengka. The leaves part of the resam was left and just the stem was used for further processes. The resam stem was converted into particle structure using knife ring flakers machine to produce small particle with many sizes. The screener was used just to separate the dust and the particles. After that, the resam particles were then dried in the oven to ensure that the average moisture content reaches below the 5 %. Usually, the oven temperature was set up to 80 °C and takes about 24 h upon complete drying.

In general, manufacturing process for the particleboard starts with the fresh raw material until they are converted into wood particles. The particles are then dried in the oven, mixed with phenol formaldehyde (PF) resin, formed into thick, pressed under cold, and hot pressed into board. There are four types of boards which are 2 single layered and 2 hybrids with a total of 12 boards. The six starting boards related to board 100 % resam, 100 % OPT, and two types of hybrid boards comprised of 50 % OPT and 50 % resam which are (25 % resam (face), 50 % OPT (core), 25 % resam (back); 25 % OPT (face), 50 % resam (core), 25 % OPT (back)), respectively. The manufacturing of the single layer and hybrid particleboard from resam and OPT, boards of size 340 mm × 340 mm × 12 mm with targeted density 600 kg/m^3 were fabricated. The particles were bonded with PF resin with 10 % resin content. The ammonium chloride (hardener) solution used had a concentration of 20 %. A weighted amount of particles was placed in the mixer and sprayed with a resin and hardener. After spraying, the particles were then manually laid in a wooden mold over a caul plate and then prepressed with cold press for 30 s. After the wooden frame was removed, two metal stops of 12 mm were placed near the side of the consolidated mat. The mat was finally pressed with hot press at 160 °C for 6 min. All the boards were conditioned in conditioning room for one week before the test samples were cut.

All the test samples were located in a conditioning room at 20 °C and 65 % relative humidity (RH) for 1 week to achieve constant weight. The mechanical and physical properties of particleboard were assessed according to Japanese Industrial Standard JIS A 5908 (2003). Modulus of rupture (MOR), modulus of elasticity (MOE), internal bond (IB), thickness swelling (TS), and water absorption (WA) properties of the boards were evaluated in this work. MOR and MOE were tested in bending with dimension 230 mm × 50 mm × 12 mm by center point loading and it was calculated. On the other hand, internal bonding with dimension 50 mm × 50 mm × 12 mm was glued with test pieces of steel blocks before testing. TS and WA were determined using specimens with the dimension of

50 mm × 50 mm × 12 mm by water immersion at 20 °C for 24 h. All specimens were weighed and measured before and after being immersed in water. The TS and WA were then calculated.

3 Results and Discussions

The mechanical and physical properties of particleboard are presented according to the type of boards (Table 1). It shows that OPT single layer gave better results for the mechanical properties while for the physical properties resam with single layer performed well. Only OPT with single layer passes the standard for the MOR and MOE. From the previous result (Bowyer et al. 2004), it shows that OPT contains high moisture content and it is favorable to absorb moisture easily because of high parenchyma cells present. Mechanical properties are usually the most important characteristics of wood product to be used in structural application. According to the obtained results of static bending test which is summarized in Table 1 for MOR and MOE strength, it showed that the average values of OPT were 22.99 and 2833 MPa, respectively. Meanwhile, the average values MOR and MOE for resam was 3.59 and 959 MPa, respectively. Single layer of resam gave better results on WA and TS compared to OPT. Both hybrid resam and OPT do not pass the MOR, MOE, and IB of the JIS standard.

Mechanical Properties

Figure 1 shows the effects of different types of boards on mechanical properties: modulus of rupture (MOR), modulus of elasticity (MOE), and internal bond (IB) of

Table 1 The mechanical and physical properties of single-layered and hybrid particleboard

Type of boards	Mechanical properties			Physical properties	
	Modulus of rupture (MPa)	Modulus of elasticity (MPa)	Internal bonding (MPa)	Water absorption (%)	Thickness swelling (%)
Resam (single layer)	3.59	959	0.048	68.42	24.13
OPT (single layer)	22.99	2833	0.35	93.67	36.19
Resam + OPT + Resam (ROR) (hybrid)	4.58	1302	0.057	100.74	60.47
OPT + Resam + OPT (ORO) (hybrid)	7.28	1344	0.035	114.11	68.39
JIS A 5908:2003	14	2000	0.5	na	12

na—not applicable

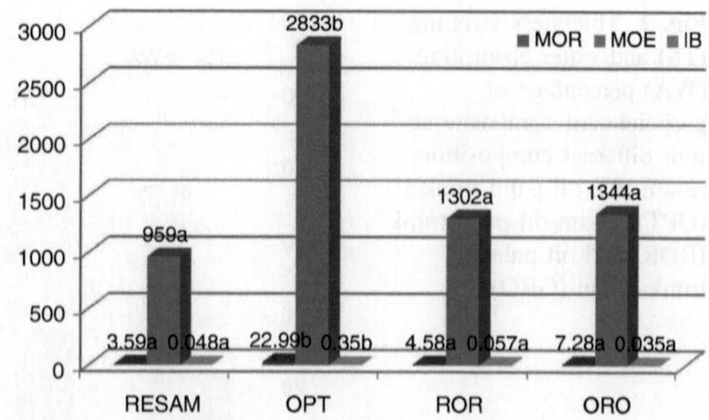

Fig. 1 Modulus of rupture (MOR), modulus of elasticity (MOE), and internal bond strength (IB) of particleboard composite at four different composition; resam (R), oil palm trunk (OPT), resam-oil palm trunk (ROR), and oil palm trunk-resam (ORO)

four different types of particleboard composite. It was obviously showed that MOR, MOE, and IB strength of panel produced using oil palm trunk (OPT) were higher compared to panel produced from resam (R). It showed no significant difference between resam, ROR, and ORO but significantly showed difference at OPT on mechanical properties. It can also be figured out that the mechanical properties of the panel with addition of OPT as core layer (ROR) and face/back layer (ORO) were improved for more than 50 %. Bending strength is the most important properties as an indicator in the manufacturing of particleboard (Mohammad et al. 2011).

Physical Properties

The physical properties: thickness swelling (TS) and water absorption (WA) of panel composed from resam (R), oil palm trunk (OPT), resam-oil palm trunk-resam (ROR), and oil palm trunk-resam-oil palm trunk (ORO) are figured out after 24 h soaking in ambient water. The result showed that, the percentage of TS and WA was higher with panel produced using OPT and lower for others panel. The TS and WA values increased in the panel composed with addition of OPT as shown in Fig. 2. Both figures disclose similar trends of water uptake and swelling in all types of particleboards. It showed that particleboard made from resam and OPT (single layered) boards shows that value has significant difference between ROR and ORO (hybrid) boards for thickness swelling. In nature, OPT contains high moisture content, nearly 300 % due to 'spongy' characteristics that can clearly be seen through the cross cut (Lim and Gan 2005). Previous study has been proven that panel prepared using OPT experienced higher WA and TS (Nurrohana et al. 2011). However, panel produced from resam was lower in both WA and TS and did not achieve JIS standard.

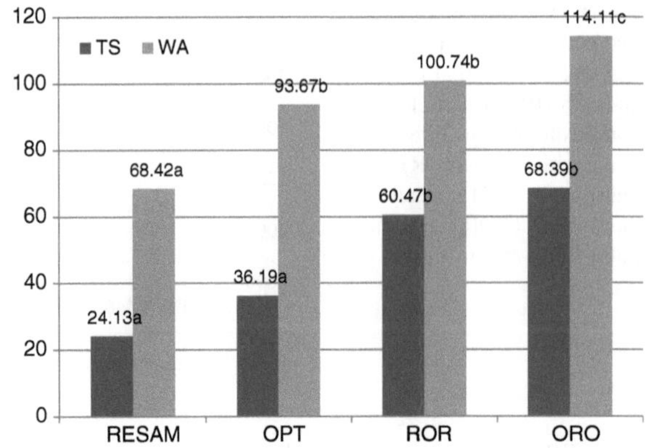

Fig. 2 Thickness swelling (TS) and water absorption (WA) percentage of particleboard composite at four different composition: resam (R), oil palm trunk (OPT), resam-oil palm trunk (ROR) and oil palm trunk-resam (ORO)

4 Conclusion

From the results, the particleboard from OPT and resam can be manufactured but it has no economical value. Resam is the type of fern which is categorized under vascular plants and nonwoody. The stem only contains the vascular tissue which distributes resources through the plants. All the boards that contained resam particle did not reach the JIS standard. Compared to OPT, resam does not have any tissue that can distribute the stress along the boards. These can clearly be seen in Table 1 of MOR and MOE result. While for the IB strength, the adhesion between the particles may not be successful. Even the physical and mechanical properties did not achieve the standard; JIS A 5908:2003. The utilization of resam to produce particleboard may be put into consideration with further research improvement.

References

Amatangelo KL, Vitousek PM (2009) Contrasting predictors of fern versus angiosperm decomposition in a common garden. Biotropica 41:154–161

Bowyer JL, Shmulsky R, Haygreen JG (2004) Forest product and wood sciences—an introduction, 4th edn, Blackwell Publishing Company

Chong TV, Ismail BS (2006) Field evidence of the allelopathic properties of *Dicranopteris linearis*. Weed Biol Manage 6:59–67

Gürü M, Tekeli S, Bilici İ (2006) Manufacturing of urea-formaldehyde-based composite particleboard from almond shell. Mater Des 27:1148–1151

Japanese Industrial Standard, JIS A 5908 (2003) Particleboards. Japanese Standards Association: 2003, Tokyo, Japan

Kochummen KM (1977) Natural plant succession after farming in Kepong. Malays For 40:61–78

Lim SC, Gan KS (2005) Characteristics and utilization of oil palm stem: Forest Research Institute. Timber Technol Bull 35:1–11

Maheswaran J, Gunatilleke IAUN (1990) Nitrogenase activity in soil and litter of a tropical lowland rain forest and an adjacent fernland in Sri Lanka. J Trop Ecol 6:281–289

Maloney TM (1996) The family of wood composite materials. For Prod J 46(2):19–20, Washington State University, Pullman, Washington
Mohammad DG, Morteza N, Ali B (2011) Influence of utilization of baggase in surface layer on bending strength of three-layer particleboard. Eur J Wood Wood Prod 69:533–535
Nadir A, Jin HK, Tae HH (2012) Effect of resin type and content on properties of composite particleboard made of a mixture of wood and rice husk. Int J Adhes Adhes 38:79–83
Nurrohana A, Jamaludin K, Siti Zalifah M, Shaikh AKY, Anis M, Nor Yuziah MY (2011) Manufacture and properties of oil palm particleboard. In: 2011 3rd international symposium & exhibition in sustainable energy & environment (ISESEE), pp 84–87
Russell AE, Vitousek PM (1997) Decomposition and potential nitrogen fixation in *Dicranopteris linearis* litter on Mauna Loa, Hawai'i. J Trop Ecol 13:579–594
Yeoh BH, Yueh FL, Luqman AC, Idrus J, Antonio P, Tahir P, Harun J (2013) Development a new method for pilot scale production of high grade oil palm plywood: effects of resin content on the mechanical properties, bonding quality and formaldehyde emission of palm plywood. Mater Des 52:828–834

Chapter 35
Effect of Tree Portion and Distance from Pith on Specific Gravity, Fiber Properties and Mechanical Properties of Kelampayan (*Neolamarckia cadamba*) Wood

Jamaludin Kasim, Siti Nadzirah Misfar, Nur Sakinah Mohamed Tamat and Nurfaizah Abd Latib

Abstract Specific gravity, fiber properties, and mechanical properties are the important factors to be considered in determining the end use of wood material. This study was conducted to determine the effect of tree portion and distance from pith on specific gravity, fiber properties (fiber length, fiber diameter), and mechanical properties of Kelampayan (Neolamarckia cadamba) wood. Three Kelampayan trees with diameter at breast height (DBH) ranging from 35 to 41 cm were harvested from UiTM Jengka, Pahang forest and were used in this study. Tree portion and distance from pith showed significant effect on fiber length, fiber diameter, modulus of rupture (MOR), compression parallel to grain, and shear parallel to grain. Specific gravity was not affected. Mechanical properties of Kelampayan wood in general increases significantly with increase in tree portion. MOR shows a positive correlation with tree portion. Modulus of elasticity also increased insignificantly with tree portion. Shear parallel to grain was also found to increase significantly with longer fiber length value. Compression parallel to grain according to tree portion was found to be significantly higher at the middle portion (25.01 MPa) than top (21.65 MPa) and bottom (20.34 MPa). All mechanical properties of Kelampayan wood increased significantly from pith to near bark. Modulus of rupture (MOR) at near bark (40.58 MPa) was found to be higher at near pith (32.11 MPa) with a positive correlation ($r = 0.293^*$).

Keywords Kelampayan · Fiber properties · Basic properties · Mechanical properties

J. Kasim
Faculty of Applied Sciences, Universiti Teknologi MARA,
Jengka, Pahang, Malaysia
e-mail: djamal@pahang.uitm.edu.my

S.N. Misfar · N.S. Mohamed Tamat (✉) · N. Abd Latib
Faculty of Applied Sciences, Universiti Teknologi MARA,
40450 Shah Alam, Selangor, Malaysia

N.A. Yacob et al. (eds.), *Regional Conference on Science, Technology and Social Sciences (RCSTSS 2014)*, DOI 10.1007/978-981-10-0534-3_35

1 Introduction

Neolamarckia cadamba is a fast-growing species with a straight and tall bole under the family of Rubiaceae. The Malaysian standard name for this timber is Kelampayan. In some places such as Sabah and Sarawak, the tree is known as Limpoh and Sempayan, respectively (Laran n.d.). The timber is white with a yellow tinge color and it darkens to creamy yellow when exposed. The sapwood and heartwood of this timber is not differentiated with each other (Choo and Lim 1999; Laran n.d.; Lim et al. 2005). In Malaysia, the timber is classified under light hardwood with its strength grouping falling under group D or SG 7 (Choo and Lim 1999; Lim et al. 2005). In terms of durability, Kelampayan is nondurable and very susceptible to wood rotting such as blue stain and fungi attack (Laran n.d.). Mechanical properties of wood are defined as the behavior of wood when such forces are applied onto it. The behavior is influenced by the types of forces exerted on the wood and the differences in basic wood properties itself (Panshin and De Zeeuw 1970).

The objectives of this study are to determine the effect of tree portion and the distance from pith on specific gravity, fiber morphology, and mechanical properties of Kelampayan wood. Correlation of tree portion and distance from pith, specific gravity, and fiber morphology to its mechanical properties will also be determined.

2 Materials and Methods

Field Procedure

Three Kelampayan trees with a diameter at breast height (DBH) ranging between 35 and 41 cm were harvested from the UiTM Pahang forest reserve and divided into three equal portions namely bottom, middle, and top. Two pieces of one inch disc were taken from each portion. Each of the discs will be marked and cut into small sample according to the distance across the disc which is near pith (NP), middle (M), and near bark (NB).

Physical Properties

Moisture content: The method used for the determination of moisture content was based on BS 373 (1957) methods of testing small clear specimen of timber. Duplicate samples were oven-dried at a temperature of 103 ± 2 °C for about 24 h until constant weight. The samples were cooled for about half an hour in a desiccator and weighed. The moisture content was calculated as follows:

$$MC = \frac{W_g - W_o}{W_o} \times 100\,\%$$

where W_o—oven-dry weight (g) and W_g—green weight (g).

Specific Gravity: Specific gravity was determined based on water displacement method. The testing was carried out according to the BS 373 (1957) testing small clear specimen of timber. The specific gravity of the wood samples was calculated as below:

$$\text{Specific gravity} = \frac{\text{oven-dried weight (g)}}{\text{weight of water displaced (g)}}$$

Fiber Length and Fiber Diameter

The maceration of fibers was carried out using a method outlined by Wilson (1954). The 30 measurement of fiber dimension (fiber length, fiber diameter) was carried out using a Motic system microscope (B1 Series) connected to a computer.

Mechanical Properties

The wood from each portion and distance from pith were cut and air-dried until the moisture content reached about 12–15 %. After that, the wood was further cut into final dimension of 20 mm × 20 mm in cross section and 300 mm in length and placed in a conditioning room at 60 ± 5 % relative humidity and 25 ± 2 °C before strength testing. The tests were carried out using an Instron universal testing machine in accordance with British Standard, BS 373 (1957) methods of testing small clear specimens of timber.

3 Results and Discussion

Moisture Content, Specific Gravity, Fiber Length, Fiber Diameter and Mechanical Properties of Kelampayan (N. cadamba) *Wood*

Table 1 gives the mean values of moisture content, specific gravity, fiber length, fiber diameter, and mechanical properties of Kelampayan (*N. cadamba*) wood according to tree portion and distance from pith. The table shows that as the tree increases in height, only moisture content (102.56–112.17 %) and fiber diameter

Table 1 Moisture content, specific gravity, fiber length, fiber diameter, and mechanical properties of Kelampayan *(N. cadamba.)* wood according to portion and distance from pith

		Properties							
		MC (%)	S.G	FL (mm)	FD (μm)	Mechanical properties (MPa)			
						Static bending		Comp	Shear
						MOR	MOE (x103)		
Portion	Bottom	112.17	0.37	1.50	34.67	32.84	4.91	20.34	6.96
	Middle	105.48	0.37	1.62	34.50	38.21	5.21	25.01	8.73
	Top	102.56	0.39	1.56	34.00	41.53	5.19	21.65	8.22
Distance from pith	Near pith	112.41	0.36	1.33	30.81	32.11	4.45	19.61	6.94
	Middle	103.23	0.37	1.63	30.51	39.89	5.07	23.66	9.24
	Near bark	94.58	0.40	1.72	35.84	40.58	5.78	23.73	7.73

Note
MC Moisture content
S.G Specific Gravity
FL Fiber length
FD Fiber diameter
MOR Modulus of Rupture
MOE Modulus of Elasticity
Comp. Compression parallel to grain

(34–34.67 μm) show a decreasing pattern. For samples further from pith, specific gravity (0.36–0.40), fiber length (1.33–1.72 mm), fiber diameter (30.81–35.84 μm), and mechanical properties of Kelampayan (MOR (32.11–40.58 MPa), MOE (4453–5779 MPa), compression parallel to grain (19.61–23.73 MPa), and shear parallel to grain (8.94–7.73 MPa) show an increasing pattern; coherent with the decrease in moisture content value (94.58–112.41 %). The results of DMRT and correlation analysis are shown in Figs. 1, 2 and Table 3, respectively.

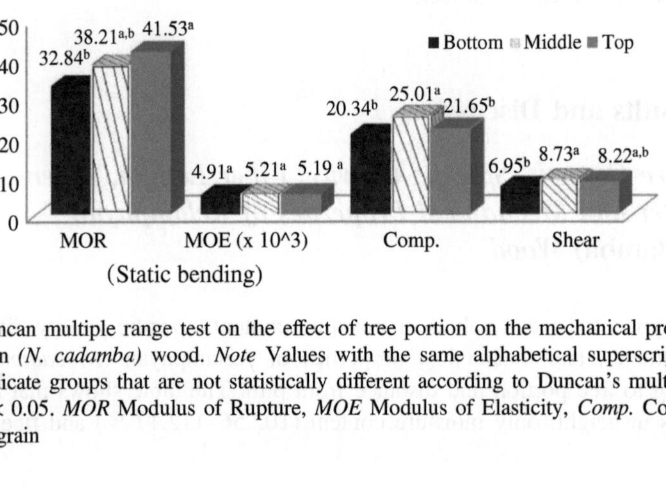

Fig. 1 Duncan multiple range test on the effect of tree portion on the mechanical properties of Kelampayan *(N. cadamba)* wood. *Note* Values with the same alphabetical superscript in each column indicate groups that are not statistically different according to Duncan's multiple range tests at $P < 0.05$. *MOR* Modulus of Rupture, *MOE* Modulus of Elasticity, *Comp.* Compression parallel to grain

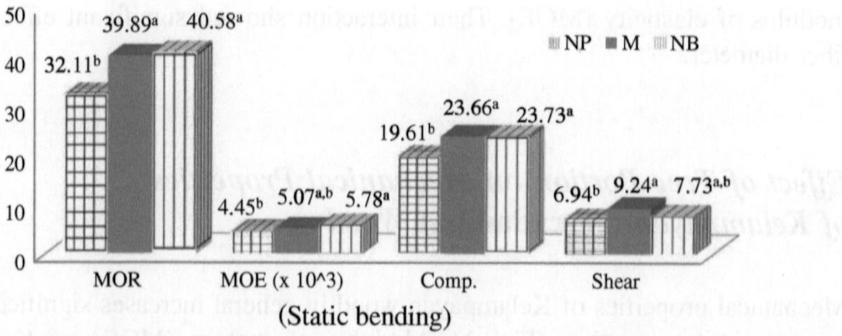

Fig. 2 Duncan multiple range test on the effect of tree distance from pith on the mechanical properties of Kelampayan *(N. cadamba)* wood. *Note* Values with the same alphabetical superscript in each column indicate groups that are not statistically different according to Duncan's multiple range tests at $P < 0.05$. *Note NP* Near Pith, *M* Middle, *NB* Near Bark, *MOR* Modulus of Rupture, *MOE* Modulus of Elasticity, *Comp.* Compression parallel to grain

Statistical Significance

The analysis of variance (ANOVA) for moisture content, specific gravity, fiber length, fiber diameter, and mechanical properties of Kelampayan wood according to portion and distance from pith are summarized in Table 2. Tree portion and distance from pith showed significant effect only on fiber length, fiber diameter, modulus of rupture (MOR), compression parallel to grain, and shear parallel to grain. Distance from pith also showed a significant effect on fiber length, moisture content, and

Table 2 Summary of ANOVA of tree portion and distance from pith on the moisture content, specific gravity, fiber length, fiber diameter, and mechanical properties of Kelampayan (*N. cadamba*) wood

Source of variation	MC (%)	S.G	FL (mm)	FD (μm)	Mechanical properties (MPa)			
					Static bending		Comp.	Shear
					MOR	MOE		
Portion	2.179^{ns}	0.805^{ns}	11.806^{*}	2.842^{**}	2.599^{**}	0.364^{ns}	5.274^{*}	2.684^{**}
Distance	18.223^{*}	1.459^{ns}	42.847^{*}	21.370^{*}	2.992^{**}	5.835^{*}	5.050^{*}	4.393^{*}
Portion × Distance	0.909^{ns}	0.096^{ns}	1.368^{ns}	5.849^{*}	0.039^{ns}	0.168^{ns}	1.759^{ns}	1.981^{ns}

Note
MC Moisture content
S.G Specific gravity
FL Fiber length
FD Fiber diameter
MOR Modulus of Rupture
MOE Modulus of Elasticity
Comp. Compression parallel to grain
*Shows F value significant at $P < 0.05$
**Shows F values significant at $P < 0.10$
^{ns}Shows F values not significant at $P < 0.05$ and $P < 0.10$

modulus of elasticity (MOE). Their interaction showed significant effect only on fiber diameter.

Effect of Tree Portion on Mechanical Properties of Kelampayan (N. cadamba) Wood

Mechanical properties of Kelampayan wood in general increases significantly with increase in tree portion (Fig. 1). Modulus of rupture (MOR) of Kelampayan according to tree portion shows a positive correlation $r = 0.301^{*}$ (Table 3). The increase in MOR value is due to the higher specific gravity ($r = 0.040^{ns}$), longer fiber length ($r = 0.213^{ns}$), smaller fiber diameter ($r = 0.145^{ns}$), and decreasing moisture content ($r = -0.308^{*}$).

Specific gravity is one of the most important wood properties in predicting the increase or decrease of wood strength. It is strongly associated with most mechanical properties (Bowyer et al. 2007; Panshin and De Zeeuw 1970). Moya and Munoz (2010) confirmed that a significant decline in specific gravity in *A. mangium, B. quinata, S. macrophylla, T. Amazonia*, and *T. oblonga* is associated with the increase in tree height based on their regression analysis. Panshin and De Zeeuw (1980) claimed that specific gravity variations with height in hardwoods show little consistency with no overall dominance of a single pattern.

There are three trends of specific gravity variation along the tree height which are (1) decreasing uniformly which is very commonly reported for conifers; (2) decreasing in the lower trunk before decreasing in the upper trunk, and (3) increasing from bottom to top in a nonuniform trend (Panshin and De Zeeuw 1970). Specific gravity value in this study follows the third trend, which is more commonly

Table 3 Correlation coefficients on mechanical properties of Kelampayan (*N. cadamba*) wood with portion, distance from pith, moisture content, specific gravity, fiber length, and fiber diameter

Properties	Static bending		Compression parallel to grain	Shear parallel to grain
	MOR	MOE		
Portion	0.301^{*}	0.094^{ns}	0.104^{ns}	0.197^{ns}
Distance	0.293^{*}	0.448^{**}	0.325^{*}	0.122^{ns}
Moisture content	-0.308^{*}	-0.374^{**}	-0.338^{*}	-0.248^{ns}
Specific gravity	0.040^{ns}	0.136^{ns}	0.159^{ns}	0.087^{ns}
Fiber length	0.213^{ns}	0.281^{*}	0.244^{ns}	0.183^{ns}
Fiber diameter	0.145^{ns}	0.081^{ns}	0.149^{ns}	0.195^{ns}

Note

*Correlation is significant at the 0.01 level (2-tailed)

**Correlation is significant at the 0.05 level (2-tailed)

^{ns}Shows correlation values are not significant at $P < 0.05$ and $P < 0.10$

reported for hardwood than any other trend. There is evidence existing to prove that wood in the crown region is basically denser however, the reasons for this have not been shown (Panshin and De Zeeuw 1970).

Modulus of elasticity of Kelampayan according to tree portion followed the order: middle (5.21×10^3 MPa) > top (5.19×10^3 MPa) > bottom (4.91×103 MPa). Although specific gravity at the top shows the highest value, however, the MOE of middle portion (5.19×10^3 MPa) was greater than top portion (5.19×10^3 MPa). This may be due to the higher value of fiber length ($r = 0.281^*$). Apart from specific gravity, fiber properties of wood are significant factors that influence the strength properties of wood. Wood fibers play an important role in determining specific gravity since their small cross section allows a great numbers of them to be massed in a small place. The specific gravity will tend to be high if the fibers are thick walled and small in lumen size and vice versa (Panshin and De Zeeuw 1970).

A few factors are said to contribute to the differences in specific gravity of wood such as (1) the size of the cells, (2) cell wall thickness and the interrelationship between (1) and (2) (Panshin and De Zeeuw 1970, 1980). Along the tree portion, fiber length increases from the base to the maximum portion up the trunk, before decreasing rapidly beyond this level. But somehow, some hardwoods and conifers do not follow the pattern mentioned above, especially for hardwoods which are heterogeneous in cell types (Horn 1978; Panshin and De Zeeuw 1970). Panshin and Zeeuw (1970) also proved that a minimal tracheid length in white fir (*Abies concolor*) occurred insignificantly at stump height, not at the top and in young growth *Pinus ponderosa* which showed no changes in tracheid length above 5-foot stem height.

Compression parallel to grain according to tree portion was found to be significantly higher at the middle portion (25.01 MPa) than top (21.65 MPa) and bottom (20.34 MPa); a similar trend as MOE with $r = 0.104^{ns}$. Panshin and Zeeuw (1970) noted that strength in compression parallel to grain varies to the specific gravity in a form of direct linear relationship. However, highest value of compression strength in middle portion may be caused by the same reason as in MOE, with correlation of compression strength to fiber length is $r = 0.244^{ns}$. Shear parallel to grain was also found to increase with the highest fiber length value in middle portion ($r = 0.183^{ns}$).

Another reason contributing to the above result could be due to a modification of general decrease from bottom to top of the tree due to the inclusion of greater proportions of high density knots in the mean specific gravity of crown wood. Variations in strength properties of wood with respect to tree height, but without regard to differences within growth increments have been shown for some other kinds of tree however, all available evidence confirms that the variations are also related with changes in the wood structure and also proportions and distributions of tissues within the tree (Panshin and De Zeeuw 1970).

*Effect of Tree Distance from Pith on Mechanical Properties of Kelampayan (*N. cadamba*) Wood*

Figure 2 shows that all mechanical properties of Kelampayan wood increase significantly from pith to near bark. Modulus of rupture (MOR) at near bark (40.58 MPa) was found to be higher near the pith (32.11 MPa) with a positive correlation ($r = 0.293^{*}$). Increase in specific gravity ($r = 0.040^{ns}$) and fiber length ($r = 0.145^{ns}$) from near pith to near bark may result in the higher MOR value. Panshin and De Zeeuw (1980) reported that high growth rate which lead into vigorous cell division might be the reason for the increase in specific gravity at the outer portion of the wood. They also claimed that among other factors, variations in wood strength and elastic properties within cross section are also a reflection of the changes in specific gravity and the length of tracheids (Panshin and De Zeeuw 1980).

A study by Zziwa et al. (2006) found the increasing pattern of density from near pith to near bark in four less utilized tropical timber species in Uganda is associated with the transition of juvenile wood into mature wood. A significant increase was also found for modulus of elasticity (MOE) from near pith to near bark ($r = 0.448^{**}$). Shanavazs and Kumar (2006) found that compression parallel to grain from inner to outer portion for all tree species also varied with wood density. Same pattern was found in this study, compression parallel to grain was increased from near pith to near bark ($r = 0.325^{*}$). Matan and Kyokong (2003) explained both the ultimate compressive stress and modulus of elasticity of rubberwood tended to increase at lower moisture content. As for shear parallel to grain, the highest value was shown from near pith to the middle, before slightly decreasing at near bark.

Besides the variability in moisture content and specific gravity, the increase in mechanical properties with distance from pith also positively correlated with the increase in fiber length. The increase of fiber length from near pith to near bark may be due to the increase in cambial initials length as well as xylem mother cells as the tree matured or aged (Jorge et al. 2000; Panshin and De Zeeuw 1970).

4 Conclusion

From the study, it can be concluded that mechanical properties of Kelampayan (*N. cadamba sp.*) wood vary significantly according to the tree portion (increasing from bottom to top) and distance from pith (increasing from near pith to near bark). Compression parallel to grain increased from near pith to near bark ($r = 0.325^{*}$) while shear parallel to grain, the highest value was shown from near pith to the middle, before slightly decreasing at near bark.

References

Bowyer JL, Haygreen JG, Shmulsky R (2007) Forest products and wood science: an introduction, Fifth edn, Blackwell Publishing

British Standard, BS 373 (1957) Methods of Testing Small Clear Specimens of Timber, British Standard Institution, ISBN 0 580 00684 0

Choo KT, Lim SC (1999) Timber Notes—Light Hardwoods VI, Timber Technology Buletin, Timber Technology Centre (ITC), Forest Research Institute Malaysia (FRIM), No. 16, ISSN:139–258

Horn RA (1978) Morphology of pulp fibre from hardwoods and influence on paper strength, Forest Products Laboratory, Forest Service, United States Department of Agriculture, pp 1–8

Jorge F, Quilhó T, Pereira H (2000) Variability of fibre length in wood and bark in *Eucalyptus globulus*. IAWA J 21(1):41–48

Laran (n.d.) Malaysian Timber Council. http://woodwizard.mtc.com.my:8888/report.asp?AttrID=79&ItemID=56. Accessed 28 Jan 2012

Lim SC, Gan KS, Thi BK (2005) Identification and Utilization of Lesser—Known Commercial Timbers in Peninsular Malaysia. 4. Kelempayan, Melembu, Membuloh and Mempari, Timber Technology Buletin, Timber Technology Centre (ITC), Forest Research Institute Malaysia (FRIM), No. 32, ISSN:139–258

Matan N, Kyokong B (2003) Effect of Moisture Content on Some Physical and Mechanical Properties of Juvenile Rubberwood *(Hevea brasiliensis Muell. Arg.*). Songklanakarin J Sci Technol 25(3):327–340 (2003)

Moya R, Munoz F (2010) Physical and mechanical properties of eight fast—growing plantation species in costa rica. J Trop For Sci 22(3):317–328

Panshin AJ, De Zeeuw C (1970) Textbook of wood technology, vol 1. McGraw Book Co., Ltd., New York

Panshin AJ, De Zeeuw C (1980) Textbook of wood technology, 4th edn. McGraw Book Co., Ltd., New York

Shanavas A, Kumar BM (2006) Physical and mechanical properties of three agroforestry tree species from Kerala. J Trop Agric India 44(1–2):23–30

Wilson JG (1954) Specific gravity of wood substances. For Prod J 16(1):55–61

Zziwa A et al (2006) Physical and mechanical properties of some less utilised tropical timber tree species growing in Uganda. Uganda J Agric Sci 12(1):29–37 (2006)

Chapter 36
Anatomical Properties of Juvenile Latex Timber Clone Rubberwood Trees

Junaiza Ahmad Zaki, Suhaimi Muhammed, Shaikh Abdul Karim Yamani, Amran Shafie and Wan Daud Wanrosli

Abstract At present, Malaysian furniture industries are dependent on the rubberwood (*Hevea brasiliensis*) species and are also highly demanded by other wood-based industry. This is because, the wood properties of rubberwood such as whitish color and good machining properties make it suitable in producing furniture and furthermore rubberwood is also known as "Malaysian Oak" or "White Mahogany" as a trade name. Due to that, Lembaga Getah Malaysia (LGM) had introduced the latex timber clone (LTC) at early years of 2000s to overcome the shortage of rubberwood. In this research, the anatomical aspect of tree heights and distances from near bark to near pith for two different types of latex timber clones (RRIM 2009 and RRIM 2024) of the same age between 3–4 years was investigated. The results for vessel frequency and vessel diameter showed no significant difference for clone and the distance from near bark to near pith except for tree height where the vessel frequency at the top portion is greater and the bottom portion gave a higher number for vessel diameter because both characteristics were related to each other. With regards to vessel frequency, RRIM 2024 clone register a higher value as compared to RRIM 2009. Overall, this study found that the RRIM 2024 clone showed better growth performance compared to RRIM 2009 clone.

Keywords Anatomical · Furniture · Latex timber clone · Rubberwood

---

J. Ahmad Zaki (✉) · S. Muhammed · S.A.K. Yamani · A. Shafie
Faculty of Applied Sciences, Universiti Teknologi MARA, Jengka, Pahang, Malaysia
e-mail: junaiza@pahang.uitm.edu.my

S. Muhammed
e-mail: prof.suhaimi@gmail.com

S.A.K. Yamani
e-mail: syamani@pahang.uitm.edu.my

A. Shafie
e-mail: amran453@pahang.uitm.edu.my

W.D. Wanrosli
School of Industrial Technology, Universiti Sains Malaysia, Gelugor, Penang, Malaysia
e-mail: wanrosli@usm.my

N.A. Yacob et al. (eds.), *Regional Conference on Science, Technology and Social Sciences (RCSTSS 2014)*, DOI 10.1007/978-981-10-0534-3_36

1 Introduction

Wood is natural resources and play very important role in human daily life. It is used as furniture, fiberboard, pulp and paper, plywood, and also has value for wood construction. Therefore, wood anatomy is a fundamental knowledge in the field of wood science and technology. Wood identification through its physical or structural features will be able to provide the information or the uniqueness of the wood that can be applied into a product.

Rubberwood species (*Hevea brasiliensis)* are from Euphorbiaceae family. This species was brought to Kuala Kangsar, Perak in 1877. The creamy to whitish color of rubberwood and its straight grain (Hong and Sim 1994) make it an important resource for the production of furniture and other products such as stairs, parquet floor, blockboard, medium density board, and wall paneling. Before this, rubberwood was only used as firewood (Salleh and Wong 1991).

In 1996, the initiative research taken by the Rubber Research Institute Malaysia (RRIM) and the Forestry Department had introduced the latex timber clone rubberwood (LTC). These clones had been identified to provide better properties because the advantages of LTC rubberwood are to produce high timber volume as well as latex in a short period of time. It is also the intention of the Malaysian government to find an alternative raw material for bio-composite materials due to depleting wood resources.

2 Methodology

The LTC rubberwood trees were RRIM 2009 and RRIM 2024 (Group II) obtained from a plantation at Jengka 8, Pahang. The age of LTC rubberwood during the time of harvesting was about 3–4 years and the average of height was about 4.1 m. Three trees were harvested for each clone. The samples were cut into wood cubes and divided into tree height according to bottom (30 cm from ground), middle (1.3 m at diameter breast height), and top portion (before the first branches) and also distance from near bark to near pith.

The wood cubes were boiled for softening. Sliding microtome was used to slice the wood cubes in transverse section, radial, and tangential. The thicknesses of each wood slices were 20–30 μm. The slice samples were then placed in the petri dishes for staining.

The purpose of staining using safranin 1 % is to differentiate the layers within the wood slices.

Next, in the dehydration process wood slices were placed in an alcohol series (90, 70, 50 and 30 %) for 2 min. The samples were washed with xylene and followed by Canada balsam for permanent slide.

The vessel frequency was evaluated on the cross section of wood species based on millimeter square and was calculated individually using 4× magnifications

Fig. 1 Transverse section of LTC rubberwood with 4× magnification

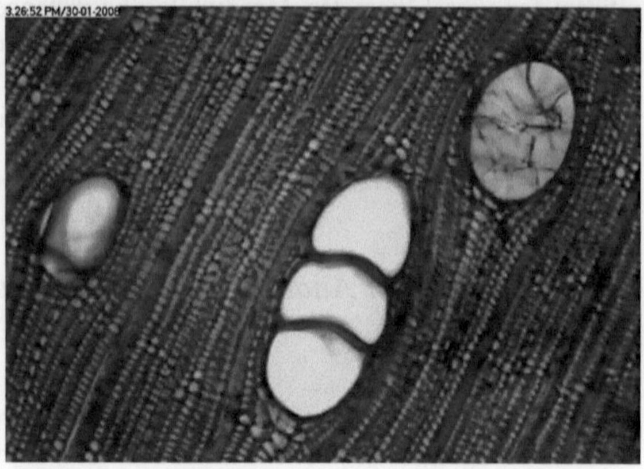

Fig. 2 Transverse section of LTC rubberwood with 10× magnification

(Fig. 1) for at least five parts of the slide with uniform sizes. The procedure is performed according to International Association of Wood Anatomist (IAWA), 1939 guideline.

Vessel's diameter was also calculated on the transverse section of wood for at least 25 vessels using 10× magnifications (Fig. 2). The vessels are selected randomly. The average vessel diameter usually is 100–200 µm. The classification of the vessel size is also based on the guideline of IAWA (1939).

3 Results and Discussion

Table 1 shows the effect of clone, tree height, distance, and also the interaction through the anatomical properties of LTC rubberwood. According to the anatomical properties that had been carried out, the results showed that clone RRIM 2009 and

Table 1 Summary of ANOVA on the anatomical properties of LTC rubberwood

SOV	Df	Vessel frequency (no/mm^2)	Vessel diameter (μm)
Clone (*C*)	1	0.37 ns	0.10 ns
Tree height (*H*)	2	3.39*	16.82**
Distance (*D*)	1	1.41 ns	1.61 ns
$C \times H$	2	0.04 ns	4.43*
$C \times D$	1	0.07 ns	0.51 ns
$H \times D$	2	0.16 ns	0.32 ns
$C \times H \times D$	2	0.04 ns	5.71**

Note: *SOV* source of variance, *ns* not significant $p > 0.05$, *significant at $p < 0.05$, **highly significant at $p < 0.01$, *Df* degree of freedom

RRIM 2024 had no significant difference on vessel frequency and vessel diameter. Tree heights show significant difference on vessel frequency and are highly significant for vessel diameter. Meanwhile, the distance from near bark to near pith showed no significant difference on vessel frequency and vessel diameter. However, the interaction between clone, tree height, and distance showed no significant difference on vessel frequency but showed significant difference on vessel diameter.

Effect of Clone on Anatomical Properties of LTC Rubberwood

Figures 3 and 4 show the effect of clones on the properties of wood anatomy of LTC *Hevea brasiliensis*. Clone RRIM 2024 recorded vessel frequency per square millimeter higher than clone RRIM 2009. The smaller the vessel diameter, the more or compact vessel in the area of 1 mm^2. Vessel frequency is closely related to the vessel diameter. However, the results of both anatomical properties attributes are not significant because both of these clones belong to the same group and age. Suhaimi et al. (1998) conducted a study on clone RRIM 600 and RRIM 623 clones and found that there are no significant differences in the vessel frequency and vessel diameter between the clones.

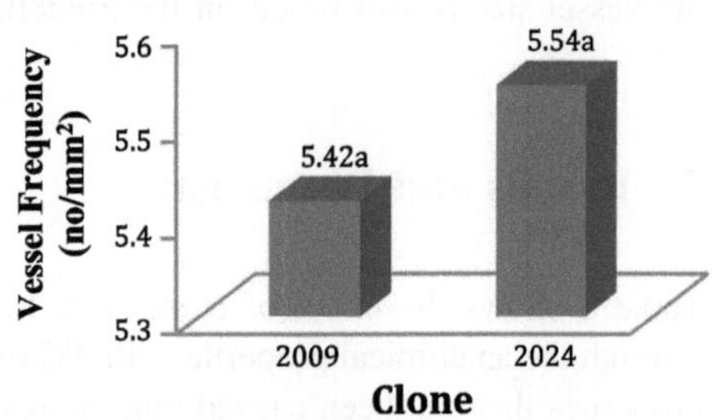

Fig. 3 Effect of clone on vessel frequency

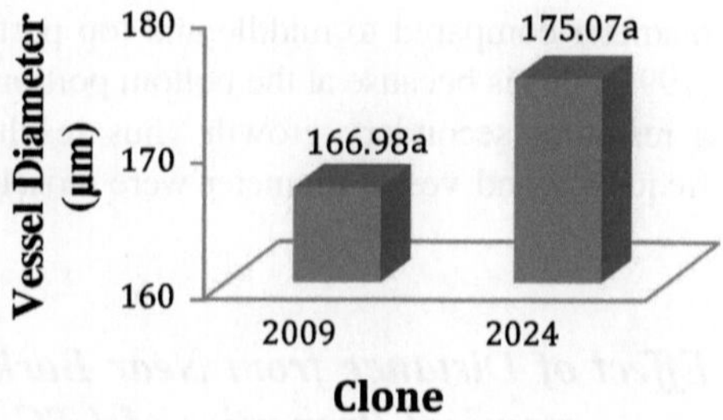

Fig. 4 Effect of clone on vessel diameter

Effect of Tree Height on Anatomical Properties of LTC Rubberwood

Figures 5 and 6 show the effect of tree heights on the anatomical properties of *Hevea brasiliensis*. The results obtained show significant difference on vessel frequency and the vessel diameter. According to Suhaimi (2006), the vessel consists of a group of cells or tissue that serves as a conducting system that drains the sap in hardwood. Based on Fig. 5, the vessel frequency had no significant difference at middle portion with the top and bottom portion of rubberwood tree. However, the bottom and top portion showed high significant difference where the top portion gives a higher value than bottom portion. According to Suhaimi et al. (1998), the higher value of vessel frequency is caused by the smaller diameter of vessel. Ramirez (2009) through his research on *Eucalyptus globulus*, obtained the same finding whereas vessel frequency increase with the increase in tree height.

On the other hand, vessel diameter showed there are highly significant difference according to the tree height. The bottom portion of the tree has greater vessel

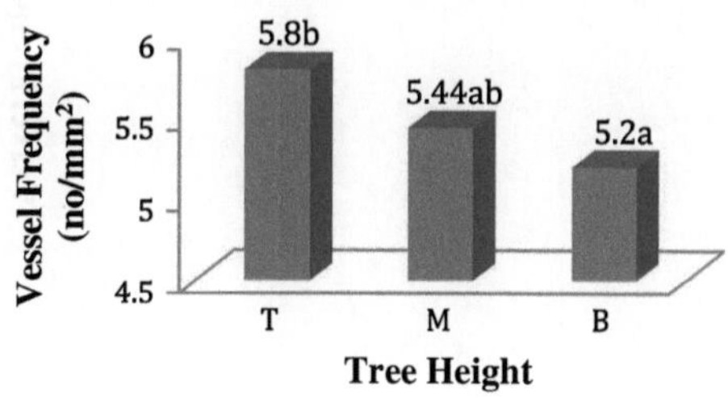

Fig. 5 Effect of tree height on vessel frequency

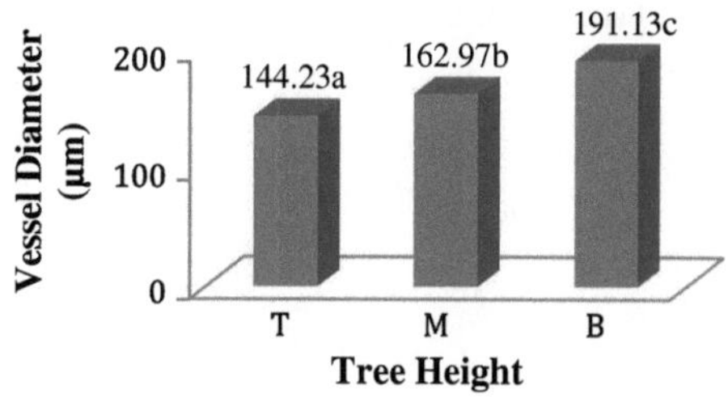

Fig. 6 Effect of tree height on vessel diameter (Note: *T* top, *M* middle, *B* bottom)

diameter compared to middle and top portion. According to Lewin and Goldstein (1991), this is because at the bottom portion of a tree, there is expansion of vessel as a result of secondary growth, thus resulting in increased tree diameter. Vessel frequency and vessel diameter were closely related to one another.

Effect of Distance from Near Bark to Near Pith on Anatomical Properties of LTC Rubberwood

The effect of the distance from near bark to near pith on the anatomical characteristics of LTC rubberwood are shown in Figs. 7 and 8. The wood portion near pith had higher vessel frequency compared to wood portion near bark. But wood portion near bark recorded a higher value of vessel diameter. The cells at wood portion near bark are active cells and a higher value of vessel diameter at this part is because the sap conducting is depends on the vessel diameter (Bowyer et al. 2007). While, Suhaimi et al. (1998) conducted study of rubberwood clones (RRIM 600 and RRIM 623) found that the vessel has high frequency at wood portion near pith and significantly reduced when approaching to near bark. This will give impacts on the vessel diameter. Study by Ramirez (2009) showed that for *Eucalyptus globulus* clone, the vessel diameter increased from wood portion near pith to near bark while the vessel frequency is decreasing.

Fig. 7 Effect of distance on vessel frequency

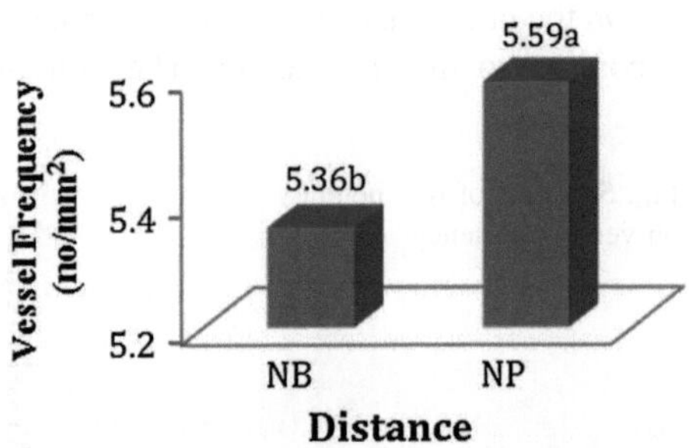

Fig. 8 Effect of distance on vessel diameter (Note: *NB* near bark, *NP* near pith)

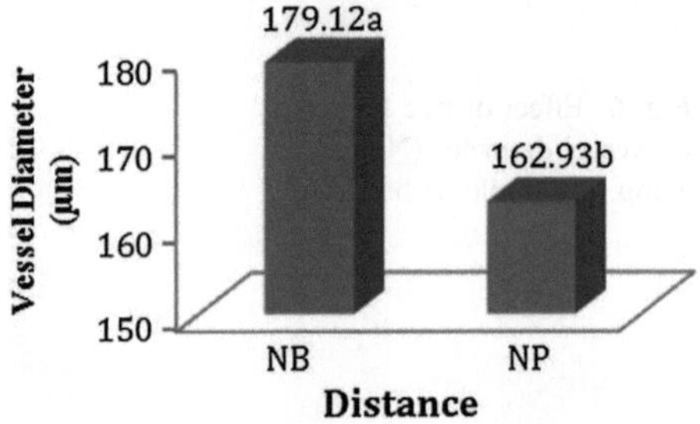

4 Conclusion

From this study, the anatomical properties of LTC rubberwood were determined. The clones of LTC rubberwood showed no significant difference on the anatomical properties. However, clone RRIM 2024 was found to have greater anatomical structure of vessel frequency and vessel diameter. Based on the different height portion of tree, both anatomical properties show significant difference. Top portion gives higher value for vessel frequency per millimeter square and bottom portion gives higher value for vessel diameter. Both of these anatomical structures are closely related to each other. The distance of wood portion near pith to wood portion near bark shows no significant difference for vessel frequency and vessel diameter. However, the vessel frequency is compact at wood portion near pith while the vessel diameter is decreasing when approaching near pith. The clones can be the alternative raw materials in supporting the furniture industries.

References

Bowyer JL, Shmulsky R, Hygreen JG (2007) Forest products and wood science: an introduction, 5th edn. Blackwell Publishing, Iowa

Hong LT, Sim HC (1994) Products from rubberwood—an overview. Rubberwood processing and utilisation. Malayan For Rec 39:173–183

International Association of Wood Anatomist (1939) Standard terms of size of vessel diameter and vessel frequency. Trop Woods 59:51–2

Lewin M, Goldstein IS (1991) Wood Structure and Composition. Marcel Dekker, Inc., New York

Ramirez M (2009) Wood anatomy and biometric parameters variation of *Eucalyptus globulus* clone. Wood Sci Technol 43(1–2):131–141

Salleh MN, Wong WC (1991) Utilization of forest plantation trees. In: Sheikh Ali Abod et al (eds) Proceedings of the of a regional symposium 1989, recent development in tree plantation of humid/subhumid tropic of Asia, Serdang, Malaysia, pp 672–684

Suhaimi M, Mohd. Hamami S, Mohd. Nazip S, Mansur A (1998) Wood quality indicators in rubberwood (*Hevea brasiliensis*). In: Proceeding Seminar of Bureau of Research and Consultancy, ITM, Shah Alam, pp 28–42

Suhaimi M (2006) Anatomi kayu – Sifat Fizikal dan Struktur kayu-kayan Malaysia. Pusat Penerbitan Universiti (UPENA), Universiti Teknologi MARA Shah Alam; Tavita Y, Polisco J, Villavalez L (1974) Fiber dimension of some Philippine woods. FORPRIDE Dig 3(3&4):66

Chapter 37
Effect of Tree Portion and Anthraquinone (AQ) on Pulp Properties from Batai (*Paraserianthes falcataria*)

Muslyza Che Hussin and Jamaludin Kasim

Abstract A study from the tree portion of Batai (*Paraserianthes falcataria*) and anthraquinone (AQ) was carried out to the pulp properties. Nowadays, demand of paper in Malaysia and the world arises year by year and need more lignocellulosic material as an alternative source. Thus, this paper describes the effect of tree portion from *Paraserianthes falcataria* and anthraquinone on pulp properties. There are two pulping conditions which are used in this study: soda pulping (without AQ) and soda-Anthraquinone pulping (with 0.1 % of AQ). The finding shows that the effects of tree portion and anthraquinone are significant to the pulp properties (pulp yield, pulp screened yield, pulp screened reject, and Canadian Standard Freeness (CSF)).

Keywords Anthraquinone · Canadian Standard Freeness (CSF) · *Paraserianthes falcataria* · Pulp properties · Tree portion

1 Introduction

Nowadays, the high demand of pulp and paper in Malaysia and the world induced researchers to find alternative wood materials that can produce pulp and paper to meet consumer needs. Agnihotri et al. (2010) reported that world demand for paper and paperboard was anticipated to grow from 300 million tonnes to over 490 million tonnes by the year 2020, which is expected to raise the cost of pulp wood. The wood material consists of fibers which are better than non-wood fibers in producing high-quality paper. Therefore, the fast growing trees such as

---

M. Che Hussin
Faculty of Applied Sciences, Universiti Teknologi MARA,
Shah Alam, Selangor, Malaysia
e-mail: outwit08@yahoo.com

J. Kasim (✉)
Faculty of Applied Sciences, Universiti Teknologi MARA,
Jengka, Pahang, Malaysia
e-mail: djamal@pahang.uitm.edu.my

N.A. Yacob et al. (eds.), *Regional Conference on Science, Technology and Social Sciences (RCSTSS 2014)*, DOI 10.1007/978-981-10-0534-3_37

Paraserianthes falcataria and *eucalyptus* are suitable as an alternative source. Large-scale cultivation of these trees could be done specifically for the pulp and paper production. Wood is largely the conventional raw material for pulp and paper production worldwide and accounts for 90 % of total fiber input (Little et al. 2003). Pulps produced from eucalyptus are dominating the world hardwood pulp markets (Patt et al. 2006).

Paraserianthes falcataria also known as batai is one of the most important pioneer multipurpose tree species in Indonesia. It has fast growth and able to grow on a variety of soils. The family and subfamily of *Paraserianthes falcataria* are Fabaceae and Mimosoideae (Krisnawati et al. 2011). *Paraserianthes falcataria* wood is generally lightweight and soft to moderately soft. The wood density is between 230 and 500 kg/m^3 at 12–15 % moisture content. This species is one of the important commercial timber species used for both the pulp and paper industry and furniture. The wood is extensively used for the manufacture of rayon and for supplying pulp for the manufacture of paper (Soerianegara and Lemmens 1993).

The main chemical pulping processes for lignocelluloses are the alkaline kraft and soda pulping processes (Haygreen and Bowyer 1982). Generally, softwoods (spruce, hemlock, and pine) and hardwoods (acacia, beech, birch, and eucalyptus) are used for the production of kraft paper pulps (Bierman 1996; Jahan et al. 2008), while non-wood material (sugarcane bagasse, reeds, straws, kenaf, bamboo, etc.) has been explored for soda pulping (De Lopez et al. 1996; Kamthai 2007; Agnihotri et al. 2010; Mossello et al. 2010). In soda pulping, the cooking liquor is composed mainly of sodium hydroxide. The application of pulping additives such as anthraquinone (AQ) has increased the extent of the delignification, so that it is comparable to the kraft method (Macloed 1983; Khristova et al. 2006).

2 Materials and Methods

In this study, matured trees of Batai (*Paraserianthes falcataria*) were collected from Donghwa plantation at Merbok, Kedah Darul Aman. Figure 1 shows the bottom, middle, and top portions of Batai tree. In this study, middle portion was mixed with top portion because their chemical and fiber properties are similar. Therefore, mixed middle and top portion in this study was known as top portion. Three concentrations of sodium hydroxide (16, 18, and 20 %) were prepared based on 500 g oven-dried (o.d.) weight of raw materials. The percentage of AQ was kept constant at 0 and 0.1 % based on oven-dried weight of raw materials. Soda pulping and soda-AQ pulping cooking conditions are given in Table 1.

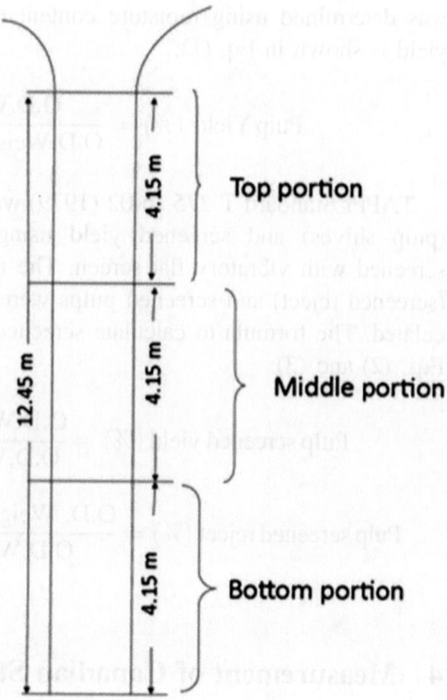

Fig. 1 The three portions of Batai

Table 1 Cooking conditions for soda and soda-AQ pulping

No.	Cooking conditions	Values
1	Tree portions (woodchips)	Bottom and Top
2	Anthraquinone (AQ)	0 and 0.1 %
3	Active alkali (as Na_2O based on o.d. fiber)	16, 18, 20 %
4	Time to maximum temperature, min	60
5	Time at maximum temperature, min	120
6	Raw material to white liquor ratio	1:6
7	Cooking temperature, °C	170
8	Raw material weight, g (based on o.d. fiber)	500
9	Rate of heating, °C/h	170

3 Determination of Pulp Yield, Pulp Screened Yield, and Pulp Screened Reject

After cooking process, the pulp slurry was disintegrated in a hydropulper for 5 min. Then, the pulps were thoroughly washed with fresh water on a fine filter to make sure that they were separated into single fiber and the black liquor was removed. The washed pulp slurry was spin dried, weighed, and moisture content of pulp yield was determined using moisture content analyser. The formula to calculate pulp yield is shown in Eq. (1):

$$\text{Pulp Yield }(\%) = \frac{\text{O.D.Weight of pulp, g}}{\text{O.D.Weight of raw material, g}} \times 100 \quad (1)$$

TAPPI Standard T 275 sp-02 (1979) was used to determine the screened rejects (pulp shives) and screened yield using Somerville Fractionator. Pulps were screened with vibratory flat screen. The residual remains on vibratory flat screen (screened reject) and screened pulps were weighed and moisture content was calculated. The formula to calculate screened yield and screened reject are shown in Eqs. (2) and (3):

$$\text{Pulp screened yield }(\%) = \frac{\text{O.D.Weight of screened pulp, g}}{\text{O.D.Weight of raw material, g}} \times 100 \quad (2)$$

$$\text{Pulp screened reject }(\%) = \frac{\text{O.D. Weight of screened reject pulp, g}}{\text{O.D.Weight of raw material, g}} \times 100 \quad (3)$$

4 Measurement of Canadian Standard Freeness (CSF)

This test method describes a procedure for measurement of the drainage rate using the Canadian Freeness Tester according to TAPPI Standard T 227 om-99. The pulps' stock was stirred homogenously and 1L of pulps stock was taken in duplicate for the test. After measuring temperature for both samples, the stock was gently poured into freeness drainage chamber containing a perforated screen plate in the bottom and a closed airtight seal underneath the plate. Then, the top lid and air cock were closed. The bottom seal was opened to allow the drainage of water. The test complete when all of the water on the chamber was released. The volume of water in the measuring cylinder was recorded while the pulps left in the chamber were oven dried at 103 °C ± 2 °C to constant weight.

5 Results and Discussion

5.1 ANOVA

Table 2 shows the statistical analysis of variance (ANOVA) of pulp properties according to tree portion and anthraquinone (AQ). The effect of tree portion and AQ was shown to be significant on CSF, pulp yield, pulp screened yield, and pulp screened reject. It was also shown that the interaction between portion and anthraquinone is significant difference on pulp properties.

Table 2 ANOVA of pulp properties according to portion and anthraquinone (AQ)

SOV	df	CSF	Pulp yield	Pulp screened yield	Pulp screened reject
Portion (P)	1	1447.95*	14.75*	18.55*	148.82*
AQ (A)	1	210.63*	307.03*	702.60*	2343.59*
P × A	1	863.41*	58.84*	8.39*	8.35*

*F value significant at $p < 0.05$

5.2 Effect of Tree Portion

Table 3 and Fig. 2 show the pulp properties of *Paraserianthes falcataria* according to tree portions (bottom and top). The CSF of bottom portion is high and significantly different to the top portion. The newsprint manufacturing industry still relies on the measurement of CSF as an indicator for the adhesion of fibers and closely related to tensile properties that can describe the nature of the pulp to the paper machine (Tessier et al. 1997). The high CSF makes low the drainage time of pulp (Muslyza 2010). The drainage time of pulp is low which results in the flow of pulps faster. The pulp yield from bottom to top portion is decreased with significant difference. The pulp screened yield from bottom to top portion is increased with significant difference. The pulp screened reject from bottom to top portion is

Table 3 The effects of portion on pulp properties

Pulp properties	Bottom	Top
CSF	418	394
Pulp yield (%)	65.31	63.88
Pulp screened yield (%)	45.75	46.82
Pulp screened reject (%)	3.60	2.96

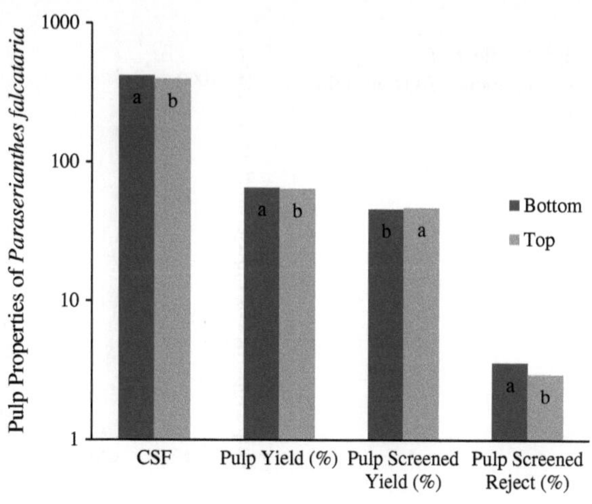

Fig. 2 Effect of tree portions (*bottom* and *top*) on pulp properties

decreased with significant difference. The bottom portion results in highest pulp yield but the top portion results in the highest pulp screened yield and the lowest pulp screened reject.

5.3 Effect of Anthraquinone

The addition of anthraquinone (AQ) to soda pulping liquor is known as an effective and simple approach that increases delignification selectivity, carbohydrate protection, and pulp yield. On the other hand, anthraquinone is an expensive material and produced alkali-AQ pulps have lower bleachability (Khristova et al. 2006; Francis et al. 2006, 2008; Jahan et al. 2008). Table 4 and Fig. 3 show the Canadian Standard Freeness (CSF), pulp yield, pulp screened yield, and pulp screened reject of *Paraserianthes falcataria* in different percentages of anthraquinone (AQ).

The CSF is increased with significant difference from 0 to 0.1 % AQ. The CSF is a test that measures the readiness of the pulp to part with its water by drainage (Bublitz 1980). In the test, the pulp is allowed to drain under standard conditions through a sieve. The resulting drainage water passes to waste through a small hole and the water over a certain level above it is taken off through an overflow and measured in a glass cylinder. A wet pulp gives a low CSF value. Freeness below

Table 4 The effect of anthraquinone (AQ) on pulp properties

Pulp properties	0 %	0.1 %
CSF	402	410
Pulp yield (%)	61.34	67.85
Pulp screened yield (%)	42.99	49.57
Pulp screened reject (%)	4.54	2.02

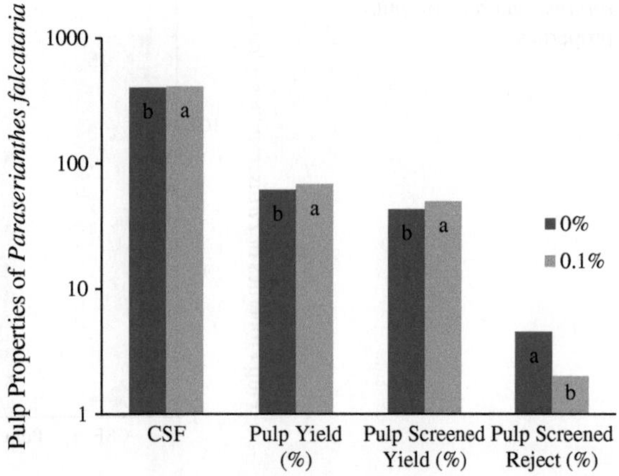

Fig. 3 Effect of anthraquinone (AQ) on pulp properties

200 ml would render the pulp "wet" resulting in poor handsheets quality (Faridah 1990). This indicates that the presence 0.1 % of AQ in pulping results in high CSF.

The pulp yield and pulp screened yield from 0 to 0.1 % AQ are increased with significant difference. This indicates that 0.1 % of AQ results in high pulp yield and pulp screened yield rather than zero percent of AQ. The addition of 0.1 % AQ dose to soda cooking results in high total yields (40.5–44.2 %) and screened yield (40.5–43.9 %) compared to soda pulping (Safaa et al. 2012). Mehmet (2003) states that the addition of 0.1 % anthraquinone increases the total pulp yield.

The pulp screened reject from 0 to 0.1 % AQ is decreased with significant difference. This indicates that 0.1 % of AQ results in low pulp screened reject rather than zero percent of AQ. The 0.1 % of AQ results in high pulp yield and pulp screened yield while low pulp screened reject. Therefore, the addition of 0.1 % of AQ results in increasing pulp yield and pulp screened yield. Furthermore, the presence of anthraquinone in the pulping decreases the pulp screened reject. Safaa et al. (2012) found that the use of 0.1 % AQ dose gave less reject amounts.

6 Conclusion

The effect of tree portion and anthraquinone on CSF, pulp yield, pulp screened yield, and pulp screened reject showed a significant difference. Bottom portion results in high CSF rather than top portion. In this study, the bottom portion results in highest pulp yield but the top portion results in the highest pulp screened yield and the lowest pulp screened reject. The presence of 0.1 % AQ in pulping results in high CSF. The 0.1 % of AQ results in high pulp yield and pulp screened yield, while low pulp screened reject compared to without using any AQ (0 %).

Acknowledgements The authors thank Universiti Teknologi MARA (UiTM) Shah Alam, Selangor, Malaysia for the financial support. Thanks also go to Amran Shafie, Rudaini Nawawi, and Sardey Idris at the Universiti Teknologi MARA (UiTM), Pahang, Malaysia for their assistance. We also acknowledge Dr. Rushdan Ibrahim, Shukri Said and Norazizi from Pulp and Paper Programme at Forestry Research Institute of Malaysia (FRIM), Kepong, Malaysia for their expertise and facilities.

References

Agnihotri S, Dutt D, Tyagi CH (2010) Complete characterization of bagasse of early species of *Saccharum officinerum-CO* 89003 for pulp and papermaking. BioResources 5(2):197–1214

Biermann CJ (1996) Handbook of pulping and papermaking, 2nd edn. Academic Press, New York

Bublitz WJ (1980) Pulp and paper chemistry and chemical technology, 3rd edn, vol 1. Wiley-Interscience, New York, pp 113–155

De Lopez S, Tissot M, Delmas M (1996) Integrated cereal straw valorization by an alkaline pre-extraction of hemicellulose prior to Soda Anthraquinone pulping. Case study of barley straw. Biomass Bioenergy 10(4):201–2011

Faridah AM (1990) Feasibility studies of utilising oil palm (*Elaeis guinensis jacquin*) leaves for paper pulp. Thesis degree of Master of Philosophy. Institute of Advanced studies, University of Malaya, Kuala Lumpur, Malaysia

Francis RC, Shin S-J, Omori S, Amidon TE, Blain TJ (2006) Soda pulping of hardwoods catalysed by AQ and methyl substituted AQs. J Wood Chem Technol 26(2):141–152

Francis RC, Bolton TS, Abdoulmoumine N, Lavrykova N, Bose SK (2008) Positive and negative aspects of soda/anthraquinone pulping of hardwoods. Biores Technol 99(17):8453–8457

Haygreen JG, Bowyer JL (1982) Forest products and wood science: an introduction, 2nd edn. Iowa State University Press

Jahan MS, Sabina R, Rubaiyat A (2008) Alkaline pulping and bleaching of *Acacia auriculiformis* grown in Bangladesh. Turk J Agric For 32:339–347

Kamthai S (2007) Comparison of AS-AQ pulping of sweet bamboo (*Dendrocalamus asperbacker*) and pulping by conventional kraft process. Chiang Mai J Sci 34(1):97–107

Khristova P, Kordsachia O, Patt R, Karar I, Khider I (2006) Environmentally friendly pulping and bleaching of bagasse. Ind Crops Prod 23(2):131–139

Krisnawati H, Varis E, Kallio M, Kanninen M (2011) Paraserianthes falcataria (L.) Nielsen: ecology, silviculture and productivity. Center for International Forestry Research (CIFOR), Bogor, Indonesia

Little KM, van Staden J, Clarke GPY (2003) The relationship between vegetation management and the wood and pulping properties of Eucalyptus hybrid clone. Ann For Sci 60:673–680)

Macloed JM (1983) Soda AQ pulp from hardwoods-physical properties and bleachability. Pulp Pap Can 84:29–32

Mehmet ST (2003) Relationship between alkaline pulp yield and the mass fraction and degree of polymerization of cellulose in pulp. Master's Thesis. The University of Maine

Mossello AA, Harun J, Tahir P, Resalati H, Ibrahim R, Shamsi SRS, Mohmamed AZ (2010) A review of literatures related of suing Kenaf for pulp production (Beating, fractionation and recycle fiber). Mod Appl Sci 4(9):21–29

Muslyza CH (2010) The influence of rice straw pulp fiber length on paper properties. Bachelor Degree Thesis. Universiti Teknologi MARA, Malaysia

Patt R, Kordsachia O, Fehr J (2006) European hardwoods versus *Eucalyptus globulus* as a raw material for pulping. Wood Sci Technol 40:39–48

Safaa HO, Tarig OK, Osman TE (2012) Pulping potential of *Ricinus communis* stems from Sudan. J Am Sci 8(6)

Soerianegara I, Lemmens RHMJ (1993) Plant resources of South-East Asia 5(1): timber trees: major commercial timbers. Pudoc Scientific Publishers, Wageningen, Netherlands

Tessier P, Broderick G, Desrochers C, Bussiere S (1997) Industrial implementation of motor load and freeness control of chemimechanical pulp refiners. Tappi J 80(12):135–142

Chapter 38
The Role of Oil Palm (*Elaeis guineensis*) Frond as Filler in Polypropylene Matrix with Relation of Filler Loading and Particle Size Effects

Nor Farhana Jasmi, Jamaludin Kasim, Mohd Shafie Ansar and Iffah Izzah Maidin

Abstract Oil palm frond seems to be among the best alternative substitution fillers in thermoplastic composite (WPC) as it is abundantly available and its lignocellulosic properties offer great usability. There is still lack of preliminary studies on the basic effects of oil palm frond's particle size and its loading in polypropylene matrix. Thus, this paper focuses on studying the mechanical and physical properties of oil palm frond–polypropylene composite in the effects of both particle size and filler loading. Composites with 10–50 % filler loading were fabricated with four different sizes from unscreen particles towards the fines which are 425, 250, and 75 μm. The findings performed a significant difference at 95 % confident level in all tested samples which are flexural MOR, flexural MOE, tensile strength, tensile modulus, elongation at break, and water absorption behavior. Largest size showed the best mechanical properties among the others, which is contradicted to previous reports on wood particle thermoplastic composite. These findings are useful as a guide throughout the future researches based on oil palm frond and polypropylene matrix.

Keywords Oil palm frond · Polypropylene · Flexural properties · Tensile properties

---

N.F. Jasmi (✉) · I.I. Maidin
Faculty of Applied Sciences, Universiti Teknologi MARA,
Shah Alam, Selangor, Malaysia
e-mail: norfarhanajasmi@yahoo.com

I.I. Maidin
e-mail: iffahizzah89@gmail.com

J. Kasim · M.S. Ansar
Faculty of Applied Sciences, Universiti Teknologi MARA,
Jengka, Pahang, Malaysia
e-mail: djamal@pahang.uitm.edu.my

M.S. Ansar
e-mail: ms_ansar@ymail.com

N.A. Yacob et al. (eds.), *Regional Conference on Science, Technology and Social Sciences (RCSTSS 2014)*, DOI 10.1007/978-981-10-0534-3_38

1 Introduction

The great expansion of oil palm plantation in this country has generated the massive amounts of residues such as oil palm frond, oil palm trunk, and empty fruit bunches which may create the problems of replanting operation. Great environmental problems may be created if the residues are left and burnt. Furthermore, there will be a problem of burning them as they consist of a lot of moisture. The residues may take up to five years to be degraded naturally and this will be something that is wasteful to the industry (MPOC 2012). Hence, implementing the wastes in fabricating thermoplastic composite may solve the problems raised. This composite commercialized its name as lignocellulosic–thermoplastic composite. The properties of this composite may be affected by many factors including the characteristics of particle as natural filler (morphology, surface chemistry, chemical composition, and crystalline contents), types of lignocellulosic species, and also the matrix properties (nature, functionality, and cross-linking agent) itself (Andrzej et al. 2010). Other basic factors that few studies have been investigated are filler content and particle size factors. These factors are indeed necessary in determining the mechanical and physical properties of thermoplastic composite.

Other than reinforcing the polymer matrix with fibers, particles are also can be used but only for filling in purpose named as filler. Increasing filler loading tends to degrade the composite properties and emerges the endurance of brittleness in the composite (Danyadi et al. 2007). Although adding filler to such content load could decrease the properties, it is somehow a way of getting better impact properties. Apart of this matter, size of filler in fabricating thermoplastic composite is also very necessary (Danyadi et al. 2007). Generally, larger specific area of the filler surface may contribute to the increase in interfacial interaction between the lignocellulosic filler and polymer matrix, thus influence the increasing of the composite strength (Rozman et al. 2003). Many studies stated that larger particle or filler size may decrease the tensile strength. This is because larger sizes of particles are easy to entangle and agglomerate to each other. The entanglement and agglomeration is then result in the lack of interfacial bonding, and the mechanical properties resulted will be decreased (Kuo et al. 2009). However, it is interesting to know that in certain cases, the larger particle size may contribute to the increasing of the strength of the thermoplastic composite (Zaini et al. 1996). The researcher believes that it is might be due to the higher aspect ratio given by the large size of the particles. This is because one of the factors of affecting the modulus of composite is being higher aspect ratio. In this study, the investigation was carried out to explore the feasibility of using oil palm frond to be implemented as filler in thermoplastic composite. This study aimed to determine the mechanical properties (flexural MOR, flexural MOE, tensile strength, tensile modulus, elongation at break, and impact) and physical properties (water absorption) with the effects of different filler loadings and particle sizes.

2 Materials and Methods

The oil palm fronds (OPF) were obtained from a local oil palm plantation at Universiti Teknologi MARA Pahang. The matrix polymer is homopolymer polypropylene (PP) in the form of pellets with a melt flow index of 33 g/10 min at 230 °C and melting temperature of 163 °C which were supplied by local manufacturer. Finest size of the ground OPF particles which is <0.5 mm was collected from vibrator screener machine and sieve into three sizes with mesh numbers 60, 40, and 30. The fines were oven dried at 80 °C for 24 h prior to compounding process for composite preparation. Particle analysis procedure in this study is necessary to be conducted in order to determine the dust distribution down to a small-scale sieve. This procedure is necessary in order to predetermine the amount of particles obtained for each size range. The sieve consists of four levels of different sizes which are ≥425, ≥250, ≥75, and <75 μm. The sieve was set to be operated within 20 min upon 200 g of oven-dried oil palm frond dust once at a time. When the screening process was finished, the dust retained on each level was weighed and the weight percentage is calculated based on the total weight of dust used. This procedure was repeated six times in a row to obtain precise reading.

OPF particles were compounded with PP matrix in a dispersion mixer with temperature of 180 °C. The mixer was then crushed to yield pellets. The pellets were then consolidated in a mold dimension of 240 mm × 150 mm × 6 mm for flexural test, and mold dimension of 300 mm × 240 mm × 2 mm for tensile test, impact and water absorption test. The composite panel was hot pressed at 1000 psi in 360 s with temperature of 180 °C followed by cold press at 1000 psi in 60 s with temperature of 25 °C. The composite was then conditioned in a room with relative humidity of 50 ± 5 % for at least 24 h. Preparation of specimens tested was done according to ASTM D 790-92 for flexural test, ASTM D 638 M-91a for tensile test, and ASTM D 570-81 for water absorption.

3 Results and Discussion

Two-way analysis of variance was conducted to determine the significance of filler loading and particle size effects on OPF–PP composite. The results of multivariate test are shown in Table 1. Filler loading variable shows significant effects on whole composite properties including both physical and mechanical performance without any exception at $p = 0.000$. On the other hand, particle size variable shows significant effects on all the composite properties except for the elongation at break at $p = 0.085$. Interaction of the two variables shows significant effect on all the composite properties at $p = 0.000$.

Table 1 Summaries of ANOVA of the filler loading and particle size effects on OPF–PP composite properties (p values)

Source of variation	MOR (MPa)	MOE (MPa)	Tstrength (MPa)	Tmodulus (MPa)	Elong (MPa)	Impact (kJ)	WA (%)
Filler loading	0.000**	0.000**	0.000**	0.000**	0.000**	0.000**	0.000**
Size	0.000**	0.000**	0.002**	0.000**	0.085NS	0.000**	0.013**
Filler loading x Size	0.000**	0.000**	0.000**	0.000**	0.000**	0.000**	0.000**

NS not significant at 0.05; *MOR* modulus of rupture; *MOE* modulus of elasticity; *Tstrength* tensile strength; *Tmodulus* tensile modulus; *Elong* elongation; *WA* water absorption; *TS* thickness swelling

Note $p < 0.05$, significant difference at the 5 % level

Flexural Properties

Flexural strength is the ability of the material to withstand the bending forces applied perpendicular to its longitudinal axis (Sreekumar et al. 2009). The mean values of flexural strength (MOR) and flexural modulus (MOE) with relation effect of filler loading are presented in a line chart as can be seen in Fig. 1. OPF content

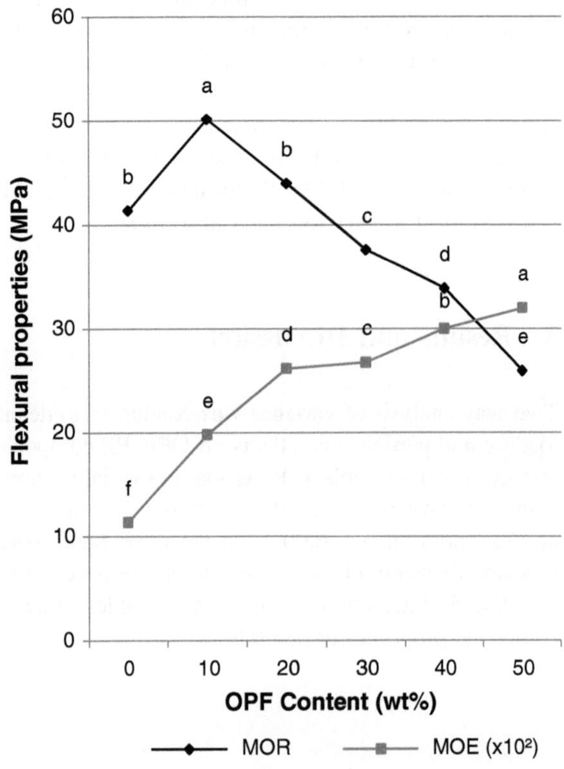

Fig. 1 Effects of filler loading on flexural properties of the composite. *a, b, c, d, e, f* different letters on top of the line chart represent significant difference among the six loadings ($p < 0.05$)

Table 2 Correlation analysis of filler loading and particle size on OPF–PP composite properties

	MOR (MPa)	MOE (MPa)	Tstrength (MPa)	Tmodulus (MPa)	Elong (MPa)	Impact (kJ)	WA (%)
Filler loading	-0.967^{**}	0.902^{**}	-700^{**}	0.390^{**}	-821^{**}	0.025	0.859^{**}
Particle size	0.057	0.019	0.003	-0.205^{*}	0.020	0.249^{**}	0.000

MOR modulus of rupture; *MOE* modulus of elasticity; *Tstrength* tensile strength; *Tmodulus* tensile modulus; *Elong* elongation; *WA* water absorption
Note **significant at 0.05

performed the best filling purpose at 10 and 20 % filler loadings which by means of better flexural strength than the neat PP. The strength was then dropping significantly as can be proved by correlation analysis in Table 2. This may be due to the conditions of the particulates that tend to agglomerate in the composite as a consequence of high filler loading. High agglomeration, hence, results in decrement strength to withstand the load forced on the sample (Danyadi et al. 2007; Luz et al. 2007). This can also be attributed to poor encapsulation of hydrophobic PP towards the hydrophilic OPF particles. A similar trend of result can be seen with other studies which show a gradual decrease in flexural strength with increasing of filler content (Rozman et al. 2003; Fuentes et al. 2007).

High filler loading influenced the composite to have higher flexural modulus as performed in other researches (Cui et al. 2008; Khalid et al. 2008). Increment of flexural modulus may be attributed to a factor of increasing in composite rigidity and strength. In addition, modulus properties are also dependence on the composite stiffness ability. Incorporation of filler in the polymer matrix is mainly intended to increase the stiffness of material. Stiffness inherited from the natural filler itself that latter may contribute to the overall composite stiffness (Rozman et al. 2003). The tendency of particles to be well wetted is likely to depend on the amount of its content reinforced in the polymer matrix. Thus, high filler content may well contribute in wetting the filler with polymer matrix. The transference of stress caused by the wettability between fiber and polymer matrix is helpful in contributing homogenous fiber distribution and hence resulting positive performance of flexural modulus (Luz et al. 2007).

The effects of particle size on flexural strength (MOR) and flexural modulus (MOE) are shown in Fig. 2. The unscreen size showed a decrement reading in flexural properties as compared to the other three size ranges. They are non-homogeneously compounded in the polymer matrix, thus resulting in a weak stress transference. Large particulate filler of 425 μm does play a positive significant effect of filling in the polymer composite in a way of high aspect ratio. Aspect ratio in this case can be interpreted as a ratio of particle length to thickness. This ratio may contribute in enhancing the transference of stress between polymer matrix and natural filler that helps to improve the composite mechanical properties (Wilczyn et al. 2012). Much studies have showed that high aspect ratio due to large particulate filler size is able to increase the composite strength. They act similar

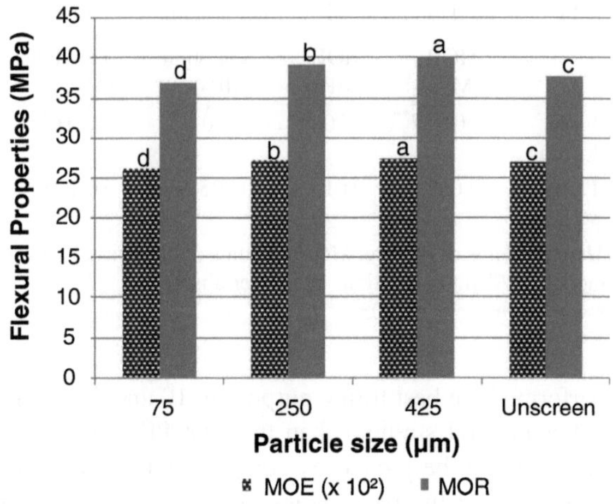

Fig. 2 Effects of particle size on flexural properties of the composite. *a, b, c, d* different letters on top of the bar chart represent significant difference among the four sizes ($p < 0.05$)

Table 3 Descriptions on particle size

Particle size (μm)	Particle analysis (%)[a]	Aspect ratio[b]
≥425	22.34	4.6
≥250	38.81	4.4
≥75	31.14	4.4
<75	7.71	≤4.2

Source [a]This study, [b]Klyosov (2007)

reaction of transferring the stress like the small particulate filler does. As a result, aspect ratio plays a significant effect on the composite board properties. Particle analysis and aspect ratio can be reffered from Table 3.

Tensile Properties

The trend of detrimental effect on tensile strength with increasing the filler content can be seen in Fig. 3 and also observed to be significantly negative correlated at $r = -700$ in Table 2. This phenomenon of negative result upon high filler content can be explained by considering the formation of clumps on the filler–matrix composite surface. The clumps of OPF filler may result in failure of transferring the stress. There is an observation of significantly increasing tensile modulus with increasing the filler loading from 40 to 50 % filler loading. The impact of increment in tensile modulus may be contributed by well distribution of the particles within composite, and hence resulting in good aggregation in the polymer matrix. Figure 3 also shows a decrement of elongation at break when the filler content increased from 10 % toward 50 %. This phenomenon is expected to happen because elongation at break is normally declined abruptly as natural particles are implemented in

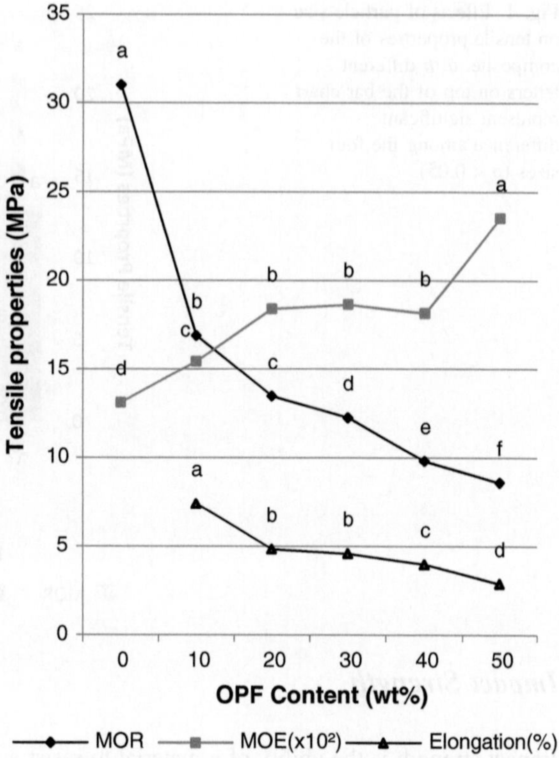

Fig. 3 Effects of particle size on tensile properties of the composite. *a, b, c, d, e, f* different letters on top of the line chart represent significant difference among the six loadings ($p < 0.05$)

the polymer matrix as compared to the elongation at break of neat polymer matrix (Premalal et al. 2002). This is attributed to the particle agglomeration in the matrix that created voids in the final composite. These situations have led a restriction by the natural filler to let plastic molecules flowing efficiently during processing and latter generated defects on the composite (Karmarkar et al. 2007; Kaewkuk et al. 2013).

Tensile properties affected by particle size can be observed in Fig. 4. Tensile modulus increases with the decrease of filler size. This result is expected to happen because the filler may react as a medium to support stress transferred from the polymer matrix and enhance the tensile modulus. Irregular shape filler supposed to have tendency to decrease the tensile modulus of composite (Zaini et al. 1996; Khalid et al. 2008). However, it is interesting to note that the unscreen size in this study has better tensile properties as compared to large particles. It may be attributed to the small particulates that have successfully clogged the pores between the large particulates and resulted in good stress transfer.

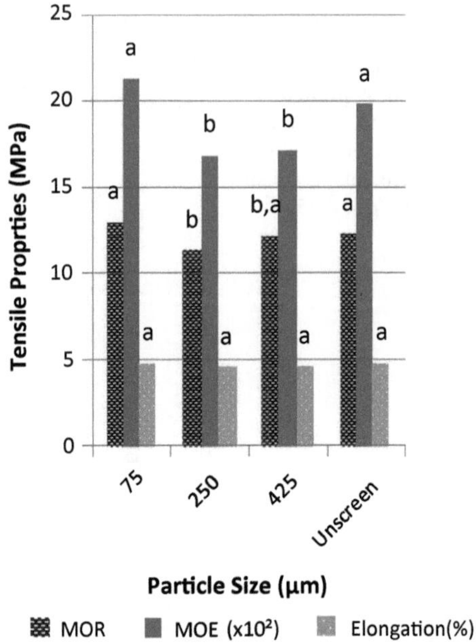

Fig. 4 Effects of particle size on tensile properties of the composite. *a*, *b* different letters on top of the bar chart represent significant difference among the four sizes ($p < 0.05$)

Impact Strength

Impact strength is the ability of a material to resist a sudden impact. It gives us an amount of energy it needed to break a specimen due to the impact forced upon it. Exceptional to the result of 10 wt% filler, Fig. 5 shows that the impact strength

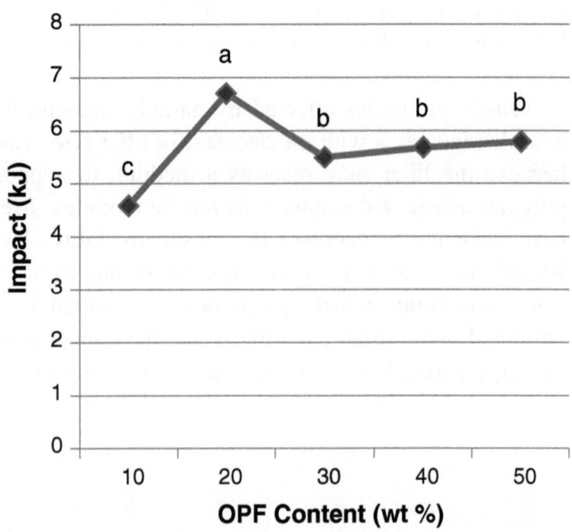

Fig. 5 Effects of filler loading on impact strength of the composite. *a*, *b*, *c* different letters on top of the line chart represent significant difference among the five loadings ($p < 0.05$)

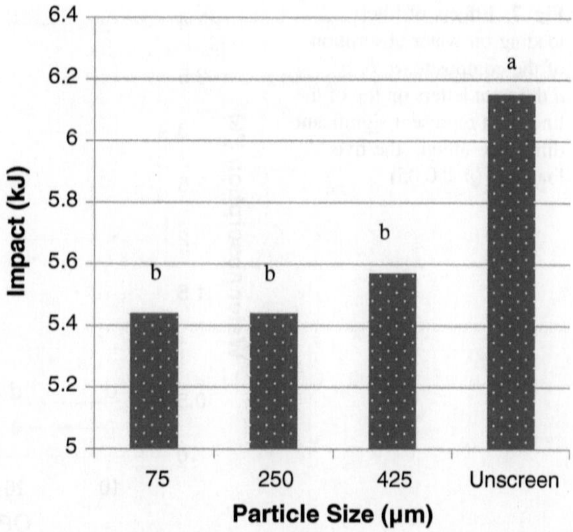

Fig. 6 Effects of particle size on impact strength of the composite. *a*, *b* different letters on top of the bar chart represent significant difference among the four sizes ($p < 0.05$)

decreases when OPF content increases. Increasing the filler content toward a polymer matrix may result in a poor interfacial adhesion between both different polar materials. The adhesion problem has led to a tendency of filler agglomeration in the matrix that latter easily propagate cracks due to an impact force. The agglomerated filler may deplete the matrix mobility and hence stiffen up the whole composite. Impact strength generally exhibits a composite stiffness which indicates a value of stress transference between filler and polymer matrix. Hence, insertion of high filler content in the polymer matrix in fabricating a composite has decreased the composite impact strength showing that a low energy is required for breaking the composite apart (Premalal et al. 2002; Kaewkuk et al. 2013).

Despite the fact of insignificant different among the three sizes (75, 250, and 425 μm), the reading shows a positive trend with relation of particle size increment as can be seen in Fig. 6. Finer particle size tends to have greater surface area that may cause the existence of more cavities. The cavities created may act as internal defects that will result in a reduction of energy transfer within the composite upon impact (Petchwattana and Covavisaruch 2013). The composite filled with unscreen particles possesses an impact strength of 6.145 kJ indicating approximately 10 % increment from the other particle sizes. The occurence of better result by unscreen particles may be caused by less cavities created on the particle surfaces.

Water Absorption

Gaps are getting more and larger when filler is inserted to the polymer matrix, thus giving the composite a tendency of absorbing water compared to neat PP. 100 % PP

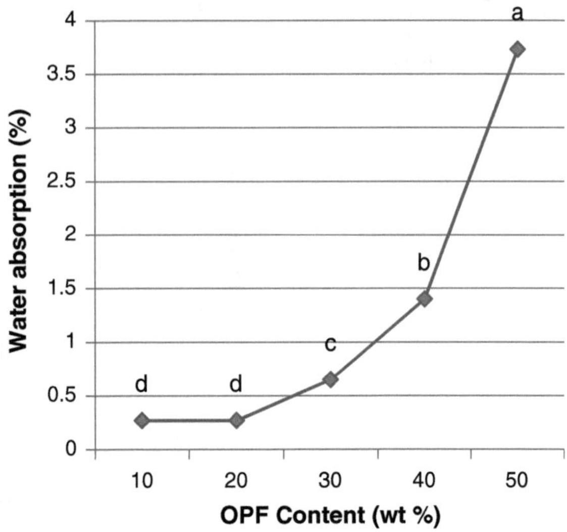

Fig. 7 Effects of filler loading on water absorption of the composite. *a*, *b*, *c*, *d* different letters on top of the line chart represent significant difference among the five loadings ($p < 0.05$)

is a control specimen which absorbs water at minimal stage. Cellulose and lignin do not give a significant effect on absorbing water. Hemicellulose, however, is likely to absorb water due to its amorphous structure and hydrophilic characteristic in nature (Bouafif et al. 2009). Hence, there is a tendency for a lignocellulosic-filled plastic composite to absorb water as observed in Fig. 7. Figure 8 shows that larger particle has led the composite to have higher water absorption. There are several reasons on that phenomenon and can be interpreted as (1) larger particle size gives low

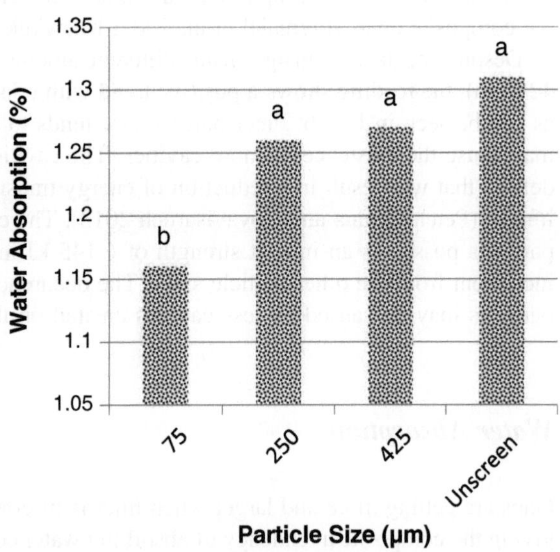

Fig. 8 Effects of particle size on water absorption of the composite. *a*, *b* different letters on top of the bar chart represent significant difference among the four sizes ($p < 0.05$)

encapsulation of polymer matrix and thus lead to void spaces; (2) weak encapsulation results in exposing hydroxyl groups and hydrophilic characteristic to absorb water (Rozman et al. 2003).

4 Conclusion

A WPC board has to reach the surpass requirement standard of strength named International Code Council Evaluation Service (ICC-ES) to eligible itself as structural decking board (Klyosov 2007). This research has found that the least strength gained by 50 wt% OPF content reached the minimum requirement by ICC code. Hence, this proved that OPF particle is feasible to be filled in thermoplastic composite where latter has the tendency to be used as structural purpose board. Large particle sizes of 250 and 425 μm are considered as the best filler in this study as it contributes the most abundant portion of a whole OPF as well as gains good properties.

References

Anonymous (2012) The oil palm tree. www.mpoc.org.my/The_Oil_Palm_Tree.aspx. Accessed 29th Jan 2012

Andrzej KB, Abdullah AM, Jürgen V (2010) Physical, chemical and surface properties of wheat husk, rye husk and soft wood and their polypropylene composites. Compos A 41:480–488

Bouafif H, Koubaa A, Perre P, Cloutier A (2009) Effects of fiber characteristics on the physical and mechanical properties of wood plastic composites. Compos A 40:1975–1981

Cui Y, Lee S, Noruziaan B, Cheung M, Tao J (2008) Fabrication and interfacial modification of wood/recycled plastic composite materials. Compos A 39:655–661

Danyadi L, Janecska T, Szabo Z, Nagy G, Moczo J, Pukanszky B (2007) Wood flour-filled PP composites: compatibilization and adhesion. Compos Sci Technol 67:2838–2846

Fuentes FJT, Guzman JAS, Richter HG, Duenas RS, Quirarte JR (2007) Effect of production variables on bending properties, water absorption and thickness swelling of bagasse/plastic composite boards. Ind Crops Prod 26:1–7

Kaewkuk S, Sutapan W, Jarukumjorn K (2013) Effects of interfacial modification and fiber content on physical properties of sisal fiber/polypropylene composites. Compos B 45:544–549

Karmarkar A, Chaulan SS, Modak JM, Chanda M (2007) Mechanical properties of wood-fiber reinforced polypropylene composites: effect of a novel compatibilizer with isocyanate functional group. Compos A 38:227–233

Khalid M, Ratnam CT, Chuah TG, Ali S, Choong TSY (2008) Comparative study of polypropylene composites reinforced with oil palm empty fruit bunch fiber and oil palm derived cellulose. Mater Des 29:173–178

Klyosov AA (2007) Wood-plastic composites. ISBN:978-0-470-14891-4(cloth)

Kuo PY, Wang SY, Chen JH, Hsueh HC, Tsai MJ (2009) Effects of material compositions on the mechanical properties of wood-plastic composites manufactured by injection molding. Mater Des. doi:10.1016/j.matdes.2009.03.012

Luz SM, Goncalves AR, Del'Arco AP (2007) Mechanical behaviour and microstructural analysis of sugarcane bagasse fibers reinforced polypropylene composites. Compos Part A 38:1455–1461

Petchwattana N, Covavisaruch S (2013) Effects of rice hull particle size and content on the mechanical properties and visual appearance of wood plastic composites prepared from Poly (vinyl chloride). J Bionic Eng 10(1):110–117. doi:10.1016/S1672-6529(13)60205-X

Premalal HGB, Ismail H, Baharin A (2002) Comparison of the mechanical properties of rice husk powder filled polypropylene composites with talc filled polypropylene composites. Polym Testing 21:833–839

Rozman HD, Saad MJ, Mohd Ishak ZA (2003) Flexural and impact properties of oil palm empty fruit bunch (EFB)-polypropylene composites-the effect of maleic anhydride chemical modification of EFB. Polym Testing 22:335–341

Sreekumar PA, Thomas SP, Saiter J, Joseph K, Unnikrishnan G, Thomas S (2009) Effect of fiber surface modification on the mechanical and water adsorption characteristics of sisal/polyester composites fabricated by resin transfer molding. Compos A 40:1777–1784

Wilczyn A, Gozdecki C, Kociszewski M, Tomaszewska J, Zajchowski S (2012) Mechanical properties of wood—polypropylene composite with industrial wood particles of different sizes. Wood Fiber Sci 44(1):1–8

Zaini MJ, Fuad MYA, Ismail Z, Mansor MS, Mustafah J (1996) The effect of filler content and size on the mechanical properties of polypropylene/oil palm wood flourcomposites. Polym Int 0959-8103/96

Chapter 39
Effects of Board Density and Resin Content on the Mechanical and Physical Properties of Oil Palm Frond Particleboard

Nur Farahin Yusoff, Nur Sakinah Mohammed Tamat and Jamaludin Kasim

Abstract The objective of this study was to determine the effect of board density and resin content on mechanical and thickness swelling properties of oil palm frond (OPF) particleboard using phenol formaldehyde (PF) as adhesive. Modulus of rupture (MOR), modulus of elasticity (MOE), internal bonding (IB), and thickness swelling (TS) of samples were tested in accordance with the Malaysian Standard for particleboard, MS 1787: 2005. Based on the results, mechanical and thickness swelling properties of board improved with increasing density and resin content. The investigation showed that OPF particleboard with density 700 kg/m^3 with 9 and 11 % resin content met and passed the minimum standard requirement for all properties.

Keywords Particleboard · Phenol formaldehyde

1 Introduction

Particleboard is an engineered board made from lignocellulosic material in the form of particles, bonded by adhesive under pressure. The idea of particleboard production was first introduced by Ernest Hubbard in 1887, which was intended as one way to overcome excessive waste and utilized wood waste produced at sawmill (Maloney 1993). This engineered product then experienced evolution in terms of

N.F. Yusoff (✉) · N.S. Mohammed Tamat
Faculty of Applied Sciences, Universiti Teknologi MARA,
Shah Alam, Selangor, Malaysia
e-mail: nurfarahin_y@yahoo.com

J. Kasim
Faculty of Applied Sciences, Universiti Teknologi MARA,
Jengka, Pahang, Malaysia
e-mail: djamal@pahang.uitm.edu.my

N.A. Yacob et al. (eds.), *Regional Conference on Science, Technology and Social Sciences (RCSTSS 2014)*, DOI 10.1007/978-981-10-0534-3_39

raw material, adhesive, end use, size, and appearance year by year. Agricultural wastes such as oil palm trunk (Boon et al. 2013), bagasse (Xu et al. 2009), rice husk (Kwon et al. 2013), kenaf core (Xu et al. 2013), and corn stalk (Akgul et al. 2010) were studied by researcher and particleboard and fiberboard manufacturer in various researches as alternative raw material to satisfy industries' demand especially furniture industries as well as reducing dependency on solid wood waste in particleboard production. High demands of particleboard are primarily because of its availability in various sizes, various thicknesses, and being cheaper than solid wood. Transforming agricultural biomass, which are unused, unutilized, and invaluable, into valuable source as in production of particleboard could reduce the environmental impact or at least utilize them as potential product. Converting agricultural waste into profitable product will reduce the cost of raw material.

Oil palm frond (OPF) is abundant and partially the only utilized agricultural waste since it has potential as new raw material for particleboard. Malaysia, as the second largest in the world behind Indonesia as palm oil exporter, covers 5.2 million hectares plantation areas by the end of 2013. This huge plantation area generates massive agriculture waste especially the fronds. It is estimated to produce 10 tons of OPF per hectare annually (MPOB 2014). The large quantity of OPF mainly left on plantation site for nutrient recycling, however, generates riotous quantity as it felled twice in a month and approximately 2–3 fronds felled for a fruit bunch.

Several researches utilizing OPF have been done including production of compressed oil palm frond composite board (Mohd Sukhairi et al. 2011), pulp (Wanrosli et al. 2007), and cement-bonded board (Hermawan et al. 2001). Each research showed comparable result to the existing products with other lignocellulosic as raw material. Although particleboard from oil palm biomass especially trunk has been investigated in several researches, studies on manufacturing particleboard using OPF especially with phenol formaldehyde (PF) as resin are generally still limited.

The objectives of this study are to investigate the mechanical and physical properties of particleboard and the effect of board density and resin content using OPF as raw material and PF as resin.

2 Materials and Methods

Oil palm frond (OPF) was collected from UiTM Pahang Plantation area during harvesting season. Phenol formaldehyde (PF) used in this study was supplied by a local adhesive manufacturer in Shah Alam, Selangor.

The leaflets of collected OPF were cut off before the samples were transported to the laboratory. The OPF chips produced using rotating chipper were then transferred into Pallmann Knives—Ring Flaker to reduce shape into particle form. The particles were then air-dried for at least 1 week to reduce its moisture content (MC) to around 8–10 % before feeding into vibrating screen to obtain 0.5 mm particle size. Screened particles were oven-dried for 24 h or until the MC reduces around 3–5 % prior to particleboard production.

Particleboards were made at three different board densities (500, 600, and 700 kg/m^3) and three levels of resin content (7, 9, and 11 %) based on oven-dried particles' weight. The particles were mixed with PF resin in the particleboard mixer and resin is sprayed at 0.35 MPa for even distribution on particle surface. The resinated particles were manually laid into 350 mm × 350 mm wooden mold lined by caul plate. Another caul plate was placed on particle mat after cold press and the mat was transferred to hot press machine at a temperature of 175 °C for 6 min. A 12-mm-thick stopper was used to control board thickness. The consolidated mat was condensed in conditioning room for at least 3 days before cut into samples for mechanical and physical testing.

Mechanical properties of samples were determined using Instron Universal Testing machine (Model 5569) and carried out in accordance with the Malaysian Standard (MS) for particleboard (MS 1787: 2005). Bending strength was tested on board samples measuring 300 mm × 50 mm and tested in accordance with MS 1787: Part 10 (2005a), while internal bond (IB) sample was carried out on square-shaped test piece measuring 50 mm × 50 mm. Thickness swelling (TS) samples with 50 mm × 50 mm size were submerged with face vertically in water with the temperature of 20 ± 1 °C for 24 h. The changes in thickness were calculated and expressed in percentage. IB and TS were tested in accordance with MS 1787: Part 11 (2005b) and MS 1787: Part 6 (2005c), respectively. Twelve samples for bending strength and 28 samples for IB and TS, respectively, were tested and analyzed.

The significance of density and resin content was statistically analyzed using the analysis of variance (ANOVA) and Duncan multiple range test (DMRT).

3 Results and Discussion

Average values of modulus of rupture (MOR), modulus of elasticity (MOE), IB, and TS of boards made from different densities and resin contents are shown in Table 1. Boards made at 700 kg/m^3 with 11 % resin content were observed to have highest values of MOR (24.86 MPa) and MOE (3382 MPa). These readings are 43 and 25 % higher than the lowest values of MOR and MOE displayed by 500 kg/m^3 board made with 7 % resin content. IB strength varies from 0.13 to 0.57 MPa and the highest value produced by 700 kg/m^3 with 11 % resin content. Based on MS, 14 and 2000 MPa are the minimum requirements for MOR and MOE, respectively, and 0.45 MPa is the minimum requirement for IB. IB of tested samples met minimum requirement at 700 kg/m^3 with 9 and 11 % resin content. The physical properties of OPF particleboard indicated poor ability to resist penetration of water as only four out of nine boards tested met the maximum thickness of swelling requirement. Only board made at 700 kg/m^3 with 9 and 11 % resin content met and passed the standard requirement.

Table 1 Average values of mechanical and physical properties

Density (kg/m^3)	Resin content (%)	Bending strength		IB (MPa)	TS (%)
		MOR (MPa)	MOE (MPa)		
500	7	6.91 (1.87)	1275 (488)	0.13 (0.03)	29.04 (1.88)
500	9	9.29 (1.04)	1458(472)	0.16 (0.02)	19.15 (0.73)
500	11	10.01 (0.99)	1967 (450)	0.27 (0.02)	24.00 (0.33)
600	7	11.50 (2.11)	2174 (342)	0.23 (0.03)	23.23 (1.30)
600	9	13.28 (2.25)	2374 (460)	0.23 (0.02)	14.68 (0.22)
600	11	14.22 (2.38)	2541 (503)	0.27 (0.02)	14.53 (0.55)
700	7	19.59 (2.09)	3027 (237)	0.33 (0.01)	27.67 (1.38)
700	9	21.34 (1.11)	3152 (253)	0.49 (0.04)	12.59 (0.81)
700	11	24.86 (3.05)	3382 (296)	0.57 (0.06)	9.26 (0.46)
MS minimum requirements furniture grade		14.00	2000	0.45	15.00

Values in parenthesis are the standard deviation

Effect of Board Density

Figure 1 displays the effect of varying board density on the bending properties. The values indicated the means and the letters (a, b, c) indicated statistical difference between samples according to Duncan multiple range test (DMRT). It seems that both MOR and MOE values positively increased and had significance effect with increasing board density. This is further supported by correlation coefficient (Table 2) with MOR ($r = 0.895^{**}$) and MOE ($r = 0.871^{**}$). Increasing board density

Fig. 1 Effect of density on mechanical properties

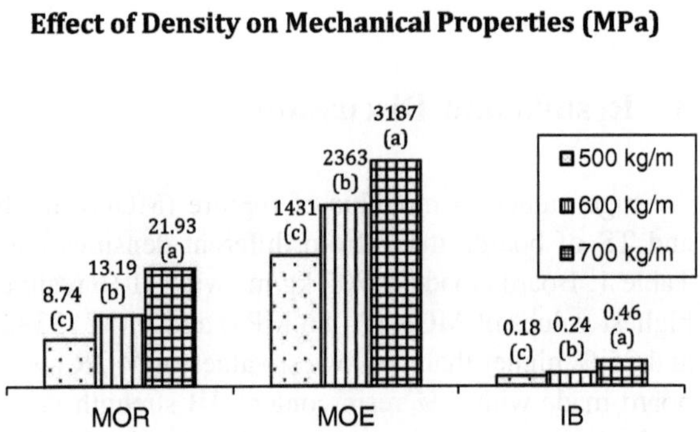

Table 2 Correlation analysis of effect of density and resin content on particleboard properties

	MOR	MOE	IB	TS
Density	0.895^{**}	0.871^{**}	0.828^{**}	-0.284^{**}
Resin content	0.251^{*}	0.167^{ns}	0.368^{**}	-0.846^{*}

*Correlation is significant at ($p < 0.05$), 2 tailed
**Correlation is significant at ($p < 0.01$), 2 tailed
ns not significant

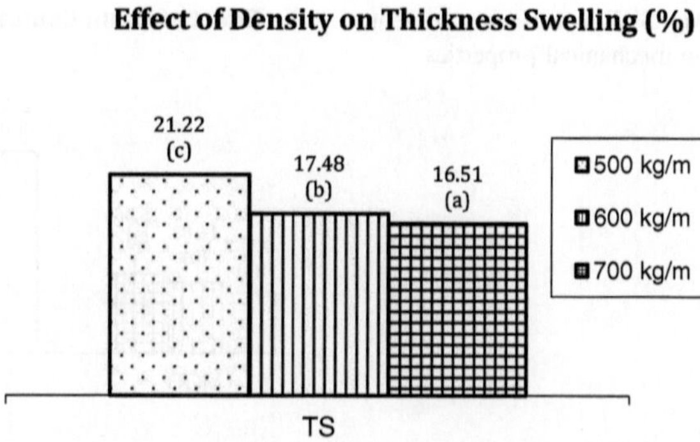

Fig. 2 Effect of density on thickness swelling

also significantly increases the IB values. Boards made at 700 kg/m^3 had highest value of IB with increment of 156 and 92 % compared to boards made at 500 and 600 kg/m^3, respectively. Result was further supported by correlation coefficient (Table 2) with IB (r = 0.828**) with significance at $p < 0.01$ level. Density improved all the mechanical properties of board due to compact arrangement of particles (Xu et al. 2013). The particles' compact arrangement brings surface of particle closer to one another and thus exhibits higher holding ability between particles (Zhang et al. 2010). This is mainly attributed by better interfacial adhesion of closely packed particles.

Effect on thickness swelling (TS) is shown in Fig. 2. Result observed that low thickness swelling was archived with higher density particleboard. This was further explained by correlation of density to TS (Table 2), showed negatively but significant (r = −0.284**). Xu et al. (2013) and Zhang et al. (2010) had similar result to this study and claim low thickness swelling at higher density because of compact arrangement. The compact arrangement of particles reduced the presence of voids that provide path for water molecules penetration to core of particleboard. Jamaludin (2006) meanwhile reported higher density resulted for better compaction and leads to stronger adhesive bond. Loh et al. (2010), however, show contradictory result. Their study on admixture of two wood species particleboard resulted in higher density which effectuates worsen thickness swelling due to the springback effect.

Effect of Resin Content

Figure 3 shows effects of resin content on the mechanical properties. Values of MOR and MOE directly increase with increasing resin content from 7 to 11 %. The increase in MOR and MOE with increasing resin content is due to the higher amount of resin comprehended at constant volume of the particle used. Thus, better coating of resin on particle surface and higher bonding site was provided between particles compared to lower resin content. MOR illustrates positive significance

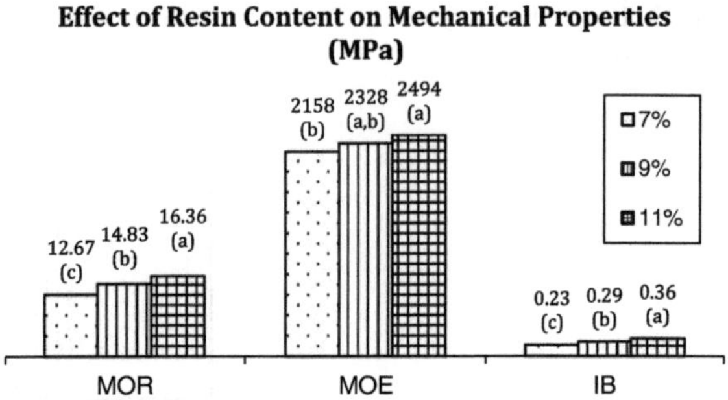

Fig. 3 Effect of resin content on mechanical properties

with correlation coefficient (Table 2) with ($r = 0.251^{*}$), while MOE shows positive effect but insignificant ($r = 0.167^{ns}$) with increasing resin content. According to Liao et al. (2013), the ability of boards to resist deflection or stiffness was determined by MOE. Higher resin content will increase the stiffness of board created by stronger bonding between particles.

Figure 3 demonstrates that IB increases with increasing resin content. IB properties of 11 % resin content exhibit the highest values; 57 % higher than IB exhibits by 7 % resin content board. The increase of IB with resin content is further revealed by correlation analysis (Table 2) with ($r = 0.368^{**}$). Higher IB strength of 11 % resin content board indicated better development of glue line between particles than lower resin content.

Table 2 shows that higher resin content negatively affected thickness swelling with correlation of coefficient with ($r = -0.846^{*}$). Figure 4 displays effect of resin content on thickness swelling. Higher resin content lowers the percentage of panels swelling in 24 h soaking. This study has similar result to previous study by Liao et al. (2013) and Bazyar et al. (2011) that stated lower resin content allows water to easily penetrate into panel through weakly formed bonded particles.

Two-way analysis of variance (ANOVA) was conducted to examine the effect of density, resin content, and their interaction on mechanical and thickness swelling properties of OPF particleboard at $p < 0.05$. The summary of ANOVA is shown in

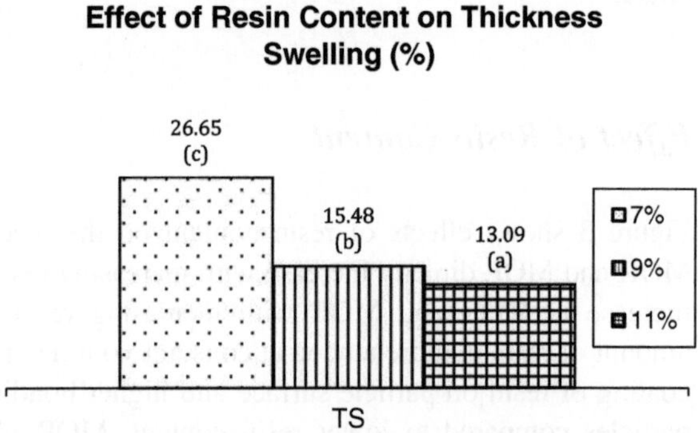

Fig. 4 Effect of resin content on thickness swelling

Table 3 Summary of ANOVA on mechanical and physical properties of particleboard

	MOR	MOE	IB	TS
Density	307.63	129.17	647.08	167.77
	(0.000)	(0.000)	(0.000)	(0.000)
Resin content	23.56	4.72	116.48	1422.98
	(0.000)	(0.92)	(0.000)	(0.000)
Density × Resin content	1.91	0.07	29.99	59.18
	(0.119)	(0.991)	(0.000)	(0.000)

The values indicate *F*—value. Numbers in parenthesis represent the *p* value

Table 3. All properties of particleboard found to be significantly affected by both density and resin content at $p = 0.000$. The interaction between variables on MOR and MOE was statistically insignificant at $F = 1.91$, $p = 0.119$ and $F = 0.07$, $p = 0.991$, respectively. Meanwhile, there were statistically significant interaction between the effect of density and resin content on IB and TS at $F = 29.99$, $p = 0.000$ and $F = 59.18$, $p = 0.000$, respectively. The means for IB and TS differ among board density for at least one level of resin content.

4 Conclusion

Effect of density and resin content on mechanical and thickness swelling properties of OPF particleboard were investigated. The experiment result indicated that increase in density and resin content directly increased the properties of particleboard. Higher density shows better compaction between particles, while higher resin content implies better coverage on particle surface that leads to stronger bond linkage formed between particles and stronger properties of board. The experimental result shows that boards made at 700 kg/m^3 with 9 and 11 % resin content meet all minimum and maximum requirements for both mechanical and physical properties.

Acknowledgement The authors thank Universiti Teknologi MARA Malaysia for UiTM Fellowship funding and Malayan Adhesives and Chemicals Sdn. Bhd for phenol formaldehyde provided.

References

Akgul M, Guler C, Copur Y (2010) Certain physical and mechanical properties of medium density fiberboards manufactured from blends of corn (*Zea mays induarata* Sturt.) stalks and pine (*Pinus nigra*) wood. Turk J Agrilc For 34:197–206

Bazyar B, Tichi AH, Rangavar H (2011) Particleboard made from fast growing aspen wood and old rail road ties. Aust J Basic Appl Sci 5(8):548–553

Boon JG, Hashim R, Sulaiman O, Hiziroglu S, Sugimoto T, Sato M (2013) Influence of processing parameters on some properties of oil palm trunk binderless particleboard. Eur J Wood Wood Prod 71:583–589

Hermawan D, Subiyanto B, Shuichi K (2001) Manufacture & properties of oil palm frond cement-bonded. J Wood Sci 47:208–213

Jamaludin K (2006) Properties of particleboard and thermoplastic board from buloh semantan (*Gigantochloa scortechinii*). University Publication Center (UPENA), Shah Alam

Kwon JH, Ayrilmis N, Han TY (2013) Enhancement of flexural properties and dimensional stability of rice husk particleboard using wood strands in face layer. Compos B 44:728–732

Liao CB, Deng YH, Wang XZ, Fan XL, Yu T, Yang Y (2013) Manufacture and mechanical properties of biocomposite made of reed and silvergrass. Appl Mech Mater 248:237–242

Loh YW, H'ng PS, Lee SH, Lum WC, Tan CK (2010) Properties of particleboard produce from admixture of rubberwood and mahang species. Asian J 3(5):310–316

Maloney, TM (1993) Modern particleboard & dry—process fiberboard manufacturing (2nd edn) Forest products society, pp 33–34. Miller Freeman, San Francisco

Mohd Sukhairi MR, Razak W, Othman S, Janshah M, Aminuddin M, Tamer AT, Izyan K (2011) Properties of composite boards from oil palm frond agricultural waste. BioResources 6 (4):4389–4403

MPOB (2014) Distribution of Oil Palm Planted Area by State: Dec 2013. Available from World Wide Web. http://www.mpob.gov.my. Accessed 1st Sept 2014

MS 1787: Part 10 (2005a) Determination of modulus of elasticity in bending and bending strength. Department of Standards Malaysia

MS 1787: Part 11 (2005b) Determination of Tensile Strength Perpendicular to Plane of the Panel. Department of Standards Malaysia

MS 1787: Part 6 (2005c) Determination of Swelling in Thickness After Immersion in Water. Department of Standards Malaysia

Wanrosli WD, Zainuddin Z, Law KN, Asro R (2007) Pulp from oil palm fronds by chemical processes. Ind Crops Prod 25:89–94

Xu X, Yao F, Wu X, Zhou D (2009) The influence of wax-sizing on dimensional stability and mechanical properties of bagasse particleboard. Ind Crops Prod 29(1):80–85

Xu X, Wu Q, Zhou D (2013) Influence of layered structure on physical and mechanical properties of kenaf core particleboard. BioResources 8(4):5219–5234

Zhang YH, Gu JY, Zuo YF, Di MW, Tan HY, Zhu LB (2010) Mechanical properties of wheat straw particleboard using composite adhesive. Adv Mater Res 113–116:2096–2099

Chapter 40
Effect of Different Pressing Times on Mechanical and Physical Properties of Phenol Formaldehyde Particleboard Made from Oil Palm Trunk

Ermadasila Mohamad and Jamaludin Kasim

Abstract The objective of this study was to examine the physical and mechanical properties of phenol formaldehyde particleboard made from oil palm trunk (OPT) at different pressing times. Three different pressing times were used in this study. The pressing times used in this research were 6, 8, and 10 min. The particle size used in this study was 2 mm and phenol formaldehyde (PF) was used as the binder. Particleboard with 10 min pressing time gave the highest mechanical (modulus of rupture (MOR), modulus of elasticity (MOE), and internal bonding (IB)) and physical properties (thickness swelling (TS) and water absorbtion (WA)). By increasing the pressing time, the properties were significantly increased. The above physical and mechanical properties were evaluated based on BS EN 310, 317, and 3261-1 standards.

Keywords Modulus of rupture · Modulus of elasticity · Internal bonding · Water absorbtion · Thickness swelling

1 Introduction

The efforts to develop particleboards had started since 1920s in the United States. At that time, failures in making particleboards were primarily due to the lack of suitable adhesives. Successful particleboard making started in the 1930s where the development of new thermosetting resin had been found. Then in 1941, the first

E. Mohamad (✉)
Faculty of Applied Science, Universiti Teknologi MARA,
Shah Alam, Selangor, Malaysia
e-mail: ermadasila21@gmail.com

J. Kasim
Faculty of Applied Sciences, Universiti Teknologi MARA,
Jengka, Pahang, Malaysia
e-mail: djamal@pahang.uitm.edu.my

N.A. Yacob et al. (eds.), *Regional Conference on Science, Technology and Social Sciences (RCSTSS 2014)*, DOI 10.1007/978-981-10-0534-3_40

industrial production of particleboards using synthetic resin started in Bremen, Germany. The importance of particleboard lies in utilizing of residue and low grade wood. Particleboard manufacturing is so flexible that it could be produced in large panel sizes with a full range of thickness (Moslemi 1974).

Nowadays, the main raw material in particleboard industry is the rubberwood, the scientific name is *Hevea brasiliensis*. However, the supply of rubber wood nowadays cannot meet demand of the particleboard industry anymore. The industry needs new materials to support the lack of these materials (Kasim et al. 2010). Rubberwood has been the main material in Malaysia to produce particleboard. The supply of rubber wood nowadays cannot meet demand of the particleboard industry anymore. The industry needs new materials to support the lack of these materials (Kasim et al. 2010).

Particleboard is the oldest composite panel that has been produced on the world and still remains until today as a furniture and structural material (Kasim et al. 2010). Particleboard has been in use since the 1940s, often used in place of the more expensive plywood as sub flooring or instead of natural hardwoods in furniture manufacturing. At the point in its history, however, particleboard was considered a costly designer material, reserved for use in exclusive homes and upscale furniture factories. Modern particleboard is now made primarily by combining discarded wood shavings, chip, and sawdust with a strong resin and pressing the mixture into serviceable boards and planks (Pizzi 1994).

However, in Malaysia particleboard commonly found in flatten pressed with board densities in the range from 600 to 800 kg m^{-3} with a thickness range of between 6 and 40 mm (Chew 1979). They can be classified into four different types according to their layers, which are single-layer homogenous, three layers, five layers, and graduated particleboard.

There is a need to search other raw materials in particleboard manufacturing due to the increasing price of wood. Therefore, an investigation of raw material other than wood and a study of particleboard manufacture without the synthetic adhesives had been carried out because of the need of formaldehyde-free particleboard and also the decreasing supply of raw material (Hashim et al. 2010).

Oil palm was first introduced in Malaysia in 1917. The growing demand for edible oil has resulted in the increase of area planted with oil palm. The oil palm plant takes 3 years to mature before it bears fruit. The economic life of the oil palm plant can extend to 25 years before being replanted. The oil palm plantation has increased from 461,791 hectares in 1975 to 4.85 million hectares in 2010, and the hectare will increase rapidly (oil palm statistics 2010). The mature oil palm plant is 7–13 m in height and has a reading diameter up to 65 cm (Teoh 2002).

Each year, a large number of oil palm biomasses were produced. These biomass yields are useful for permanent feedstock and renewable resource for higher value products. Biomass has a high potential to be one of the raw materials to supply of finished product which is traditionally been made from wood. Nevertheless, the biomass still awaits the commercial exploitation for the development of system for handling and processing (Kamaruddin et al. 1997).

2 Materials and Methods

The raw material used in this study was oil palm trunk. The oil palm trunk use was harvested from the oil palm plantation in FELDA Jengka 25, FELDA Jengka 24, and FELDA Ulu Jempol, Pahang. Then it was transported to UiTM Pahang using a lorry. The adhesive used was phenol formaldehyde (PF) which was made available by Malaysian Adhesive and Chemical Sdn. Bhd. (MAC) in Shah Alam, Selangor. The oil palm tree was harvested after 25 years old for replacement planting.

Phenol formaldehyde (PF) was used as the binder in this study. PF was supplied by the private company at Shah Alam Selangor. The oil palm trunk particle was mixed together with PF using the laboratory mixer. 11 % resin content has been added to the mixture. Then the mixture was formed into mat size of 350 mm × 350 mm using forming box and then prepressed by a cold press. The mat was then hot-pressed at 175 °C. Three different pressing times were used in this study, and the hot-press times were 6, 8, and 10 min. 12-mm stopper was used in this study to form 12-mm-thick boards. The board was conditioned in a climate chamber at temperature of 20 °C and relative humidity of 65 %. The board was cut into the sample size based on BS EN standard. The testings conducted in this study were internal bonding (IB), modulus of rupture (MOR), modulus of elasticity (MOE), thickness swelling (TS), and water absorption (WA).

3 Results and Discussion

Mechanical Properties

The effect of pressing time on mechanical properties of particleboard is shown in Fig. 1. The MOR value significantly increases with increasing pressing time. By increasing the pressing time from 6 to 10 min, the MOR values increase about 24.71 %. Table 1 shows that the correlation analysis is significant at $R = 0.262^*$. At higher pressing time, more heat is spread through the thickness of the board providing enough heat to cure the resin thus giving the higher MOR value.

The effect of pressing time on modulus of elasticity (MOE) of particleboard is also shown in Fig. 1. The MOE values significantly when pressing time increase. By increasing the pressing time from 6 to 10 min, the MOE increases about 15.66 %. Table 1 shows that the correlation analysis is significant at $R = 0.220^*$. At higher press time, more heat is spread through the thickness of the board providing enough heat to cure the resin, thus giving the higher MOE value. The bond between resin and the particle will be improved, and the resin will cure perfectly.

The effect of pressing time on internal bonding (IB) properties is also shown in Fig. 1. The IB value significantly increased with increment in hot-pressed time. By increasing the board density from 6 to 10 min, the IB increases about 39.62 %. 6 min press time to show 0.53 Mpa of IB and 8 min press time to show the strength

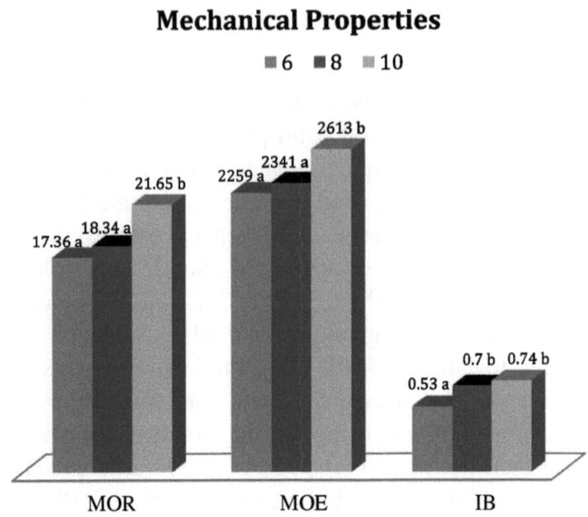

Fig. 1 Effect of pressing time on mechanical properties of particleboard

Table 1 Correlation analysis

SOV	MOR	MOE	IB	TS	WA
Press time	0.262*	0.220*	0.320*	−0.19ns	−0.024ns

ns not significant
*Significant

are increased to 0.7 Mpa and 10 min press time to show 0.74 MPa. This significant result for IB was supported by the correlation analysis (R = 0.320*) as shown in Table 1.

Increasing the press time will improve the mechanical properties of the particleboard. This is because of the increasing bond between the particles and hardening of the resin efficiency during hot pressing. Similar result was recorded in the previous studies by Alireza and Amir (2008).

Physical Properties

The effect of pressing time on thickness swelling (TS) of particleboard is also shown in Fig. 2. The TS properties were decreased with decrease in board density. By increasing the press time from 6 to 10 min, the TS value improved about 1.81 %. 6 min press time shows 13.29 % of TS, but 8 min press time shows that the percentage is decreased to 12.87 %. However, the percentage is slightly increased to 13.05 % for 10 min press time. All the TS values failed to meet the minimum value of BS EN standards of 14 %. This improvement is future reviled by the correlation analysis in Table 4.3 where $R = -0.19$ns.

The effect of pressing time on water absorption (WA) of particleboard is also shown in Fig. 2. The WA values decrease with increase in press time. By increasing

Fig. 2 Effect of pressing time on physical properties of particleboard

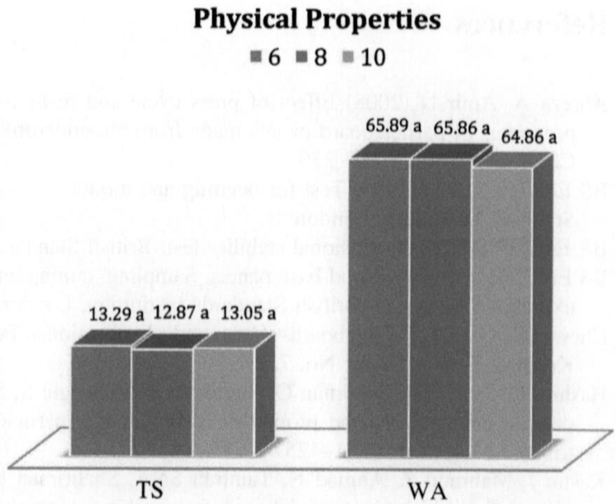

the press time from 6 to 10 min, the WA improved about 1.56 %. 6 min press time boards to show 65.89 % of WA, and 8 min pressing time to show that the percentage is decreased to 65.86 %. But in 10 min press time, the percentage of WA is slightly dropped to 64.86 %. This was further supported by the correlation analysis significant of $R = 0.024$ns.

Board containing the high amount of particle increases the particle bonding and higher compaction ratio and with the present, the less the voids, the better the board stability is created. Kasim et al. (2010), Xu and Suchsland (1999) also reported similar relationship between board density and TS.

4 Conclusion

Effect of hot-pressed time on some of the basic properties of phenol formaldehyde particleboard panels made from oil palm trunks was investigated. The results showed that the hot-press time cycle has an influence on the properties of the samples. Overall, the panels met strength requirements based on BS standards. However, thickness swelling of the samples needs to be improved using either some kind of chemical, steam treatment, or adding of some wax in the furnish. The study shows that the best hot-pressed time is 10 min.

Acknowledgement We would like to acknowledge Universiti Teknologi MARA Shah Alam and Universiti Teknologi MARA Pahang for the support and assistance for this study.

References

Alireza A, Amir N (2008) Effect of press cycle and resin content on physical and mechanical properties of particleboard panels made from the underutilized low-quality raw materials. Ind Crops and Prod 28:225–230

BS EN 310 & 319 (1993) Test for bending and modulus of elasticity, and internal bond. British Standard Institution, London

BS EN 317 (1993) Dimensional stability test. British Standard Institution, London

BS EN 3261-1 (1994) Wood base panels. Sampling, cutting and inspection. Part 1. Sampling and expression of results. British Standards Institution, London

Chew LT (1979) Particleboards–Uses and Applications. Forest Research Institute Malaysia Kepong. Timber Digest No. 7.3

Hashim R, Saari NH, Sulaiman O, Sugimoto T, Hiziroglu S, Sato M, Tanaka R (2010) Effect of particle geometry on the properties of binder less particleboard manufacture from oil palm trunk. J Mater Des 4251–4257

Kasim J, Mahmud Z, Ahmad N, Tamiran SNA, Shahriman N, Razak NA (2010) Properties of phenol formaldehyde particleboard from oil palm trunk particles. Paper Presented at the XXI IUFRO conference in Seoul, South Korea

Kamaruddin H, Mohamad HD, Arifin D, Jalani S (1997) An estimated availability of oil palm biomass in Malaysia: No. 37, Occasional Paper, PORIM

Malaysia Oil Palm Statistic (2010) Malaysia Palm Oil Board

Moslemi AA (1974) Particleboard Volume 1: Materials. Southern Illionis University Press, USA, pp 785–863

Pizzi A (1994) Advanced wood adhesives technology. Marcel Dekker Inc, New York

Teoh CH (2002) The palm oil industry in Malaysia. From Seed to Frying Pan, WWF Switzerland

Xu W, Suchsland O (1999) Within-panel variability and selected property relationships of particleboard from single- and mixed-species processes. Forest Prod J 49: 36–40

Chapter 41
Properties of Particleboard from Kelempayan (*Neolamarckia cadamba*) Wood

Nur Sakinah Mohamed Tamat, Nur Farahin Yusoff, Jamaludin Kasim and Wan Mohd Nazri Wan Abdul Rahman

Abstract Growing demand for wood based panels has led to a shortage of wood, especially rubberwood, which has been the main raw material in the production of the Malaysian furniture and panel board industries. Kelempayan has potential to be used as alternative resources for the coming years because it is a fast growing species and can be cultivated in Malaysia as tropical country. This study was conducted to explore the potential of kelempayan for particleboard manufacture and to characterize water resistance and mechanical properties of kelempayan particleboard as affected by various resin contents and board densities. Single layer particleboard was fabricated from 2 mm particle size and bonded with phenol formaldehyde (PF) adhesive at 145 °C of hot pressing temperature. Three different levels of resin contents (7, 9, 11 %) and board densities (500, 600, 700 kg/m^3) were used as variable factors. The experimental panels were tested for modulus of elasticity (MOE), modulus of rupture (MOR), internal bonding strength (IB) and water resistance according to the procedures defined by Malaysian Standard. Overall results showed that samples made from density 700 kg/m^3 and 11 % resin content had the highest MOE and MOR value compare to others. Internal bonding strength also indicated the similar pattern. However, thickness swelling of the panels were very poor than requirements. We concluded that kelempayan can have potential for particleboard manufacture and promising better qualities especially for exterior purposes. Dimensional stability need to be enhanced by using various treatments or addition of chemical such as wax.

Keywords Kelempayan · Particleboard · Phenol formaldehyde

N.S. Mohamed Tamat (✉) · N.F. Yusoff · J. Kasim · W.M.N. Wan Abdul Rahman
Faculty of Applied Sciences, Universiti Teknologi MARA, Shah Alam, Selangor, Malaysia
e-mail: sakinah292@yahoo.com

J. Kasim
e-mail: djamal@pahang.uitm.edu.my

N.S. Mohamed Tamat · N.F. Yusoff · J. Kasim · W.M.N. Wan Abdul Rahman
Faculty of Applied Sciences, Universiti Teknologi MARA, Jengka, Pahang, Malaysia

N.A. Yacob et al. (eds.), *Regional Conference on Science, Technology and Social Sciences (RCSTSS 2014)*, DOI 10.1007/978-981-10-0534-3_41

1 Introduction

Neolamarckia cadamba is locally known as Kelempayan belongs to the family of *Rubiaceae*. It is a fast-growing tree species with a tall and straight bole. The timber is light creamy yellowish colour and classified under Light Hardwood in Malaysia (Ismail et al. 1995). Lim et al. (2005) stated that, this species are distributed in lowlands to mountain forest up to 1000 m altitude. They are also frequently found by streams and in open sites in the forest. The timber is widely use in moulding, general utility furniture, veneer and plywood (Lim et al. 2005). As it is a fast-growing species, it could become alternative resources to support the wood industry. Recently, short rotational plantations and shorter forest management periods have been practiced all over the world to balance between supply and demand of wood (Guler et al. 2007). These methods also help to protect natural resources like soil, water and wildlife (Nourbakhsh 2008).

Particleboard consists of discrete wood particles of various sizes that are bonded together with synthetic glue under heat and pressure (Colak et al. 2010). It is used in furniture, wall and ceiling panels, office dividers, bulletin boards, flooring, cabinets, counter tops and desk tops (Wang et al. 2007). The demand for particleboard has increased especially for both construction and industrial production (Sari et al. 2012a). Nemli et al. (2008) mentioned that the strength properties of particleboard are more consistent compare with natural lumber. Particleboard can utilize low-grade logs such as thinning, bowed and twisted logs. Its properties also can be engineered.

There are many factors which can affect the performance of particleboard. Ayrilmis et al. (2012) in their study found that one of the most successful ways to improve dimensional stability and mechanical properties of wood-based panels is to increase resin content. Study by Ashori and Nourbakhsh (2008) also proved that resin content is the main parameter which can create significant impact on the dimensional stability and mechanical properties of particleboard. One of the major considerations in the manufacturing particleboard is to maintain necessary physical and mechanical properties upon where it will use. Siti Norbaini et al. (2013) carried out experiments on mechanical properties of homogeneous and heterogeneous three layered particleboard composite in relation to different resin content. It was found that density is one of the parameter that indicated the properties of wood based panel. Another previous work (Rachtanapun et al. 2012) also reported that physical and mechanical properties of particleboard increased with increasing board density.

The present study emphasizes on utilization of kelempayan as alternative resources for manufacturing particleboard. Furthermore the study relating on that is very rare and has not been study in detail. This work also conducted to study the properties of particleboard made from kelempayan wood focusing on the effect of resin content and board density. Both mechanical and physical properties of such composite panels were evaluated.

2 Materials and Methods

Sample Preparation and Particleboard Manufacturing

Kelempayan trees with the diameter of breast height (DBH) of 35–45 cm were used in this study. The trees were harvested from UiTM Pahang Forest Reserve. The felled trees were cross-cut into eight-foot bolts before they were sawn into 1 in. × 1 in. × log length of planks. Later the logs were fed into wood chipper to produce chips and then flaking into small particles using knife ring flaker. The particles were air dried for a week in a shaded area. Kelempayan particles then screened into 2 mm particle size using a vibrating screener. The screened particles were dried in an oven having temperature of 60 °C for 24 h to reduce the moisture content to less than 5 % moisture content. Single layered particleboard was fabricated at density 500, 600 and 700 kg/m^3. Phenol Formaldehyde (PF) was used as adhesive. The adhesive was supplied by Malayan Adhesive Company Sdn. Bhd., Shah Alam, Selangor. The properties of the PF adhesive are given in Table 1.

A weighted amount of kelempayan particles were blended with 7, 9 and 11 % of PF adhesive. The resulting mat was formed manually to a size 35 × 35 × 1.2 cm. The mat was pre-pressed with a cold press at 1000 psi for 1 min before being hot pressed to the required thickness for 6 min at 145 °C. The hot press used removable steel stopper to achieve a constant thickness of particleboard. All boards were kept at 20 °C and 65 % relative humidity in a conditioning room for 1 week before they were cut into various sizes for property evaluation.

Mechanical and Physical Testing

Finished particleboards were cut into required sizes for testing modulus of elasticity (MOE), modulus of rupture (MOR), internal bond strength (IB) and dimensional stability. The samples were evaluated based on Malaysian Standards. Three samples with dimensions 30 × 5 × 1.2 cm from each board were used for bending strength test according to MS1787: part 10 (2005a) and MS1787: part 11 (2005b). Bending strength included MOE and MOR. Seven samples with dimensions 5 × 5 × 1.2 cm from each type of panel were used for internal bond test. Both tests were conducted using an Instron Universal Testing Machine Model UiTM-5569 with movable crosshead speed 10 mm/min for bending strength test and 1.5 mm/min for IB test.

Table 1 Properties of phenol formaldehyde (PF) adhesive

Property	Phenol formaldehyde
pH at 30 °C	12.62/m
Viscosity at 30 °C	0.43p
Specific gravity at 30 °C	1.182
Solid content (%)	40.3

The thickness swelling tests were carried out on seven samples cut from each particleboard with dimensions 5 × 5 × 1.2 cm. Dimensional stability of the composite was measured following the MS1787: part 6 (2005c). The thickness of the samples was measured before soaking in water for 24 h. The water was changed after each test. After immersion time has elapsed, the samples were taken out of water and excess water was removed. The thickness of each sample was measured within 10 min after removal from the water. The thickness swelling was calculated as shown in Eq. (1).

$$\text{Thickness Swelling, TS } (\%) = \frac{t_2 - t_1}{t_1} \times 100 \quad (1)$$

where:

t_1 the thickness of test piece before immersion, in mm and

t_2 the thickness of test piece after immersion, in mm

Data for each test were statistically analysed. The effects of resin content and board density on the particleboard properties were evaluated by analysis of variance (ANOVA). When the ANOVA indicated a major difference among factors and levels, a comparison of the means was done employing Duncan Multiple Range Test to identify which groups were considerably different from other groups at 95 % confidence level.

3 Results and Discussion

The average values of mechanical and physical properties of particleboard from kelempayan are presented in Table 2. The data obtained in this research showed that improved MOE, MOR, IB and water resistance with increasing resin content. Most panels made with 11 % resin content had the highest values of MOE, MOR and IB compared with 7 and 9 % resin content. Thickness swelling data ranged from 10 to 37 % after soaking for 24 h. Better dimensional property was indicated by the sample of 11 % resin content while the panel with 7 % resin content showed poor dimensional stability. As shown clearly in Table 2, the MOE, MOR and IB values of the boards increased with increased in board density. However, increased in panel density resulted in poor thickness swelling property. Particleboard with a density 700 kg/m^3 exhibited superior mechanical properties compared to the particleboard with the density 500 and 600 kg/m^3. However, samples made from 700 kg/m^3 gave the highest TS values compared with 500 and 600 kg/m^3. As stated on Malaysian Standards 1036 (2006), particleboard should have a maximum TS value of 15 % for 24 h immersion. According to the test results, all of the samples were higher than the Malaysian Standard requirements except for the panels of 500 kg/m^3 with 11 % resin content and 600 kg/m^3 with 9 % resin content were

Table 2 Mechanical properties and water resistance of the particleboard from kelempayan

Board Density	Resin (%)	MOE (MPa)	MOR (MPa)	IB (MPa)	TS (%)
	7	1512	10	0.36	21
500	9	1607	11	0.35	19
	11	2022	16	0.41	13
	7	2297	18	0.45	33
600	9	2196	17	0.48	10
	11	2849	22	0.67	17
	7	2984	26	0.66	37
700	9	3216	27	0.58	26
	11	3122	27	0.60	21
Furniture grade for use in humid condition (MS Standard 1036:2006)		2000	14	0.45	15

Notes MOE modulus of elasticity, *MOR* modulus of rupture, *IB* internal bonding, *TS* thickness swelling

Table 3 Analysis of variance (ANOVA) on the effect of board density and resin content on the particleboard properties

SOV	Df	MOE (MPa)	MOR (MPa)	IB (MPa)	TS (%)
Density (D)	2	318.00*	412.25*	425.79*	320.73*
Resin content (RC)	2	34.78*	36.65*	65.99*	611.51*
D × RC	4	8.52*	6.44*	53.03*	97.46*

Notes *F-values are significant at $p < 0.05$
SOV source of variance, *Df* degree of freedom, *MOR* modulus of rupture, *MOE* modulus of elasticity, *IB* internal bonding and *TS* thickness swelling

found to comply with the standard requirements. Table 3 illustrates the ANOVA on the effect of board density and resin content on the particleboard properties. Resin content (RC) and board density (D) significantly affect all board properties. Their interaction (D × RC) also shows significant effects on all the board properties.

Effect of Board Density

Figures 1 and 2 indicated the Duncan Multiple Range Test (DMRT) on the effect of board density on mechanical and physical properties of kelempayan particleboard. It is well known fact that mechanical properties of particleboard have improved with increasing panel density. It is evident that MOE, MOR and IB values increase significantly with increasing board density as can be seen in Fig. 1. The reason for this behaviour is attributed to the tighter structure, more compact and low porosity of the particleboard at high-density compared to the lower density boards (Nemli et al. 2005). High density board consist of high amount of wood material. In

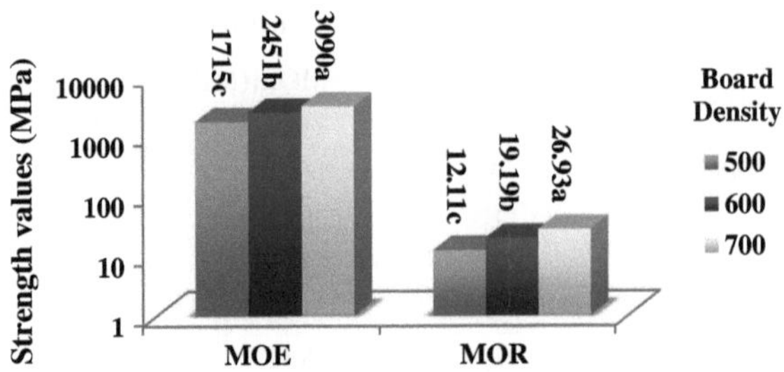

Fig. 1 Effects of board density on MOE and MOR

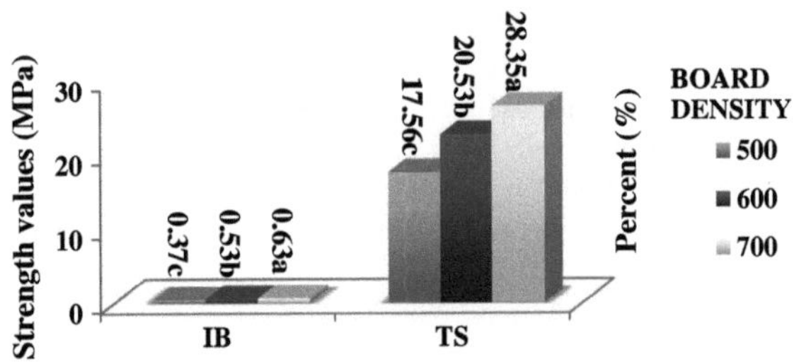

Fig. 2 Effects of board Density on IB and TS

addition, the resin is used based on the weight of the particles. For this reason, specimens at 700 kg/m^3 density had higher amount of resin compared to panels at 500 and 600 kg/m^3 which resulted in tighter and more compact structure of the panels. Previous work also found that high density particleboards had lower porosity so that particles and adhesive can interact with each other more easily to form stronger crosslink compared to low density particleboards (Zheng et al. 2007). Youngquist (1999) mentioned that choosing a proper particleboard density is a very important step in particleboard industry so that proper density can be determined based on the intended application requirements.

Figure 2 shows the effect of varying board density on thickness swelling properties. Results revealed that the increase in board density leads to the increase in thickness swelling. Thickness swelling of the samples showed significant difference from each other. Based on the findings of this work, particleboard with the density 700 kg/m^3 showed the greatest board swelling. However all the samples produced failed to meet the maximum TS value of 15 % for 24 h immersion as stated in the Malaysian Standards. The negative influence of the board density on the TS values of particleboard might be due to larger amount of wood particles required to produce 700 kg/m^3 particleboard compare to those 500 and 600 kg/m^3 boards. This trend is agrees with previous studies by (Sari et al. 2012b; Mohd Hazim et al. 2013). Bowyer et al. (2007) also stated that dimensional stability may be adversely affected by increased board density. Wood is a porous structure and tends to absorb liquids. Colak et al. (2010) in their study found that high density wood absorbs more water than wood of low density because of more woody cell membrane presence in the high density wood. Moreover, in this study no hydrophobic substance or wax has been added during panel manufacturing.

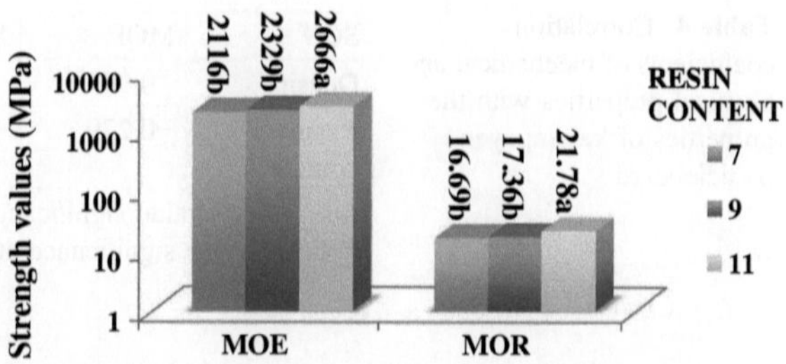

Fig. 3 Effect of resin content on MOE and MOR

Taramian et al. (2007) mentioned that the addition of wax (0.5–1 %) to the mixture of adhesive and particles during manufacturing process may improve water resistance. Heat treatment and chemical treatments also could be considered to enhance thickness swelling and water absorption (Mohd Hazim et al. 2013).

The effects of resin content on mechanical and dimensional properties are depicted in Figs. 3 and 4. It seems that using high resin content enhanced overall mechanical properties as compared to those with low resin content. The qualities of particleboards were significantly improved with the increase of resin content. It is assumed that the bonding between wood particles of 7 % resin content was not as strong as 9 and 11 % resin amount, therefore resulted in lower strength values of the samples. Compared to low percentage of resin content, high percentage of resin content had better bonding as more resin is available for inter-particle bonding (Jamaludin 2006). Research conducted by (Ashori and Nourbakhsh 2008; Jamaludin et al. 2010) also found similar resin content relationship. No significant differences were determined between MOE, MOR and IB values of particleboard with 7 and 9 % of resin content. However, the MOE, MOR and IB values from specimens made of 11 % resin content were significantly different from board made with 7 and 9 % resin content. This observation support the theory of resin content increment will increase the mechanical properties of wood panels (Colak et al. 2010). Higher resin content means higher production cost. Therefore, proper levels of resin content should be determined for specific applications with considerations of reasonable cost.

The effects of resin content on the physical properties are shown in Fig. 4. Resin content was found to be significantly effective on the TS of panels. The quantity of resin used to manufacture a board is the major factor determining strength and

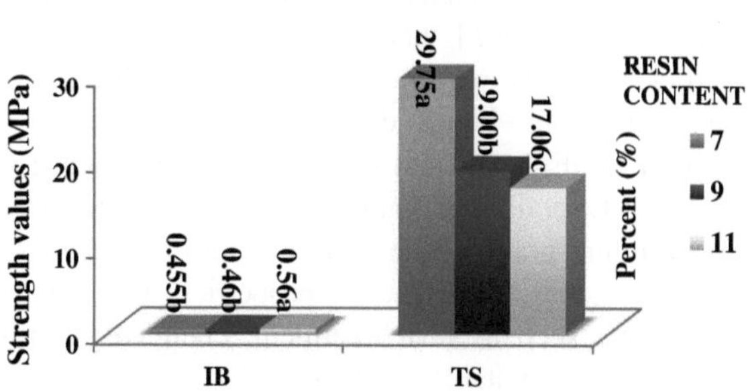

Fig. 4 Effect of resin content on IB and TS

Table 4 Correlation coefficients of mechanical and physical properties with the properties of kelempayan particleboard

SOV	MOE	MOR	IB	TS
Density	0.903**	0.907**	0.817**	0.461**
Resin content	−0.279*	−0.244	−0.182	0.637**

Note *shows value significance at $P < 0.05$
**shows value significance at $P < 0.01$

dimensional properties. It was observed that higher resin content has been significantly improved thickness swell behaviour of particleboard. This is related to a higher degree of cross linking resulting in retarded water penetration (Rachtanapun et al. 2012). Based on the tested ranged, samples with 7 % resin content were less dimensional stability compared to samples with 9 and 11 % resin content. The thickness swelling values in this work were relatively high owing to the porous structure of the lower resin content board, which absorbed more water compared to the compressed higher resin content board. Nur Farahin et al. (2014) reported that low amount of resin leads to lack of bonding site between particles. Consequently, weaker bond form and allow water molecule to penetrate into the specimens.

The correlation coefficients of mechanical and physical properties with the properties of board are tabulated in Table 4. The correlation analysis revealed that all mechanical and physical properties showed positive correlation and significantly correlated with the density of the board. As can be seen from Table 4, strength properties negatively correlated with the resin content of the board produced. However dimensional stability showed a positive correlation ($r = 0.637^{**}$) and significantly correlated with the resin content.

4 Conclusion

Kelempayan could offer a great potential as new materials for the production of particleboard composites with high performance and suitable for exterior purposes or where the water resistant are needed. The study indicated that the board density and resin content were the main parameters influencing the physical and mechanical properties of the panels. According to the test results, all mechanical and physical properties of the panels were improved when the resin content increased. The results also showed that board density have an effect on strength properties of particleboard. However, the thickness swelling of the panels was very poor and suggesting that board density adversely influences dimensional stability. The highest MOE, MOR and IB were reached at density 700 kg/m^3 and a resin content of 11 %. Overall, most panels made in this study satisfied the strength requirements for furniture grade use in humid condition as stated in the Malaysian Standards. In further studies, dimensional properties need to be enhanced by using various treatments such as heat treatment, chemical modification or addition of wax to become a more stable product.

Acknowledgements The authors would like to thank Institute of Graduate Studies of Universiti Teknologi MARA Malaysia for providing funding support. In addition, the authors thank Malayan Adhesive Company (MAC) Sdn. Bhd. Shah Alam Selangor for providing adhesive for this research. The authors also acknowledged Universiti Teknologi MARA Pahang for providing raw materials, necessary facilities and much guidance.

References

Anonymous (2005a) Malaysian Standard MS 1787: part 10: 2005, Wood-based panels—determination of modulus elasticity in bending and bending strength. Department of Malaysian Standard, Putrajaya

Anonymous (2005b) Malaysian Standard MS 1787: part 11: 2005, Wood-based panels—determination of tensile strength perpendicular to the plane of the panel. Department of Malaysian Standard, Putrajaya

Anonymous (2005c) Malaysian Standard MS 1787: part 6: 2005, Wood-based panels—determination of swelling in thickness after immersion in water. Department of Malaysian Standard, Putrajaya

Anonymous (2006) Malaysian Standard MS 1036: 2006, Wood-based panels—particleboards-specification (First Revision). Department of Malaysian Standard, Putrajaya

Ashori A, Nourbaksh A (2008) Effect of press cycle time and resin content on physical and mechanical properties of particleboard panels made from the underutilized low-quality raw materials. Ind Crops Prod 28:225–230

Ayrilmis N, Kwon JH, Han TH (2012) Effect of resin type and content on properties of composite particleboard made of a mixture of wood and rice husk. Int J Adhes Adhes 38:79–83

Bowyer JL, Shmulsky R, Haygreen JG (2007) Forest products and wood science, an introduction. Blackwell Publishing, USA

Colak S, Nemli G, Demirkir C, Aydin I, Demiral S (2010) Utilization potential of waste from window joints for particleboard. J Compos Mater 45(1):29–37

Guler C, Copur Y, Akgul M, Buyuksari U (2007) Some chemical, physical and mechanical properties of juvenile wood from black pine (*Pinus nigra Arnold*) plantations. J Appl Sci 7 (5):755–758

Ismail J, Jusoh MZ, Mohd Sabri H (1995) Anatomical variation in planted kelempayan (*Neolamarckia cadamba rubiaceae*). IAWA J 16(3):277–287

Jamaludin K (2006) Properties of particleboard and thermoplastic board from buluh semantan (*Gigantochloa scortechinii*). University Publication Centre (UPENA), Shah Alam

Jamaludin K, Zalifah M, Nurrohana A, Siti Nor Ain T, Nor Suziana S, Nor Ashikin R (2010) Properties of phenol formaldehyde particleboard from oil palm trunk particles. Paper Presented at theXXI IUFRO Conference in Seoul, South Korea

Lim SC, Gan KS, Thi BK (2005) Identification and utilization of lesser-known commercial timbers in Peninsular Malaysia. For Res Inst Malaysia 32:139–258

Mohd Hazim MA, Rokiah H, Hiziroglu S, Nurul Syuhada S, Othman S (2013) Properties of particleboard made from rubberwood using modified starch as binder. Compos Part B 50:259–264

Nemli G, Ozturk I, Aydin I (2005) Some of the parameters influencing surface roughness of particleboard. Build Environ 40:1337–1340

Nemli G, Yildiz S, Gezer ED (2008) The potential for using the needle litter of scotch pine (*Pinussylvestris* L.) as a raw material for particleboard manufacturing. Bioresour Technol 99:6054–6058

Nourbakhsh A (2008) Mechanical and thickness swelling of particleboard composites made from three-year-old poplar clones. J Reinf Plast Compos. doi:10.1177/0731684408097771

Nur Farahin Y, Jamaludin K, Hazwani L, Muslyza CH, Nor Farhana J (2014) Evaluation of 3-layer oil palm frond particleboard. Int J Latest Res Sci Technol 3(6):183–186

Rachtanapun P, Sattayarak P, Ketsamak N (2012) Correlation of density and properties of particleboard from coffee waste with urea-formaldehyde and polymeric methylene diphenyl diisocyanates. J Compos Mater 46(15):1839–1850

Sari B, Ayrilmis N, Nemli G, Baharoglu M, Gumuskaya E, Bardak S (2012a) Effects of chemical composition of wood and resin type on properties of particleboard. Lignocellulose 1(3):174–184

Sari B, Nemli G, Baharoglu M, Bardak S, Zekovic E (2012b) The role of solid content of adhesive and panel density on the dimensional stability and mechanical properties of particleboard. J Compos Mater. doi:10.1177/0021998312446503

Siti Norbaini S, Shaikh Abdul Karim YZ, Jamaludin K (2013) Mechanical properties of homogeneous and heterogeneous three layered particleboard composite in relation on different resin content. Mater Sci Chem Eng 699:637–640

Taramian A, Doosthoseini K, Sayyed Ahmad M, Faezipour M (2007) Particleboard manufacturing: an innovative way to recycle paper sludge. Waste Manage 27:1739–1746

Youngquist JA (1999) Wood Based Composites and Panel Products. In Wood Handbook: Wood as an Engineering Material. Gen. Tech. Rept. FPL-GRT-113. USDA Forest Serv., Forest Prod. Lab., Madison, WI, pp. 1–31 (Chapter 10)

Wang SY, Yang TH, Lin LT, Lin CJ, Tsai MJ (2007) Properties of low-formaldehyde emmision particleboard made from recycled wood-waste chips sprayed with PMDI/PF resin. Build Environ 42:2472–2479

Zheng Y, Pan Z, Zhang R, Jenkins BM, Blunk S (2007) Particleboard quality characteristics of saline jose tall wheat grass and chemical treatment effect. Bioresour Technol 98:1304–1310

Chapter 42
Effect of Different Portion on Calorific Value, Ash Content, and Specific Gravity of *Leucaena leucocephala* Wood

Nur Saidah Nordin, Junaidah Md Sani, Jamaludin Kasim and Wan Mohd Nazri Wan Abdul Rahman

Abstract The effect of tree portion on the calorific value, ash content, and specific gravity of *Leucaena leucocephala* wood was investigated using 2-year-old tree. Wood samples of *Leucaena leucocephala* were taken from three portions, namely top, middle, and bottom. Sampling was carried out according to the standard of Technical Association of the Pulp and Paper Industry (TAPPI) for ash content and calorific value was measured using bomb calorimeter. The results indicated that calorific value was found to be higher at the bottom and the trend increased from top to bottom portion. Specific gravity shows similar pattern to calorific value. However, the ash content of *Leucaena leucocephala* wood decreased from bottom to top portion. This study reveals that tree portion showed no significant effect on the ash content, specific gravity, and calorific value. This implied that the whole tree stem of 2-year-old *Leucaena leucocephala* wood at 2 years old could be used for source of biomass for renewable energy material.

Keywords Ash · Calorific · Energy · Renewable · Thermochemical

N.S. Nordin (✉)
Faculty of Applied Sciences, Universiti Teknologi MARA, Shah Alam, Selangor, Malaysia
e-mail: anursaidahnordin@gmail.com

J. Md Sani · J. Kasim · W.M.N. Wan Abdul Rahman
Faculty of Applied Sciences, Universiti Teknologi MARA, Jengka, Pahang, Malaysia
e-mail: ajun@pahang.uitm.edu.my

J. Kasim
e-mail: djamal@pahang.uitm.edu.my

W.M.N. Wan Abdul Rahman
e-mail: wmdnazri@pahang.uitm.edu.my

N.A. Yacob et al. (eds.), *Regional Conference on Science, Technology and Social Sciences (RCSTSS 2014)*, DOI 10.1007/978-981-10-0534-3_42

1 Introduction

Fossil fuels namely oil, coal, and natural gas are currently the primary sources of energy used to generate electricity and power the transport sector. However, these fuel resources are considered nonrenewable on the ground that natural processes that produce them take millions of years. Therefore, in order to sustain the comfort of modern life and power the high-end needs of transportation industry, there should be efforts to replace them with other energy sources that are renewable. Undeniably, the burnings of some fossil fuels come with a package of hazardous gasses such as carbon monoxide, sulfur dioxide, and nitrate aerosol that would go into the atmosphere, gradually polluting the environment as well as increasing the risk of lung diseases (Rom and Ryon 2011). Biomass is the most common form of renewable energy source. Its capability to be utilized directly as heat, even though is always underestimated, cannot be belittled. In fact, the biomass could actually be manipulated to produce mechanical energy using both thermochemical and biochemical conversion technologies. Furthermore, it generates comparatively lower levels of atmospheric pollutant. Woody biomass has long been used as firewood especially in the remote civilization and even by the most isolated community in the world. In developing countries, woody biomass is one of the major sources of energy. The use of woody biomass as firewood is even mentioned in the holy Qur'anic verses Chapter 36: 81, "He Who produces for you fire out of the green tree, and behold, you kindle from it."

Leucaena leucocephala is a tree that can be found in most tropical countries. It is a fast growing woody tree and produced plentiful of seeds, thus is a potential biomass source for renewable energy. On the other hand, the planting of these trees would protect from gradual wearing of soil. Compared to fossil fuels, plants grown for energy crops absorb the amount of carbon dioxide (CO_2) released during their combustion. Hence, the reducing of CO_2 will assist in controlling the air pollution problems arising from the greenhouse effect (Yazdani et al. 2012). Economically, the planting of *Leucaena leucocephala* will put underutilized agricultural land to use, thus benefits the farmers by providing additional source of income since *Leucaena leucocephala* can be used to generate electricity via steam boiler, gasification, and combustion. The potential of *Leucaena leucocephala* to be a good energy source depends on its thermochemical properties such as high calorific value and also physical properties such as lower moisture content and less ash content (Yazdani et al. 2012). In this study, the physical and thermochemical of 2-year-old *Leucaena leucocephala* wood would be determined according to the tree height portions.

2 Materials and Methods

The 2-year-old *Leucaena leucocephala* trees were acquired from Tawau, Sabah. The sampling was done according to the standard of Technical Association of the Pulp and Paper Industry (TAPPI). The trees were segregated into top, middle, and bottom portions. A total of nine tree samples for each section were used for the determination of specific gravity and ash content. The calorific value was evaluated using the same batch of samples. All the samples were chipped and grinded before undergoing drying process in the oven at the temperature of 70 °C for 24 h.

Determination of Specific Gravity

The determinations of specific gravity of the samples were done using conventional method according to TAPPI: T258 om-94 (1996).

Determination of Ash Content

For the ash content determination, each sample was allowed to carbonize by placing 2 g of sample on a porcelain crucible and heated in the furnace at the temperature of 100 °C. The temperature was slowly raised to 525 °C in 3 h. The crucible was then taken out, placed in a desiccator, and left to cool to room temperature of 27 °C. The final weight was then measured. The ash content was determined by calculating the fractional percent of the weight of carbonizing sample after burning to the weight of sample before burning.

Determination of Calorific Value

The calorific value was taken via combustion process in the oxygen bomb calorimeter (Model C5000 Adiabatic Calorimeter, IKA, Germany D79019). Prior to that, a calibration to get the energy equivalent of the calorimeter was done using the benzoic acid standard reference sample according to the National Bureau of Analyzed Standard. An amount of 0.5 g sawdust of the sample was loaded onto crucible. They were then placed on the crucible holder to be ignited in the combustion chamber insulated with an isothermal jacket. The platinum ignition wire and the sample were connected with a cotton thread. The pressure is set to 30 bars. The samples were ignited at 298 K in oxygen with 1 cm^3 of water added to the decomposition vessel. The combustion process was completed in 20 min. The heating value in terms of a lower (LHV) or higher (HHV) values were recorded digitally and the calorific value in joule per grams were then evaluated from the enthalpy change between reactants and products.

3 Results and Discussion

Calorific value, ash content, and specific gravity of 2-year-old *Leucaena leucocephala* according to tree sections are shown in Table 1. The specific gravity plays an important role in determining the quality of heat produced. The higher it is, the more energy per volume it has and the longer time the process of burning takes to be completed. In other words, the parts with higher density may provide fires that are sustainable in terms of energy produced. In this study, the specific gravity of *Leucaena leucocephala* wood shows no significant difference between the top and middle section but there is a sudden increase from middle to bottom part. The increase in specific gravity of wood from top to bottom portion could be attributed to the thicker cell wall in the bottom part (Lemenih and Bekele 2004).

There is a noticeable discrepancy between the ash content and the specific gravity of the 2-year-old *Leucaena leucocephala* in this study. The highest percentage of ash content comes from the upper part of the tree. According to Kumar et al. (2011), ash content in general is higher in the top portions of the trees. The ash content of a whole tree is also positively correlated with tree height. There was a slightly decreasing pattern in the calorific value of 2-year-old *Leucaena leucocephala* from bottom to top, unlike the ash content which shows rising value in the opposite direction. However, there is a visibly positive correlation between the calorific value and the specific density of the wood which suggested that there is a locality of chemical wood content such as extractive materials in the 2-year-old *Leucaena leucocephala*. The ash as a by-product of wood burning could be contributed by the inorganic extractives that are negatively correlated with calorific value (Moya and Tenorio 2013). This is in line with the result which shows that calorific value decreases with ash content. Nevertheless, there are other chemical contents such as cellulose, hemicelluloses, and lignin that also give a hand in affecting the calorific value. The bottom of the tree shows the highest calorific value, suggested due to its highest specific gravity figures, related to the more calories it holds per cubic meter (Kumar et al. 2011). However, from the statistical point of view, there were no significance differences between the mean values of top, middle, and bottom portion. Thus, this means that three portions of the tree are suitable for firewood.

Table 1 The calorific value, ash content and specific gravity of 2-year-old *Leucaena leucocephala*

Age	Portion/Parameter	Top	Middle	Bottom
2 years	Calorific value (J/Kg)	18196a	18278a	18427a
	Ash content (%)	0.99a	0.91a	0.77b
	Specific gravity	0.64a	0.65a	0.81b

Note Means with the same letter down the column are not significantly different at $p < 0.05$

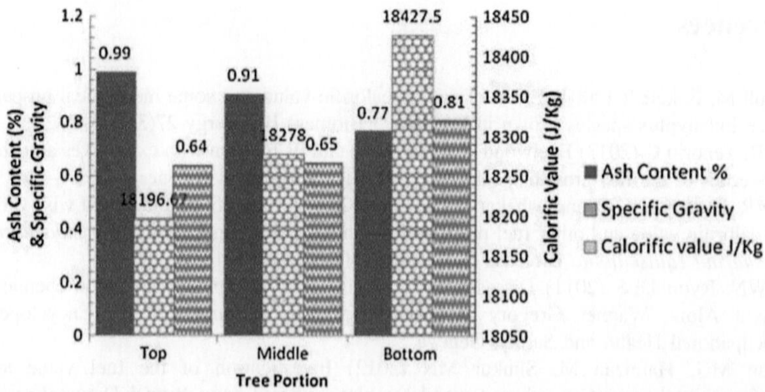

Fig. 1 The calorific value, ash content and specific gravity for 2-year-old *Leucaena leucocephala* vary with different portion of tree

Figure 1 further illustrated the relation between the sectional calorific value, ash content, and specific gravity for the 2-year-old *Leucaena leucocephala.* Looking at the chart, the top and middle section of the 2-year-old *Leucaena leucocephala* is unlikely to be a choice for firewood due to the highest content of ash. The bottom section stands the highest in terms of calorific value but lowest in ash content, hence has the necessary characteristic for use as firewood.

4 Conclusion

The result shows that the calorific value of *Leucaena leucocephala* decreases along the tree height toward the top. The result also shows that the bigger the calorific value, the lower the ash content. The ash content is found highest at the top section while the specific gravity increases toward the bottom part. However, this study reveals that the height portion of the tree has slight significant effect on the ash content, specific gravity, and calorific value. This means that the whole stem of *Leucaena leucocephala* wood could be harvested to be used as firewood but the most suitable part should be the bottom portion. The *Leucaena leucocephala* wood, a fast growing species is suitable to be planted alongside other crops for a sustainable environment. Developing and increasing the consumption of *Leucaena leucocephala* for energy biomass could provide new markets for the wood and increased profits for its planters. Furthermore, it may reduce our dependency on the nonrenewable fossil energy sources.

Acknowledgments The authors wish to thank Dr. Ahmad Rafizan, Miss Hajatun Rabani, and the Deans of the Faculty of Chemical Engineering, UiTM Malaysia for their help and support in completing this study.

References

Lemenih M, Bekele T (2004) Effect of age on calorific value and some mechanical properties of three Eucalyptus species grown in Ethiopia. J Biomass Bioenergy 27(3):223–232

Moya R, Tenorio C (2013) Fuelwood characteristics and its relation with extractives and chemical properties of ten fast-growth species in Costa Rica. J Biomass Bioenergy 56:14–21

Kumar R, Pandey KK, Chandrashekar N, Mohan S (2011) Study of age and height wise variability on calorific value and other fuel properties of Eucalyptus hybrid, *Acacia auriculaeformis* and *Casuarina equisetifolia*. Biomass Bioenergy, 35(3):1339–1344

Rom WN, Ryon DLS (2011) Diseases caused by respiratory irritants and toxic chemicals. In: David Alois, Wagner Gregory R (eds) Respiratory system, vol 10. Encyclopedia of Occupational Health and Safety, Geneva

Yazdani MG, Hamizan M, Shukur MN (2012) Investigation of the fuel value and the environmental impact of selected wood samples gathered from Brunei Darussalam. Renew Sustain Energy Rev 16(7):4965–4969

Part VII
Plantation and Agrotechnology

Chapter 43
Composting of Empty Fruit Bunch Treated with Palm Oil Mill Effluent and Decanter Cake

Salwa Adam, Syed Saiful Nashrizam Syd Ahmad, Nur Masriyah Hamzah and Noor Azimah Darus

Abstract Malaysia is a major exporter of palm oil however it created abundance of oil palm waste particularly empty fruit bunch (EFB) and palm oil mill effluent (POME). Composting of EFB and POME is one of the alternative ways to reduce the amount of by-product and towards the zero emission programs in palm oil mill industry. These by-products are good sources of plant nutrients and could be recycled into the fields as organic fertilizers. This study was conducted to determine the optimum rate of POME plus oil palm decanter cake (OPDC) on composting EFB. Composted EFB with different rate of POME and OPDC was carried out in composting plant at FELCRA Mill, Maran. Shredded EFB was partially treated with POME from anaerobic pond in four different ratio (1:0, 1:2, 1:3 and 1:3:0.2) with 60 % moisture content. The composting materials were also mixed on a weekly basis. The 10-weeks old composts were dark in colour with earthy smell and the pH of compost keep increasing from moderate alkaline to strongly alkaline. The temperature and C/N ratio were found to decrease significantly while N, P and K content increased significantly. The compost formed from the addition of OPDC (1:3:0.2) found to be the most optimum compost with low C/N ratio (23.64), pH 8.4 and high nutrient content (N 1.57 %, P 0.21 %, K 0.65 %). In conclusion, the addition of OPDC as a nitrogen source can speed up composting process by reducing the C/N ratio value. The use of these oil palm wastes could reduce environmental damage and enhance economic benefits in the oil palm industry.

S. Adam (✉)
Faculty of Plantation and Agrotechnology, Universiti Teknologi MARA,
Merlimau, Melaka, Malaysia
e-mail: salwa@melaka.uitm.edu.my

S.S.N. Syd Ahmad · N.M. Hamzah · N.A. Darus
Faculty of Plantation and Agrotechnology, Universiti Teknologi MARA,
Jengka, Pahang, Malaysia
e-mail: nashrizam@gmail.com

N.M. Hamzah
e-mail: nurmasriyah@pahang.uitm.edu.my

N.A. Darus
e-mail: azimah_yute@yahoo.com.my

N.A. Yacob et al. (eds.), *Regional Conference on Science, Technology and Social Sciences (RCSTSS 2014)*, DOI 10.1007/978-981-10-0534-3_43

Keywords Composting · Decanter cake · Empty fruit bunch · Palm oil mill effluent

1 Introduction

Malaysia has more than 3.9 million hectares of land that were cultivated with oil palm and this resulting large amount of oil palm waste being produced particularly Empty fruit bunch (EFB) and Palm oil mill effluent (POME). In the year 2005, it was identified more than 51 million tonnes of both waste has been produced (MPOB 2006). The waste keep increased due to increase in production through research and development activity. Since the large amount of EFB is being produced EFB were recycled and applied in many industrial enterprise and agricultural activities. EFB contain high in nutrient (Table 1) thus normally it was further processed into organic fertilizer. The use of organic fertilizer could reduce production cost as well as conserve our environment. Instead of that, EFB were applied into various industries such as use in the production of wood-based goods, paper and pulp production, stabilizes the soil material and used in the field of horticulture (Kala et al. 2009). In plantation sector, EFB were used widely as mulching because it promote microbial activity and conserve soil moisture that help in improve soil structure and organic matter content and subsequently improved plant growth (Hoong and Nadarajah 1988) as quoted Lim and Zaharah (2002).

Palm oil mill effluent (POME) is colloidal suspensions that contain 95–96 % water, 0.6–0.7 % oil and 4–5 % total solids including 2–4 % suspended solids (Singh et al. 1999). It generated from the final stage of palm oil processing where it consists of washing and cleaning processes in the mill and these effluents contains cellulosic material, fat, oil and grease (Agamuthu 1995). Basically, the effluents produced from the mill were in liquid form because to produce one tonne of crude palm oil it requires about 50–75 % of water thus it ends up in liquid form (Singh et al. 1999). The effluents will be discharge into water bodies but it requires a proper treatment. POME that not properly treated will result in water depletion and aquatic pollution (Hwang 1978) as cited by Igwe and Onyegbado (2007). It also will cause

Table 1 Nutrient content of empty fruit bunch from oil palm

Nutrient	Concentration (%)
Nitrogen	0.35
Phosphorus	0.03
Potassium	2.29
Calcium	0.18
Magnesium	0.15

Source Lim and Zaharah (2002)

water logging of the soil and kills the vegetation. Similar with EFB, POME also benefit to agricultural activities where it could be converted into useful materials through microbial process (Habib et al. 1997). This is because it contains high concentrations of protein, nitrogenous compounds, carbohydrate, lipids and minerals that act as food source for microorganisms. For example, through anaerobic system where microorganisms break down biodegradable material in the absence of oxygen POME can be processed into bio-gas. About 400 m^3 of bio-gas produced from 100 tonnes of POME, of which this amount of POME have been released during processing of 20 tonnes of fresh fruit bunches (Habib et al. 1997).

Oil palm decanter cake also one of the waste material that being produced at palm oil mills. The increasing of the OPDC is becoming the problem (Nafis et al. 2012). Even though OPDC is categorized as a waste but it contain higher nutrient such as nitrogen, phosphorus and potassium compared to POME (Haron et al. 2008). The OPDC also better in the moisture content than POME. As stated by Kala et al. (2009) moisture content is a vital factor to ensure acceleration in compost formation as well as conserve moisture on planting media. Proper waste management are required to ensure less environmental damage and nutrient conservation derived from oil palm mill. Recently, there has been a great interest in converting oil palm waste into composts through microbial activity. Composting EFB treated with POME was one of the options for waste utilization and could offer many environmental and economic benefits. Baharuddin et al. (2009) reported that compost obtained amount of calcium, magnesium, phosphorus, potassium and other micronutrients, therefore it might be suitable to be used as organic fertilizer for plantation purpose.

The use of organic fertilizer is more eco-friendly when compared to inorganic fertilizer. There are many formulation in compost preparation either use water or POME. But most of producer used POME as main material in producing compost. Therefore, this study was conducted to determine the optimum rate of POME plus oil palm decanter cake (OPDC) on composting EFB to get the optimum nutrient status for plant requirement. The findings from this study provide valuable information about composting formulation of empty fruit bunch (EFB). On the other hand, this study also promotes more economic benefit through producing organic fertilizer as well as promotes sustainable agriculture practices through eco-friendly method.

2 Materials and Methods

The EFB, POME and OPDC were collected from composting plant in FELCRA Mill Maran, Pahang. The EFB were shredded into loose fibrous material into 15–20 cm in size using shredder machine. A total of 12 heaps of shredded EFB are incubated at the composting site. The heaps of composting were constructed with the size

150 cm × 150 cm × 45 cm. The treatments for this study consisted EFB treated with different ratio of POME and OPDC (1:0, 1:2, 1:3 and 1:3:0.2) with three replications. The ratios of the treatment were referred to the weight of composting material (kg). This study was conducted on field scale basis and arranged in a complete randomized design (CRD). Each of composting heap has 60 cm clearance in between to allow the turning process. Plastic cover was placed to prevent raining in order to avoid the loss of nutrient. The plastic cover is used throughout the composting period except during spraying of POME and turning process. POME and OPDC were mixed manually into shredded EFB according to treatment design. The same amount of water was sprayed to each of the composting heaps every 3 days to keep the compost moist about 60 % moisture and was composted for 10 weeks. Application of water is stopped 1 week prior to harvesting in order to avoid the final product from being too wet. Turning process was done three times per week to maintain an even distribution of moisture and prevent the build-up heat. The composting heaps were monitored for changes in physical (colour, odour and temperature) and chemical properties (pH, C/N ratio, N, P and K) during the composting process. Temperature reading of each composting heap was recorded every week using thermometer probe trough insert the probe at 15 cm deep. A sample of each compost heap were sampled weekly and tested for their nutrients content, pH and C/N ratio. The pH was determined in the suspension of 1:5 (w/v) compost: deionized water using pH-meter (Mettler MP 225). The total nitrogen was determined using Kjeldhal method (wet digestion) (Bremner and Mulvaney 1982) and organic C was determined according to the combustion method (Kala et al. 2009). Total analysis of macronutrient (P and K) was determined using the aqua-regia method and were analyzed using UV Visible Spectrophometer. All the data were analyzed statistically using analysis of variance (ANOVA). Tukey's Simultaneous Test was used for means comparison of treatment.

3 Results and Discussion

Compost Characteristic

Based on field observation there were changes in colour and texture of empty fruit bunch (EFB) compost along the 10 weeks of composting process. The matured compost of T4 (EFB + POME + DC) at the ratio of (1:3:0.2) was exhibited blackish in colour and fined texture as compared to the first week of composting (Fig. 1). EFB contain relatively large amounts of cellulose calluses and lignin material where it required more time to decompose. However, the application of POME can accelerates decomposition in compost materials due to high moisture content. Moisture content has been described as a vital factor that ensures and speed up decomposition in compost materials (Chang and Chen 2010).

Fig. 1 Physical changes of T4 during composting **a** 1st week of composting **b** 10th weeks of composting

Temperature Pattern

The temperature pattern of composting process was shown in Fig. 2. At the early stage of decomposition process, it was found that T1, T2, T3 and T4 showed a slightly increased in the temperature. This reflecting was due to the intensive microbial activity during the decomposition process which resulted in higher composting rate (Kala et al. 2009). The temperature were keep increased until 6th week however all of the treatments did not reach the thermophilic phase (>40 °C). This might be due to a small size of composting heap that used to build-up heat. Yet, on 7th week onwards the temperature decreased gradually until the final temperature reached at 31–33 °C with the T1 shows the lowest temperature (31.5 °C).

pH of the Compost

Figure 3 illustrates the pH of composts that was fluctuating between the ranges of 7.8–9.0 within 10 weeks of composting process. As mention by Jusoh et al. (2013), the alkaline condition indicate a good quality compost and within suggested range of 6.0–8.5. However as stated by Miller et al. (1991), the high pH of compost can cause loss of nitrogen as NH_3 and lead to odour problems as shown in T2, T3 and T4 that were exceed 8.5 pH values. As refer to Fig. 3, the pH values of T1, T2, T3

Fig. 2 Changes in temperature during composting process

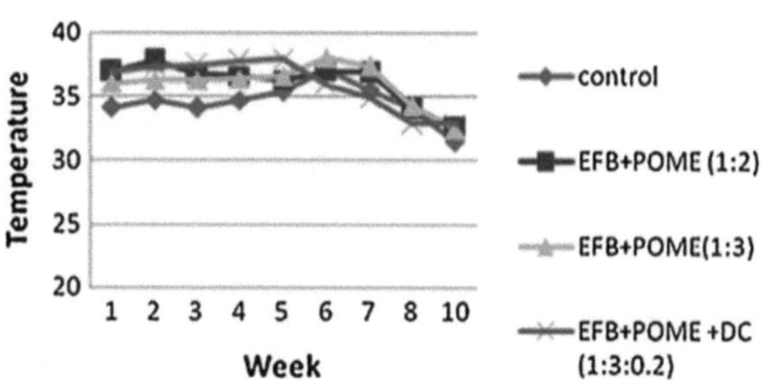

Fig. 3 Changes of pH during composting process

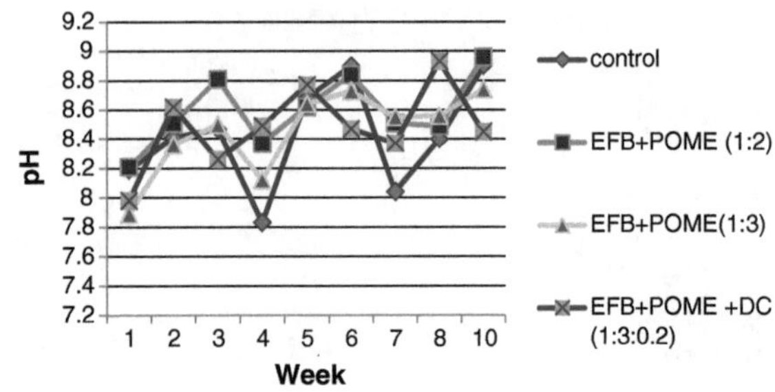

and T4 had increased at the initial stage of composting process. According to Satisha and Devarajan (2007), the increases of pH value in the composting process was likely to be the consequences of rapid metabolic degradation of organic acids and intense proteolysis of liberating alkaline ammonia compound due to protein degradation. Further, T4 (EFB + POME + DC) had found to provide the most desirable compost with 8.45 pH value that was within the suggested range.

C/N Ratio

Initial C/N ratio of the compost were higher than 70 (Fig. 4). The ratio available of carbon to nitrogen is very important to determine the decomposition rate of compost as well as reflect the maturity of the compost. The C/N ratio of all treatments was decreased until week ten with the final reading at 23.64–45.5. As in Fig. 4, the lowest rate of POME applied (T2) was resulted in high C/N ratio as compared to high rates of POME practise (T3 and T4). The application of POME were accelerates mineralization of organic matter by microorganisms (Kala et al. 2009). On the other hand, the addition of decanter cake to the compost (T4) showed the lowest C/N ratio (23.64) among all of treatments. This was because the source of nitrogen in decanter cake could further enhance the microbial activity and accelerates the rate of compost formation (Yahya et al. 2010). Compost material with decanter cake could be said near to its maturity stage because according to Kala et al. (2009)

Fig. 4 Changes in C/N ratio during composting process

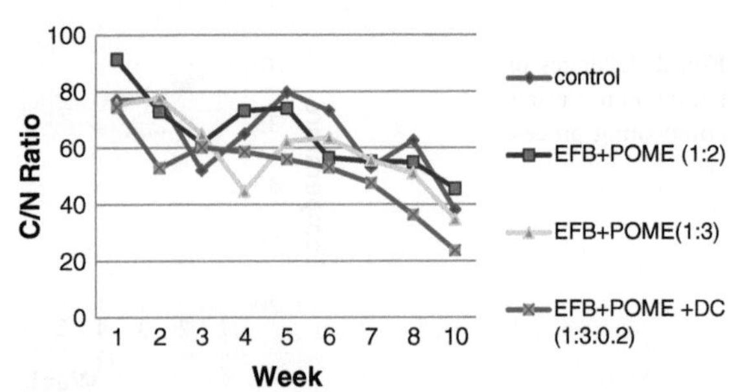

C/N ratio ≤ 20 is considered as matured compost and can be used without any restriction.

Total NPK Availability in Matured Compost

The availability of total NPK is presented in Table 2, while ANOVA table for NPK was presented in Table 3. Expressly, the NPK contents were increased throughout the composting process (Table 2). The increasing of these nutrient contents was due to degradation of organic matter and carbon content (Schuchardt et al. 2002). Carbon degradation was reflecting mineralization process that makes all of the nutrients available in compost materials. As refer to Table 2, the mean value for N of the compost material with decanter cake (T4) was greater than others compost material without decanter cake (T1, T2 and T3). According to Haron et al. (2008), decanter cake remains known as rich in N, P_2O_5, K_2O, CaO and MgO where could further enhanced the nutrient value of resultant compost. Besides, the increase of N value at the end of composting process was due to the nitrogen-fixing bacteria activity commonly occurs at the end of composting (Jusoh et al. 2013). However, the ANOVA table for N indicates there were no significant difference in the nitrogen content among four treatments ($p = 0.097$).

The total P content was increased greatly with high ratio of POME (T3) as well as in T4 that contain decanter cake. The P content at initial composting process was between 0.04–0.08 % and between 0.13–0.22 % at the end of composting stage for

Table 2 Total NPK availability in matured compost

Nutrient	Composts ratio			
	Control 1:0 (%)	EFB + POME 1:2 (%)	EFB + POME 1:3 (%)	EFB + POME + DC 1:3:0.2 (%)
Nitrogen (N)	1.36[a]	1.14[a]	1.37[a]	1.57[a]
Phosphorus (P)	0.13[a]	0.15[a]	0.22[b]	0.21[b]
Potassium (K)	0.85[a]	0.57[b]	0.63a	0.65[a]

[a,b]Tukey grouping showing different letters indicates that the mean elements of treatments are significantly at 5 % significant level

Table 3 Analysis of variance (ANOVA) table for NPK analyzed in matured compost, F-values with degrees of freedom and probabilities for each nutrient analyzed with significant differences ($p \leq 0.05$) highlighted in bold

Nutrient	$F_{(3, 8)}$	p
Nitrogen (N)	2.98	0.097
Phosphorus (P)	7.03	0.012
Potassium (K)	4.88	0.032

all treatments. Therefore, both treatments had the potential to produce compost with a high value of P content. The ANOVA table of P shows there were significant difference in the phosphorus content among the treatment (p = 0.012). Whereas, the potassium content shows differently trend when compared to N and P content where the K content was slightly decreased in time. Control treatment (T1) showed the highest percentage of K content (0.85 %) among others and continued to decrease over time. The ANOVA table of K shows there was significant difference in the potassium content between the treatments (p = 0.032). Potassium is known as the element that easily leached out (Jusoh et al. 2013) and probably the application of POME in liquid form resulting in high leaching rate of K for those treatments applied with POME. Moreover, the high moisture content derived from POME might encourage the leaching process to occur and reduced the K value of compost materials.

4 Conclusion

It can be concluded that the rate of nutrients had increased among the treatments meanwhile decreased in the temperature and C/N ratio within 10 weeks of composting process. The compost formed from the addition of OPDC (1:3:0.2) was found to be the most optimum compost with low C/N ratio (23.64), pH 8.4 and high nutrient content (N 1.57 %, P 0.21 %, K 0.65 %). In this study, it was revealed the potential of oil palm mill wastes could be recycled to form compost which is suitable for the sources of organic fertilizer as well as reduced environmental damage thus enhance economic benefits in the oil palm industry.

References

Agamuthu P (1995) Palm oil mill effluent treatment and utilization. In: Sastry CA, Hashim MA, Agamuthu P (eds) Waste treatment plant. Narosa Publishing House, New Delhi, pp 338–360

Baharuddin AS, Nakamura M, Tabatabaei M (2009) Co-composting empty fruit bunches and partially treated palm oil mill effluents in pilot scale. Int J Agric Res 4(2):69–78

Bremner JM, Mulvaney CS (1982) Nitrogen-total N. In: Miller RH, Keeney DR (eds) Methods of soil analysis (Part 2), p 70–73

Chang JI, Chen YJ (2010) Effects of bulking agents on food waste composting. Bioresour Technol 101:5917–5924

Habib MAB, Yusoff FM, Phang SM, Ang KJ, Mohamed S (1997) Nutritional values of chironomid larvae grown in palm oil mill effluent and alga culture. Aquaculture 158:95–105

Haron K, Mohammed AT, Halim RM, Din AK (2008) Palm-based bio-fertilizer from decanter cake and boiler ash of palm oil mill. Malaysian Palm Oil Board (MPOB). Information Series. MPOB TT No. 412, 1–4

Hoong HW, Nadarajah MN (1988) Mulching of empty fruit bunches of oil palm. SLDB/PORIM Workshop on Palm Oil Milling Technology. Kota Kinabalu, Sabah p 38–50

Hwang TK (1978). Chemical composition of palm oil mill effluents. Planter 54:749–756

Igwe JC, Onyegbado CC (2007) A review of palm oil mill effluent (pome) water treatment, global. J Environ Res 1(2):54–62
Jusoh MLC, Manaf LA, Latiff PA (2013) Composting of rice straw with effective microorganisms (EM) and its influence on compost quality. Iran J Environ Health Sci Eng 10:17
Kala DR, Rosenani AB, Fauziah CI, Thohirah LA (2009) Composting oil palm wastes and sewage sludge for use in potting media of ornamental plants. Malays J Soil Sci 13:77–91
Lim KC, Zaharah AR (2002) The effects of oil palm empty fruit bunches on oil palm nutrition and yield, and soil chemical properties. J Oil Palm Res 14:1–9
Miller FC, Harper ER, Macauley BJ (1991) Investigation of various gases, pH and redox potential in mushroom composting phase I stacks. Aust J Exp Agric 31:415–425
MPOB (2006) Malaysian palm oil board. Available online at http://www.mpob.gov.my. Accessed Oct 2006
Nafis M, Razak A, Ibrahim MF, Yee PL, Hassan MA, Abd-aziz S (2012) Utilization of oil palm decanter cake for cellulase and polyoses production. Biotechnol Bioprocess Eng 17:547–555
Satisha GC, Devarajan L (2007) Effect of amendments on windrow composting of sugar industry press mud. Waste Manag 27:1083–1091
Schuchardt F, Darnoko D, Guritno P (2002) Composting of empty oil palm fruit bunch (EFB) with simultaneous evaporation of oil mill waste water (POME). In: Proceedings of the international oil palm conference. Organized by Indonesian Agency for Agricultural Research and Development, Nusa Dua, Bali, Indonesia, July 8–12, pp 1–9
Singh G, Lim KH, Leng T, David LK (1999) Oil palm and the environment: a Malaysian perspective. Malaysia Oil Palm Growers' Council, Kuala Lumpur, pp 113–126
Yahya A, Sye CP, Ishola TA, Suryanto H (2010) Effect of adding palm oil mill decanter cake slurry with regular turning operation on the composting process and quality of compost from oil palm empty fruit bunches. Bioresour Technol 101:8736–8741

Chapter 44
The Effect of Lateral Shoot Number Manipulation on the *Capsicum frutescens* var Centel Growth and Yield

Yaseer Suhaimi Mohd and Mohamad Abd Manas

Abstract Bird's eye chilli (*Capsicum frutescens*), widely grown in Malaysia, is a high value cash crop. Cultivation of chilli had been done using conventional method and fertigation system. There were increasing suggestion and demands for improvement in cultural practices, including growing techniques, nutrition, pest and disease control, irrigation, planting media and seed availability to increase crop yields. Therefore, this study was done to increase the yield the *Capsicum frutescens* var Centel by manipulating the number of lateral shoot. This study was carried on both conventional and fertigation planting systems. The treatments used in the study were as follows: control—all lateral shoots were removed; T1—two lateral shoots left; T2—four lateral shoots left; T3—five lateral shoots left; T4—seven lateral shoots left, and T5—no lateral shoot was removed. Results indicated that there were similar patterns in term of yields between treatment in both conventional planting and fertigation system after 6 months of planting periods. The highest yields were obtained in T2 from both fertigation system (1600 ± 150 g) and conventional planting (1200 ± 110 g). However, there were no significant difference between T2 and T3 in both cultivation systems. The lowest yields were recorded in T5 for fertigation system (800 ± 80 g) and in control for conventional planting (500 + 65 g). This study revealed that by removing and leaved 4–5 lateral shoots can increase the yield of *Capsicum frutescens* in conventional planting than fertigation system.

Keywords Lateral shoot · Soilless culture · *Capsicum*

Y.S. Mohd (✉) · M. Abd Manas
Technology Promotion and Development Centre, Malaysian Agricultural Research and Development Institute (MARDI), Serdang, Selangor, Malaysia
e-mail: ysuhaimi@mardi.gov.my

M. Abd Manas
e-mail: matnoha@mardi.gov.my

N.A. Yacob et al. (eds.), *Regional Conference on Science, Technology and Social Sciences (RCSTSS 2014)*, DOI 10.1007/978-981-10-0534-3_44

1 Introduction

Capsicum frutescens belongs to the kingdom plantae, order solanales, family Solanaceae, and genus *Capsicum*. *Capsicum frutescens* can be annual or short-lived perennial plants. The fruits are usually very small and pungent, growing 1–3 cm long and 0.3–0.7 cm in diameter. The fruit typically grows a pale green and matures to a bright red, but can also be other colors depending on the varieties. *Capsicum frutescens* or commonly known as Bird's eye chilli is widely grown in Malaysia. The total area of bird's eye chilli cultivation in Malaysia was 276 ha (DOA 2011). It is one of the major condiments widely used in every household in Malaysia. With the increasing demand for bird's eye chilli for fresh consumption as well as for processing and with its favorable prices, there is tremendous potential for growers to venture into bird' eye chilli cultivation in Malaysia.

In Malaysia, cultivation of chilli had been done using conventional method or in soil cultivation and soilless system. Soilless culture technology had been proven for its efficiencies and effectiveness for plant growth and yields. Demands for this technology had been increased significantly throughout the years. Production of chillies had showed an increase in yields up to three to five times higher than in soil planting. However, the conventional method had showed a longer harvesting period and better cost-effectiveness compared to soilless system (Vimala et al. 2007). Despite the availability of planting technology, the main consideration is the production cost.

The cost of agriculture inputs such as fertilizer, pesticide, fungicide, and seeds are increasing very year. High input price will lead to the increase of production cost. At certain point, the yield obtained cannot cover the production cost, thus will lead to loss. There were increasing suggestion and demands for improvement in cultural practices, including growing techniques, nutrition, pest and disease control, irrigation, planting media, and seed availability to increase crop yields. However, improved in the agronomy practices that involve addition of agriculture input will also increase the production cost. Experimental studies have revealed that the increased productivity associated with fertilizer application is with a lot of negative effects which include pollution of ground water after harvest and does not improve soil tilth (Gordon et al. 1993), non-readily available, and economically not feasible (Lar and Kang 1982). Several studies had showed that shoot number manipulation by pruning could result in yield increase (Cebula 1995; Guo et al. 1990; Yaseer Suhaimi et al. 2012; Yadvinder et al. 2014). However, frequent pruning can lead to decrease in production and slow death of the plants (Duguma et al. 1988; Romero et al. 1993).

Initially, as input increases, output increases but eventually, a point of maximum efficiency will be reached and additional input will lead to diminished increase in yield. There is a need in increasing the yields while maintaining the overall production cost with less need for expensive and environmentally damaging inputs. Therefore, this study was done to increase the yield of *Capsicum frutescens* var Centel by manipulating the number of lateral shoots. This study was carried on both conventional and fertigation planting systems.

2 Materials and Methods

Planting Materials

Capsicum frutescens var Centel (Chiap Hup Trading Sdn. Bhd.) was selected and used in the study. Chilli seeds were sown in the peat moss for germination. The seedlings were transferred to nursery to minimize the pest attack after germination. After 21 days of germination, the chilli seedlings about 10–15 cm tall were transferred to the field based on the treatments.

Field Experimental Design

Two trials were conducted at farm located at Mambau, Negeri Sembilan for 6 months duration, from January 1, 2010 until July 31, 2010. Trials were conducted separately in different production systems, namely plants cultivated using conventional method and soilless culture. Both conventional method and soilless production system were conducted in the open field. In the soilless production system, the distance between polybags was 0.45 m and between lines was 1.5 m. Meanwhile, the planting distance in conventional method was 0.45 m × 0.9 m. The treatments used in the study were as follows: control—all lateral shoots were removed; T1—two lateral shoots left; T2—four lateral shoots left; T3—five lateral shoots left; T4—seven lateral shoots left; and T5—no lateral shoot was removed. The lateral shoots were removed after 3 weeks until 6 weeks of planting periods. For each treatment there were 100 samples and the experiment was replicated three times.

Soilless Production System

The seedlings for soilless production system were transferred into the 0.33 m × 0.53 m black polyethylene bag filled with coir dust (coco peat) as growth media. Each polyethylene bag was placed on sides of the tertiary irrigation pipes and individually irrigated with fertilizer solution via a 1 mm micro tube equipped with arrow dripper. Fertilizer used was based on MARDI Formulation solution (Mahamud et al. 2002, 2007). All of the fertilizer components were water soluble. The fertilizer stocks were prepared according to Yaseer Suhaimi et al. (2011). The macro and micro nutrients were prepared separately as A and B stock solutions, respectively, for 100× dilution. Solution A contained calcium nitrate and iron, while solution B contained all others components. All components were added one by one to ensure that they dissolved completely in water. In preparing stock A solution, calcium nitrate was added into container containing tap water (pH 5.5–6.5) and stirred until it dissolved, then the solution was poured into 100 L vessel.

Then, iron powder was added into the container that contained tap water, stirred until it dissolved completely, and added into the vessel. Same procedure was applied in preparing stock B solution.

The irrigation solutions were prepared in a 1500-L tank. Stock A and stock B were added into the tank at 1:1 ratio until the needed electricity conductivity (EC) was achieved. The electrical conductivity (EC) of the irrigation solutions was between 1.8 and 2.3 μS. The irrigation scheduling was automatically implemented by digital timer, four times per day (0700, 1000, 1200, and 1600 h). The duration of irrigation was 3 min and an identical amount of fertilizer solution was applied to all polybags. The daily irrigation volume per plant was 1200 ml. The pH was measured and maintained within a range of pH 5.8–6.1.

Conventional Method

Seedlings were transplanted to 0.3 m high and 1.2 m wide soil ridges. The ridges were, 0.9 m apart and seedlings were transplanted in a single row on the ridges. Application of chicken manure (60 kg/ha) as preplant fertilizer was applied based on soil analysis results. Plants where then irrigated with drip tape (discharge rate of 2.1 L/h) placed on the soil next to each plant. The plants were irrigated twice a day on 8.00 am and 5.00 pm, and the duration of irrigation was 20 min. Inorganic fertilizer ($N:P_2O_5:K_2O:MgO$ = 12:12:17:2) was applied at monthly intervals.

Plant Growth Analysis

Parameters that were observed in both trials include, plant height, fruit numbers, total yields, and harvesting patterns. Routine horticultural practices for pest, disease, and weed were followed. Insecticide (Malathion) and fungicide (Benlate) were applied once in every 2 weeks. Data obtained were statistically analyzed using analysis of variance procedures to test the significant effects of all variables investigated using SAS version 9.1. Means were separated using the Duncan multiple range test (DMRT) as the test of significance at $p \leq 0.05$.

3 Results and Discussion

Plant Growth Performance

The seedling in the soilless cultivation received nutrient solution via microtube second day after transplanting is compared to seedling in the conventional planting that only received water and the granular fertilizer was applied at day seven after

Table 1 Plant height between treatments in both soilless culture and conventional planting

No	Treatments	Plant height (cm)	
		Soilless culture	Conventional planting
T0/control	All lateral shoots were removed	153^{a}	148^{a}
T1	2 lateral shoots left	143^{b}	138^{b}
T2	4 lateral shoots left	145^{b}	140^{b}
T3	5 lateral shoots left	147^{b}	142^{b}
T4	7 lateral shoots left	137^{c}	130^{c}
T5	No lateral shoot removed	138^{c}	135^{b}

transplanting (DAT). The plant in the conventional planting system received nutrient from three source; granular NPK fertilizer, processed poultry manure, and soil. The average plant height in the soilless was higher (143 ± 21) cm compared to conventional planting (138 ± 19 cm) after 6 months of cultivation period (Table 1). However, the highest plant height was observed in the control plant for both planting systems. The shortest plant was observed in the T5 for both planting systems. Prunning or removing the lateral or side shoot created taller plant compared to plant that left with a few lateral shoots and not removed the lateral shoot at all. Prunning the lateral shoot allows the plant to concentrate the nutrition to the main shoot and resulted in taller plant. Plant with excess or many lateral shoots had to divide the nutrient to every shoot for growth and development. Prunings have made substantial contributions to the development of a cost-effective mechanism for optimizing crop yields for efficient and stable crop production (Kang 1997; Young 1997). The avaibility of nutrient in the root zone plays a significant role in the plant performance. The soilless culture plant received fertilizer in the liquid form every day that results in greater plant performance compared to soil planting that depends on the granular fertilizer at 2 weeks frequency. Liquid fertilizer also made nutrients ready for plant to absorb compared to granular fertilizer.

Yields Performance

There were significant differences in terms of yield between each treatment from both soilless and conventional planting (Table 2). The highest average yield was obtained in T2 from both soilless culture (1600 ± 150 g) and conventional planting (1200 ± 110 g). However, there was no significant difference in terms of average yields between T2 and T3 in soilless culture. Similar patterns were also observed in conventional planting system. The average yield in the soilless culture was higher in all treatments and control than plant in soil planting. The lowest yields were recorded in T5 for fertigation system (800 ± 80 g) and in control for conventional planting (500 + 65 g). The highest fruit number was observed in T2 from both soilless culture (1081) and conventional planting (813). There were also no

Table 2 Plant yield between treatments in both soilless culture and conventional planting

No	Treatments	Planting system			
		Soilless culture		Conventional planting	
		Average yields (g)	Number of fruit	Average yields (kg)	Number of fruit
T0/control	All lateral shoots were removed	890 ± 85^{b}	605^{b}	500 ± 65^{b}	338^{e}
T1	2 lateral shoots left	1000 ± 105^{b}	676^{b}	700 ± 70^{b}	473^{c}
T2	4 lateral shoots left	1600 ± 150^{a}	1081^{a}	1200 ± 110^{a}	813^{a}
T3	5 lateral shoots left	1550 ± 145^{a}	1045^{a}	1000 ± 100^{a}	670^{b}
T4	7 lateral shoots left	900 ± 97^{b}	608^{c}	600 ± 68^{b}	405^{d}
T5	No lateral shoot removed	800 ± 80^{b}	540^{d}	580 ± 66^{b}	393^{d}

^{a,b,c,d}Mean values in the same column followed by the same letter are not signicantly different at $p < 0.05$ using DMRT

significant difference in terms of fruit number between T2 and T3 in soilless culture and conventional planting system. Plant cultivated using soilless culture shows higher fruit number than plant in soil planting in all treatments. The fruit number per plant showed the same trend as plant average yield. The fruit number is proportional to the average yield which is the highest fruit number that gives rise to highest average yield. Studies showed that lateral shoot is contributed to the fruits number and plant average yield. Number of shoots per plant are factors that affect light interception by the canopy. As solar radiation varies through the year, the management of these variables to obtain high yield and fruit quality might be different for crops grown in different seasons (Papadopoulos and Ormrod 1990; Lorenzo and Castilla 1995). Mean fruit weight was 1.5 g and did not differ significantly with treatments. Fruit length ranged from 3.5 to 4 cm. Though there were significant effects, no trend was discernable. However, the fruit length to fruit weight ratio was 2:1. This study revealed that 1 cm of fruit length gives rise to 0.5 g of fruit weight in all treatments from both planting systems.

4 Conclusion

The manipulation of lateral shoot number gave a significant effect on the *Capsicum frutescens* var Centel growth and yield. Prunning or removed lateral shoot can increase plant height, fruit number, and average yield. By removing and leaved 4–5 lateral shoots can optimum the yield of *Capsicum frutescens* var Centel in conventional planting than fertigation system.

References

Cebula S (1995) Optimization of plant and shoot spacing in greenhouse production of sweet pepper. Acta Hortic 412:321–328

DOA (2011) Hectareage of species by state Malaysia. Department of Agriculture, Ministry of Agriculture, Malaysia

Duguma B, Kang BT, Okali DUU (1988) Effect of pruning intensities of three woody leguminous species grown in alley cropping with maize and cowpea on an alfisol. Agrofor Sys 6:19–35

Gordon WR, Whitney DA, Raney RJ (1993) Nitrogen management in furrow irrigated, ridge-tilled corn. J Prod Agric 6:213–217

Guo F-C, Fujime Y, Hirose T, Kato T (1990) Effects of the number of training shoots, raising period of seedlings and planting density on growth, fruiting and yield of sweet pepper. J Jpn Soc Hort Sci 59:763–770

Kang BT (1997) Alley cropping—soil productivity and nutrient recycling. For Ecol Manag 91:75–82

Lal R, Kang BT (1982) Management of organic matter in soils of the tropics and subtropics. Int Soil Soe 3:153–178

Lorenzo P, Castilla N (1995) Bell pepper yield response to plant density and radiation in unheated plastic greenhouse. Acta Hortic 412:330–334

Mahamud S, Jamaludin S, Mohamad Roff MN, Ab Halim AH, Mohamad AM, Suwardi AA (2002) Manual teknologi penanaman cili, rockmelon dan tomato secara fertigasi di tanah rendah. MARDI, Serdang

Mahamud S, Jamaludin S, Mohamad Roff MN, Ab Halim AH, Mohamad AM, Suwardi AA (2007) Manual teknologi penanaman cili, rockmelon dan tomato secara fertigasi. MARDI, Serdang

Papadopoulos AP, Ormrod DP (1990) Plant spacing effects on yield of the greenhouse tomato Can. J Plant Sci 70:565–573

Romero F, Montenegro J, Chana C, Pezo D, Borel R (1993) Cercas vivas y bancos de proteína de Erythrina beteroana manejados para la produccíon de biomasa comestible en el trópico húmedo de Costa Rica. In: Westley SB, Powell MH (eds), Erythrinas in the New and Old Worlds. p 205–210

Vimala P, Melor R, Ahmad Shokri O, Balasubramaniam P (2007) Effect of organic and inorganic fertilizers on growth, yield and nutrient content of bird chilli (*Capsicum frutescence*). J Trop Agric Food Sci 35(1)

Yadvinder SS, Kukal S, Mangi LJ, Harminder SS (2014) Improving water productivity of wheat-based cropping systems in South Asia for sustained productivity, vol 127, pp 157–258

Yaseer Suhaimi M, Mohamad AM, Mahamud S, Rezuwan K, Fadlilah Annaim Huda H, Azman J (2011) Effects of temperature gradient generated by fan-pad cooling system on yield of cabbage grown using fertigation technique under side netted rain shelter. J Trop Agric Food Sci 39(1):93–101

Yaseer Suhaimi M, Mahamud S, Mohamad AM, Abdul Kadir Y (2012) Penanaman cili menggunakan sistem fertigasi terbuka. Buletin Teknologi MARDI 1:89–96

Young A (1997) Agroforestry for soil management. UK, CAB International, Wallingford, p 320

Chapter 45
Effects of Light Intensity and Mycorrhiza Association on the Growth Performance of *Capsicum annum*

Anisah Mohammed and Mohamad Amir Shah Yusop

Abstract The study was conducted to examine the effects of light intensity and mycorrhiza inoculation toward the growth and yield performance of *Capsicum annum*. Split-plot design was employed as an experimental design consists of 13 treatments including control with three replications. Four different levels of shade percentage: 0, 30, 50, 70 % and four different rates of mycorrhiza: 0, 50, 60, 70 g were used in this study. Shading was designed as main plot and mycorrhiza as subplot. Data of plant height, size of leaves, number of fruits, fresh weight of fruits, fresh weight of roots, and dry weight of roots were collected. 70 % shading and 60 g mycorrhiza gave the highest plant height with 40.70 cm. 50 % shading and 40 g mycorrhiza was highest in leaf size with 14.50 cm. Plant grown under 30 % shading and 60 g mycorrhiza was the highest in fruit number and fresh fruit weight. Application of 50 % shading and 60 g mycorrhiza contribute to the highest in root weight. Shading was contributed to the significant effect on growth and yield response of *Capsicum annum*.

Keywords *Capsicum annum* · Mycorrhiza · Shading

1 Introduction

Capsicum annum belongs to the dicotyledonous group of flowering plants. It was grown as rain fed crop but too much rain resulted in poor fruit set and rotting of fruit (Bosland and Votava 1999). The optimum temperature for seed germination is

A. Mohammed (✉) · M.A.S. Yusop
Faculty of Plantation & Agrotechnology, Universiti Teknologi MARA, Jengka, Pahang, Malaysia
e-mail: anisahm@pahang.uitm.edu.my

M.A.S. Yusop
e-mail: amirshah@pahang.uitm.edu.my

N.A. Yacob et al. (eds.), *Regional Conference on Science, Technology and Social Sciences (RCSTSS 2014)*, DOI 10.1007/978-981-10-0534-3_45

25–30 °C whereas the temperature recommended for the growth performance is 24–28 °C but these temperatures depend on the varieties of plant. The growth of chilli will be retarded if temperature is too low and resulting to lower yield and quality of fruits. *Capsicum annum* require 80 % of humidity and 5.5–6.5 of soil pH for best performance (Wahyu 2005).

Sunlight is the main factor for plant photosynthesis especially in producing yield. Extreme heat from sunlight contribute to leaf necrosis which reduces the value of plants and at the same time the production of fruits (Menzel and Simpson 1988) will be reduced. According to Schwarz et al. (2010), high temperature and radiation reduce the growth, decrease the photosynthesis rate, reduce water uptake, and increase respiration of plant. The application of shading is an alternative to overcome the heat problem. According to Marin-Lopez et al. (2011), shading screens were reducing solar radiation, heat load inside the greenhouse, and maintain the good level of plant humidity. Shading influences the changes of plant morphology such as stem elongation, leaf size, flower size, and number of flower. Other than that, application of shading will conserve soil humidity and improve the quality of root in uptake of nutrients and water from the soil (Syed Kamaruddin 1983). Wilawan (1998) reported that Prikkeenusuan (varieties of chilli) only can be grown under shade area. Poor yield was obtained when it is grown in large scale without any shade being applied.

Mycorrhiza is a symbiotic association fungus that colonizes the roots in order to improve the uptake of nutrients for the host plant. Myrorrhiza is associated with plant by absorbing nutrients that can be up taken by the plant and in return the plant supply carbon to myrorrhiza (Jaya and Umesh 2012). Application of mycorrhiza will improve the nutrient uptake by external hyphae that can reach outside depletion zone. Besides that, mycorrhiza can increase the volume of soil exploited due to the extent and high surface area of the extrametrical mycelium (Strack et al. 2003). Apart from beneficial effects on the nutrition and water uptake, mycorrhiza has been proven to increase the resistance of plant due to infection of pathogens in roots system (Godbold and Sharrock 2003).

Light availability had been recognized as the main factor in controlling physiological function of plant and known as the critical factor in producing high quality of crop (Matsumoto et al. 2000). Mycorrhiza hyphae improve plant absorption efficacy for water and nutrient uptake for boost plant growth and maximizing crop yield (Koide and Mosse 2004). Since, *Capsicum annum* are considered as a sensitive plant that cannot tolerate to adverse effects, this study aims to examine the effect of light intensity and mycorrhiza association toward the growth performance of this crop.

2 Materials and Methods

Materials Preparation

Capsicum annuum of F1 Hybrid with 90 % of germination rate was used as planting material in this study. It needs to be sowing in germination tray before transplanting into the polybag. The size of polybag that been used in this study is 12 in. × 15 in. and filled with 10 kg of soil mixture with the ratio of (3 parts of top soil: 2 parts of sand: 1 part of organic matter) as planting medium. Shades from nylon mass netting material were used in this study. This shading was bought from local agriculture shop with different percentage level of light intensity that are 30, 50, and 70 %. Mycorrhiza that was used in this study has been obtained from Malaysia Agri Group Company under the name of Myco Gold. Every 10 g of Myco Gold consisted of endomycorrhiza or arbuscular mycorrhiza that contains 200–250 spores of *Acaulospora*, *Gigaspora*, *Glomus*, and *Scutellospora.*

Treatment and Experimental Design

Black shade netting with different percentage of light intensity used as the main factors in this study are 30, 50, and 70 % with 0 % (uncovered) as control. The second factors focused in this study are rate of mycorrhiza that contains 0 g as control, 40, 50, and 60 g. The combinations of shade and mycorrhiza were used as treatments in this study. Split-plot design was employed as experimental design where shading is the main plot and mycorrhiza is a subplot conducted with three replications. The treatments of study were shown in Table 1.

Table 1 Treatments of study

No	Treatment
1	Control (0 % Shading; 0 g Mycorrhiza)
2	0 % Shading 40 g Mycorrhiza
3	0 % Shading 50 g Mycorrhiza
4	0 % Shading 60 g Mycorrhiza
5	30 % Shading 40 g Mycorrhiza
6	30 % Shading 50 g Mycorrhiza
7	30 % Shading 60 g Mycorrhiza
8	50 % Shading 40 g Mycorrhiza
9	50 % Shading 50 g Mycorrhiza
10	50 % Shading 60 g Mycorrhiza
11	70 % Shading 40 g Mycorrhiza
12	70 % Shading 50 g Mycorrhiza
13	70 % Shading 60 g Mycorrhiza

Data Collections and Statistical Analysis

Transplanting of *Capsicum annum* was carried out after 2 weeks of sowing. The seedlings were irrigated using drip irrigation system twice a day for 15 min per irrigation. Plant height and size of leaves were measured in weekly basis. Data on number and fresh weight of fruits was collected after harvesting. Wet and dry weights of roots were recorded at the end of experiment. Data were analyzed using MINITAB and interpreted using Microsoft Excel. Effect of treatments on plant growth performance was evaluated using ANOVA. Variances and separation of means were analyzed using Duncan new multiple range test (DNMRT) for post hoc comparisons at alpha = 0.05.

3 Results and Discussion

The performance of plant height was constantly increased starting from week one until week four. Plant height for each treatment was slightly different starting from week five. In week six, 50 % shading and 60 g mycorrhiza was leading the performance of plant height. 70 % shading and 60 g mycorrhiza gave the highest plant height with 40.70 cm at the end of study. The lowest plant height with 23.17 cm goes to uncovered shading and 40 g mycorrhiza (Fig. 1). According to the result of the variance analysis, plant height was affected by only shading (p = 0.041). Mycorrhiza and interaction between shading and mycorrhiza had no notable effect. According to Riadh et al. (2013), studying the effect of different shading levels on growth and yield of hot pepper in Tunisia reported that the value of plant height is

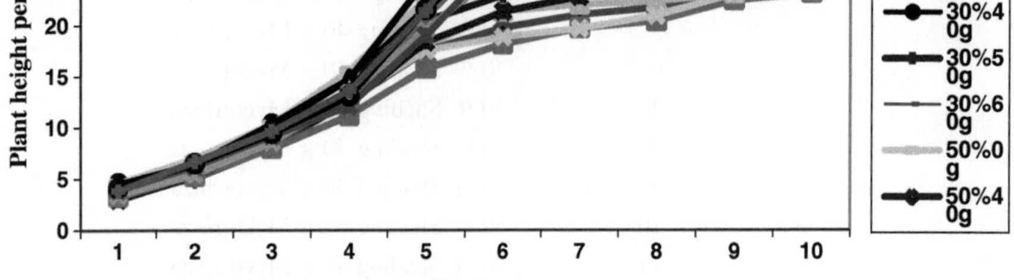

Fig. 1 Plant height performance of *Capsicum annum*

97 cm grown under 100 % shading, 82.45 cm grown under 50 % shading, and the lowest is 68.67 cm grown under 0 % shading. Planted tomato under 80 % shading were significantly taller than plant grown without shading (Syed Kamaruddin 1983). Highest value for plant height of *Leucojum aestivum* was observed in shaded area compared to uncovered area (Ali et al. 2004). The decreased light intensity due to increased percentage of shading resulted in increase of plant height due to promotion of cell elongation and enlargement resulting in taller plants (Schoch 1972; Wilawan 1998).

Leaf size from 50 % shading and 40 g mycorrhiza was highest with 14.50 cm. 0 % shading and 40 g mycorrhiza contribute to the lowest size of leaf with 6.01 cm (Fig. 2). Based on the variance analysis, only shading affected the leaf size with $p = 0.006$. Similar to plant height, mycorrhiza and interaction between shading and mycorrhiza did not affect the leaf size. Marin-Lopez et al. (2011) reported that leaf area of sweet pepper values ranged from 15.14^2 cm planted without shade and 18.30^2 cm planted under shaded condition. Under low light condition, plant morphology was changed in order to maximize the light utilization. Crops that are planted under shade condition have greater foliar surface and leaf area compared with crops in open field area (Larcher 1995). According to Diaz-Perez (2013), studying on bell pepper crop is affected by shade level: microenvironment, plant growth, leaf gas exchange, and leaf mineral nutrient concentration reported that bell pepper planted under shaded condition have longer internode, large leaves, greater whole plant leaf area, and thinner leaves. Wilawan (1998) stated that application of 50 % shading contributes to the highest size of leaves. Reduction of light causes endogenous changes such as reduction in net assimilation or change in hormone level that affect the stem elongation, leaf size, flower size, and flower number (Syed Kamaruddin 1983).

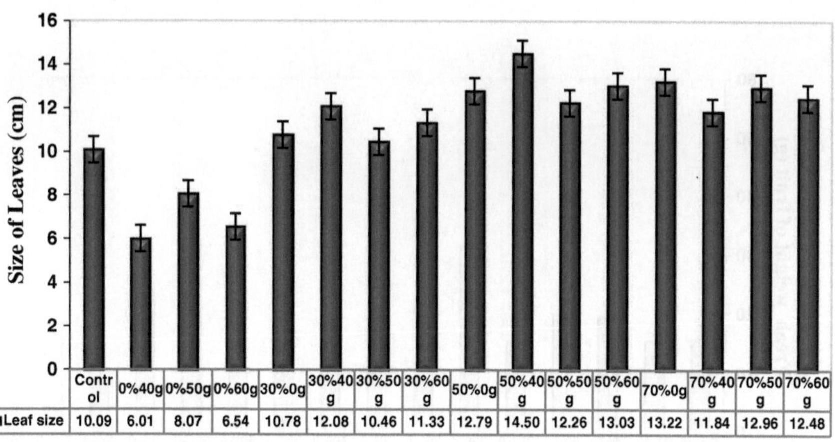

	Control	0%40g	0%50g	0%60g	30%0g	30%40g	30%50g	30%60g	50%0g	50%40g	50%50g	50%60g	70%0g	70%40g	70%50g	70%60g
■Leaf size	10.09	6.01	8.07	6.54	10.78	12.08	10.46	11.33	12.79	14.50	12.26	13.03	13.22	11.84	12.96	12.48

Fig. 2 Leaf size of *Capsicum annum*

Different yield parameters including fruit number and fresh fruit weight are shown in Figs. 3 and 4. Plant grown under 30 % shading and 60 g mycorrhiza was the highest fruit number and fresh fruit weight. Result of 70 % shading and 40 g mycorrhiza indicated the lowest value for both parameters. Results from analysis of variance, only shading ($P < 0.05$) affected the fruit number and fresh fruit weight. There were no mycorrhiza and shading-mycorrhiza interaction for any of plant yield variable. Plant grown under intermediate level of light intensity can produce more branches, enhance fruit setting, favor early flowering, and reduce time of harvesting (Menzel and Simpson 1988). Total yield planted under 26 % shading gave the highest yield and quality of fruits (Riadh et al. 2013). Application of 50 % shading

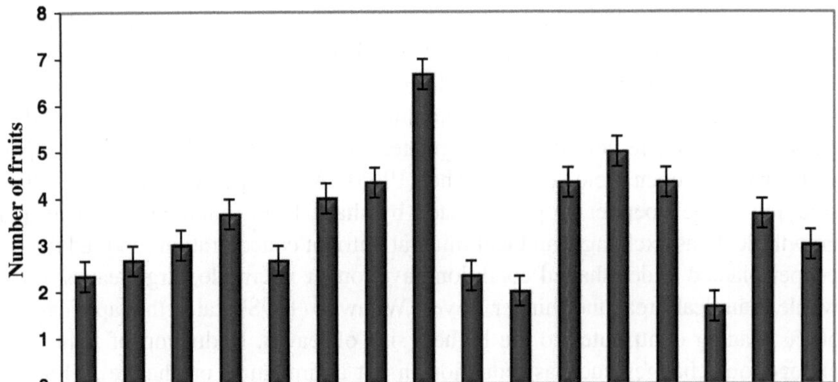

	Contro l	0%40g	0%50g	0%60g	30%0g	30%40 g	30%50 g	30%60 g	50%0g	50%40 g	50%50 g	50%60 g	70%0g	70%40 g	70%50 g	70%60 g
■Fruit No	2.33	2.67	3.00	3.67	2.67	4.00	4.33	6.67	2.33	2.00	4.33	5.00	4.33	1.67	3.67	3.00

Fig. 3 Fruit number of *Capsicum annum*

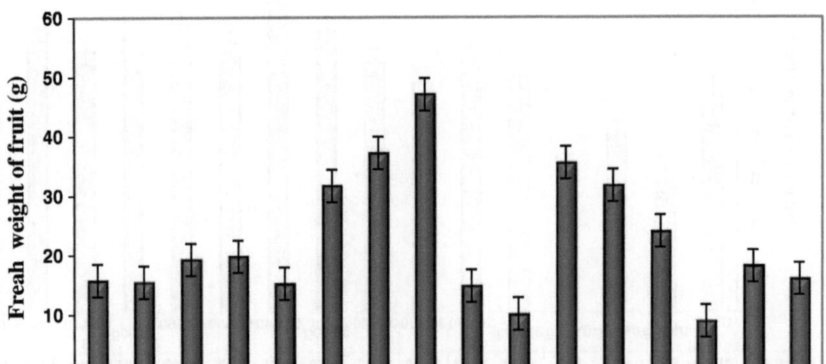

	Contr ol	0%40 g	0%50 g	0%60 g	30%0 g	30%4 0g	30%5 0g	30%6 0g	50%0 g	50%4 0g	50%5 0g	50%6 0g	70%0 g	70%4 0g	70%5 0g	70%6 0g
■Fresh weight	15.80	15.47	19.33	19.80	15.20	31.67	37.20	47.10	14.77	10.07	35.53	31.73	24.00	8.80	18.10	15.97

Fig. 4 Fresh fruit weight of *Capsicum annum*

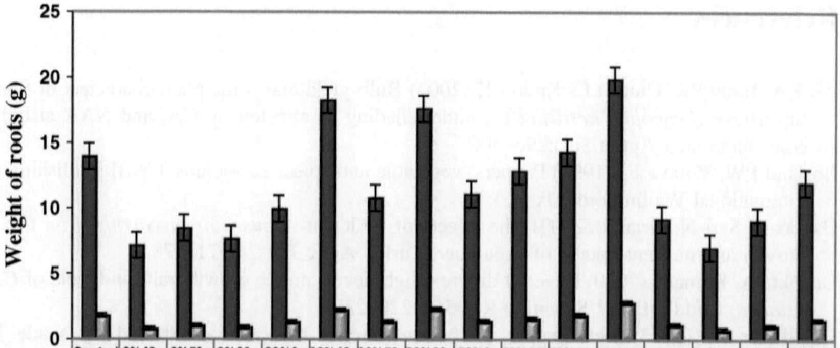

	Control	0%40g	0%50g	0%60g	30%0g	30%40g	30%50g	30%60g	50%0g	50%40g	50%50g	50%60g	70%0g	70%40g	70%50g	70%60g
Fresh Weight	13.93	7.17	8.50	7.67	9.97	18.27	10.80	17.60	11.10	12.87	14.30	19.87	9.20	7.03	9.07	12.00
Dry Weight	1.80	0.80	1.03	0.90	1.30	2.23	1.33	2.30	1.37	1.57	1.87	2.83	1.13	0.77	1.03	1.50

Fig. 5 Root weight of *Capsicum annum*

gave the greatest result on marketable fruit fraction (Gent 2007). The fruit production of pepper was best under 30 % shading where it increased number of branches, enhanced fruit setting, and favored early flowering (Wilawan 1998). Devokta and Kumar (2010) studied the effects of different light intensity on the growth traits and yield of *Centell asiatica* reported that the plant was grown well when exposed to 30 % shading.

Application of 50 % shading and 60 g mycorrhiza contribute to the highest root weight (Fig. 5). Based on the variance analysis, there were no significant difference among the treatments in root weight. Cagras et al. (2000) reported that mycorrhiza inoculation significantly increased the root biomass of plant. Ortas (2010) studied the effect of mycorrhiza application on the growth and nutrient uptake in cucumber production under field conditions. They found that mycorrhiza inoculation was underlined as the main problem in the results where mycorrhiza dependency was reduced. The effects from isolates used were still unpredictable after several periods of controlled experiment. The overall yield was changed each year although mycorrhiza inoculation always gave the benefits to the plant growth.

4 Conclusion

The use of shading in controlling light intensity gave an effect to the plant growth and yield of *Capsicum annum*. It is interesting to note that use of shading significantly affects the plant height, leaf size, fruit number, and fresh fruit weight. The application of mycorrhiza in this study does not show any significant effect among the treatments. 30 % shading looks as a very useful practice to enhance crop yield due to significant effect of yield in this study.

References

Ali KA, Ertan SK, Cuneyt C, Kudret K (2004) Bulb yield and some plant characters of Summer Snowflake (*Leucojum aestivum* L.) under shading as affected by GA, and NAA at different concentration. J Agron 3(4):296–300

Bosland PW, Votava EJ (1999) Peppers: vegetable and spices capsicums. CABI Publishing CAB International Wallingford, Oxon, UK

Cagras S, Sari N, Ortas I (2000) The effects of vesicular *Arbuscular mycorrhizae* on the plant growth and nutrient uptake of cucumber. Turk J Agric For 24:571–578

Devokta A, Kumar J (2010) Effect of different light levels on the growth traits and yield of *Centell asiatica*. Middle-East J Scientific Res 5(4):226–230

Diaz-Perer JC (2013) Bell Pepper (*Capsicum annum* L.) crop as affected by shade level; microenvironment, plant growth, leaf gas exchange and leaf mineral nutrient concentration. Hortic Sci 48(2):175–182

Gent MPN (2007) Effect of degree and duration of shade on quality of greenhouse tomato. Horticultural Science 42(3):514–520

Godbold DL, Sharrock R (2003) Mycorrhizas: In trees, crop and soil fertility. CABI Publishing, United Kingdom

Jaya SP, Umesh BK (2012) Study of arbuscular mycorrhiza associated with important medicinal plants in suburban area of Mumbai. Online International Interdisciplinary Research Journal 2 (2):116–127

Koide RT, Mosse B (2004) A history of research on *Arbuscular mycorrhiza*. Mycorrhiza 14: 145–165

Larcher W (1995) Physiological plant ecology: ecophysiology and stress physiology of functional group. Springer, Berlin. Germany, 48–54

Marin-Lopez J, Gonzalez A, Perez-Alfocea XY, Egea-Galibert C, Fernandez JA (2011) Effect of shade on quality of greenhouse peppers. Acta Horticulturae 893:895–900

Matsumoto J, Hiroyuki M, Izumi W (2000) Whole plant carbon gain of an endangered herbaceous species *Aster kantoensis* and the influence of shading by an Alien Grass *Eragrostis curvula* in its Gravelly Floodplain Habitat. Ann Bot 86:789–797

Menzel CM, Simpson DR (1988) Effect of continuous shading on growth, flowering and nutrient uptake of passionfruit. Horticultural Science 35:77–88

Ortas I (2010) Effect of mycorrhiza application on the growth application on plant growth and nutrient uptake in cucumber production under field conditions. Spanish Journal of Agricultural Research 8(1):116–122

Riadh I, Thouraya R, Imen T, Hager J (2013) Effect of different shading levels on growth and yield parameters of a Hot Pepper (*Capsicum annum* L.) Cultivar 'Beldi' grown in Tunisia. Global Science books 7(1):32–35

Schoch PG (1972) Effect of shading on structural characteristics of the leaf and yield of fruit in *Capsicum annum* L. Journal of the American Society of Horticultural Science 97:461–464

Schwarz D, Rouphael Y, Colla G, Venema GH (2010) Grafting as a tool to improve tolerance of vegatbles to abiotic stresses; Theral stresses, water stresses and organic pollutants. Sci Hortic 127:162–171

Strack D, Thomas F, Betinna H (2003) *Arbuscular mycorrhiza*: Biological, chemical and molecular aspects. J Chem Ecol 29(9):1955–1979

Syed Kamaruddin SW (1983) The effect of shade on growth of tomato (*Lycopersicon esculentum*) and ease of rooting of its cuttings. MARDI Res. Bull 11(2):187–192

Wahyu WBT (2005) Bertanam cabai pada musim hujan. Agro Media Pustaka. Indonesia

Wilawan, K. (1998). Effect of shading on the growth and yield of a local hot pepper variety. AVRDC-ARC. Research Report Thailand

Chapter 46
Inhibition of Egg Hatching of the Golden Apple Snail by Synthetic Molluscicides

Mohd Hafezan Sisa, Firdaus Aspani, Rosdiyani Massaguni, Hazmi Awang Damit and Hendry Joseph

Abstract Golden apple snail is an invasive freshwater snail species that originates from South America. This snail is considered a threat to paddy rice which has caused great attention all over the world. In Malaysia, synthetic molluscicides have been used to control the golden apple snail without much success. Hence, it is imperative to find a practical method and effectively respond to golden apple snail invasions as early and efficiently as possible. This study was initiated to control the snails by inhibiting the egg hatchling through spot spraying the egg clutches using synthetic molluscicides. The eggs of the golden apple snail belonging to 1 day-old were sprayed with metaldehyde (5 %) and niclosamide (81.4 %) at five different concentrations. Then, the hatching rates of eggs within 20 days after treatment were evaluated. The analysis of variance test showed that the molluscicides significantly affected on hatchability of golden apple snail eggs ($p < 0.05$). The results demonstrated that increasing the concentration of the molluscicides were decreasing the hatchability of snail eggs, where the lower hatchability was found at 6.87 and 0.36 mg/ml for metaldehyde (5 %) and niclosamide (81.4 %), respectively. Since the treatment is confined to snail egg clutches, the amount of molluscicides used is relatively small and makes the control method become economically feasible to use.

M.H. Sisa (✉) · F. Aspani · H. Awang Damit · H. Joseph
Faculty of Plantation and Agrotechnology, Universiti Teknologi MARA,
Kota Kinabalu, Sabah, Malaysia
e-mail: mohdh445@sabah.uitm.edu.my

F. Aspani
e-mail: daus92@gmail.com

H. Awang Damit
e-mail: hazmiad@sabah.uitm.edu.my

H. Joseph
e-mail: hendry@sabah.uitm.edu.my

R. Massaguni
Faculty of Plantation and Agrotechnology, Universiti Teknologi MARA,
Shah Alam, Selangor, Malaysia
e-mail: naniefrezy@yahoo.com

N.A. Yacob et al. (eds.), *Regional Conference on Science, Technology and Social Sciences (RCSTSS 2014)*, DOI 10.1007/978-981-10-0534-3_46

Keywords Egg hatchability · Golden apple snail · Molluscicides · Paddy rice · Spot spraying

1 Introduction

Golden apple snail is an invasive freshwater snail species that is considered as a threat to rice production in Malaysia. They are causing significant economical damage to the farmers due to the loss of the plants. These pests prefers young rice seedlings and are able to eliminate completely the leaf and stem from plant bases, which causes the death to the plant (Massaguni and Latip 2012). Its damage could range from 20–100 % from the germinating to the transplanting stage. The extent of the damage can be determined by rice crop stage, snail size, and snail density. After 30 days of transplanting, medium-sized snail with 2–3 cm of shell height at a density of one and eight snails per square meter reduced the number of rice tillers by 19 and 98 %, respectively (Basilio 1991; Naylor 1996). In addition, several characteristics owned by golden apple snail have led to its rapid multiplication and significant damage such as voracious appetite, high fecundity, able to aestivate in soil during dry season, and have enormous reproductive capacity (Du et al. 2007). In the year round warmth of Malaysia, golden apple snail is able to reproduce three to four times per year with eggs ranging from 20 to 500 eggs per cluster at one time (Wang et al. 2012). A female snail starts to lay eggs when two months old with 80 % hatchability rate and the incubation ranges from 10 to 15 days (Joshi 2005).

Recently, a diversity of control strategies have been developed to control the infestation of golden apple snail in paddy field including chemical, biological, physical, cultural, and natural products approach. However, chemical methods are widely used by farmers to protect paddy plants from golden apple snail damage. Metaldehyde ($C_8H_{16}O_4$) and niclosamide ($C_{13}H_8C_{12}N_2O_4$) are the most common synthetic molluscicides used in Malaysia paddy field. Metaldehyde acts mainly through ingestion with symptoms caused by the irreversible damage of mucous cells leading to excessive mucous production and ultimately to death of snails (Coloso et al. 1998). Meanwhile, niclosamide kills snail by affecting the respiration and the carbohydrate metabolism which probably disturbs the oxidation processes by inhibiting oxygen uptake (Chalida et al. 2009). However, both of these chemicals are recommended for controlling the juvenile and adults of golden apple snail only. Thus, excessive amount of synthetic molluscicides is required to control these growth stages of snail which leads to a problem of environmental pollution and increase costs of production (Devanand and Pathipati 2008; Prakash et al. 2008; Baskar et al. 2011). San Martin et al. (2008) have reported that 0.5–1.0 mg a.i./L of niclosamide is the effective rate recommended for the control of golden apple snail, but the LC_{50} for carps is only 0.14 mg a.i./L. This means that no fish must be present in the rice fields while the product is applied. In addition, the use of pesticide to control golden apple snail at adult stage is difficult as snails bury themselves in the soil to avoid exposure from the pesticide spray (Dela Cruz and Joshi 2001).

Hence, it is imperative to find a practical method and effectively respond to golden apple snail invasions as early and efficiently as possible. According to Wu et al. (2005), the mature female of golden apple snail lays egg masses at night by depositing them on paddy stalks, weeds, concrete walls, and stones above the water level. The bright pink colour of egg masses has made it highly visible and easy to find in the fields. Studies in the past demonstrated the eggs hatchability rate and that the offspring development are very sensitive to the availability of oxygen (O_2) and carbon dioxide (CO_2) concentration. Decuypere et al. (2006) have stated that the O_2 and CO_2 exchange is a fundamental importance for eggs embryonic development during incubation. They may not only affect liveability of the embryo, but also affect hatchability as well as the later development and functioning of the eggs. These findings have a theoretical value which suggests that the hypothesis of the study where the golden apple snail eggs will be affected by the availability of O_2 and CO_2 during the incubation.

Thus, a study about the ability of chemical pesticides (niclosamide and metaldehyde) to form the diffusion barrier on snail's egg masses surface which is attributed to the inhibition of O_2 and CO_2 exchange should be conducted. If these pesticides have a potential to be used as an inhibitory to the hatching of snail eggs, the golden apple snail in paddy fields can be decreased and its damage is limited by spot spraying the egg masses. Since this approach is confined to egg masses, it will have minimum environmental impact and gives low cost of production to farmers. This study was conducted to evaluate the effectiveness of metaldehyde and niclosamide in inhibiting eggs hatching of golden apple snail at recommended dosage using spot spraying approach.

2 Method

Egg Production, Collection and Sorting of Eggs

Adults of golden apple snail were collected from the paddy field in Kota Belud, Sabah and were brought to the laboratory. They are subsequently left in a plastic aquarium to acclimatize under lab condition and allowed to mate. The aquariums were filled with distilled water and 10 cm layer of soil which was also gathered from the paddy field to let golden apple snail produce more eggs. Four opposite sidewall inside the aquarium were covered with transparent plastic for deposition of egg masses which simulated the walls of irrigation ditches in the paddy fields. During acclimatization period, the snails were given newly sprouted rice seedlings as a food source. The collection of snail eggs were done randomly by collecting the eggs masses laid by snails reared in the aquarium. 1 day-old eggs which were released during acclimation were taken as an experimental unit.

Treatment Application

In assessing the ability of the synthetic molluscicides, Mostox® (metaldehyde 5 %) was used at five different concentrations ranging from 1.72 to 6.87 mg/ml, while that Bayluscide® (niclosamide 81.4 %) at a range from 0.09 to 0.36 mg/ml which is a recommended rate used by the farmer in Malaysian paddy field. Distilled water was used to dissolve the tested molluscicides. Five clusters of eggs were used for each treatment and placed in a rectangular plastic aquarium that was filled with tap water. Then, about 1.0 ml test chemical was sprayed on each egg mass in each set using a spray bottle (Fig. 1a). An aquarium without chemical treatment was also set up using 1.0 ml of distilled water and marked as a control treatment. After treatment, treated egg masses were left at room temperature to allow the eggs to hatch for a period of 20 days (Fig. 1b), after which the number of unhatched eggs were determined. Golden apple snail eggs usually take 14 days for completion of hatching, so unhatched eggs in each egg mass were separated with a needle and counted after 20 days (Wu et al. 2005).

(a) (b)

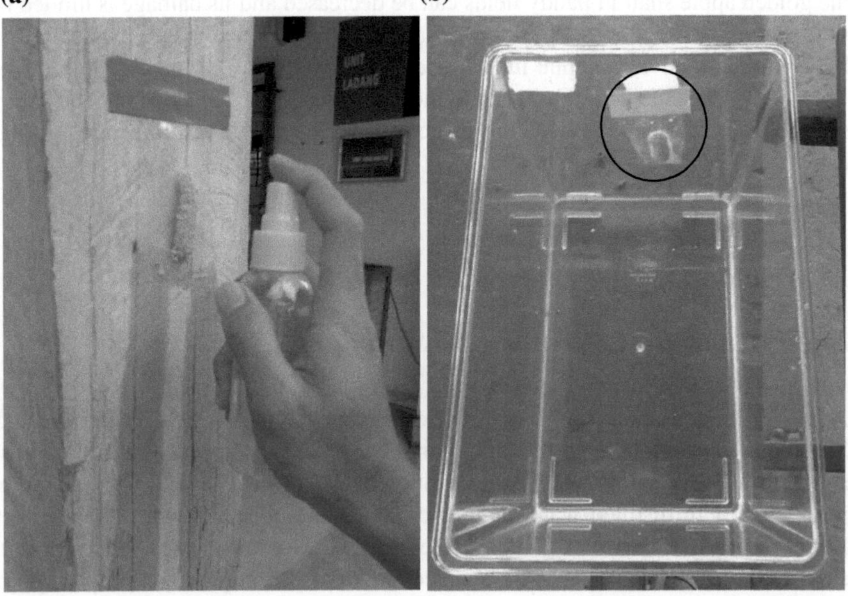

Fig. 1 An approach of spot spraying the egg masses (**a**); Treated egg masses were left for 20 days (**b**)

Data and Statistical Analysis

The efficiency of synthetic molluscicide solution at different concentrations against the golden apple snail eggs were determined by one way analysis of variance (ANOVA) test. Data were analyzed using the statistical software package of Minitab® Release 14.1 (Minitab, Inc.).

3 Results and Discussion

Effect of Metaldehyde (5 %) and Niclosamide (81.4 %) on Hatching Rate of Golden Apple Snail Eggs

The sensitivity of the golden apple snail eggs to the most common molluscicides applied by farmers in paddy field, metaldehyde, and niclosamide was estimated at five different concentrations. After 20 days of observation, the result of non hatching eggs of golden apple snail was pooled across all the exposures time. The analysis of variance among treatments in terms of concentrations of metaldehyde and niclosamide were analyzed statistically by determination of one way ANOVA test. The ANOVA test demonstrated that both molluscicides significantly affected the hatchability of golden apple snail eggs at all concentrations compared to control treatment. In this study, exposure of golden apple snail eggs to metaldehyde at concentration ranges of 1.72–6.87 mg/ml and niclosamide at 0.09–0.36 mg/ml caused a significant decrease in the hatchability ability of eggs which were $F = 4.76$ ($p < 0.05$) and $F = 3.78$ ($p < 0.05$), respectively (Table 1).

The relationship between the hatchability rate of snail eggs with the concentration of tested molluscicides are summarized in Figs. 2 and 3. The figures showed that increasing the concentration of metaldehyde (Fig. 2) and niclosamide (Fig. 3) were increasing the non hatching rate of golden apple snails. This is in agreement with Musman et al. (2013) who reported that the number of golden apple snail eggs hatched was decreased when the concentration of *Barringtonia racemosa* extracts increasing. They had assumed that the decreasing of egg hatchability was due to the active substances in the *Barringtonia racemosa* extracts.

Table 1 Analysis of Variance (ANOVA) of the non-hatching rate eggs on the different treatments

Groups	Variation source	Df	Sum of squares (SS)	Mean square (MS)	F-value	Significance (p-value)
Treatment	Metaldehyde	4	67.84	16.96	4.76	0.021
	Niclosamide	4	200.7	50.2	3.78	0.040

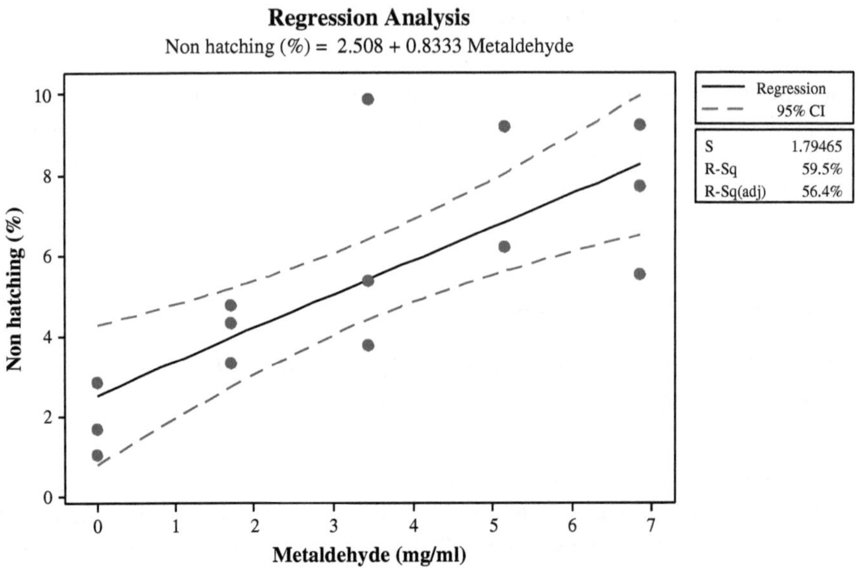

Fig. 2 Non hatching rate (%) of golden apple snail eggs treated with different concentrations of metaldehyde

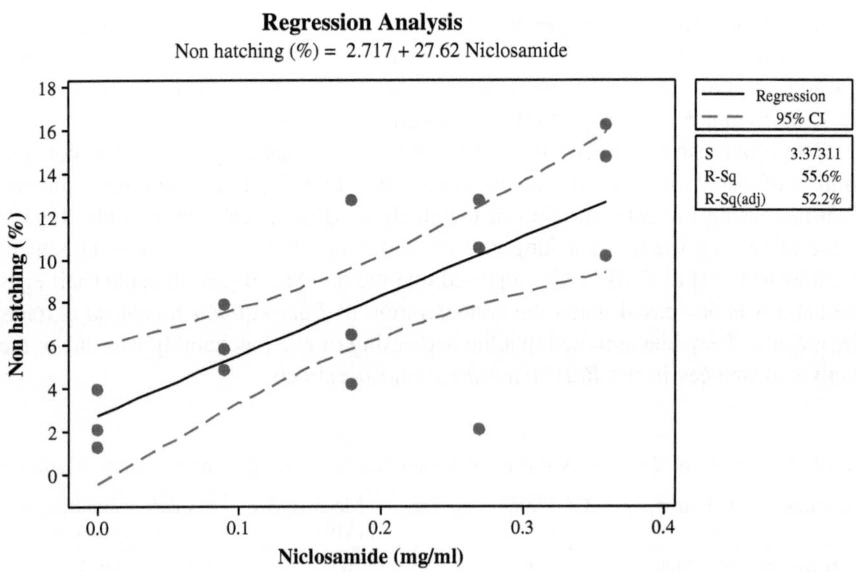

Fig. 3 Non hatching rate (%) of golden apple snail eggs treated with different concentrations of niclosamide

Effect of metaldehyde on mean percentage reduction of the eggs hatchability ranged from 4.1 to 7.4 % at concentrations of 1.72 and 6.88 mg/ml, while 6.1–13.7 % for 0.09 and 0.36 mg/ml of niclosamide (Figs. 2 and 3). It is noticed that the high percent of non hatching eggs was at the highest concentration of tested molluscicides indicating that a double rate of recommended concentration of tested molluscicides are required to control golden apple snail eggs. San Martin et al. (2008) have reported that it only required 0.5–1.0 mg a.i./L of niclosamide to control juvenile and adult stage golden apple snail effectively, which is lower than the rate required to control eggs hatchability in this study. Results showed that the eggs of golden apple snail had a limited sensitivity to tested molluscicides as the mean of non hatching eggs were less than 15 % for both molluscicides. It appears that the egg shells of golden apple snail are permeable to the toxic material contained in the metaldehyde and niclosamide used in this study. Wang et al. (2012) had reported that the surface of newly laid eggs of golden apple snail contain colloid that serves as a protector and help eggs to achieve normal hatchling progress. This colloid was solidified within 24 h of egg laying by forming the calcareous fraction of the egg capsule. Once the calcareous capsules of the snail's egg are formed, it is able to withstand the water spraying and submersion effectively. Musman et al. (2013) also had mentioned that the egg capsule of golden apple snail is formed by a calcium carbonate layer that acts as a membrane for gas exchanges. It suggests that the application of tested molluscicide on golden apple snail eggs should be done within 24 h after egg being laid for more non hatching rate of eggs.

Comparison of Metaldehyde and Niclosamide on Hatchability Eggs

The comparison between metaldehyde and niclosamide on the efficacy in inhibiting the egg hatching could be visualized by plotting the non hatching rate against its frequency as illustrated in Fig. 4. As the difference between non hatching rate changes with the concentration level, it showed that the egg masses in the niclosamide treatments had higher rates of non hatching eggs than metaldehyde. Approximately 7.7 % ± 4.9 mean of egg cluster produced non hatchling eggs in the niclosamide treatments versus 5.4 % ± 2.7 in the metaldehyde treatments. This result concurs with the statement of Dela Cruz et al. (2000), who observed a quick knock down effect of niclosamide 250 EC on mortality of golden apple snail adult compared to metaldehyde formulations.

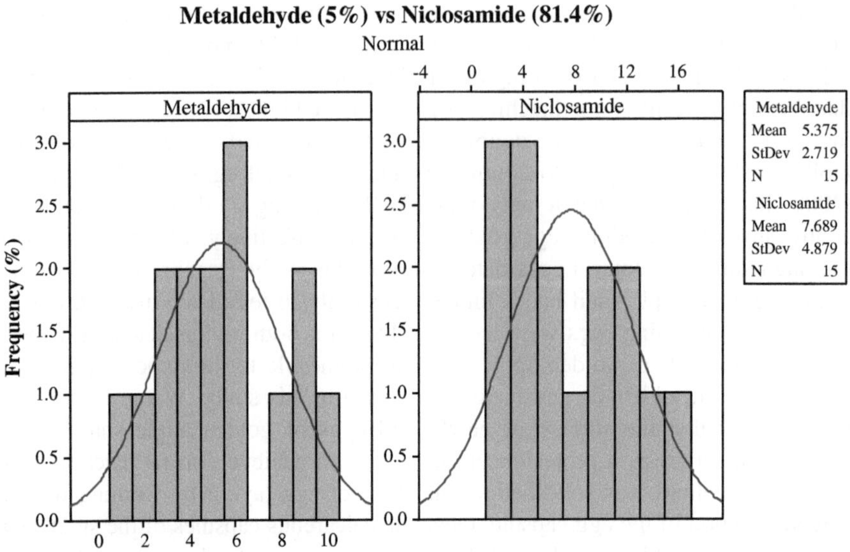

Fig. 4 Comparison on the efficacy of metaldehyde (5 %) and niclosamide (81.4 %) on hatching rate of golden apple snail eggs

4 Conclusions

Based on this study, it is concluded that niclosamide and metaldehyde significantly affected on the hatchability of golden apple snail eggs at all concentrations. It also showed that niclosamide treatments are more effective in inhibiting the hatchability of snail eggs compared to metaldehyde due to its quick knock down effect. Non hatching rate of snail eggs were directly proportional to the increase of concentration for both the tested molluscicides, which is attributed to that highest concentration gives the highest reduction of the eggs hatchability. Even though, the highest concentration for both molluscicides have been applied, it only caused less than 15 % of non hatching eggs which revealed that the eggs have a limited sensitivity to tested molluscicides. This observation has suggested that the earlier treatment which is within 24 h of egg laying could affect more on the hatchling process. Nevertheless, further research is needed to determine the development of offspring after application and mode of action for both molluscicides on golden apple snail eggs. In addition, it should be assessed for field efficacy.

Acknowledgement This work was supported by the RAGS Grant 600-RMI/RAGS 5/3 (192/2013), UiTM Sabah. Author also thank the paddy field staffs Kota Belud, Sabah which really helpful in golden apple snail sampling for this work.

References

Basilio R (1991) Problem of golden snail infestation in rice farming. In: Acosta BO, Pullin RSV (eds) Environmental impact of the golden snail (*Pomacea* sp.) on rice farming systems in the Philippines. Manila, ICLARM, pp 11–12

Baskar K, Maheshwaran R, Kingsley S, Ignacimuthu S (2011) Bioefficacy of plant extracts against Asian army worm *Spodoptera litura* Fab. (Lepidoptera: Noctuidae). J Agric Technol 7(1):123–131

Chalida T, Smarn T, Apiporn S, Supawadee P, Panita K, Jutharat K, Sattrachai P, Wipaporn R, Varima W, Pairat T (2009) Molluscicidal effects of Bayluscide (niclosamide) on *Bithynia siamensis goniomphalos*, First Intermediate Host of Liver Fluke, Opisthorchis viverrini. KKU Res J (GS) 9(2):107–117

Coloso RM, Borlongan IG, Blum RA (1998) Use of metaldehyde as a molluscicide in semicommercial and commercial milkfish ponds. J Crop Prot 17(8):669–674

Decuypere E, Onagbesan O, De Smit D, Tona K, Everaert N, Witters A, Debonne M, Verhoelst V, Buyse J, Hassanzadeh M, Debaerdemaeker J, Arckens L, Bruggeman V (2006) Hypoxia and hypercapnia during incubation of chicken eggs on development and subsequent performance. In: Proceedings of the XII European Poultry Congress, Verona, Italy, 10–14 Sept, pp 486–487

Dela Cruz MS, Joshi RC (2001) Efficacy of commercial molluscicide formulations against golden apple snail *Pomacea canaliculata* (Lamarck). Philipp Agric Sci 84(1):51–55

Dela Cruz MS, Joshi RC, Martin EC (2000) Potential effects of commercial molluscicides used in controlling golden apple snails on the native snail *Vivipara costata* (Quoy & Gaimard). Philipp Ent 14(2):149–157

Devanand P, Pathipati UR (2008) Biological potency of certain plant extracts in management of two lepidopteran pests of *Ricinus communis* L. J Biopesticides 1(2):170–176

Du LN, Jonathan D, Chen XY, Cui GH, Yang JX (2007) A record of the invasive Golden Apple Snail, *Pomacea canaliculata* (Lamarck 1819) at Black Dragon Spring, Dianchi Basin. Zool Res 28(3):325–328

Joshi RC (2005) Managing invasive alien mollusc species in rice. Int Rice Res Notes 30(2):5

Massaguni R, Latip SNH (2012) Neem crude extract as potential biopesticide for controlling golden apple snail, *Pomacea canaliculata*, pesticides—advances in chemical and botanical pesticides. In: Soundararajan RP (ed), ISBN: 978-953-51-0680-7, InTech. doi:10.5772/48626. http://www.intechopen.com/books/pesticides-advances

Musman M, Kamaruzzaman S, Karina S, Rizqi R, Arisca F (2013) A preliminary study on the anti hatching of freshwater golden apple snail *Pomacea canaliculata* (Gastropoda: Ampullariidae) eggs from *Barringtonia racemosa* (Magnoliopsida: Lecythidaceae) seeds extract. AACL Bioflux 6(4):394–398

Naylor R (1996) Invasions in agriculture: assessing the cost of the golden apple snail in Asia. Ambio 25(7):443–448

Prakash A, Rao J, Nandagopal V (2008) Future of botanical pesticides in rice, wheat, pulses and vegetables pest management. J Biopesticides 1(2):154–169

San Martin R, Ndjoko K, Hostettmann K (2008) Novel molluscicide against *Pomacea canaliculata* based on quinoa (*Chenopodium quinoa*) saponins. J Crop Prot 27:310–319

Wang Z, Tan J, Tan L, Liu J, Zhong L (2012) Control the egg hatchling process of *Pomacea canaliculata* (Lamarck) by water spraying and submersion. Acta Ecol Sinica 32:184–188

Wu DC, Yu JZ, Chen BH, Lin CY, Ko WH (2005) Inhibition of egg hatching with apple wax solvent as a novel method for controlling golden apple snail (*Pomacea canaliculata*). J Crop Prot 24:483–486

Part VIII
Sports Science and Recreation

Chapter 47
Kinematics and Kinetics of High and Low Velocity Resistance Training Equated by Time Under Tension: Implications for Hypertrophy Training

Nur Ikhwan Mohamad, Kazunori Nosaka and John Cronin

Abstract Time under tension (TUT) is thought as an important mechanical stimulus for strength and hypertrophic adaptation. The purpose of this study was to determine if two training loads (35 and 70 % 1RM) equated by TUT, differed in terms of their kinematic and kinetic characteristics. Twelve recreationally trained men were recruited in this acute randomized within-subject cross-over design. The 35 and 70 % 1RM loading schemes were equated by TUT based on the ground reaction forces data of a half squat exercise. TUT, average force, peak force, average power, peak power, total work and total impulse were calculated for the eccentric and concentric phases of the squat. Of the 12 variables that were found to be significantly different between loads, 10 of the variables were greater in the lighter 35 % 1RM loading scheme. The major findings were that significantly ($P < 0.05$) greater (41.5–59.5 %) total session average and peak power were found for both the eccentric and concentric phases for the 35 % 1RM loading scheme. However, significantly greater (32.3–34.3 %) eccentric and concentric total impulses were found for the heavier loading scheme. It would seem that when equated by TUT, that lighter loading schemes offer similar peak and average forces but superior velocity and power outputs which may have interesting implications for high-velocity hypertrophic adaptation.

N.I. Mohamad (✉)
Faculty of Sports Science & Coaching, Universiti Perguruan Sultan Idris,
Tanjung Malim, Perak, Malaysia
e-mail: nur.ikhwan@fsskj.upsi.edu.my

K. Nosaka
School of Exercise and Health Sciences, Edith Cowan University, Perth,
Western Australia, Australia
e-mail: k.nosaka@ecu.edu.au

J. Cronin
Sports Performance Research Institute New Zealand, Auckland University
of Technology, Auckland, New Zealand
e-mail: john.cronin@aut.ac.nz

N.A. Yacob et al. (eds.), *Regional Conference on Science, Technology and Social Sciences (RCSTSS 2014)*, DOI 10.1007/978-981-10-0534-3_47

Keywords Heavy load · Hypertrophy · Kinematics · Kinetics · Light load · Time under tension

1 Introduction

Training to increase the cross-sectional area (CSA) of muscle is of great importance to many populations, from the atrophied muscle of the injured and elderly to increasing muscle size of athletes. In terms of increasing CSA, there is no doubt that there are many interacting factors responsible for hypertrophic adaptation. To ensure that protein synthesis exceeds protein degradation, appropriate loading needs to optimize: the muscles mechanical, metabolic and hormonal responses; signalling pathways; and nutrition and recovery. In terms of mechanical loading, Toigo and Boutellier (2006) identified the distinct role of active tension in generating muscle hypertrophy, reporting that time under tension (TUT) is one of the important stimuli in promoting cross-sectional or radial growth of skeletal muscle. Furthermore, in terms of the magnitude of the active tension it is thought that loading the muscle with high loads (high forces/tensions) is fundamental to the development of maximal strength and an important stimulus for muscle hypertrophy (Hakkinen and Pakarinen 1993; Kraemer et al. 1990, 2002; Linnamo et al. 2005; Ratamess et al. 2009). For example, typical loading parameters used to increase muscle hypertrophy are: moderate-to-high repetitions and intensities (6–12 RM or 70–85 % RM) of slow to moderate contraction velocity with moderate-to-short rest intervals (60–90s), and a high volume of training (Abe et al. 2003; Ahtiainen et al. 2003; Crewther et al. 2008; Denton and Cronin 2006; Ratamess et al. 2009).

Although these loading parameters have proven to be effective in increasing the CSA of muscle (Abe et al. 2003; Ahtiainen et al. 2005; Cureton et al. 1988; Dons et al. 1979; Goto et al. 2004; Hurley et al. 1995; McCall et al. 1996; Narici et al. 1996; Rønnestad et al. 2010) the duration of each contraction is long and subsequent movement velocity of such training is slow. This may decrease power production typified by high-velocity explosive movement (Baker 2003), which is very important for sporting and athletic performance. Adaptation of muscle contractile properties via training is dependent on what kind of exercise and protocol is performed (Aagaard et al. 2001; Dons et al. 1979; Kurokawa et al. 2003) as muscle fibre-type shift occurs according to the stimulus applied, which is normally from faster to slower myosin isoforms (Costill et al. 1979). In terms of hypertrophy for sporting performance, it would appear advantageous for hypertrophic adaptation to occur side-by-side with high-velocity adaptation. Using lighter loads would seem interesting in this regard, as there is evidence that training at higher velocities particularly during lengthening contractions increases protein remodelling and preferential hypertrophy of fast fibres (Farthing and Chilibeck 2003; Shepstone et al. 2005).

The use of lighter loads therefore is intuitively appealing, however, as stated previously, heavier loads (higher forces) and TUT are thought critical stimuli for hypertrophic adaptation and therefore the longer eccentric and concentric phases and subsequent slower movement velocities associated with the higher loads used in traditional hypertrophic training would seem fundamental to adaptation. Such an assumption however, could be misleading especially if TUT can be equated in some manner to disentangle the efficacy of heavy load-low velocity (70 % 1RM) as opposed to light load-high velocity (~35 % 1RM) loading schemes in increasing strength and/or hypertrophy. A study by Moss et al. (1997) did exactly this by equating training volume by matching the total TUT (as measured by EMG) and trained the elbow flexors of the non-dominant arm in three groups using loads of 15 % (10 reps), 35 % (7 reps) and 90 % (2 reps) of 1RM (Moss et al. 1997). After 9 weeks of training significant maximal strength gains were recorded (6–7 %) but the differences between the light (15 % 1RM) and heavy (90 % 1RM) groups were non-significant. It was concluded by the authors that the results were not in conflict with the idea that high tension may be the stimulus for increased maximal strength, albeit the high forces were of shorter duration. From these findings it may be surmised that a minimum load or threshold tension is not an important training stimulus as initially proposed and that there may be a case for light load-high velocity training for strength and hypertrophic adaptation.

What seems obvious from this brief treatise is that if we are to progress understanding into the adaptational effects of resistance strength training, we need to have a greater appreciation of the mechanics (kinematics and kinetics) of various loading schemes. Adaptation of muscle tissue is determined by the mechanobiological stimuli imposed upon the muscle and when they are wrongly manipulated will reduce the desired training effects such as regulation of skeletal muscle girth, length and fibre type (Toigo and Boutellier 2006). Given the preceding information and the interest in the possible adaptational effects of light load-high velocity loading schemes for the development of strength and hypertrophy, it would seem prudent to develop a better understanding of the mechanics associated with such loading as apposed to more traditional heavy load-low velocity loading parameters. Furthermore given the importance of TUT in strength and hypertrophic adaptation it would seem beneficial to equate loading schemes by total contraction durations, i.e. TUT. Such an approach should enable a better mechanical understanding of the potential benefits of two divergent loading schemes, i.e. light load-high velocity versus heavy load-low velocity. Consequently, the purpose of this study was to determine two training loads (35 and 70 % 1RM) equated by TUT, differed in terms of their kinematic and kinetic characteristics. The findings will clarify the magnitude of the mechanical stimuli associated with heavy load-low velocity as opposed to light load-high velocity loading schemes, and may provide insight into the adaptational effects with repeated application of these schemes.

2 Methods

Experimental Approach to the Problem

In this acute randomized within-subject cross-over design, 12 recreational trained male student–athletes were recruited to investigate the effects of load on set and session squat kinematics and kinetics. Two loading schemes (70 % 1RM vs. 35 % 1RM) were equated by TUT, by increasing the number of repetitions in the 35 % 1RM loading scheme to match the eccentric and concentric durations of the 70 % 1RM condition. The dependent variables of interest that were quantified during the eccentric and concentric phases included: TUT, average velocity, peak velocity, average force, peak force, average power, peak power, work and total impulse. The total set and session values of each of the variables were then used for statistical analysis.

Subjects

Twelve recreationally trained men with at least 6 months resistance strength training experience volunteered to participate in this research. Subject mean ± SD age, height and mass were 26.3 ± 3.0 y, 175.3 ± 4.0 cm and 80.3 ± 11.2 kg, respectively. All subjects recruited were considered injury free as indicated by no lower limb and spine injury record for the past 2 years. Subjects completed an informed consent form prior to the experiment. Ethics approval from the Human Research Committee of Edith Cowan University was obtained prior to commencing the study.

Equipment

Subjects performed the squat on a force plate (400 Series, Fitness Technology, Australia) supported by a power cage (FT 700, Fitness Technology, Australia). The Olympic bar was interfaced with the Ballistic Measurement System (BMS, Fitness Technology, Australia), which consisted of a linear position transducer (Celesco, PT5A-0150-V62-UP-1K-M6, USA), computer interface (XPV Interface, Fitness Technology, Australia) and the BMS software (BMS, Version 2007.2.3, Innervations, Australia). The sampling frequency of the BMS system (force plate and position transducer) was set at 200 Hz.

Procedures

The procedures involved one familiarization and two testing sessions. The testing sessions were randomized to eliminate any learning, order or fatigue effects that could confound the outcome measures. A minimum of 72 h rest was given between all sessions to ensure full recovery. Participants were asked to replicate exercise and dietary intake 24-h prior to each testing occasion.

Preliminary assessments and familiarization. During the first session, technique and squat maximal strength (1RM) were assessed and anthropometric measurements were taken. The anthropometric variables of interest included standing height (cm) and body mass (kg). Technique for the half squat was analysed and corrections were made as necessary. A 5 minute general warm-up was undertaken prior to lifting. Each participant was then required to perform two warm-up sets of 8 reps at 50 % of estimated 1RM and 3 reps at 70 % of estimated 1RM, respectively. The 1RM assessment followed the warm-up, the maximal load a subject could lift once found over 3–5 trials (4–5 min rest in between 1RM tests).

Squat Technique. The squat movement began from a standing position with the feet approximately shoulder width apart. The squat was initiated by a controlled downward eccentric knee bend until the tops of the thighs became parallel with the floor, which was followed by a concentric phase. The depth of the eccentric phase for each participant was marked and the rubber band was used to regulate depth in subsequent contractions.

Testing Procedures. Subjects were randomly allocated to one of the two testing sessions (35 and 70 % of 1RM). They warmed up as described previously prior to each testing session (e.g. jogging and warm-up sets). TUT was equated quantifying the eccentric and concentric durations for each individual performing the 70 % 1RM intervention (i.e. 36 repetitions—3 sets of 12 reps). Thereafter, the required number of repetitions at the 35 % 1RM load (on average about 54 repetitions) was used to match the eccentric and concentric TUT of the heavier loading scheme. A 90s inter-set rest period was used for both conditions.

Data Analysis

Ground reaction forces were sampled using a portable force plate and interface (Fitness Technology BMS v2008.1.2, Australia) at a rate of 200 Hz. The force plate was synchronized with a linear position transducer to measure the various dependent variables and information was stored in the BMS data acquisition program. The dependent variables of interest were eccentric and concentric; TUT, average velocity, peak velocity, average force, peak force, average power, peak power, work and total impulse.

Statistical Analysis

Means and standard deviations were used to represent centrality and spread of data. Paired sample t-test comparisons were used to determine if significant differences existed for the dependent variables (eccentric and concentric TUT, average velocity, peak velocity, average force, peak force, average power, peak power, work and total impulse) between the two loading schemes. The percent difference between loading schemes were calculated (% Difference = (1 − Lowest Variable/Highest Variable)*100). An alpha level of 0.05 was set to assess statistical significance for all tests.

3 Results

Tables 1 and 2 compare the total session eccentric and concentric outputs (i.e. the sum of all repetitions completed for the 35 and 70 % 1RM loading schemes) when equated by TUT. For the eccentric phase, significant differences were found to be greater (25.7–59.5 %) in total session average and peak power, and total work of the 35 % 1RM loading scheme, while total impulse (32.3 %) was significantly greater in the 70 % 1RM loading scheme (see Table 1).

As for the concentric phase significantly greater (29.2–46.7 %) total session peak force and, average and peak power were associated with the 35 % 1RM loading scheme, whilst significantly greater (34.3 %) total impulse was associated with the 70 % 1RM loading scheme (see Table 2).

Table 1 Total eccentric values (mean ± SD) for the 35 and 70 % 1RM schemes equated by TUT

Variable	35 % 1RM	70 % 1RM	% Difference	Significance
TUT (s)	37.2 ± 6.46	37.1 ± 6.44	0.27	0.826
Average velocity (m s^{-1})	0.66 ± 0.16	0.35 ± 0.08	47.0	0.000
Peak velocity (m s^{-1})	1.08 ± 0.24	0.63 ± 0.13	41.7	0.000
Average force (N)	49549 ± 28313	40929 ± 13695	17.4	0.123
Peak force (N)	22154 ± 15325	26142 ± 13574	15.3	0.260
Average power (W)	34406 ± 22606	14340 ± 5472	58.3	0.003
Peak power (W)	68327 ± 41299	27642 ± 9687	59.5	0.002
Total work (J)	19884 ± 11970	14772 ± 5432	25.7	0.038
Total impulse (N s^{-1})	29185 ± 13925	43108 ± 16180	32.3	0.000

Table 2 Total concentric values (mean ± SD) for the 35 and 70 % 1RM schemes equated by TUT

Variable	35 % 1RM	70 % 1RM	% Difference	Significance
TUT (s)	31.2 ± 5.92	31.1 ± 5.96	0.32	0.483
Average velocity (m s^{-1})	0.69 ± 0.12	0.42 ± 0.07	39.1	0.000
Peak velocity (m s^{-1})	1.14 ± 0.18	0.77 ± 0.13	36.8	0.000
Average force (N)	41164 ± 23030	40226 ± 13684	2.28	0.824
Peak force (N)	72860 ± 30670	51580 ± 14087	29.2	0.005
Average power (W)	28914 ± 17641	16909 ± 6352	41.5	0.010
Peak power (W)	60515 ± 31022	32280 ± 11984	46.7	0.002
Total work (J)	1056 ± 913	700 ± 375	33.7	0.163
Total impulse (N s^{-1})	23227 ± 11880	35339 ± 12936	34.3	0.000

4 Discussion

This study quantified whether two different loading schemes that were termed high load-low velocity (70 % 1RM) and low load-high velocity (~35 % 1RM) loading schemes, differed in terms of their kinematics and kinetics when equated by TUT. The interest in this analysis was prompted by the critical importance of TUT as a hypertrophic stimulus. Typical hypertrophic loading schemes increase TUT by decreasing movement velocity therefore extending the duration of the eccentric and concentric contractions. Given the findings of research that has found significant hypertrophy with lighter loads (Campos et al. 2002; Goldspink et al. 1991; Shepstone et al. 2005) and the contention that slower velocity hypertrophic adaptation may be suboptimal for sport specific adaptation, it may be better to manipulate TUT by increasing the number of repetitions with a lighter load and then move this load at a higher velocity. This contention formed the basis of the experimental design and is discussed subsequently.

The kinematics and kinetics associated with the 35 % 1RM loading scheme were associated with an additional 18 repetitions on average, to equate TUT between loading schemes. Using the lighter loads allowed greater eccentric (~41–47 %) and concentric (~36–40 %) movement velocities as compared to the heavier loading scheme. The effect of these higher movement velocities on total session peak and average forces was for the most part non-significant as compared to the 70 % 1RM loading scheme, the exception being the concentric peak forces (29.2 % greater in the lighter scheme). Though greater forces would have been associated with the individual repetitions of the heavier scheme the additional repetitions used to equate TUT no doubt had a balancing out effect in terms of the total force output for the session.

Whilst the forces were relatively similar between the two loading schemes, significantly greater (41.5–59.5 %) total session average and peak power were found for both the eccentric and concentric phases for the 35 % 1RM loading scheme. The combination of the higher movement velocity and mechanical power outputs associated with the lighter load, as well as the additional repetitions to equate TUT between schemes, offered a substantial mechanical benefit. These findings were similar to Crewther et al. (2008) who used an equivolume approach to study the kinematics and kinetics of a power load (30 % 1RM) and a hypertrophy load (60 % 1RM) on a supine squat machine. They found that greater total power output (40–69 %) was associated with the 30 % 1RM compared to the 60 % 1RM condition.

In contrast to the power results, significantly greater (32.3–34.3 %) eccentric and concentric total impulses were seen for the heavier loading scheme. Despite equating by TUT and the subsequent additional repetitions, when the product of force and time were compared between schemes, the heavier loading scheme was superior. Given that the total forces between schemes were for the most part similar it would seem that the extended contraction durations were responsible for the greater impulses. Thus, it would seem that heavy loads maximize the integration of force and time compared with lighter conditions of equal TUT. If impulse production were the most important stimulus for hypertrophic adaptation, then heavy-load training would seem superior to lighter load training. This finding is not novel and has been reported in research that has studied the kinematic and kinetics of equivolume loading schemes (Crewther et al. 2008; Cronin and Crewther 2004). Unfortunately, there is very little literature that has investigated the importance of impulse as a training stimulus and therefore understanding in this area is rudimentary.

Greater total eccentric work (25.7 %; $p = 0.038$) was evident for the lighter loading scheme, whereas there were non-significant differences between the concentric phase between conditions. It is widely accepted that the eccentric or lengthening phase in resistance exercise training is an important contributor to muscle growth (Farthing and Chilibeck 2003; Goldspink 1999; Higbie et al. 1996; Hortobagyi et al. 1996). Furthermore, hormone release (e.g. growth hormone, testosterone), which is important for hypertrophy, is also associated with the amount of work performed across a workout (Craig and Kang 1994; Kraemer et al. 1990, 1991). Greater work would generally be associated with increased load intensity on a single-repetition basis (i.e. 70 % 1RM load), but the accumulative responses of the additional repetitions in the 35 % 1RM condition when equated by TUT were superior.

A direct comparison between the kinetics of this study to other research and integration into the discussions has been problematic given that this study was the first to equate free squat loading by TUT and then analyse total session kinematics and kinetics. To the knowledge of these authors, the only other study to equate loading schemes by TUT was Moss et al. (1997), however, the loads (90, 35 and 15 % 1RM) and movement (elbow flexion) used were very different to that investigated in this study and the research was longitudinal in nature. The research of Crewther et al. (2008) and Cronin and Crewther (2004) most likely provide the best comparative data, however, direct comparisons are difficult given that: (1) these studies equated volume and not TUT, while hypertrophy (75 % 1RM) and power (45 % 1RM) loading schemes in Crewther et al. (2008) were not equated in any manner; (2) training status such that Crewther et al. (2008) used well-trained athletes who were lifting twice a week as compared to the recreational athletes used in this study; (3) different maximal strength (1RM) therefore each sample will be lifting different relative masses, which will equate to different forces, impulses, etc.; (4) some research used only position transducer data (Crewther et al. 2008; Cronin and Crewther 2004)—the dual use of force plate and position transducer a better option as used in this study; (5) different methodological approaches, e.g. use of mass versus system mass in the calculations and whether using a ballistic or non-ballistic technique (Cormie et al. 2008; Crewther et al. 2008; Cronin and Crewther 2004; Sheppard et al. 2008); (6) different machinery as in the utilization of a Smith machine, supine squat machine and a free weight squat, which will undoubtedly affect kinematics and kinetics (Crewther et al. 2008; Cronin and Crewther 2004); and, (7) different comparative loads, e.g. Crewther et al. 2008 used power and hypertrophy loads of 45 and 75 % 1RM, respectively. Nonetheless, some comparisons have been made throughout the discussion and the reader needs to be cognizant of these limitations.

Understanding the mechanical stimuli associated with various repetitions, set and session loading schemes and their interaction with hormonal and metabolic factors are of paramount importance, if we are to prescribe and understand the adaptational effects of resistance exercise to better effect. In addition undertaking cross-sectional mechanical analyses as in this paper should underpin longitudinal research.

With regard to hypertrophy, it is unlikely that hypertrophic adaptation is caused by one single mechanism and more likely it depends on the integration of multiple local and systemic factors, which are influenced by the mechanical stimuli presented to the muscle and the resultant hormonal and metabolic responses. The tension developed within a session (TUT) appears to be a critical mechanical stimuli in this process. Our understanding of how to develop tension throughout the work period during resisted strength training has been well researched, i.e. using slow contraction velocities (slow tempos), and maximizing both the eccentric and concentric contraction duration. However, it has been speculated by these authors that the slower movement velocities may be suboptimal for sport specific adaptation and that the use of lighter loads with higher movement velocities may be a better choice to invoke "high-velocity" hypertrophic adaptation. Certainly, there is some support in the literature for the use of lighter loads. The findings of this study seem to support such an approach to a degree given the similar forces, superior movement velocities, power outputs and eccentric work of the lighter loading scheme when equated by TUT. However, greater eccentric and concentric impulses were associated with the heavier loading scheme. If impulse production was the most important stimulus for hypertrophic adaptation, then heavy-load training and the associated slower movement velocities would seem the training method of choice. It may be that periodising both light- and heavy-loading schemes equated by TUT may provide the best options for hypertrophic adaptation.

To conclude, low load-high velocity loading equated by TUT may offer a better or alternative training stimulus for athletes who want to maximize hypertrophic adaptation while at the same time maintaining or increasing movement velocity.

However, longitudinal research is needed to validate such a contention. Also future research needs to investigate the effect of impulse and other mechanical stimuli across different loading parameters and profile the subsequent effects on the metabolic and hormonal milieu, and ultimately adaptation, e.g. cross-sectional area.

References

Aagaard P, Andersen JL, Dyhre-Poulsen P, Leffers A-M, Wagner A, Magnusson PS, Halkjaer-Kristensen J, Simonsen EB (2001) A mechanism for increased contractile strength of human pennate muscle in response to strength training: changes in muscle architecture. J Physiol 534:613–623

Abe T, Kojima K, Kearns C, Yohena H, Fukuda J (2003) Whole body muscle hypertrophy from resistance training: distribution and total mass. Br J Sports Med 37:543–545

Ahtiainen JP, Pakarinen A, Alen M, Kraemer WJ, Hakkinen K (2003) Muscle hypertrophy, hormonal adaptations and strength development during strength training in strength-trained and untrained Men. Eur J Appl Physiol 89:555–563

Ahtiainen JP, Pakarinen A, Alen M, Kraemer WJ, Hakkinen K (2005) Short vs. long rest period between the sets in hypertrophic resistance training: influence on muscle strength, size, and hormonal adaptations in trained men. J Strength Cond Res 19:572–582

Baker D (2003) Acute negative effect of a hypertrophy-oriented training bout on subsequent upper-body power output. J Strength Cond Res 17:527–530

Campos GER, Luecke TJ, Wendeln HK, Toma K, Hagerman FC, Murray TF, Ragg KE, Ratames NA, Kraemer WJ, Staron RS (2002) Muscular adaptations in response to three different resistance-training regimens: specificity of repetition maximum training zones. Eur J Appl Physiol 88:50–60

Cormie P, McBride JM, McCaulley GO (2008) Power-time, force-time, and velocity-time curve analysis during the jump squat: impact of load. J Appl Biomech 24:112–120

Costill DL, Coyle EF, Fink WF, Lesmes GR, Witzman FA (1979) Adaptations in skeletal muscle following strength training. J Appl Physiol 46:96–99

Craig BW, Kang H-Y (1994) Growth hormone release following single versus multiple sets of back squats: total work versus power. J Strength Cond Res 8:270–275

Crewther BT, Cronin J, Keogh JWL (2008) The contribution of volume, technique, and load to single-repetition and total-repetition kinematics and kinetics in response to three loading schemes. J Strength Cond Res 22:1908–1915

Cronin JB, Crewther B (2004) Training volume and strength and power development. J Sci Med Sport 7:144–155

Cureton KJ, Collins MA, Hill DW, MCelhannon FMJ (1988) Muscle hypertrophy in men and women. Med Sci Sports Exerc 20:338–344

Denton J, Cronin JB (2006) Kinematic, kinetic and blood lactate profiles of continous and intraset rest loading schemes. J Strength Cond Res 20:528–534

Dons B, Bollerup K, Bonde-Petersen F, Hancke S (1979) The effect of weight-lifting exercise related to muscle fiber composition and muscle cross-sectional area in humans. Eur J Appl Physiol 40:95–106

Farthing JP, Chilibeck PD (2003) The effects of eccentric and concentric training at different velocities on muscle hypertrophy. Eur J Appl Physiol 89:578–586

Goldspink G (1999) Changes in muscle mass and phenotype and the expression of autocrine and systemic growth factors by muscle in response to stretch and overload. J Anat 194:323–334

Goldspink G, Scutt A, Martindale J, Jaenicke T, Turay L, Gerlach G (1991) Stretch and force generation induce rapid hypertrophy and myosin isoform gene switching in adult skeletal muscle. Biochem Soc Trans 19:368–373

Goto K, Nagasawa M, Yanagisawa O, Kizuka T, Ishii N, Takamatsu K (2004) Muscular adaptations to combinations of high- and low-intensity resistance exercise. J Strength Cond Res 18:730–737

Hakkinen K, Pakarinen A (1993) Acute hormonal responses to two different fatiguing heavy-resistance protocols in male athletes. J Appl Physiol 74:882–887

Higbie EJ, Cureton KJ, Warren GL III, Prior BM (1996) Effects of concentric and eccentric training on muscle strength, cross-sectional area, and neural activation. J Appl Physiol 81:2173–2181

Hortobagyi T, Hill JP, Houmard JA, Fraser DD, Lambert NJ, Israel RG (1996) Adaptive responses to muscle lengthening and shortening in humans. J Appl Physiol 80:765–772

Hurley BF, Redmond RA, Pratley RE, Treuth MS, Rogers MA, Goldberg AP (1995) Effects of strength training on muscle hypertrophy and muscle cell disruption in older men. Int J Sports Med 16:378–384

Kraemer WJ, Adams K, Enzo C, Dudley GA, Dooly C, Feigenbaum MS, Fleck SJ, Franklin B, Fry AC, Hoffman JR, Newton RU, Potteiger J, Stone Michael H, Ratamess NA, Triplett-McBride T (2002) Progression models in resistance training for healthy adults. Med Sci Sports Exerc 34:364–380

Kraemer WJ, Gordon SE, Fleck SJ, Marchitelli L, Mello R, Dziados JE, Friedl K, Harman E, Maresh CM, Fry AC (1991) Endogenous anabolic hormonal and growth factor responses to heavy resistance exercise in males and females. Int J Sports Med 12:228–235

Kraemer WJ, Marchitelli L, Gordon SE, Harman E, Dziados JE, Mello R, Frykman P, McCurry D, Fleck SJ (1990) Hormonal and growth factor responses to heavy resistance exercise protocols. J Appl Physiol 69:1442–1450

Kurokawa S, Fukunaga T, Nagano A, Fukashiro S (2003) Interaction between fascicles and tendinous structures during counter movement jumping investigated in vivo. J Appl Physiol 95:2306–2314

Linnamo V, Pakarinen A, Komi PV, Kraemer WJ, Hakkinen K (2005) Acute hormonal responses to submaximal and maximal heavy resistance and explosive exercises in men and women. J Strength Cond Res 19:566–571

McCall GE, Byrnes WC, Dickinson A, Pattany PM, Fleck SJ (1996) Muscle fiber hypertrophy, hyperplasia, and capillary density in college men after resistance training. J Appl Physiol 81:2004–2012

Moss BM, Refsnes PE, Abildgaard A, Nicolaysen K, Jensen J (1997) Effects of maximal effort strength training with different loads on dynamic strength, cross-sectional area, load-power and load-velocity relationship. Eur J Appl Physiol 75:193–199

Narici M, Hoppeler H, Kayser B, Landoni L, Claassen H, Gavardi C, Conti M, Cerretelli P (1996) Human quadriceps cross-sectional area, torque and neural activation during 6 months strength training. Acta Physiol Scand 157:175–186

Ratamess NA, Alvar BA, Evetovich TK, Housh TJ, Kibler WB, Kraemer WJ, Triplett NT (2009) Progression models in resistance training for healthy adults. Med Sci Sports Exerc 41:687–708

Rønnestad B, Hansen E, Raastad T (2010) Effect of heavy strength training on thigh muscle cross-sectional area, performance determinants, and performance in well-trained cyclists. Eur J Appl Physiol 108:965–975

Sheppard JM, Doyle TLA, Taylor DC (2008) A methodological and performance comparison of free weight and smith-machine jump squats. J Aust Strength Cond 16:5–9

Shepstone TN, Tang JE, Dallaire S, Schuenke MD, Staron RS, Phillips SM (2005) Short-term high- vs. low-velocity isokinetic lengthening training results in greater hypertrophy of the elbow flexors in young men. J Appl Physiol 98:1768–1776

Toigo M, Boutellier U (2006) New fundamental resistance exercise determinants of molecular and cellular muscle adaptations. Eur J Appl Physiol 97:643–663

Chapter 48
Relationship Between EMG Activity and Endurance Time of the Biceps Brachii During Isokinetic Contraction

S.A.M. Matiur Rahman, Nizam Uddin Ahamed, Mahdi Alqahtani, Omar Altwijri, Kenneth Sundaraj and N. Ahmed

Abstract Studies of the surface electromyographic or EMG signal on the skeletal muscles have provided valuable information on the physiology to understand the motor unit activity strategy during contraction. This paper investigated the relationship between EMG and time during isokinetic contraction (a type of exercises or movement to increase muscular strength) from the upper limb's biceps brachii (BB) muscle. Five right-hand-dominated male subjects (age 23.2 ± 1.6 years) participated in the study. The muscle activation during load lifting (using a standard 6-kg dumbbell) was determined as the root mean square (RMS) electromyographic signal normalized to the peak RMS EMG signal of a maximal contraction for 10 s. For the statistical analysis, the slope of the regression relationship was used to test the relationships between EMG and time from the maximal voluntary isokinetic

S.A.M. Matiur Rahman
College of Computer Science and Information System,
Najran University, Najran, Saudi Arabia
e-mail: matiurrahman7363@gmail.com

N.U. Ahamed (✉)
Faculty of Manufacturing Engineering, Universiti Malaysia Pahang,
Pekan, Pahang, Malaysia
e-mail: nizamuddin@ump.edu.my

M. Alqahtani · O. Altwijri
Biomedical Technology Department, College of Applied Medical Sciences,
King Saud University, Riyadh, Saudi Arabia
e-mail: amahdi@ksu.edu.sa

O. Altwijri
e-mail: oaltwijri@ksu.edu.sa

K. Sundaraj
AI-Rehab Research Group, School of Mechatronics Engineering,
Universiti Malaysia Perlis, Arau, Perlis, Malaysia
e-mail: kenneth@unimap.edu.my

N. Ahmed
School of Computer and Communication Engineering,
Universiti Malaysia Perlis, Arau, Perlis, Malaysia
e-mail: nasim@unimap.edu.my

N.A. Yacob et al. (eds.), *Regional Conference on Science, Technology and Social Sciences (RCSTSS 2014)*, DOI 10.1007/978-981-10-0534-3_48

contraction (MVC). The measurement duration (10 s) was divided into four phases for analyzing the relationship. The results of the study showed a linear but poor relationship between EMG and time during such contraction, where $r^2 = 0.11$ and $F = 3.74$. Also, no significant difference exits from the analysis where, $P = 0.062$. These results provided insight toward applications of biomedical engineering methods in the analysis and control of the neuromuscular system, ergonomics field of research, rehabilitation engineering, and movement biomechanics during such type of dynamic contraction.

Keywords Biceps brachii · EMG-time relationship · Isokinetic contraction · Surface EMG · Muscle activity

1 Introduction

Surface electromyography (EMG) is only the noninvasive technique which offers straight information on muscle involvement during movement and contraction (Merletti et al. 2004; Rainoldi et al. 2004). It has been widely accepted that the processing of EMG signals is essential for understanding how muscular activity (during contraction) generates electrical signals that control the joint movements in the human body (Benedetti et al. 2001). According to Putatunda, each muscle in the human muscular system is composed of thousands of tiny fibres and cells, and these control the movement of the various parts of the human body (Putatunda 2011).

In general, EMG signals are generated from these skeleton muscles in the human body as a result of the contraction of the muscles fibres, and these signals are always stochastic (random) (Komi and Viitasalo 1976). Biceps brachii (BB) is one of the active skeleton muscles in the upper extremity, and EMG signals are repeatedly generated by this muscle during contraction due to the elbow flexion relaxation phenomenon (Clark et al. 2003). As a result, these random raw and integrated EMG signals need to be recorded for further analysis. There are two types of recording processes that are mainly used to record the signals generated by muscle contraction: needle (invasive) and surface (noninvasive) EMG procedures. The surface EMG sensor (electrode) is used more frequently because it is preferred by the subjects, a noninvasive and painless technique that minimizes the signal interference (Mandryk et al. 2006; Pullman et al. 2000). Based on these advantages of the surface EMG technique, this research work has used surface EMG sensor to record the EMG signals from the subjects. It is therefore important to understand how well the surface EMG signals of motor units representing the muscular activity of the BB during.

The muscular contraction velocity remained constant, while force was allowed to be varied during the isokinetic or isovelocity contraction (Kellis and Baltzopoulos 1996). This type of contraction showed muscle lengthening and shortening at a constant speed Mercer et al. (2006). Several investigators have used the amplitude of the raw EMG signal (in terms of volts) as a measure of BB muscle activation during contractions. For example, Horita and Ishiko (1987) investigated the relationship between muscle lactate accumulation and surface EMG activities during isokinetic contractions (knee extensions) from the male subjects. Perry-Rana et al. (2002) discussed the EMG responses of the superficial quadriceps muscles during isokinetic contraction with different velocities. Then Fernández-Pena et al. (2009) calculated the EMG activity for 6-s maximal isokinentic pedalling sprints. Another research by Jingyi et al. (1995) showed the EMG characteristics of quadriceps muscle during different range of motion (ROM) and velocities from the ongoing contractions. One article reported the EMG-angle relationship during isokinetic exercise from lower limb muscle during agonist and antagonist moment (Howatson and van Someren 2005; Kellis and Baltzopoulos 1997). Likewise, other researchers reported that the EMG effect of human body muscle during contractions (Zhang and Sun 2011). Most published studies on isokinetic contraction have used different normalization techniques and relationship status among the parameters for the purpose of EMG signal processing. However, we could not find the EMG effect of BB as well as the relationship between time and EMG amplitude during isokinetic contraction. Therefore, the main aim of this research is to investigate the relationship of BB muscular activity during isokinetic contraction.

2 Methods

Participants

Five right-hand-dominated male subjects, with mean ± standard deviation, age: 23.2 ± 1.6 yrs, height: 1.68 ± 0.19 m and weight: 67.8 ± 13.6 kg, were recruited from the university students. None of the participants had any upper extremity musculoskeletal disorders or other medical problems. Prior to the test, the purposes and procedures of the experiment were explained to the participants. All subjects gave written informed consent before entering the study. All experimental procedures conformed to the declaration of Helsinki and were approved by the Human Research Ethics Committee at the university.

Apparatus

A wireless, named Shimmer^{TM} (Real-time Technologies Ltd., Ireland) was used to record the signal (EMG activity in microvolt (mV)). The positive and negative input electrodes were used for the EMG recording on the BB muscle, and a third was used as the reference channel. The EMG signals were detected from the belly of the BB muscle using two foam adhesive electrodes (silver-silver chloride). The reference electrode was attached to the lateral epicondyle of the humerus of the right arm (approximately 1 in. on the olecranon of the elbow). Some protocols such as inter sensor distance, sensor placement procedures, and skin preparation suggested by Hermens et al. (1999); and SENIAM (2012) were followed. The raw EMG signals were recorded at a sampling rate of 1 kHz before their A-D conversion and stored on a compatible computer. The fourth-order band pass Butterworth filter was used to remove any skin movement artifacts as well as high frequency noises (cutoff frequency between 10–500 Hz). The recorded digitized EMG data sets were processed offline by filtering, windowing, and extracting signal. Signal processing was performed with MATLAB (MathWorks, USA) software.

Experimental Protocol

Before the test, the subject was told not to move and sit on a chair as relax as possible. The isokinetic contraction (up and down with same velocity) was performed by lifting and lowering a 6-kg Dumbbell. The elbow moved (flexion and extension) at a speed, generates a pendulum arm motion within 0°–90° angles. This angle measurement was considered from the shoulder to elbow and then elbow to palm which was calculated by a digital inclinometer. Three trials were performed for 10-s each and a rest period of 5-min was provided between each trial. Figure 1 depicts the experimental protocol from the subject. Here, the up and down arrows indicating the lifting load (6-kg) of constant weight at same speed.

Data Processing

EMG data were divided into 4 time slags and analyzed for each 2500-ms time window. The sampled waveform of EMG signals (raw and without filtering) were presented in Fig. 2. Normalization procedure was applied to analyse the EMG amplitude. Thus, all data have been normalized in terms of the root mean square (RMS) values (i.e, the individual RMS values of the tests during the contraction

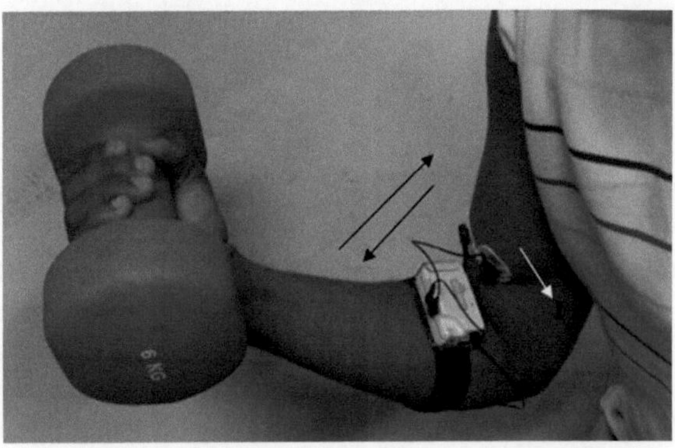

Fig. 1 Schematic representation of contractility of BB muscle during isokinetic contraction

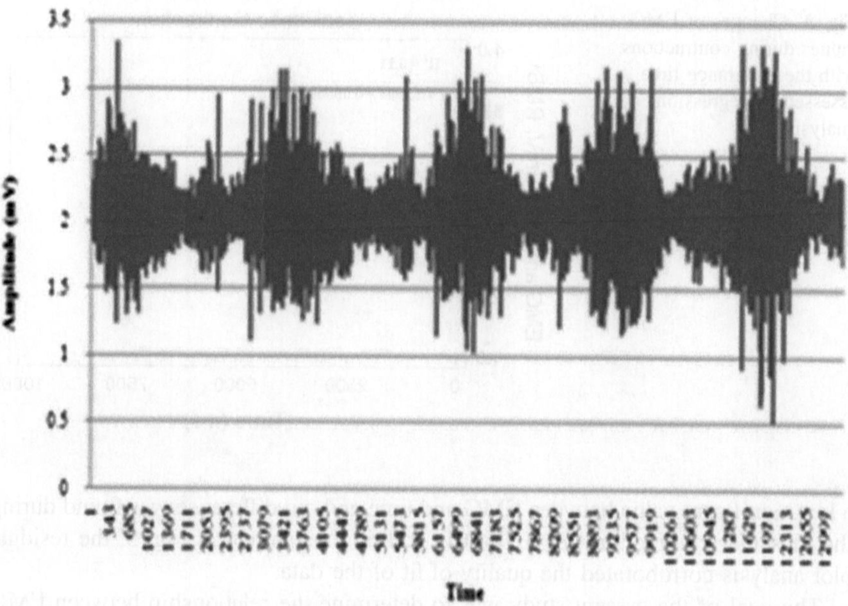

Fig. 2 Sampled waveforms of the surface EMG signal during Isokinetic contractions

were taken as 100 % MVC). The filtered EMG activity was normalized within each subject by dividing the observed EMG value for BB by the maximum value recorded during the three maximal tests. The mean (RMS) normalized EMG activity was then calculated as the mean of the sum of the normalized EMG percentages from the subject in each contraction. The EMG associated with 100 % MVC was designated as 100 % and fractions thereof. The maximum peak-to-peak value of the EMG was considered as a relative measure of motor activity (the positive portion of the peak is defined as the peak-maximum). Statistical analysis was performed using Minitab™ software (version 13.32). Significant differences in three contractions between the time lag and EMG amplitudes (RMS) were detected using repeated measures with analysis of variance (ANOVA), and post hoc tests were applied to test on the differences (between trials) with the significance level set at $\alpha = 0.05$, 95 % confidence intervals for all variables. Linear regression (r^2) analysis was used to analyze the relationship between time and the EMG variables.

3 Results and Discussion

The relationship between the EMG activity (amplitude) and time during isokinetic contraction has been demonstrated in Fig. 3. The slope of the regression line between EMG and time is linear but, the predictability among the two parameters are significantly inferior ($r^2 = 0.11$). Also, a low F-ration of 3.73 was observed with

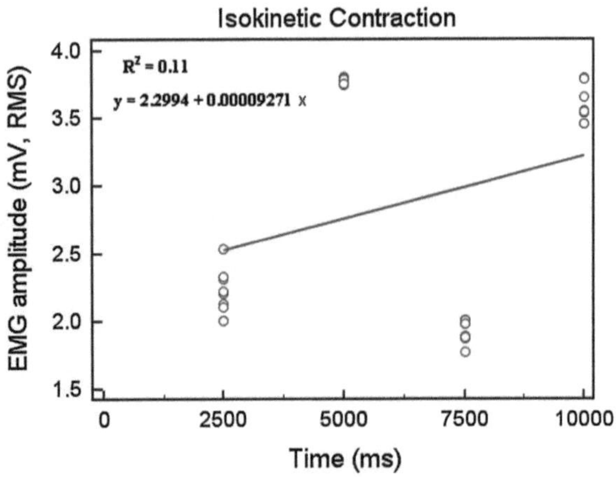

Fig. 3 Changes of EMG values during contractions with the endurance time assessed by regression analysis

a low significant value between EMG and time and, no difference was found during the muscular contraction (p = 0.062) In addition, as shown in Fig. 3, the residual plot analysis corroborated the quality of fit of the data.

The goal of the present study was to determine the relationship between EMG amplitude and time lag during isokinetic contraction. Isokinetic tests are often applied to evaluate muscular strength and EMG activity from the different subjects (Oliveira et al. 2012), but its behavior in terms of time series is still a matter for discussion. To well define the individual muscle activities around the BB area during contractions, previous studies attempted to use regression-based techniques to predict muscle activities. For example, Beck et al. (2005) found the BB muscle activity based on inter-electrode distance and established the torque relationship between isokinetic and isometric using EMG amplitude and mean power frequency. In this study the quantitative relation between the EMG activity and endurance time has been investigated. The EMG activity in respect with the time relationship shows inferior activity during the contraction.

The outcomes from this research may be helpful to researchers who are interested in EMG analysis in terms of time, sensor placement, and isokinetic contraction. In addition, our results can be used for different practical applications for the noninvasive evaluation in BB. Our findings may also help to develop EMG-assisted biomechanical modeling techniques as well as prosthetic devices. Surface EMG is commonly used to control prosthetic devices, while in most industrial devices the utilized force is estimated proportionally to muscle activity (Kamavuako et al. 2013). Further studies are needed to define the EMG-force relationship in terms of different sensor placement locations during isokinetic contraction. Such studies must include different anthropometric parameters variations (e.g., age, sex, and muscle thickness) during movements of the upper limbs at different angles and various external conditions that influence the signal amplitude (e.g., heat versus. cold). There were some limitations in this study with regard to the tests performed. First, the method which used to normalize the EMG data and sensors throughout the study were not capable to record the activity from the deep muscle tissues of the BB. Second, all subjects were given verbal instruction during the MVC trials and seated in identical testing positions in order to ensure maximal effort. However, without the use of the superimposed burst technique (short or long period), there is no way to determine whether a subject afforded maximal effort during the procedure (Kent-Braun and Le Blanc 1996; Norcros et al. 2010).

4 Conclusion

Considering the outcomes and the above discussion, we found that EMG activity is variable and less significant in terms of time lags during isokinetic contraction. The results from the current study outline a valuable guideline of understanding the BB muscle activity during such type of dynamic contraction. Our findings can also be applied to biomedical engineering methods for the analysis and control of the neuromuscular system, ergonomics field of research, rehabilitation engineering, and movement biomechanics.

5 Acknowledgements

This research was supported by the Universiti Malaysia Pahang (UMP) Research Grant Scheme, 2014 (No. RDU1403157).

References

Beck TW, Housh TJ, Johnson GO, Weir JP, Cramer JT, Coburn JW, Malek MH (2005) The effects of interelectrode distance on electromyographic amplitude and mean power frequency during isokinetic and isometric muscle actions of the biceps brachii. J Electromyogr Kinesiol 15(5):482–495

Benedetti MG, Ebenbichler G, Loisel P, Roy SH (2001) Clinician's view: dynamic EMG. Eng Med Biol Mag IEEE 20(6):33–37. doi:10.1109/51.982273

Clark BC, Manini TM, Doldo NA, Ploutz-Snyder LL (2003) Gender differences in skeletal muscle fatigability are related to contraction type and EMG spectral compression. J Appl Physiol 94(6):2263–2272

Fernández-Pena E, Lucertini F, Ditroilo M (2009) A maximal isokinetic pedalling exercise for EMG normalization in cycling. J Electromyogr Kinesiol 19(3):e162–e170

Hermens HJ, Freriks B, Merletti R, Stegerman D, Block J, Gre A (1999) SENIAM: European recommendations for surface electromyography Roessingh Research and Development, Enschede. http://www.seniam.org/

Horita T, Ishiko T (1987) Relationships between muscle lactate accumulation and surface EMG activities during isokinetic contractions in man. Eur J Appl Physiol 56(1):18–23

Howatson G, van Someren K (2005) The reproducibility of peak isokinetic torque and EMG activity in unfamiliarised subjects on repeated days. Isokinetics Exerc Sci 13(3):179–186

Jingyi Y, Ruiyuan W, Kaiyu X, Anping H (1995) The EMG determination and analysis of quadriceps isokinetic concentric contraction. J Beijing Univ Phys Educ 4:004

Kamavuako EN, Rosenvang JC, Bøg MF, Smidstrup A, Erkocevic E, Niemeier MJ, Farina D (2013) Influence of the feature space on the estimation of hand grasping force from intramuscular EMG. Biomed Sig Process Control 8(1):1–5

Kellis E, Baltzopoulos V (1996) The effects of normalization method on antagonistic activity patterns during eccentric and concentric isokinetic knee extension and flexion. J Electromyogr Kinesiol 6(4):235–245

Kellis E, Baltzopoulos V (1997) Agonist and antagonist moment and EMG-angle relationship during isokinetic eccentric and concentric exercise. Occup Health Ind Med 36(4):180–181

Kent-Braun JA, Le Blanc R (1996) Quantitation of central activation failure during maximal voluntary contractions in humans. Muscle Nerve 19(7):861–869

Komi PV, Viitasalo JHT (1976) Signal characteristics of EMG at different levels of muscle tension. Acta Physiol Scand 96(2):267–276

Mandryk RL, Atkins MS, Inkpen KM (2006) A continuous and objective evaluation of emotional experience with interactive play environments. In: Paper presented at the proceedings of the SIGCHI conference on human factors in computing systems

Mercer J, Bezodis N, DeLion D, Zachry T, Rubley M (2006) EMG sensor location: Does it influence the ability to detect differences in muscle contraction conditions? J Electromyogr Kinesiol 16(2):198–204

Merletti R, Bottin A, Cescon C, Farina D, Gazzoni M, Martina S, Mesin L, Pozzo M, Rainoldi A, Enck P (2004) Multichannel surface EMG for the non-invasive assessment of the anal sphincter muscle. Digestion 69(2):112–122

Norcross MF, Troy Blackburn J, Goerger BM (2010) Reliability and interpretation of single leg stance and maximum voluntary isometric contraction methods of electromyography normalization. J Electromyogr Kinesiol 20(3):420–425

Oliveira A, Corvino R, Gonçalves M, Caputo F, Denadai B (2012) Maximal isokinetic peak torque and EMG activity determined by shorter ranges of motion. Human Mov 13(2):102–108

Perry-Rana SR, Housh TJ, Johnson GO, Bull AJ, Berning JM, Cramer JT (2002) MMG and EMG responses during fatiguing isokinetic muscle contractions at different velocities. Muscle Nerve 26(3):367–373

Pullman SL, Goodin DS, Marquinez AI, Tabbal S, Rubin M (2000) Clinical utility of surface EMG Report of the therapeutics and technology assessment subcommittee of the american academy of neurology. Neurology 55(2):171–177

Putatunda R (2011) Human muscular system. http://www.buzzle.com/articles/human-muscular-system.html. Accessed on 28 Nov 2011

Rainoldi A, Melchiorri G, Caruso I (2004) A method for positioning electrodes during surface EMG recordings in lower limb muscles. J Neurosci Methods 134(1):37–43

SENIAM (2012) http://www.seniam.org/. Accessed on 15th Mar 2012

Zhang J, Sun D (2011) Surface EMG observation and isokinetic test on pressing-kneading manipulations for exercise fatigue of anterior tibial muscle. J Acupunct Tuina Sci 9(1):62–66

Chapter 49
The Effects of Eight Weeks Consecutive Swimming Exercise and Estrogen Therapy on Cardiovascular Risk Factors in Ovariectomised Rat Model

Wan Mohd Norsyam Wan Norman, Asok Kumar Ghosh, Chen Chee Keong and Siti Amrah Sulaiman

Abstract Menopause plays an important role in lipid distribution that leads to the apparent acceleration in cardiovascular diseases in women. The aim of this study is to investigate the effectiveness of combination of regular exercise and estrogen therapy in reducing cardiovascular risk factors in postmenopausal animal model. Forty-eight Sprague Dawley female rats aged 3 months were used in the study and randomly divided into two main groups which were ovariectomy and sham operation groups. Later, four groups which were nontreatment, estrogen therapy, exercise and combination estrogen therapy and exercise derived from ovariectomy controlled group and nontreatment and exercise groups derived from sham-controlled group. Each group consists of seven to nine rats. Bilateral ovariectomy was done for all experimental and ovariectomised controlled groups while no ovary removal was performed to sham controlled groups. Estrogen therapy and swimming exercise were given for consecutive 8 weeks and body weight

W.M.N. Wan Norman (✉)
Faculty of Sports Science & Recreation, Universiti Teknologi MARA, Jengka, Pahang, Malaysia
e-mail: norsyam@pahang.uitm.edu.my

A.K. Ghosh
Faculty of Medicine and Health Sciences, Universiti Tunku Abdul Rahman, Petaling Jaya, Selangor, Malaysia
e-mail: asok@utar.edu.my

C.C. Keong
Sports Science Unit, School of Medical Sciences, Universiti Sains Malaysia, Kota Bharu, Kelantan, Malaysia
e-mail: ckchen@kck.usm.my

S.A. Sulaiman
Department of Pharmacology, School of Medical Sciences, Universiti Sains Malaysia, Kota Bharu, Kelantan, Malaysia
e-mail: sbsamrah@usm.my

N.A. Yacob et al. (eds.), *Regional Conference on Science, Technology and Social Sciences (RCSTSS 2014)*, DOI 10.1007/978-981-10-0534-3_49

progress was measured weekly. Rats were sacrificed after a period of 8 weeks intervention and blood was taken for lipid analysis. The body weight progression of the rats was analyzed by using paired *t*-test and repeated measures ANOVA. The difference on plasma lipid profile was analyzed by using one-way ANOVA. The results showed that there was a significant ($P < 0.05$) increase in the body weight 4 weeks after ovariectomy. No significant difference ($P > 0.05$) detected in the body weight after 8 weeks intervention. Estrogen therapy and swimming exercise treatment lowered the total plasma cholesterol and low-density lipoprotein (LDL) cholesterol levels as compared with estrogen therapy only and a higher plasma high-density lipoprotein (HDL) cholesterol was observed in exercise treatment group. In conclusion, estrogen therapy and exercise treatment might be beneficial on lipid and lipoprotein metabolism which led to an increase in cardiovascular health in ovariectomised rat model.

Keywords Cardiovascular disease · Estrogen therapy · Menopause · Ovariectomy · Swimming exercise

1 Introduction

Cardiovascular disease (CVD) is less common in premenopausal women than men of the same age or postmenopausal women. CVD includes coronary heart disease (CHD), cerebrovascular disease, and/or peripheral arterial disease which attribute more than 30 % of global death (World Health Organization (WHO) 2011). Furthermore, statistics had shown that 52.6 % died due to CVD and CVD had become a leading cause of death in American women (Zhang 2010). A study had shown that postmenopausal women exhibit higher risk for CVD as compared to nonmenopausal women in Malaysia (Ariffin et al. 2012).

Estrogen replacement therapy seems to improve the estrogen levels in postmenopausal women, and thereby may reduce the risk of having cardiovascular disease (Chu et al. 2006). This may be due to the fact that estrogen replacement therapy improves blood flow by releasing a potent vasodilator known as nitric oxide (Moien-Afshari et al. 2003). Nevertheless, recent study had indicated that the usage of hormone replacement therapy (HRT) may ameliorate coronary heart disease or stroke (White et al. 2010) and yet there is no clear enlightenment for the divergence. Hence the type, dose, and route of administration of HRT are suggested being the main factors of the conflicting results (Khalil 2013).

Regular exercise elicits a lot of benefit to both men and women ranging from increased longevity and decreased risk of cardiovascular disease. Thus, endurance exercise helps to reduce the metabolic risk variables and improves the lipid profile as had been seen in postmenopausal women (Wang et al. 2014). Therefore, combination of both estrogen replacement therapy and regular exercise may reduce the cardiovascular risk factors in postmenopausal women. However, the effectiveness of combination of estrogen therapy and regular exercise in reducing cardiovascular risk factors in postmenopausal women is scanty in the literature. Moreover, the specific effect on weight-bearing type of exercise, particularly swimming exercise to the postmenopausal women is not well understood.

It had been shown that menopause alters the lipid profile in both humans either via natural menopause (Kuh et al. 2005) or surgical menopause (Baksu et al. 2007) and in an animal model (Kishida et al. 2003). Rats have been used as model system to study substance with estrogenic effects on plasma lipid levels and ovariectomised rats resemble the decline in estrogen levels in postmenopausal women (Gallo et al. 2005). Hence, the present study will investigate the effect of menopause, estrogen replacement therapy and exercise on body weight, blood pressure, lipid profiles, and blood glucose in animal model. The outcome of this study will demonstrate the role of exercise and ERT in manipulating metabolic syndrome parameters in postmenopausal state.

2 Materials and Methods

Animals Preparation

All procedures were carried out in accordance with the institutional guidelines for animal research of the Universiti Sains Malaysia. Forty-eight adult female *Sprague Dawley* rats, 3-month-old, were used in the experiment. The rats were housed at normal room temperature with adequate ventilation and normal 12-h light–dark cycle with free access to food (commercial laboratory rat food) and water and were divided into two main groups: Sham operation and ovariectomised. After 4 weeks of postsurgery, sham operation groups were divided into sham-controlled (Sham, $n = 8$) and sham-controlled with exercise (Sham-Ex, $n = 8$). The ovariectomised group was divided into ovariectomised-controlled (Ovx, $n = 8$), estrogen therapy treated (Ovx-ERT, $n = 8$), exercise treated (Ovx-Ex, $n = 8$), and combination of exercise and estrogen therapy (Ovx-ERT-Ex, $n = 8$).

Ovariectomy

Bilateral ovariectomy was performed under anesthesia using a ventral approach. Rats were anesthetized via intramuscular injection of Xylazil 20 (Troy Laboratories, Australia) 10 mg/kg and Ketamil (Troy Laboratories, Australia) 90 mg/kg. After the surgery, rats were randomly allocated to the nominated groups.

Interventions

The intervention started 1 month after the surgery in the following manner: estrogen therapy (120 μg/kg/day) (Al-Wahaibi et al. 2008) was given via gavage technique. The exercise groups swam in a large acrylic tank filled with water 30 cm in depth in which water temperature was maintained 28 ± 2 °C and changed weekly. Each rat swam in its own space and dried before being returned to the cage. The exercise rats were submitted to continuous moderate swimming exercise for five consecutive times during a week and 8 weeks period (Guerra et al. 2007). Adaptation time in accordance with the volume principles of training was as follows: on day-1 the rats swam for 30 min; on day-2 for 60 min; on day 3 and throughout the remaining days for an hour and a half. All rats adapted to the conditions imposed.

The body weight of the rats were measured by using digital weighing scale (OHAUS Navigator, Switzerland) every week throughout the study to monitor the body weight changes. At the end of the experimental period, the blood was collected from the rats via decapitation methods and stored in an appropriate anticoagulant tube before centrifuged to separate the plasma. An amount of 200 μl plasma was used and analyzed for total cholesterol, low-density lipoprotein (LDL) and high-density lipoprotein (HDL) cholesterol by using automated chemistry analyser (Vitalab, USA).

Statistical Analysis

The mean differences in body weight among all groups were analyzed using repeated measures analysis of variance (RM-ANOVA) with post hoc by using Bonferroni test. The difference of each cholesterol parameter among all groups were measured by using one-way analysis of variance (ANOVA) with LSD post hoc test. All analysis was made using statistical package for social science (SPSS) version 20.0. Significant value was reported when $P < 0.05$.

3 Results and Discussion

Body Weight

Rats show an increase in body weight significantly ($P < 0.05$) after 2 weeks of ovariectomy and became highly significant ($P < 0.01$) after 4 weeks in ovariectomised groups (Fig. 1). However, no significant difference in body weight existed after second and fourth week after surgery in sham operated rats as compared to their baseline body weight. Therefore, it is suggested that ovariectomy-induced

Fig. 1 Body weight changes upon surgery ($n = 8$ for each group). *$P < 0.05$ versus 0 week

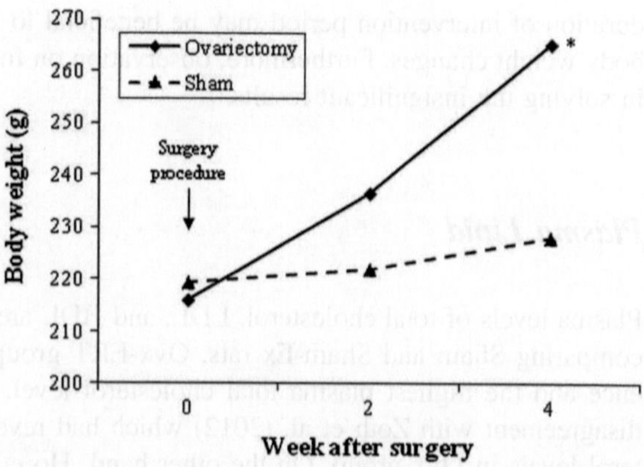

weight gain may be due to increase in food intake among ovariectomised rats (Fazlina et al. 2009).

Current study on ovariectomised rat models revealed that neither estrogen replacement therapy nor exercise treatment nor even combination of estrogen therapy and exercise treatment were sufficient enough to reduce the body weight even after 8 weeks of intervention period, as it is reported that $P > 0.05$ among all groups. However, the mean difference in body weight changes occurs in all ovariectomised groups regardless of the treatment given with the Sham group (Fig. 2). This finding was lead to the inconsistency results in body weight changes with regard to the effect of estrogen therapy and exercise treatment for ovariectomised animal model. Moreover, Figard et al. (2006) did not find any significant difference in body weight after 8 weeks of swimming exercise. Therefore, a longer

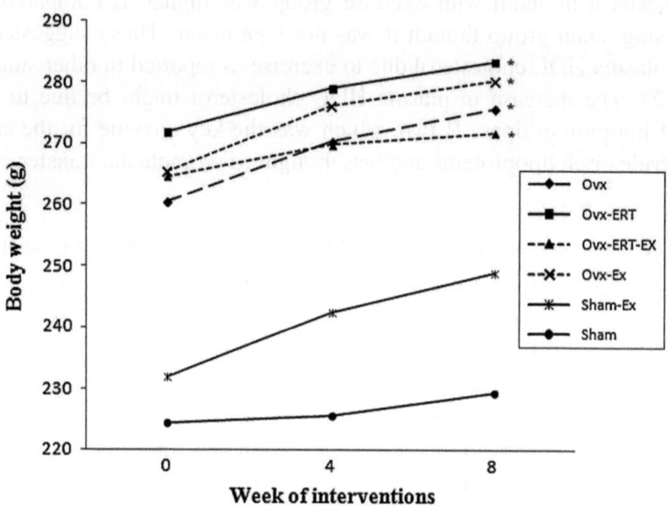

Fig. 2 Body weight changes throughout 8 weeks of intervention. *$P < 0.05$ versus Sham group

duration of intervention period may be beneficial to see a significant difference in body weight changes. Furthermore, observation on food intake also may be helpful in solving the insignificant results.

Plasma Lipid

Plasma levels of total cholesterol, LDL, and HDL are demonstrated in Table 1. On comparing Sham and Sham-Ex rats, Ovx-ERT group exhibits a significant difference and the highest plasma total cholesterol level. Hence, current study was in disagreement with Zoth et al. (2012) which had revealed the highest total cholesterol levels in Ovx group. On the other hand, Ho et al. (2010) found higher total cholesterols in ovariectomised rats which received only estrogen therapy compared to other groups. Therefore, the increment of total plasma cholesterol upon estrogen therapy was still not well understood.

Swimming exercise seems to be an effective way of lowering LDL cholesterol. A significant low level of LDL cholesterol in Sham and Sham-Ex group was compared with either of the received estrogen therapy . It was well documented that ovariectomy altered the plasma lipid due to the removal of estrogen sources. Consequently, LDL can be lowered by taking exogenous estrogen thus, reducing the risk of cardiovascular disease in humans (Bittner 2001), which was not parallel with this study. In the present study, highest LDL level was observed in Ovx-ERT group. However, Sham-Ex exhibited lowest LDL level and thus suggested the promising effect of exercise.

Plasma high-density lipoprotein cholesterol did not reveal significant difference between sham-controlled and sham with exercise group. However, the mean plasma HDL cholesterol in sham with exercise group was higher as compared with the nonexercising sham group though it was not significant. These suggested the elevation of plasma HDL cholesterol due to exercise as reported in other studies (Zoth et al. 2012). The increase in plasma HDL cholesterol might be due to increased activity of lipoprotein lipase (LPL), which was the key enzyme for the catabolism of triglycerides-rich lipoproteins and was thought to promote the transfer of lipids to

Table 1 Effects on total cholesterol, LDL, and HDL plasma in rats after intervention

Intervention group	Total cholesterol (mmol/L)	LDL (mmol/L)	HDL (mmol/L)
Sham	1.47 ± 0.27	0.13 ± 0.03	0.31 ± 0.14
Sham-Ex	1.50 ± 0.42	0.11 ± 0.03	0.40 ± 0.24
Ovx	1.67 ± 0.23	0.18 ± 0.04^{b}	$0.25 \pm 0.08^{b,c}$
Ovx-ERT	$2.00 \pm 0.20^{a,b}$	$0.21 \pm 0.05^{a,b}$	0.26 ± 0.10
Ovx-Ex	1.75 ± 0.10	0.17 ± 0.06	0.40 ± 0.12
Ovx-ERT-Ex	1.60 ± 0.27	0.14 ± 0.04	$0.20 \pm 0.05^{b,c}$

Values express as mean ± SD

$^{a}P < 0.05$ versus Sham; $^{b}P < 0.05$ versus Sham-Ex; $^{c}P < 0.05$ versus Ovx-Ex

HDL from chylomicrons and a very low-density lipoproteins as reported in Imamura et al. (2000). Other potential mechanisms included augmented HDL cholesterol production by exercise-induced increases in lecithin-cholesterol acyl-transferase (LCAT) activity and diminished HDL cholesterol clearance by exercise-related decreases in hepatic lipase activity (Pronk 1993).

4 Conclusion

It was concluded from this study that ovariectomy exhibited similar symptoms due to the loss of estrogen function which increased the body weight after 3 weeks of surgery. Nevertheless, 8 weeks of estrogen therapy and exercise treatment was insufficient to reduce the body weight of ovariectomised rats. However, a longer intervention period might be needed in order to show an improvement in body weight reduction. The estrogen therapy and swimming exercise treatment lowered the total plasma cholesterol and LDL levels. Meanwhile a higher plasma HDL was observed in ovariectomy with exercise treatment group.

Therefore, the current study had demonstrated that estrogen therapy and exercise treatment might be beneficial on lipid and lipoprotein metabolism which led to an increase in cardiovascular health in ovariectomised rats. However, clinical trials are warranted to investigate their effects on the target population who are post-menopausal women.

Acknowledgments This work was supported by the USM Short Term Grant (304/PPSP/6131545).

References

Al-Wahabi A, Wan Nazaimoon WM, Norsyam WN, Farihah HS, Azian AL (2008) Effect of water extract of*Labisia pumila* var *alata* on aorta of ovariectomized Sprague Dawley rats. Pak J Nutr 7(2):208–213

Ariffin F, Hamid HA, Daher AM, Safura RA, Krishnapillai ADS, Miskan M, Keat NK, Razak SA, Nawawi H, Yusoff K (2012) Relationship between menopause and cardiovascular risk factors in Malaysia. J Hypertens. doi:10.1097/01.hjh.0000420071.05038.35

Baksu B, Davas I, Agar E, Akyol A, Uluocak A (2007) Do different delivery systems of estrogen therapy influence serum lipids differently in surgical menopause women? J Obstet Gynaecol Res 33(3):346–352

Bittner V (2001) Estrogens, lipids and cardiovascular disease: no easy answers. J Am Coll Cardiol 37(2):431–433

Chu MC, Cosper P, Nakhuda GS, Lobo RA (2006) A comparison of oral and transdermal short-term estrogen therapy in postmenopausal women with metabolic syndrome. Fertil Steril 86(6):1669–1675

Fazlina M, Wan Nazaimoon WM, Gu HF, Ostenson CG (2009) Labisia pumila extract regulates body weight and adipokines in ovariectomized rats. Maturitas 62:91–97

Figard H, Mougin F, Gaume V, Barthelot A (2006) Combined intervention of dietary soy bean proteins and swim training: effect on bone metabolism in ovariectomized rats. J Bone Min Metab 24:206–212

Gallo D, Zannoni GF, Apollonio P, Martinelli E, Ferlini C, Passetti G, Riva A, Morazzoni P, Bombardelli E, Scambia G (2005) Characterization of the pharmacologic profile of a standard soy extract in the ovariectomized rat model of menopause: effect on bone, uterus and lipid profile. Menopause 12(5):589–600

Guerra RLF, Prado WL, Cheik NC, Viana FP, Botero JP, Vendramini RC, Carlos IZ, Rossi EA, Dâmaso AR (2007) Effect of 2 or 5 consecutive exercise days on adipocyte area and lipid parameter in Winstar rats. Lipids Health Dis 6(16):1–24

Ho L, Wang Y, Duan Y, Bu S (2010) Effects of treadmill exercise training on liver fat accumulation and estrogen receptor alpha expression in intact and ovariectomized rats with or without estrogen replacement treatment. Eur J Appl Physiol 109:879–886

Imamura H, Katagiri S, Uchida K, Miyamoto N, Nakano H, Shirota T (2000) Acute effects of moderate exercise on serum lipids, lipoproteins and apolipoproteins in sedentary young women. Clin Exp Pharmacol Physiol 27:975–979

Khalil RA (2013) Estrogen, vascular estrogen receptor and hormone therapy in postmenopausal vascular disease. Biochem Pharmacol 86:1627–1642

Kishida T, Miyazato S, Ogawa H, Ebihara K (2003) Taurine prevent hypercholesterolemia in ovariectomized rats fed corn oil but not in those fed coconut oil. J Nutr 133:2616–2621

Kuh D, Langenberg C, Hardy R, Kok H, Cooper R, Butterworth S, Wadsworth MEJ (2005) Cardiovascular risk at age 53 years in relation to the menopause transition and use of hormone replacement therapy: a prospective British birth cohort study. Int J Obstet Gynaecol 112: 476–685

Moein-Afshari F, Kenyon E, Choy JC, Battistini B, MacManus BM, Laher I (2003) Long term effect of ovariectomy and estrogen replacement treatment on endothelial function on mature rats. Mauritas 45:213–223

Pronk NP (1993) Short term effects of exercise on plasma lipids and lipoproteins in humans. Sports Med 16:431–448

Wang CH, Chung MH, Chan P, Tsai JC, Chen FC (2014) Effects of endurance exercise training on risk components for metabolic syndrome, interleukin-6, and the exercise capacity of postmenopausal women. Geriatr Nurs 35:212–218

White RE, Gerrity R, Barman SA, Han G (2010) Estrogen and oxidative stress: a novel mechanism that may increase the risk for cardiovascular disease in women. Steroids 75:788–793

World Health Organization (2011) Global status report on non-communicable diseases 2010. World Health Organization, Geneva

Zhang Y (2010) Cardiovascular disease in American women. Nutr Metab Cardiovasc Dis 20: 386–393

Zoth N, Weigt C, Zengin S, Selder O, Selke N, Kalicinski M, Piechotta M, Diel P (2012) Metabolic effects of estrogen substitution in combination with targeted exercise training on the therapy of obesity in ovariectomized Wistar rats. J Steroid Biochem Mol Biol 130:64–72

Part IX
Health and Medicine

Chapter 50
Differentiating Benign from Malignant Adnexal Masses: Comparison of Two-Dimensional and Three-Dimensional Ultrasound Imaging

Marlina Tanty Ramli, Anushya Vijayananthan, Gan Gek Choo, Ghana Kumar and Lim Boon Kiong

Abstract This is a prospective study to compare two-dimensional (2D) and three-dimensional (3D) ultrasounds in differentiation of benign from malignant adnexal masses with histopathology as the gold standard. In this study, we performed 2D and 3D US on 94 adnexal masses from 76 women prior to surgery. Two radiologists independently analysed the 2D and 3D US images to differentiate benign from malignant adnexal lesions based on clinical judgment and ultrasound sassone scoring system. The inter-observer agreement and the sensitivity, specificity, positive predictive value (PPV), negative predictive value (NPV), and accuracy were calculated for both the techniques. There were 82 benign (87 %) and 12 malignant lesions (13 %). The inter-observer agreement was substantial to almost perfect with kappa value ranging from 0.825 to 1. Compared to 2D and 3D US is more superior ($p < 0.01$) in the detection of wall thickness (49 % vs. 68 %), thick papillary projections (14 % vs. 43 %) and thick septations (43 % vs. 51 %). The sensitivity, specificity, PPV, NPV and accuracy were 92 %, 77 %, 36 %, 98 % and 78 %, respectively, for 2D US and 92 %, 61 %, 26 %, 98 % and 65 %, respectively, for 3D US. In conclusion, 3D US imaging is superior to 2D US in defining septal structures and papillary projection on the inner surface of the

M.T. Ramli (✉)
Medical Imaging Unit, Faculty of Medicine, University Teknologi MARA, Shah Alam, Selangor, Malaysia
e-mail: marlina352@salam.uitm.edu.my

A. Vijayananthan · G.G. Choo · G. Kumar
Department of BioMedical Imaging, University Malaya Medical Centre, Kuala Lumpur, Malaysia

L.B. Kiong
Department of Obstetric & Gynaecology, University Malaya Medical Centre, Kuala Lumpur, Malaysia

N.A. Yacob et al. (eds.), *Regional Conference on Science, Technology and Social Sciences (RCSTSS 2014)*, DOI 10.1007/978-981-10-0534-3_50

tumour. However, it does not have a better diagnostic accuracy than 2D US in the diagnosis of ovarian cancer.

Keywords Adnexal masses · Sassone scoring · Two-dimensional · Three-dimensional · Ultrasound

1 Introduction

Adnexal mass is a commonly encountered problem in clinical practice and an accurate diagnosis is essential to establish optimal treatment (Parker et al. 1994). Ultrasound is, without doubt, the most commonly utilised diagnostic modality in the differential diagnosis of adnexal masses (Granberg 1993). Morphological analyses with 2D ultrasound help in narrowing the differential diagnosis of adnexal masses, however, the variable macroscopic features of these masses make a precise diagnosis from 2D ultrasound features alone difficult.

Studies using (2D) grayscale ultrasound ± colour Doppler have reported that the presence of intra-tumoral papillae, solid parts and thick septa suggests malignancy (Alcazar et al. 2003; Bonilla-Musoles et al. 1995). The use of 3D ultrasound will allow for more detail evaluation of the internal structures and surfaces of these masses, thus enabling earlier diagnosis of malignancy.

This study aims to determine if the use of 3D US will improve the diagnostic accuracy of adnexal masses; hence, the earlier detection of ovarian cancer.

2 Methods

This was a prospective study of 94 adnexal masses in 76 female patients that were scheduled for surgery (18 patients had bilateral adnexal masses). 2D and 3D US scans were performed preoperatively using a Philips iU22 Ultrasound system. No age or ethnic constraints were applied. All patients underwent surgery and definitive histological diagnoses were obtained. The masses were classified according to the World Health Organization criteria. Primary ovarian carcinomas were staged surgically according to International Federation of Gynaecology and Obstetrics (FIGO) criteria. Morphologic parameters, operative staging and histopathological diagnoses of the masses were correlated retrospectively. This study was approved by the Medical Ethics Committee of University Malaya Medical Centre.

The 2D and 3D US were performed via transabdominal approach by an ultrasonographer with 10 years experience. 2D US was done using a C5-2 Philips convex transducer (2–5 MHz) and the 3D US using Philips xMatrix x3-1 transducer. The x3-1 transducer consists of an array of 2D transducers that generate ultrasound pulses and process the echoes to generate 3D information in real time.

The 2D and 3D US images were saved into an external workstation with Philip QLAB quantification software for analysis. The images were analysed by two radiologists who were blinded to the histopathological results. Ovarian volume was calculated using the prolate ellipsoid formula (length × width × height × 0.523). An ovarian volume greater than or equal to 20 cm^3 in pre-menopausal women and greater than or equal to 10 cm^3 in post-menopausal women was considered abnormal (Kurjak et al. 2003).

2D US diagnoses of malignancy were done using Sassone et al. scoring system (Sassone et al. 1991). Scores equal or greater than 4 were regarded as malignant. Table 1 demonstrated the adnexal masses features that were being analysed. Surface rendering mode of 3D US allows the study of the inner wall surfaces of the masses. The diagnostic criteria for malignancy by 3D US were the presence of thick irregular septa, irregular/thick papillary projection, high echogenicity, irregular inner wall and solid component. Adnexal masses with two or more of these criteria were defined as malignant. Histopathological diagnoses were used as the gold standard.

The sets of data were analysed collectively using statistical software package (SPSS version 17.0). A p-value < 0.05 is considered statistically significant. Using the definitive histopathological diagnoses as the reference standard, sensitivity, specificity, positive and negative predictive value and accuracy, with its 95 % CI, were calculated.

Table 1 2D and 3D US morphological criteria for diagnosis of ovarian malignancy

Morphology	Criteria	2D	3D
Volume			
Pre-menopausal	<20 cm^3	0	0
	>20 cm^3	1	1
Post-menopausal	<10 cm^3	0	0
	>10 cm^3	1	1
Wall thickness	Thin < 3 mm	0	0
	Thick > 3 mm	1	1
Papillary projection	No	0	0
	Thin < 3 mm	1	1
	Thick > 3 mm	2	2
Septation	No	0	0
	Thin < 3 mm	1	1
	Thick > 3 mm	2	2
Solid area	No	0	0
	Yes	1	1
Echogenicity	Purely cystic	0	0
	Mixed echogenicity	1	1
	Solid and cystic	2	2
	Purely solid	2	2

3 Results

There were a total of 76 patients with age ranging from 17 to 76 years old (mean 40.9 ± 13.4). 13 patients were post-menopausal (17 %) and 63 patients were pre-menopausal (83 %). There were a total of 82 benign (87 %) and 12 malignant lesions (13 %). The most common ovarian malignancy was serous cystadenocarcinoma which was detected in six patients (6 %). The most common benign lesion was endometrioma with a total of 38 lesions (40 %). Table 2 listed the detailed histopahological diagnosis of the adnexal masses. All but one of the histologically confirmed malignant lesions was detected by both 2D and 3D US. The lesion that was missed was a borderline mucinous cystadenocarcinoma in a 20-year-old patient with 2D and 3D US appearance of a large benign simple cyst.

Kappa coefficent used for the intraobserver agreement between the two radiologists demonstrated substantial to almost perfect agreement with values ranging from 0.825 to 1 on both the 2D and 3D US. Wilcoxon signed-rank test ($p < 0.01$) also demonstrated no significant interobserver variability in the assessment of the morphological criteria of the adnexal masses between the two radiologists in the interpretation of both the 2D and 3D US images.

Table 3 summarises the detailed morphology criteria of the ovarian lesions that was detected in both the 2D and 3D US. When the results of the image analysis from both modalities are compared and analysed using the Wilcoxon signed-rank test, there is a significant difference in the detection of the wall thickness, septations and papillary projections between 2D and 3D US ($p < 0.01$).

Table 2 Histopathology diagnosis of the adnexal masses

Histopathological diagnosis	N (94)
Benign ovarian tumours ($n = 78$)	
Simple ovarian cyst	4
Teratoma	11
Benign serous cystadeno(fib)ma	10
Benign mucinous cystadeno(fib)ma	2
Endometrioma	38
Corpus luteal cyst	5
Haemorhagic cyst	6
Fimbrial cyst	2
Mature dermoid cyst	1
Follicular cyst	2
Primary ovarian cancer ($n = 9$)	
Serous (papillary) cystadenocarcinoma	6
Clear cell adenocarcinoma	1
Endometrioid adenocarcinoma	1
Poorly differentiated carcinoma	1
Borderline ovarian tumour ($n = 2$)	
Mucinous cystadeno(fib)oma	2
Metastatic tumour ($n = 1$)	
Adenocarcinoma of the bowel	1
Others (TB, Hydrosalpinx, Leiomyoma)	4

Table 3 Detailed morphological criteria of the ovarian lesions detected by 2D and 3D US

Morphology	Criteria	2D US (N)	2D US (%)	3DUS (N)	3D US (%)
Volume					
Pre-menopausal	<20 cm^3	6	7	6	7
	>20 cm^3	68	72	68	72
Post-menopausal	<10 cm^3	2	2	2	2
	>10 cm^3	18	19	18	19
Wall thickness	Thin < 3 mm	48	51	30	32
	Thick > 3 mm	46	49	64	68
Papillary projection	No	80	85	54	57
	Thin < 3 mm	1	1	0	0
	Thick > 3 mm	13	14	40	43
Septation	No	36	38	32	34
	Thin < 3 mm	18	19	14	15
	Thick > 3 mm	40	43	48	51
Solid area	No	57	61	53	57
	Yes	37	39	41	43
Echogenicity	Purely cystic	13	14	12	13
	Mixed echogenicity	72	77	72	77
	Solid and cystic	7	8	8	9
	Purely solid	2	1	2	1

Table 4 2D US diagnosis and histopathology diagnosis (n = 94)

		HPE	
		Benign	Malignant
2D US	Benign	62	1
	Malignant	20	11

Table 5 3D US Diagnosis with histopathology diagnosis (n = 94)

		HPE	
		Benign	Malignant
3D US	Benign	50	1
	Malignant	32	11

Tables 4 and 5 demonstrate the 2D and 3D US diagnosis of the adnexal masses compared to histopathology. The accuracy, sensitivity, specificity, positive predictive value (PPV) and negative predictive value (NPV) of 2D and 3D US in the diagnosis of ovarian malignancy against histopathology were 92 %, 77 %, 36 %, 98 % and 78 % for 2D US and 92 %, 61 %, 26 %, 98 % and 65 % for 3D US, respectively. Wilcoxon signed-rank test demonstrated significant difference between the predictive diagnosis of ovarian malignancy by 2D and 3D US as compared to the gold standard.

4 Discussion

Although the sample size was adequate, the number of malignant lesions was small (n = 12). There were only 17 % (n = 13) post-menopausal patients and 12 of them confirmed malignant lesions, 50 % presented in post-menopausal patients.

Inter-observer Agreement

There were substantial to almost perfect inter-observer agreement in the assessment of the morphological criteria of the adnexal masses between the radiologists. This provides evidence to the reproducibility, diagnostic precision and accuracy in determining the exact nature of the adnexal masses on both modalities by the two readers.

Comparison Between 2D and 3D US

Morphological evaluation of adnexal masses on US has been shown to be useful for the prediction of ovarian cancer. Most ovarian cancers demonstrate complex sonographic appearance with solid components, thick papillary projections and thick septations (Hatta et al. 2006; Geomini et al. 2007; Laban et al. 2007; Valentin et al. 2006). Compared to 2D US, 3D US is more superior in the detection of wall thickness (49 % vs. 68 %), thick papillary projections (14 % vs. 43 %) and thick septations (43 % vs. 51 %). There is no significant difference however, between 2D and 3D US, in the rate of detection of solid areas and degree of echogenicity of the masses. 2D US missed the papillary projectons in three malignant lesions which were clearly depicted on 3D US.

Comparison Between 2D and 3D US with Histopathology

The low specificity and positive predictive values in both modalities were due to the high false positive (FP). These were caused by complex lesions such as teratoma, dermoid tumours and fibroma that although benign, scored high in the scoring system. There were a total of 38 endometriomas, 11 teratomas and 1 dermoid tumour. All the teratomas and 12 of the endometriomas were diagnosed as malignant in our study. This was intensified on the 3D US as its capability for multiplanar detailed cross sectional imaging increased the detection rate of papillary projections and thick septations within a tumour (Figs. 1 and 2).

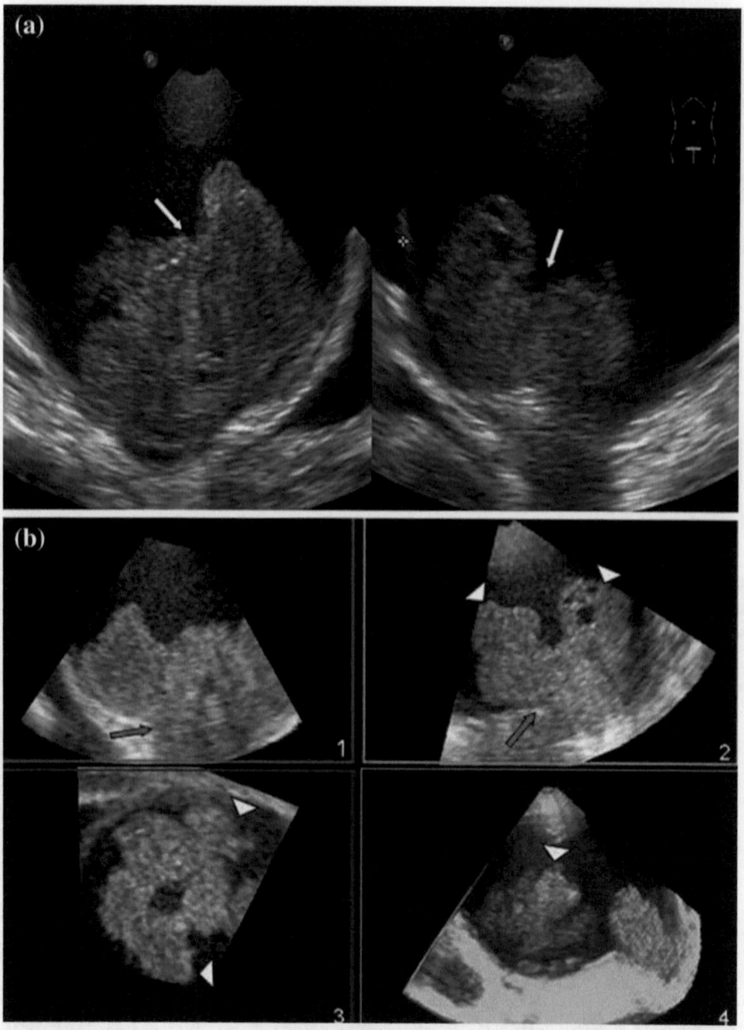

Fig. 1 **a** 2D US images showing the solid component (*white arrow*) within the tumour in a case of a 52-year-old pre-menopausal female with endometrioid carcinoma. **b** The corresponding 3D volumetric display clearly improved the definition of the solid component of the tumour, demonstrated the tumoral extension through the capsule (*red arrow*) and the multiple thick papillary projections (*white arrowhead*)

A borderline mucinous cystadenocarcinoma in a 20-year-old patient that had sonographic appearance of a simple cyst resulted in a false negative result across both modalities (Fig. 3).

Borderline mucinous ovarian tumour makes up about 10 % of mucinous ovarian neoplasm. A borderline tumour is an intermediate between a *mucinous cystadenoma* and a *mucinous cystadenocarcinoma* and is considered to be precursor of

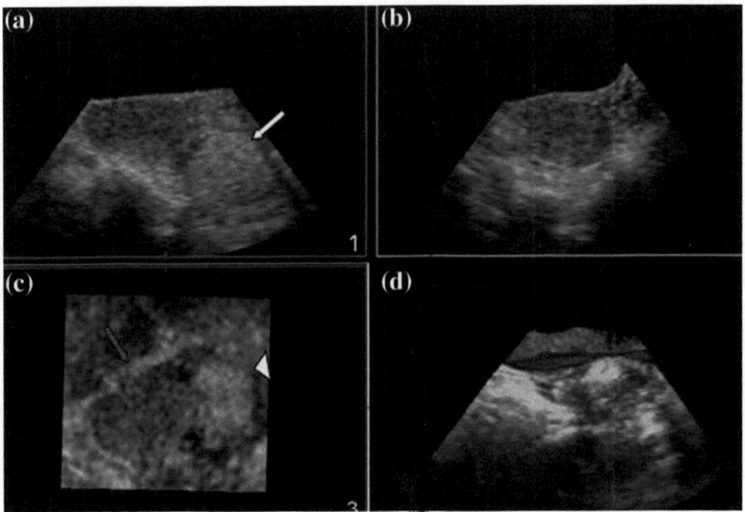

Fig. 2 The 3D feature of a benign mature teratoma in a 45-year-old pre-menopausal female. This complex lesion with solid component (*white arrow*), thick septation (*red arrow*—Fig. 2**c**) and papillary projections (*white arrowhead*—Fig. 2**c**) scored high in the scoring system, resulting in false positive result

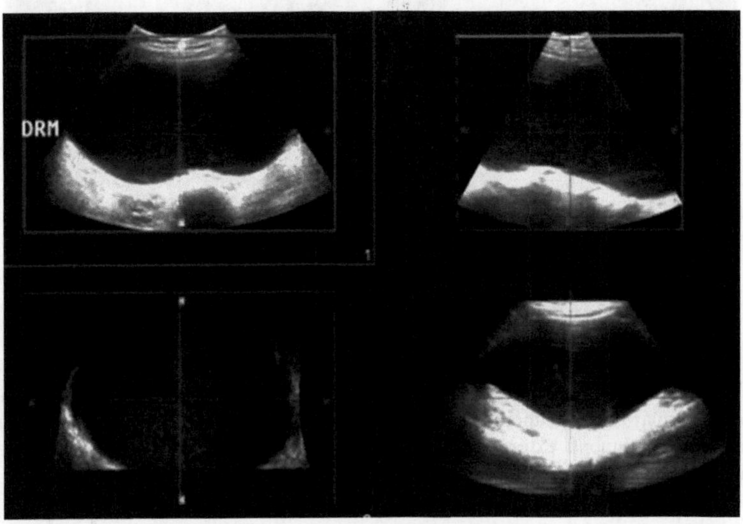

Fig. 3 The 3D volumetric display of the borderline mucinous cystadenocarcinoma in a 20-year-old patient that resulted in a false negative result in both modalities. There was no demonstrable suspicious feature to suggest malignancy

low-grade ovarian cancers (Lalwani et al. 2010; Jones 2006). Ultrasonography is the primary screening imaging technique in the evaluation of any suspected adnexal mass. Mucinous cystadenomas usually present as a large, unilateral, smooth surfaced, multilocular or unilocular cystic masses containing watery or viscous mucoid material (Zanetti et al. 2001; Brown et al. 1998; Alcazar et al. 1999). A borderline tumour often presents with identical gross features and are indistinguishable from a simple mucinous cystadenoma on ultrasound. Hence, this explains the false negative result that we had in our study. MRI and CT can characterise adnexal masses into benign and malignant up to 93 % and 89 % of the cases, respectively. Sonographically, indeterminate lesions would benefit from MRI for further characterisation (Sohaib et al. 2003; Tsili et al. 2008; Bazot et al. 2006). However, final diagnosis and staging of borderline ovarian tumours will still require pathologic evaluation after surgical excision (Van Nagell and DePriest 2005).

5 Conclusion

In conclusion, our study demonstrated that 3D US imaging was superior compared to 2D US, in defining septal structures and papillary projection on the inner surface of the tumour. This superiority is a drawback as it leads to overclassification and overdiagnosis of benign adnexal masses into malignant lesions that affect the outcome of the study.

References

Alcázar JL, Galan MJ, García-Manero M, Guerriero S (2003) Three-dimensional sonographic morphologic assessment in complex adnexal masses: preliminary experience. J Ultrasound Med 22(3):249–254

Alcázar JL, Errasti T, Zornoza A, Minguez JA, Galan MJ (1999) Transvaginal color doppler ultrasonography and CA-125 in suspicious adnexal masses. Int J Gynecol Obstet 66(3):255–261

Bazot M, Nassar-Slaba J, Thomassin-Naggara I, Cortez A, Uzan S, Darai E (2006) MR imaging compared with intraoperative frozen-section examination for the diagnosis of adnexal tumors: correlation with final histology. Eur J Radiol 16:2687–2699

Bonilla-Musoles F, Raga F, Osborne NG (1995) Three-dimensional ultrasound evaluation of ovarian masses. Gynecol Oncol 59(1):129–135

Brown DL, Doubilet PM, Miller FH, Frates MC, Laing FC, DiSalvo DN, Benson CB, Lerner MH (1998) Benign and malignant ovarian masses: selection of the most discriminating gray-scale and Doppler sonographic features. Radiology 208(1):103–110

Granberg S (1993) Macroscopic characterization of ovarian tumors and the relation to the histological diagnosis: Clinical Obstetrics & Gynecology. 36(2):363–374

Geomini PMAJ, Coppus SFPJ, Kluivers KB, Bremer GL, Kruitwagen RFPM, Mol BMJ (2007) Is three-dimensional ultrasonography of additional value in the assessment of adnexal masses? Gynecol Oncol 106(1):153–159

Hata T, Yanagihara T, Hayashi K, Yamashiro C, Ohnishi Y, Akiyama M, Manabe A, Miyazaki K (2006) Ultrasonographic evaluation of adnexal masses. Ultrasound Med Biol 32(5,1):1228–1269

Jones MB (2006) Borderline ovarian tumors: current concepts for prognostic factors and clinical management. Clin Obstet Gynecol 49:517–525

Kurjak A, Kupesic S, Sparac V, Prka M, Bekavac I (2003) The detection of stage I ovarian cancer by three-dimensional sonography and power Doppler. Gynecol Oncol 90(2):258–264

Laban M, Metawee H, Elyan A, Kamal M, Kamel M, Mansour G (2007) Three-dimensional ultrasound and three-dimensional power Doppler in the assessment of ovarian tumors. Int J Gynecol Obstet 99(3):201–205

Lalwani N, Shahbhogue AK, Vikram R, Nagar A, Jagirdar J, Prasad SR (2010) Current update on borderline ovarian neoplasm. Am J Roentgenol 194(2):330–336

Parker WH, Levine RL, Howard FM, Sansone B, Berek JS (1994) A multicentre study of laparoscopic management of cystic adnexal masses in post menopausal women. J Am Coll Surg 179:733–737

Sassone AM, Timor-Tritsch IE, Artner A (1991) Transvaginal sonographic characterization of ovarian disease: evaluation with a new scoring system to predict ovarian malignancy. Am J Obstet Gynaecol 78:70–76

Sohaib SA, Sahdev A, Van Trappen P, Jacobs IJ, Reznek RH (2003) Characterization of adnexal mass lesions on MR imaging. Am J Radiol 180:1297–1304

Tsili AC, Tsampoulas C, Charisiadi A, Khalef-Ezra J, Dousias V, Paraskevaidis E, Efremidis SC (2008) Adnexal masses: accuracy of detection and differentiation with multidetector computed tomography. Gynecol Oncol 110:22–31

Valentin L, Ameye L, Testa A, Lecuru F, Bernard JP, Paladini D, Van Huffel S, Timmerman D (2006) Ultrasound characteristics of different types of adnexal malignancies. Gynecol Oncol 102(1):41–48

Van Nagell JR, DePriest PD (2005) Management of adnexal masses in postmenopausal women. Am J Obstet Gynaecol 193(1):30–35

Zanetta G, Rota S, Lissoni A, Meni A, Brancatelli G, Buda A (2001) Ultrasound, physical examination, and CA 125 measurement for the detection of recurrence after conservative surgery for early borderline ovarian tumors. Gynecol Oncol 81:63–66

Chapter 51
Prevalence of Iron Deficiency Anemia (IDA) Among Medical Laboratory Technology Students in UiTM Puncak Alam

Mazura Bahari, Mohd Kamil Ariff Md Fiah, Wan Mazlina Md Saad and Safura Ramli

Abstract Approximately, two billion people are anemic based on the hemoglobin concentrations below recommended thresholds suggested by the World Health Organization (WHO). The most common cause of anemia is iron deficiency anemia (IDA) which is also known as nutritional anemia. The aim of this study is to identify the prevalence of IDA among MLT students in UiTM Puncak Alam through laboratory investigations. This cross-sectional study was conducted in UiTM Puncak Alam. Five milliliters of blood samples were drawn from each student into ethylenediaminetetraacetic acid (EDTA) tube. The blood samples were used to analyze the full blood count (FBC) and full blood picture (FBP). The serum ferritin test was performed if necessary for confirmation of IDA. The data were analyzed using the SPSS program for Windows version 18.0 and using the Mann–Whitney U test with 95 % confidence interval. This study concludes that the prevalence of IDA among MLT students in UiTM Puncak Alam is 5.6 %. Out of 162 students involved in this study, nine female students were detected as IDA and none of male student with IDA. This study provided the evidence of IDA among UiTM students in Puncak Alam which could be of importance to overcome the IDA problem in UiTM Puncak Alam.

Keywords Iron deficiency anemia (IDA) · Full blood count (FBC) · Full blood picture (FBP)

M. Bahari (✉) · M.K.A. Md Fiah · W.M.M. Saad · S. Ramli
Faculty of Health Sciences, Universiti Teknologi MARA,
Puncak Alam, Selangor, Malaysia
e-mail: mazurabahari@puncakalam.uitm.edu.my

N.A. Yacob et al. (eds.), *Regional Conference on Science, Technology and Social Sciences (RCSTSS 2014)*, DOI 10.1007/978-981-10-0534-3_51

1 Introduction

Anemia is a common disease that occurs in every country of the world. Milman (2011) reported that two billion people suffer from anemia worldwide and most of them have iron deficiency anemia (IDA), where 40–50 % of children under 5 years old. World Health Organization (WHO) estimates that approximately two billion people are anemic based on hemoglobin concentrations below recommended thresholds (WHO and UNICEF 2001). Anemia is referred as a condition of low hemoglobin (Hb) concentration in blood which is less than 12 g/dL and it can be caused by several factors (Siti Noor et al. 2006). IDA is the most important cause of microcyctic hypochromic anemia (Siti Noor et al. 2006). Morphology of red blood cells in blood film shows that the red blood cell is small (microcytic) and pale in color (hypochromic). Several studies have been done in Malaysia based on affected group and factor causing IDA. Nadarajan and Eow (2002) concluded that iron deficiency mostly occurred with the regular blood donation as a result of regular iron loss from each donated blood unit. Ferritin level was found to be significantly lower among regular blood donor rather than first time blood donor. A study among young nonpregnant women aged 20–40 years old in Kuala Lumpur reveals that the prevalence of IDA was the highest among the Indians (26.4 %) as compared to the Malays (16.4 %) (Loh and Khor 2010). Azma et al. (2012) reported that in the study which focuses on the university students concluded that out of 400 samples, there were about 14.5 % or 58 students had hypochromic microcytic indices that consisted of 11 % of thalassemia red cell indices while the other 3.5 % of all IDA were red cell indices. IDA is one of the most common public health concerns in Malaysia. Therefore, the aim of this study is to determine the prevalence of iron deficiency anemia (IDA) among Medical Laboratory Technology (MLT) students at MARA University of Technology (UiTM) Puncak Alam.

2 Material and Method

The study was conducted at Faculty of Health Sciences, UiTM Puncak Alam from February 2013 to June 2013. One hundred and sixty two (162) volunteered students involved in this study. Sample size is according to previous study done by Krejcie and Morgan. Consent forms for this study were sent out before the blood sample was taken and no refusals to participate. All blood samples were collected in ethylenediaminetetraacetic acid (EDTA) blood tubes by the qualified phlebotomist. This study was approved by the Research and Ethic Committee Faculty of Health Science, UiTM Puncak Alam. Full blood count (FBC) was measured using Beckman Coulter (Coulter LH 500) machine in the hematology laboratory at UiTM Puncak Alam. Blood samples which were showing red cell indices result of characteristic of anemia was preceded to full blood picture (FBP) examination. Thin blood smear was prepared using Leishman stain. The FBP slide was examined under microscope for analysis of RBC morphology. The abnormal hematology parameters results of participants blood were then tested with serum ferritin (SF) at the Biochemistry laboratory of Faculty Medicine at UiTM Sungai Buloh and measured using Hitachi analyzer. The data were analyzed using the SPSS program for Windows version 18.0. The data were analyzed using the Mann–Whitney U test with 95 % confidence interval for comparison between normal and IDA parameters of hemoglobin, hematocrit, MCV, MCH, MCHC, and RDW.

3 Results

The volunteered students consisted of 132 female (82 %) and 30 male (18 %) students. All the blood samples were analyzed for hematological parameters included Hemoglobin (Hb), Mean Cell Volume (MCV), Mean Cell Hemoglobin (MCH), Mean Cell Hemoglobin Concentration (MCHC), and Red cell Distribution Width (RDW) done by Beckman Coulture LH500 analyzer. Samples with low Hb, MCV, MCH, MCHC, and high RDW were analyzed for serum ferritin to diagnose IDA. 16 out of 162 student's blood samples were tested further test which are full blood picture (FBP) and ferritin in order to differentiate thalassaemia and iron deficiency anemia (IDA). Seven students were excluded from this study due to thalassemia trait, megaloblastic anemia, and other anemia.

Full Blood Count and Serum Ferritin Result

Full blood count results are summarized in Table 1, for level of hemoglobin more than 12 g/dl (>12 g/dl) had 29 males (96.7 %) and 117 females (88.7 %). Whereas only 1 male (3.3 %) and 15 females (11.4 %) having hemoglobin level below 12 g/dl (<12 g/dl). Approximately, about 4:1 female students were noted to have low levels of hemoglobin. The main criteria in analyzing IDA were hemoglobin <12 g/dl, MCV < 80 fl, MCH < 27 pg, MCHC < 31.5 %, and RDW > 15 %. Level of MCV < 80 fl had also 1 male (3.3 %) and 17 females (12.9 %). The MCH < 27 pg shows that male students consist of 3 (30 %) and female students 27 (20.5 %). The MCHC < 31.5 % results show that only 1 (3.3 %) male student and 22 (16.7 %) female students. The last parameter for IDA in CBC is RDW. It showed that RDW > 15 % had 1 male (3.3 %) and 15 females (11.3 %). Based on FBC result only 16 students were suspected to be IDA, therefore all of them proceed with serum ferritin test. All 16 anemic subjects were identified based on hematological parameters. Number of students having lower serum ferritin <12 is 9 female students (60 %) and no male students meanwhile students with higher serum ferritin >12 is 1 male student (100 %) and 6 female students (40 %).

Table 1 Full blood count and serum ferritin result

Reference range	Male N (%)	Female N (%)	Total N (%)
Hb (g/dl)			
<12	29 (96.7)	117 (88.7)	146 (90.1)
>12	1 (3.3)	15 (11.36)	16 (9.8)
MCV (fl)			
<80	29 (96.7)	115 (87.1)	144 (88.9)
>80	1 (3.3)	17 (12.9)	16 (11.1)
MCH (pg)			
<27	27 (90)	105 (79.6)	16 (9.9)
>27	3 (30)	27 (20.5)	146 (90.1)
RDW (%)			
<15	1 (3.3)	15 (11.4)	16 (9.9)
>15	29 (96.7)	117 (88.7)	146 (90.1)
MCHC (%)			
<31.5	29 (96.7)	110 (83.3)	139 (85.8)
>31.5	1 (3.3)	22 (16.7)	23 (14.2)
For the 16 anemic students the test was continued with serum ferritin (SF) test			
SF (μmol/L)			
<12	1 (100)	6 (40)	7 (43.8)
>12	–	9 (60)	9 (56.3)

4 Full Blood Picture Result

Based on FBP results, 16 subjects with abnormal results in FBC have abnormal characteristics in their blood smear especially hypochromic microcytic cell, tear-drop cell, and target cell shown in Fig. 1.

Comparison between normal students and IDA students in the investigated hematological parameters and comparison between IDA group and excluded group based on serum ferritin test result.

Overall FBC and SF profile results divided subjects into two groups: normal and IDA. Tables 2 and 3 summarize FBC and SF results in each group. The median hemoglobin level of the first group (normal) was 13.10 g/dl (1.40) while the median hemoglobin level of the second group (IDA) was 10.40 g/dl (1.30). Then the median of MCV was 90.30 fl (5.13) for normal group whereas 75.10 fl (3.45) for IDA group. Also, the median of MCH was 29.05 pg (1.90) for normal group and 23.40 pg (2.80) for IDA group. Meanwhile the median of MCHC was 32.10 (0.72) for IDA group and 30.80 (1.20) for IDA group. Then the median of RDW was 12.70 % (1.10) for normal group whereas 16.40 % (5.55) for IDA group. Based on FBC result, 16 students were suspected IDA. Table 3 showed the median level of SF in IDA and student. The SF test was done to confirm whether the student was

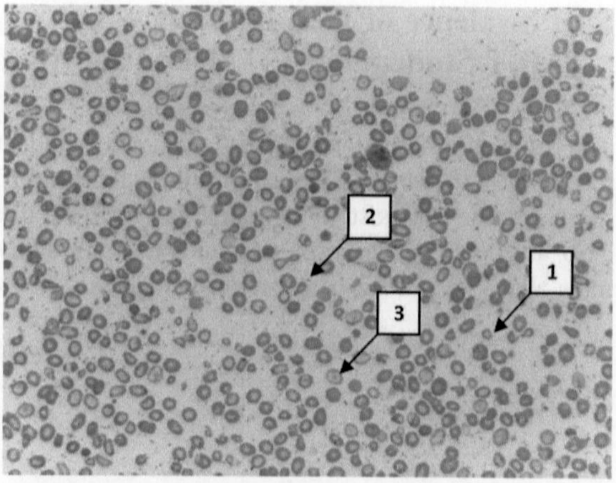

Fig. 1 shows result of FBP with characteristics of (*1*) hypochromic microctic cell, (*2*) teardrop cell, and (*3*) target cell (codocyte)

Table 2 Comparisons between normal ($n = 146$) and (IDA) ($n = 9$) in the investigated hematological parameters value

	Normal (n = 146)		IDA (n = 9)		Z statistic	*p*-value
	Median	IQR	Median	IQR		
Hb (g/dl)	13.10	1.40	10.40	1.30	−5.031	0.000
MCV (pg)	90.30	5.13	75.10	3.45	−4.909	0.000
MCH (pg)	29.05	1.90	23.40	2.80	−-4.945	0.000
MCHC (%)	32.10	0.72	30.80	1.20	−4.596	0.000
Hct (%)	41.20	4.33	32.70	7.30	−4.897	0.000
RDW (%)	12.70	1.10	16.40	5.55	−4.917	0.000

**p*-value < 0.05: Mann–Whitney U Test

Table 3 Comparison between IDA (n = 9) and excluded students (n = 7) in the investigated iron study of serum ferritin

	IDA (n = 9)		Excluded (n = 7)		Z statistic	*p*-value
	Median	IQR	Median	IQR		
SF (μmol/l)	8.10	3.72	76.70	113.52	−3.334	0.001

**p*-value < 0.05: Mann–Whitney U test

having IDA. The result shows that nine out of sixteen students have a low SF result and the median of the result was 8.10 μmol/l (3.72) whereas seven students have a normal value of SF and the median was 76.7 μmol/l (113.52). Therefore, the criteria of ferritin in IDA are low and for thalassemia cases show that serum ferritin level is normal. The samples with positive criteria of IDA were confirmed by hemoglobin electrophoresis.

5 Prevalence of Iron Deficiency Anemia (IDA) Among MLT Students in UiTM Puncak Alam

The result shows that the case of IDA in MLT student, only from female student, is 5.6 % (n = 9) whereas there is no prevalence of IDA among male student. The normal status in this study revealed that for male it is 18.5 % (n = 29) and female 71.6 % (n = 117). The excluded student in this study was only 0.62 % (n = 1) for male and 3.70 % (n = 6) for female students.

6 Discussion

The purpose of this study is to find out the prevalence of IDA among MLT students in UiTM Puncak Alam. This study had shown that the prevalence of IDA is 5.6 %. The previous study done by Azma et al. (2012) reported that among medical students of Universiti Kebangsaan Malaysia Medical Centre (UKMMC) in 2012 the prevalence of IDA was 13.5 %. Loh and Khor (2010) stated that the prevalence of IDA is the highest among Indians (18.0 %) followed by Chinese (9.9 %) and Malays (4.3 %). According to Hassan et al. (2005), the major health risk in pregnancy is anemia and the prevalence of anemia in a study done in Singapore was 15.3 % and the highest prevalence of anemia was among Malays. Al-Mekhlafi et al. (2008) reported that the prevalence of IDA was 34 % and accounted 70.1 % of the anemic cases. Earlier Foo et al. (2004) found that the prevalence of anemia and IDA among rural adolescents in Sabah was 20 and 17 %, respectively.

All subjects were Malay ethnicity with an age between 20 and 26 years old. All blood sample collections were tested for FBC. Among 162 students, 16 were found anemic. The 16 samples were then tested continuously for FBP and SF. After performing the test, nine out of sixteen students were confirmly IDA. Table 1 shows the results of hemoglobin among MLT students of UiTM Puncak Alam. Results from this study show that there were 29 males (96.7 %) and 117 females (88.64 %) having the level of hemoglobin >12 g/dl. Whereas only 1 male (3.3 %) and 15 females (11.36 %) were having a hemoglobin level at 12 g/dl. Only 16 students (9.88 %) were suspected anemic.

Table 1 shows the Hb level of normal (n = 146) students and IDA (n = 9) students showed a statistical significant difference (P = 0.000). Other FBC parameter also showed a statistical significant difference (P = 0.000) between normal (control group) and IDA such as MCV, MCH, MCHC, and RDW. This study shows that this parameter was lower among IDA patients. Besides that, for the RDW level of IDA patient, the result would be higher than the normal range. This finding was consistent with studies done by Al-sayes et al. (2011) and Loh and Khor (2010).

Based on Table 3, the SF test is only done with the students who were having the FBC result of low in parameter of Hb, MCV, MCHC, MCHC, and high RDW.

Table shows that there is a statistical significant difference (P = 0.001) of SF between IDA group (n = 9) and excluded groups (n = 7). Based on the median of IDA group, the result was 8.10 which is low from the normal reference to SF level whereas the median of excluded groups was 76.7 which is within the normal reference range of SF. The prevalence of IDA among female students is 5.6 % (n = 9), whereas no IDA was detected in male students.

Based on Fig. 1, the blood smear shows abnormal morphology of RBCs that consists of target cells, teardrop cells, and microcytic hypochromic RBCs. In this study the prevalence of IDA among MLT students of UiTM Puncak Alam maybe due to several factors. Harris (2007) stated that IDA is a condition where the reduced heme protein, reduced by iron-containing enzyme and the reduced function at a cellular level are influenced by iron and therefore the deficiency of iron will have various clinical manifestations. Al-sayes et al. (2011) reported that there is a statistical significant difference between IDA and IDA with consuming no or infrequent amount of red meat. Furthermore, heme iron from meat usually provides 10–20 % of iron while nonheme iron which is from vegetable, cereals, and fruits provides 80–90 %. According to Al-sayes et al. (2011) IDA is the most prevalent nutritional problem in the world and major cause of anemia, especially among female. Heavy menstrual blood loss is an important risk factor for IDA among women, Al-sayes et al. (2011).

7 Conclusion

This study concludes that the prevalence of IDA among MLT students in UiTM Puncak Alam is 5.6 %. Out of 162 students involved in this study, nine female students were detected as IDA whereas none from male students. This study had shown the evidence of IDA among UiTM students in Puncak Alam which could be of importance to overcome the IDA problem in Malaysia. For improvement of IDA problem, educational programs should be held to improve public awareness of IDA and its causes. Furthermore, nutrition education programs should be conducted especially for students to advocate healthy dietary habits.

References

Al-Mekhlafi MH, Surina J, Atiya AS, Ariffinc W, Mohammed Mahdya A (2008) Anaemia and iron deficiency anaemia among aboriginal schoolchildren in rural Peninsular Malaysia: an update on a continuing problem. R Soc Trop Med Hyg 102(10):1046–1052

Al-sayes F, Gari M, Qusti S, Bagatian N, Abuzenadah A (2011) Prevalence of iron deficiency and iron deficiency anemia among females at university stage. J MedLab Diagn 2(1):5–11. http://www.academicjournals.org/JMLD

Azma RZ, Ainoon O, Azlin I, Hamenuddin H, Hadi NA, Tatt WK, Syazana IN, Asmaliza AM, Das S, Hamidah NH (2012) Prevalence of iron deficiency anaemia and thalassaemia trait among undergraduate medical students. Clin Ther 163(4):287–91

Foo LH, Khor GL, Tee E-S, Dhanaraj P (2004) Determinants of iron status in Malaysian adolescents from a rural community. Int J Food Sci Nutr 55(6):517–25. http://www.ncbi.nlm.nih.gov/pubmed/15762316

Harris RJ (2007) Iron deficiency anaemia: does it really matter? Paediatr Child Health 17(4):143–146. doi:10.1016/j.paed.2007.01.019

Hassan R, Wan ZA, Hussain NHN (2005) Anemia and iron status of malay women attending an antenatal clinic in kubang kerian, Malaysia. Southeast Asia J Trop Med Public Health 30 (5):1304–130

Loh SP, Khor GL (2010) Iron Intake and Iron Deficiency Anaemia among Young Women in, *6* (January). *Malaysian Joulrronna l Ionft aMkee daincidn eI raonnd D Heefiaclitehn cSyc ieAnncaeesm Vioal .a 6m(oln) gJ aYnouuanrgy W20olm0e: n6 3*

Milman N (2011) Anemia–still a major health problem in many parts of the world! Ann Hematol 90(4):369–377. doi:10.1007/s00277-010-1144-5

Nadarajan VS, Eow GI (2002) Anaemia and iron status among blood donors in a blood transfusion unit in Malaysia. Malays J Pathol 24(2):99–102. http://www.ncbi.nlm.nih.gov/pubmed/12887168

Siti Noor AS, Wan Maziah WM, Narazah MY, Quah B (2006) Prevalence and risk factors for iron deficiency in Kelantan se, 47(11):935–939

WHO/UNICEF/UNU (2001) Iron deficiency anemia: assessment, prevention and control a guide for programme managers. Geneva, World Health Organization, (Document WHO/NHD/01.3)

Chapter 52
Malay Cupping Therapy: A Haematological Analysis Pilot Study

Siti Aishah Abdullah, Mohd Nadzri Mohd Najib, Ahmad Fauzi Dali and Suraya Sulaiman

Abstract Cupping is an ancient practice used for healing as well as health maintenance. It can be simply explained by the act of suction and removal of blood after some superficial incisions were made on the skin at various specific points on the body. Though cupping therapy has been used in the Malay community as well as throughout the world till today, there have been only few researches attempted at uncovering its secret. This study is aimed to evaluate haematological changes of the blood drawn intravenously from respondents at pre- and post-cupping session. Ten healthy male and female individuals between the age of 26–46 years old were randomly selected for this pilot study. Five millilitres (5 ml) blood samples (1 h interval before for pre-cupping and 1 h interval after for post-cupping of each session) were collected from the median cubital vein of each individual for the analysis of some haematological parameters. Fifteen (15) haematological parameters were examined. There were some changes in almost all parameters tested. Nevertheless, the changes were found to be not statistically significant. In conclusion, small number of samples may have contributed to the parameter changes being negative. Another reason is possibly due to the fact that healthy individuals were selected; hence their body's ability to maintain internal haematological stability is optimised.

Keywords Blood · Cupping · Haematology · Malay · Healthy

S.A. Abdullah · M.N. Mohd Najib (✉) · A.F. Dali · S. Sulaiman
Faculty of Pharmacy, Universiti Teknologi MARA, Kepala Batas
Pulau Pinang, Malaysia
e-mail: mohdna2857@ppinang.uitm.edu.my

S.A. Abdullah
e-mail: aisha_humair@yahoo.com

A.F. Dali
e-mail: fauzi.dali@ppinang.uitm.edu.my

S. Sulaiman
e-mail: suraya.sulaiman@ppinang.uitm.edu.my

N.A. Yacob et al. (eds.), *Regional Conference on Science, Technology and Social Sciences (RCSTSS 2014)*, DOI 10.1007/978-981-10-0534-3_52

1 Introduction

Cupping therapy is an ancient medical treatment practiced throughout the world either in Eastern or Western. This therapy is claimed as the oldest medicine practiced as the ancient Greek and Roman artefacts prove the long history of this unique therapy (Christopoulou-Aletra and Papavramidou 2008). In China, cupping therapy was recorded in an ancient book in Han Dynasty (Chirali 1999). Positive perception towards this therapy has made most people continuously practicing it thus making it one of the main complementary medical treatments nowadays.

There are various types of cupping practiced in China namely retained cupping, bleeding cupping, moving cupping, needle cupping, herbal cupping, water cupping and empty cupping (Cao et al. 2010). Other types of cupping therapy recorded are light cupping, medium and strong cupping, flash cupping and Moxa cupping (Qasim 2012). The varied types of cupping depend on the methods of cupping itself. However, the most eminent cupping types are dry cupping and wet cupping. Wet cupping involves skin bleeding which is applied in this study, while dry cupping does not apply any skin laceration or pricking to withdraw the blood.

Recently, some trial study showed the effectiveness of cupping towards chronic non-specific neck pain (Lauche et al. 2012), and some also showed positive effects on hemiplegic shoulder pain, hemiplegic hand edema and also aphasia accompanied by several terms (Lee et al. 2010). The favourable effect from cupping was also seen in treating migraine, headache, tension and hypertension (Ahmadi et al. 2008; Zarei et al. 2012). Another study also reported that cupping therapy benefited patients with osteoarthritis of the knee (Teut et al. 2012). Although these studies showed some effectiveness of cupping therapy, but their attempt to show how cupping therapy works to treat the diseases are inconclusive.

The aim of this study is to investigate the differential of haematological parameters, pre- and post-cupping. Similar studies were done on Iraq men and resident with diabetes (Al-Kazazz et al. 2014), but none has been done in Malaysia. Furthermore, the methods used varied which contribute to contrasting result. Blood parameters test is crucial as a baseline study to analyse any blood disorder that might be present.

2 Method

Subjects Consent

This study was carried out from 1 May 2014 to 29 May 2014. Ten healthy subjects were conveniently selected (walk-into the selected certified cupping centre) for cupping therapy. Their agreement to participate in this study was obtained through the consent form. However, subjects will be excluded if they have these following conditions; haemophilic, having menstruation and pregnant women, having bleeding disorder history, ongoing anticoagulating therapy, having acquired immune deficiency syndrome (AIDS) and hypotensive patient.

Bleeding

The subjects' blood was obtained 1 h before and 1 h after cupping therapy by the phlebotomists through the median cubital veins.

Cupping Procedures

The type of cupping used in this research was wet cupping. Ten cupping points were identified throughout the body. The duration taken for each cupping session was about 10–15 min per person. The cupping points were initially disinfected using 70 % alcohol. Next, cups were placed on the cupping points and the air in the cups was sucked out. The cups were left on the skin for 1–2 min. Then, the pins of the cups were removed to let the air out. Next, the cupping points were pricked using disposable needles. After that the cups were placed on the same cupping points and the air was sucked out again. After 3–5 min, the cups were removed from the cupping points. The blood that remained on the skin was cleaned using disinfectant.

Blood Profiling

The withdrawn blood was transferred into tubes and analysed using haematological instrument Sysmex XE2100 by Sysmex America, USA and Abbott CELL-DYN Ruby by Abbott Diagnostics, USA. Changes of the blood profile before and after cupping were recorded.

Data Analysis

Paired t-test was applied to compare pre- and post-cupping results. Wilcoxon signed-rank test was used to determine the association between variables. All tests were carried out with a 95 % level of confidence using SPSS version 20.0.

3 Results and Discussion

The subjects in this study were considered healthy as their haematological parameters values fall in the normal range as indicated in Table 1.

Table 2 shows red blood cell (RBC) haematological parameters of blood samples for pre- and post-cupping. The result shows some increment in haemoglobin and total RBC count after cupping session. However, a paired-samples t-test was conducted to compare haemoglobin count for pre- and post-cupping and no significant difference in the scores for pre (M = 135.67, SD = 15.42) and post (M = 136.33, SD = 16.56) values; $t(14) = -0.80$, $p = 0.44$. These results suggest that

Table 1 Normal range of haematological parameters (Lewis 2006)

Parameters	Normal range
Haemoglobin (gm/L)	M 125–175, F 11–155
Total RBC (x10^12/L)	M 4.5–6.0, F 4.0–5.5
PVC (L)	M 0.4–0.5, F 0.37–0.45
MCV (fl)	82–98
MCH (pg)	27–33
MCHC (g/L)	310–350
RDW (%)	11.0–16.0
Total WBC (x10^9/L)	4–11
Polymorphs (%)	50–70
Lymphocytes (%)	20–40
Monocytes (%)	<6
Eosinophils (%)	<4
Basophils (%)	<1
ESR (mm/hr)	M 0–15, F 1–20
Platelets (x10^9/L)	150–400

M male, *F* female, *RBC* red blood cell, *PVC* packed cell volume, *MCV* mean corpuscular volume, *MCH* mean corpuscular haemoglobin, *MCHC* mean corpuscular haemoglobin concentration, *RDW* red blood cell distribution width

Table 2 RBC haematological parameters of blood samples for pre- and post-cupping

Parameter	Pre-cupping Mean ± SD	Post-cupping Mean ± SD	p-value
Haemoglobin (gm/L)	135.667 ± 15.4211	136.333 ± 16.5644	0.439
Total RBC (x10^12/L)	4.753 ± 0.5986	4.780 ± 0.6002	0.301
PVC (L)	0.4187 ± 0.03739	0.4240 ± 0.03979	0.135
MCV (fl)	88.8667 ± 6.54508	88.9333 ± 6.15823	0.882
MCH (pg)	28.8667 ± 2.41622	28.9333 ± 2.43389	0.582
MCHC (g/L)	322.6667 ± 17.51190	324.6667 ± 18.46490	0.189
RDW (%)	15.1867 ± 5.41187	14.5467 ± 2.28875	0.475

SD standard deviation

cupping therapy does not affect haemoglobin count. Therefore, it dissmisses the misconception that cupping therapy may lead to the reduction of haemoglobin count. However, contrasting results were reported where the haemoglobin level decreased after cupping therapy (Al-Kazazz et al. 2014), but it was also not significant. Similar trend was seen in the total RBC count as well as other RBC haematological parameters such as packed cell volume (PVC), mean corpuscular volume (MCV), mean corpuscular haemoglobin (MCH) and mean corpuscular haemoglobin concentration (MCHC). The t-test of each parameter shows no significant difference for pre- and post-cupping ($p > 0.05$).

The result also shows a slight decrease in red blood cell distribution width (RDW). However, a Wilcoxon signed-rank test showed that cupping therapy did not elicit a statistically significant change in RDW value in our subjects ($Z = -0.597$, $p = 0.550$). Therefore, the decline of RDW is not significant.

Table 3 shows white blood cell (WBC) haematological parameters of blood samples for pre- and post-cupping. The result shows a slight decrease in total WBC, lymphocytes and eosinophils percentage after cupping session. However, a Wilcoxon signed-rank test showed that cupping therapy did not elicit a statistically significant change in WBC, lymphocytes and eosinophils value in our subjects ($Z = -0.53, p = 0.59$), ($Z = -1.12, p = 0.26$) and ($Z = -0.23, p = 0.82$), respectively. A contrasting result was seen for WBC count after 2 weeks cupping session (Al-Kazazz et al. 2014).

The result shows some increment in polymorphs, monocytes and eosinophils percentage after cupping session. However, a paired-samples t-test was conducted to compare polymorphs percentage for pre- and post-cupping and no significant difference in the scores for pre (M = 55.40, SD = 6.82) and post (M = 57.07, SD = 5.71) values; $t(14) = -0.84$, $p = 0.41$. These results suggest that cupping therapy does not affect polymorphs percentage. Similar trend was seen in the monocytes percentage. However, t-test also revealed that no significant difference for pre- and post-cupping was seen, pre (M = 3.07, SD = 1.75) and post (M = 3.20, SD = 1.08) values; $t(14) = -0.25$, $p = 0.81$, respectively. The polymorphs, monocytes and eosinophils percentage increase in tissue or cell with the infection and inflammation (Marya 2003) so the number will decrease in the blood vessel as seen in this study though the value is not significant.

Table 3 White blood cell haematological parameters of blood samples for pre- and post-cupping (for white blood cell)

Parameter	Pre-cupping Mean ± SD	Post-cupping Mean ± SD	ρ-value
Total WBC (x10^9/L)	7.4333 ± 2.14032	7.3467 ± 1.95954	0.695
Polymorphs (%)	55.4000 ± 6.82223	57.0667 ± 5.71298	0.414
Lymphocytes (%)	39.3333 ± 6.65117	37.3333 ± 5.86353	0.268
Monocytes (%)	3.0667 ± 1.75119	3.2000 ± 1.08233	0.806
Eosinophils (%)	2.2667 ± 1.22280	2.3333 ± 0.89974	0.869

WBC white blood cell, *ESR* erythrocyte sedimentation rate

Platelet count (10^9/L) result shows a slight increase, however, a paired-samples t-test was conducted to compare platelet count for pre- and post-cupping and no significant difference in the scores for pre (M = 236.00, SD = 75.16) and post (M = 254.20, SD = 75.16) values; $t(14) = -0.98, p = 0.35$. Even though the result is not significant, the slight increase may be due to laceration of the skin during wet cupping therapy, the blood vessel was injured and triggered platelet release (Marya 2003).

Erythrocyte sedimentation rate (ESR) (mm/hr) is the point of reference for inflammation and trauma. It is widely used throughout the world in clinical practice since it is a simple and inexpensive test (Ismailov et al. 2005). Erythrocyte sedimentation rate (ESR) count result shows a slight decrease, however, a paired-samples t-test was conducted to compare platelet count for pre- and post-cupping and no significant difference in the scores for pre (M = 14.93, SD = 14.45) and post (M = 12.07, SD = 10.47) values; $t(14) = 1.84$, $p = 0.09$. This may be due to the changes of ESR which is low rate, therefore 1 h blood withdrawing after cupping is not sufficient for the ESR changes. Similar result was reported for slow ESR change during initial stage of inflammation (Ismailov et al. 2005). It is also reported that ESR change can only be observed after 3 months conventional treatment (Ahmed et al. 2005), thus the duration needed for ESR changes in cupping therapy might be longer than 1 h. In addition the minor change in ESR may be due to the fact that the subjects involved were healthy and do not have any acute or chronic inflammation.

4 Conclusion

Changes in almost all parameters tested were found to be not significant. Several factors were indentified that may contribute to this result which are; small number of subjects, insufficient interval time for post-cupping blood withdrawal and involvement of healthy individual as subjects. Nevertheless, this preliminary test is crucial as references for other researchers to explore more on cupping therapy.

Acknowledgements This study was supported by grants from Research Acculturation Grant Scheme (600-RMI/RAGS 5/3 (96/2012)) and UiTM (Pulau Pinang) Excellence Fund.

References

Ahmadi A, Schwebel DC, Rezaei M (2008) The efficacy of wet-cupping in the treatment of tension and migraine headache. Am J Chin Med 36(1):37–44. doi:10.1142/S0192415X08005564

Ahmed SM, Madbouly NH, Maklad SS, Abu-Shady EA (2005) Immunomodulatory effects of blood letting cupping therapy in patients with rheumatoid arthritis. Egypt J Immunol/Egypt Assoc Immunol 12(2):39–51

Al-kazazz FF, Abdulsattar SA, Mohammed K (2014) Study effect of wet cupping on hematological parameters and inflammatory proteins of healthy Iraqi men. Am J Phytomed Clin Ther 1–6 Iraq

Cao H, Han M, Li X, Dong S, Shang Y, Wang Q, Liu J (2010) Clinical research evidence of cupping therapy in China: a systematic literature review. BMC Complement Altern Med 10:70. doi:10.1186/1472-6882-10-70

Chirali I (1999) The cupping procedure in traditional chinese medicine cupping therapy. Churchil Livingstone, London, p 3

Christopoulou-Aletra H, Papavramidou N (2008) Cupping: an alternative surgical procedure used by Hippocratic physicians. J Altern Complement Med (New York, N.Y.) 14(8):899–902. doi:10.1089/acm.2008.0238

Ismailov RM, Shevchuk NA, Khusanov H (2005) Mathematical model describing erythrocyte sedimentation rate. Implications for blood viscosity changes in traumatic shock and crush syndrome. Biomed Eng Online 4:24. doi:10.1186/1475-925X-4-24

Lauche R, Cramer H, Hohmann C, Choi K-E, Rampp T, Saha FJ, Dobos G (2012) The effect of traditional cupping on pain and mechanical thresholds in patients with chronic nonspecific neck pain: a randomised controlled pilot study. Evid Based Complement Altern Med. doi:10.1155/2012/429718

Lee MS, Choi TY, Shin BC, Han C, Ernst E (2010) Cupping for stroke rehabilitation: a systematic review. J Neurol Sci 294(1–2):70–73. doi:10.1016/j.jns.2010.03.03

Lewis SM (2006) Reference ranges and normal values. In: Lewis SM, Bain BJ, Bates I (eds) Dacie and lewis—practical haematology. Churchill Levingstone, New York, USA, pp 11–24

Marya R (2003) Medical physiology. CBS Publishers & Distributors, New Delhi

Qasim AAK (2012) The clinical and histological skin changes after the cupping therapy (al-hijamah). J Turk Acad Dermatol 6(1):1–7. doi:10.6003/jtad.1261a1

Teut M, Kaiser S, Ortiz M, Roll S, Binting S, Willich SN, Brinkhaus B (2012) Pulsatile dry cupping in patients with osteoarthritis of the knee—a randomized controlled exploratory trial. BMC Complement Altern Med 12(1):184. doi:10.1186/1472-6882-12-184

Zarei M, Hejazi S, Javadi SA, Farahani H (2012) The efficacy of wet cupping in the treatment of hypertension. ARYA Atheroscler J 8(1):3–6

Chapter 53
Pharmaceutical Manipulation of Chronic Anal Fissure

Kadhim Jawad, Waqar Al-Kubaisy, Ali Al Shaham, Suneet Sood and Yahya Mohammed Arpuin

Abstract Physiologically, it is the resting tone of the internal anal sphincter that chiefly interferes with the healing of the fissure. An anal fissure which does not heal with sitz baths and laxatives is a chronic anal fissure. Till now, the treatment for chronic anal fissures has been surgery in the form of anal dilatation or lateral sphincterotomy. Fissures heal in most patients, but a few develop transient or even permanent incontinence. There are exciting new advances in the form of chemical sphincterotomy, by the application of drugs that relax the sphincter. Glyceryl trinitrate and isosorbide dinitrate relax smooth muscle. They promote healing in about half of patients, but often cause headaches. Consequently compliance is a problem. The calcium antagonists, nifedipine and diltiazem, also reduce anal pressure by 28 %, but healing rates are low. Botulinum toxin is the most promising of the agents used for chemical sphincterotomy. This toxin can be used to weaken smooth muscle in the gastrointestinal tract, for example in achalasia and infantile hypertrophic pyloric stenosis. Botulinum toxin injection near the fissure reduces anal pressure lasts for about 3 months. This allows the fissure to heal, thus eliminating the need for surgery. After injection of botulinum toxin, there is a decrease in resting anal pressure by 18–30 %. The injection relieves the pain almost immediately. Cure rates are over 60 %, and the procedure can be repeated. Botulinum toxin

K. Jawad (✉) · Y. Mohammed Arpuin
Faculty of Medicine, Universiti Teknologi MARA, Shah Alam, Malaysia
e-mail: kadhim_jawad@yahoo.com

Y. Mohammed Arpuin
e-mail: mayahya3@gmail.com

W. Al-Kubaisy
Faculty of Medicine, Universiti Teknologi MARA, Shah Alam, Malaysia
e-mail: waqar_abd@yahoo.co.uk

A. Al Shaham
General Surgery, Al Kindy Medical School, Baghdad, Iraq
e-mail: alialshaham@yahoo.com

S. Sood
Faculty of Medicine, Monash University, Melbourne, Australia
e-mail: suneetsood@yahoo.com

N.A. Yacob et al. (eds.), *Regional Conference on Science, Technology and Social Sciences (RCSTSS 2014)*, DOI 10.1007/978-981-10-0534-3_53

is a reasonable first-line alternative to surgery in the management of chronic anal fissure.

Keywords Botulinum toxin · Anal fissure · Glyceryl trinitrate · Lateral sphincterotomy

1 Background

An anal fissure is a longitudinal tear at the anal verge. It is caused by hard stools, and presents with fresh bleeding and anal pain during defecation. Acute fissures may respond to conservative management with sitz baths and laxatives (Ellis et al. 2006). Until now, surgery has been necessary for acute fissures that do not respond to conservative therapy and for chronic fissures. Fissures, therefore, are typically managed by surgeons. However, there are exciting new advances in nonoperative management that may well see this condition becoming a primarily medical disorder. This review discusses the current status of nonsurgical treatment for anal fissures.

2 Pathophysiology and Pathogenesis

In a patient with constipation, the passage of hard stools can injure the skin in the anal canal. Such injuries in most parts of the body would heal. Therefore, why does this injury not heal easily? The cause of this difficulty in healing is the internal anal sphincter (IAS). The IAS is a dense condensation of circular muscle fibers that form a complete ring around the anal canal. The IAS is in a state of tonic contraction state (except when defecating). It is the single most important contributor to anal resting pressure. In contrast, the external sphincter contributes little to resting anal pressure. The anal injury triggers a spasm of the IAS, increasing the resting anal pressure. This sphincteric spasm is fundamental in the pathophysiology of anal fissure. The degree of spasm in an anal fissure is sufficient to cause regional ischemia and hinder healing. Ninety percent of anal fissures occur at the posterior anus (Maria et al. 1998a, b) this part of the anus is less perfused than other sections of the anal canal. A permanently elevated resting pressure impairs the intrasphincteric blood flow. This results in ischemia to the anoderm and therefore hinders healing of the fissure (Fig. 1).

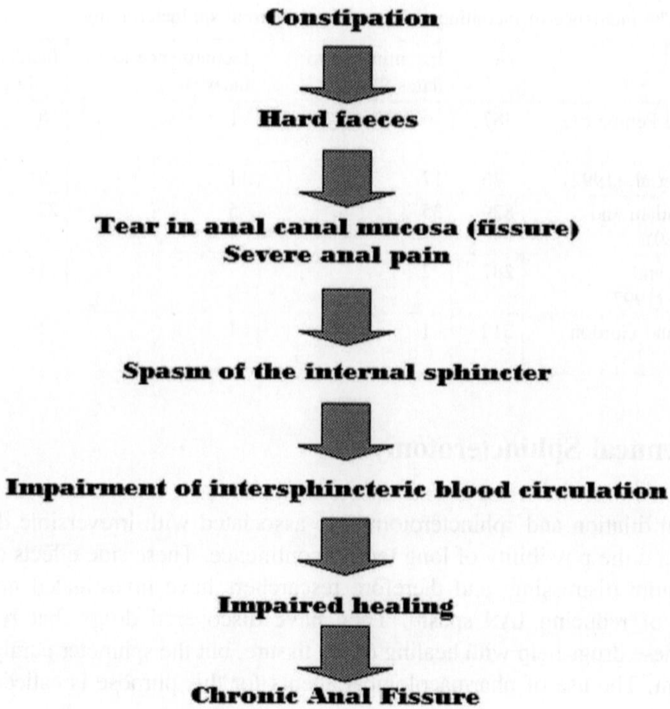

Fig. 1 The pathophysiology of development of a chronic anal fissure

3 Surgical Sphincterotomy

It follows that a relief of IAS spasm will improve the blood supply to the fissure area and result in healing. Indeed, the surgery of anal fissure is based on this concept. The two most popular procedures are anal dilatation and lateral sphincterotomy. In an anal dilatation, the surgeon stretches the sphincter so much that the IAS becomes paralyzed for several days. During a lateral sphincterotomy, the surgeon divides the lowermost fibers of the IAS. These fibers lose their spasm, and improve the blood supply with consequent healing of the fissure. After anal dilatation, fissures heal in 93 % of patients. Nevertheless, because of the sphincter damage, as many as 38 % of patients can have transient incontinence, and a small number of patients have a degree of permanent incontinence (Nyam and Pemberton 1999) Since anal dilatation may cause incontinence, many surgeons prefer lateral internal sphincterotomy, often considered the procedure of choice for anal fissure. The sphincterotomy is simple to perform, and provides immediate relief of pain. Healing occurs in 96 % of patients. Although incontinence is rare, it does occur, and may be permanent (Nyam and Pemberton 1999). Many studies show Some patients develop incontinence to flatus or faeces (Khubchandani and Reed 1989), (Littlejohn and Newstead 1997), (Hananel and Gordon 1997), (Mélange et al. 1992) as in table 1.

Table 1 The incidence of incontinence after lateral internal sphincterotomy

Authors	No.	Incontinence to flatus %	Incontinence to faces %	Incidence of soiling %
Nyam and Pemberton (1999)	487	6	1	8
Melange et al. (1992)	76	17	11	9
Khubchandani and Reed (1989)	829	35	5	22
Littlejohn and Newstead (1997)	287	2		1
Hananel and Gordon (1997)	312	1	1	1

4 Chemical Sphincterotomy

Both anal dilation and sphincterotomy are associated with irreversible damage to the IAS and the possibility of long term incontinence. These side effects of surgery can be quite distressing, and therefore researchers have investigated nonsurgical methods of reducing IAS spasm. They have discovered drugs that reduce this spasm: these drugs help with healing of the fissure, but the sphincter paralysis is not permanent. The use of pharmacological agents for this purpose is called chemical sphincterotomy.

Glyceryl Trinitrate and Isosorbide Dinitrate

Glyceryl trinitrate (GTN) (Fig. 2) and isosorbide dinitrate relax smooth muscle. They reduce resting anal pressure, and consequently improve blood flow. Their vasodilatory effect may also help. The resting anal pressure falls after local anal application of GTN ointment. Immediately after the application of 0.2 % GTN, a significant rise to normal blood flow levels occurs (Kua et al. 2001). GTN ointment

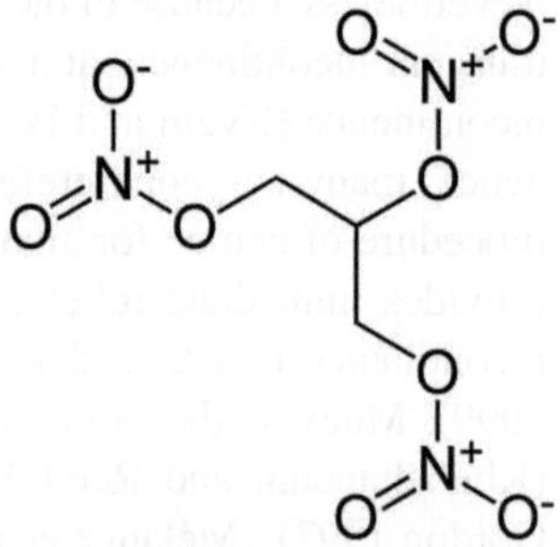

Fig. 2 Glyceryl trinitrate (GTN)

promotes healing of anal fissures in 33–88 % of patients in different trials (Lund and Scholefield 1997). Moderate to severe headaches are common, and occur in about 80 % of patients (Richard et al. 2000) and compliance is a major issue (Altomare et al. 2000; Palazzo et al. 2000). These side effects, along with tachyphylaxis, limit the utility of topical GTN ointment in the treatment of anal fissure (Richard et al. 2000). The problems encountered with isosorbide dinitrate are similar to those with GTN.

Calcium Antagonists

Nifedipine and diltiazem are calcium channel blockers, and are effective in the management of selected patients with hypertension. Topical formulations of both nifedipine (Nyam and Pemberton 1999) and diltiazem, have been evaluated. The reduction in anal pressure is a modest 28 %, and the effect lasts for hours after application (Ellis et al. 2006). At present these drugs are not used in routine clinical practice for managing anal fissure (Antropoli et al. 1999).

Botulinum Toxin

Toxin is a lethal biologic neurotoxin released by *Clostridium botulinum*. It binds rapidly and strongly to presynaptic cholinergic nerve terminals, and prevents the release of acetylcholine into the neuromuscular junction (Westfall and Westfall 2006). This decreases the activity within parasympathetic and sympathetic cholinergic synapses, and produces a flaccid paralysis of skeletal muscle (Fig. 3). Botulinum toxin is a versatile tool and is used in disorders of striated muscles, e.g., spasmodic torticollis, strabismus, blepharospasm, and hemifacial spasm. The toxin can be used to weaken smooth muscle in the gastrointestinal tract' as in achalasia and infantile hypertrophic pyloric stenosis. In the treatment of anal fissure, botulinum toxin has greater clinical potential than GTN or the calcium antagonists. Figure 3 shows the normal nerve function, the synaptic vesicle adheres to the cell membrane at the synaptic cleft and releases acetylcholine by exocytosis. The process of release of acetylcholine requires a complex of proteins, commonly known as SNARE proteins, consisting of VAMP—synaptobrevin, SNAP-25, and syntaxin. When botulinum toxin is administered, it attaches itself to the presynaptic nerve terminal and enters the cell by endocytosis. Botulinum toxin cleaves SNAP-25, and prevents the release of acetylcholine into the synaptic cleft, paralyzing the nerve.

The advantage of using botulinum toxin injections in patients with anal fissure is that the ensuing reduction in anal pressure lasts for 3 or more months. This prolonged relaxation of the IAS allows the fissure to heal, thus eliminating the need for surgery. When botulinum toxin is used for the management of other disorders, as in disorders of the face, its disadvantage is that the effect wears off, necessitating

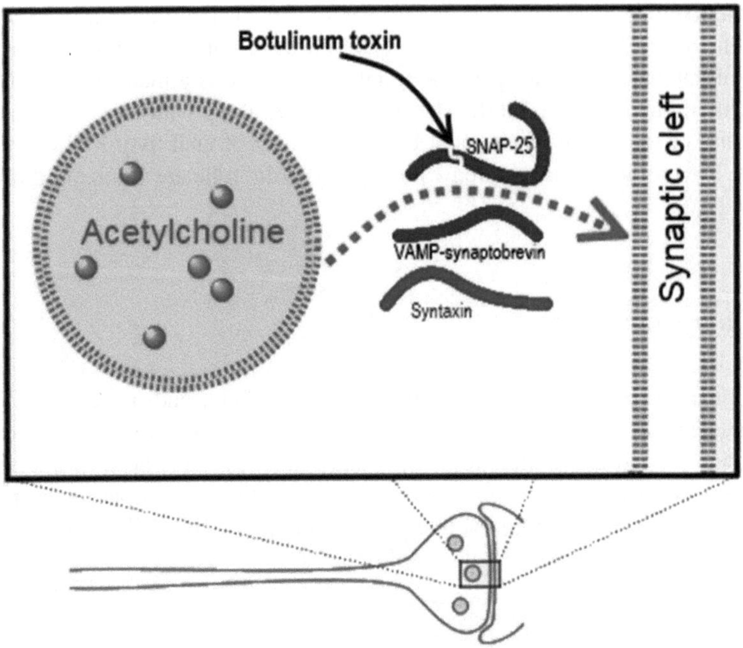

Fig. 3 The effect of botulinum toxin

reapplication. In the management of anal fissure this property is actually an advantage, since there is no risk of permanent incontinence (Mason et al. 1996). Typically, 0.4 ml of botulinum toxin is injected around the fissure in a saline solution containing 50 units/ml. Some patients may need re-injection after 2 months (Maria et al. 1998a). After injection of botulinum toxin, there is a significant decrease in resting anal pressure by 18–30 % (Maria et al. 1998b). The therapeutic effect of sphincter relaxation occurs within a few hours after injection, relieving pain almost immediately. Another benefit over surgery is that no sedation or local anesthesia is required during the procedure. Botulinum toxin should be used with caution in certain situations. The presence of abscesses in the area of injection constitutes a definite contraindication. Coagulation disorders and intake of anticoagulants are relative contraindications (Fernandez et al. 1999). A transitory incontinence for flatus occurs in about 3 % of patients (Maria et al. 1998b). It is natural to assume that botulinum toxin injection is dangerous because it is a toxin, but in fact it is very safe. Others show no temporary incontinence (Minguez et al. 1999) as shown in Table 2. When used for cosmetic purposes (e.g., facial disorders), side effects include bruising at the site of the injection are headache, respiratory infection, temporary eyelid droop, nausea, and an influenza-like syndrome. Injection for fissure in ano seems very safe and lacks systemic side effect.

In a randomized trial, Nasr et al. (2010) showed that fissure healing occurred in 25/40 patients after use of botulinum toxin, as compared to 36/40 patients after

Table 2 Treatment of chronic anal fissure using botulinum toxin injection in three large studies

Author	No.	Temporary incontinence (%)	Healing rate (%)
Maria et al. (1998a, b)	57	2	68
Fernandez et al. (1999)	76	3	67
Minguez et al. (1999)	69	0	63

lateral sphincterotomy. Recurrence rates were also lower following surgery. Unfortunately, five patients had incontinence, which persisted in two. Permanent incontinence does not seem to occur with the use of botulinum toxin.

We believe that botulinum toxin injection should be tried before surgery in patients with chronic anal fissure. Over 60 % of patients will heal without the morbidity or side effects of surgery. The injection can be repeated, and surgery is always an option in case the fissure still does not heal.

5 Conclusion

Although surgical sphincterotomy reduces anal tone and sphincter spasm and promotes fissure healing and it is effective treatment in fissure in ano, it is a surgical procedure, and is associated with cases of fecal incontinence and soiling. Therefore, pharmacological means to treat chronic anal fissure is an interesting alternative. GTN 0.2 % ointment has an efficacy of up to 68 % in healing chronic anal fissures, but it is associated with headache as a major side effect. Botulinum toxin injected into the anal sphincter healed over 60 % of chronic anal fissures.

The use of botulinum toxin appears to be effective for the treatment of chronic anal fissure. It is less expensive and easier to perform as an outpatient procedure and does not require anesthesia. No adverse effects or permanent sphincter damage resulted from the injections of the toxin. It might be considered also for patients with high surgical risks or in selected cases with high risk of future incontinence or patients to avoid surgery.

References

Altomare DF, Rinaldi M, Milito G (2000) Glyceryl trinitrate for chronic anal fissure–healing or headache? Results of a multicenter, randomized, placebo-controlled, double-blind trial. Dis Colon Rectum 43:174–179

Antropoli C, Perrotti P, Rubino M, Martino A, De Stefano G, Migliore G et al. (1999) Nifedipine for local use in conservative treatment of anal fissures: preliminary results of a multicenter study. Dis Colon Rectum 42:1011–1015

Ellis H, Calne R, Watson C (2006) Anal fissure. Lecture notes: general surgery, 11th edn. Blackwell, Massachusetts, pp 219–231

Fernandez LF, Conde FR, Rios RA (1999) Botulinum toxin for the treatment of anal fissure. Dig Surg 16:515–518

Hananel N, Gordon PH (1997) Lateral internal sphincterotomy for fissure- in-ano–Revisited. Dis Colon Rectum 40:597–602

Khubchandani IT, Reed JF (1989) Sequelae of internal sphincterotomy for chronic fissure in ano. Br J Surg 76:431–434

Kua KB, Kocher HM, Kelkar A (2001) Effect of topical glyceryl trinitrate on anodermal blood flow in patients with chronic anal fissures. Aust N Z J Surg 71:548–550

Littlejohn DR, Newstead GL (1997) Tailored lateral sphincterotomy for anal fissure. Dis Colon Rectum 40:1439–1442

Lund JN, Scholefield JH (1997) A randomised, prospective, doubleblind, placebo-controlled trial of glyceryl trinitrate ointment in treatment of anal fissure. Lancet 349:11–14

Maria G, Brisinda G, Bentivoglio AR, Cassetta E, Gui D, Albanese A (1998a) Botulinum toxin injections in the internal anal sphincter for the treatment of chronic anal fissure: long-term results after two different dosage regimens. Ann Surg 228:664–669

Maria G, Cassetta E, Gui D (1998b) A comparison of botulinum toxin and saline for the treatment of chronic anal fissure. N Engl J Med 338:217–220

Mason PF, Watkins MJG, Hall HS, Hall AW (1996) The management of chronic fissure in ano with botulinum toxin. J R Coll Surg Edinb 41:235–238

Melange M, Colin JF, Van Wymersch T (1992) Anal fissure: correlation between symptoms and manometry before and after surgery. Int J Colorectal Dis 7:108–111

Minguez M, Melo F, Espi A (1999) Therapeutic effects of different doses of botulinum toxin in chronic anal fissure. Dis Colon Rectum 42:1016–1021

Nasr M, Ezzat H, Elsebae M (2010) Botulinum toxin injection versus lateral internal sphincterotomy in the treatment of chronic anal fissure: a randomized controlled trial. World J Surg 34:2730–2734

Nyam DC, Pemberton JH (1999) Long-term results of lateral internal sphincterotomy for chronic anal fissure with particular reference to incidence of fecal incontinence. Dis Colon Rectum 42:1306–1310

Palazzo FF, Kapur S, Steward M (2000) Glyceryl trinitrate treatment of chronic fissure in ano: one year's experience with 0.5 % GTN paste. J R Coll Surg Edinb 45:168–170

Richard CS, Gregoire R, Plewes EA (2000) Internal sphincterotomy is superior to topical nitroglycerin in the treatment of chronic anal fissure: results of a randomized, controlled trial by the Canadian colorectal surgical trials group. Dis Colon Rectum 43:1048–1057

Westfall TC, Westfall DP (2006) Neurotransmission: the autonomic and somatic motor neuron systems. In: Brunton LL, Lazo JS, Parker KL (eds) Goodman & Gillman's, the pharmacological basis of therapeutics, 11th edn. McGraw-Hill, New York, pp 137–181

Part X
Biology

Chapter 54
Genotyping the Exon 10 of Low-Density Lipoprotein Receptor: Discovery of New Single Nucleotide Polymorphism

J.S.K. Shia, Cannilia Kerine, K.L. Teh, M.Z. Salleh, S.N. Hussin, I. N. Ismail and N.J. Abdul Wahab

Abstract Familial hypercholesterolemia (FH) is an autosomal dominant inherited disease characterized by elevated cholesterol levels. Studies have shown that polymorphism in the *LDLR* gene is one of the contributors of FH. The aim of this study is to investigate the role of polymorphisms in exon 10 of *LDLR* gene in relation to LDL levels, using both biochemical test and allele-specific polymerase chain reaction. Two different single nucleotide polymorphisms (SNPs) were detected. The first SNP 29209G > A was detected in 33 % of the samples. And the second SNP 29326A > G (novel SNP) was detected in only 8 % of the samples. However, both SNPs 29209G > A and 29326A > G were synonymous with no change in the resultant amino acids. Moreover, they lack significant phenotypic changes in relation to the LDL levels in blood. Therefore, further studies need to be done on metabolomics aspects to uncover the clinical significance of these two SNPs.

Keywords Familial hypercholesterolemia · Low-density lipoprotein · Polymerase chain reaction · Single nucleotide polymorphism

1 Introduction

Familial hypercholesterolemia (FH) is an autosomal dominant inherited disease which is characterized by elevated cholesterol level (Goldstein and Brown 2009). It is associated with the presence of xanthomas and early development of coronary heart diseases (CHD). FH can be classified into heterozygous FH and homozygous

J.S.K. Shia (✉) · C. Kerine · S.N. Hussin
Faculty of Pharmacy, Universiti Teknologi MARA, Puncak Alam, Selangor, Malaysia
e-mail: johnshia@puncakalam.uitm.edu.my

K.L. Teh · M.Z. Salleh · I.N. Ismail · N.J. Abdul Wahab
Integrative Pharmacogenomics Institute (IPROMISE),
Universiti Teknologi MARA, Puncak Alam, Selangor, Malaysia

N.A. Yacob et al. (eds.), *Regional Conference on Science, Technology and Social Sciences (RCSTSS 2014)*, DOI 10.1007/978-981-10-0534-3_54

FH (Austin et al. 2004). Homozygous FH expresses more severe phenotype compared to heterozygous FH, such as much higher cholesterol levels. However, they rarely occur usually one in every million (Goldberg et al. 2011). Heterozygous FH is more common with prevalence of one in every 500 individuals (Goldstein and Brown 2009). Although many studies have been done about FH, most of them were done in the Western population, there are not many studies done among the Asian population especially among Malaysians (Livy and Lye 2011).

FH is caused by mutation in the low-density lipoprotein receptor (*LDLR*), Apolipoprotein-B (*APO-B*) or in Convertase Subtilisin/Kexin type 9 (*PCSK9*). Studies done on the Asian population showed that most of mutations occur in the *LDLR* (Livy and Lye 2011). This also applies to the Malaysian population in which most of the mutations were detected in exon 7 to exon 9 of the *LDLR*. Low occurrence of mutation in *APO-B* among Malaysians maybe a factor of lower incidence of xanthomas and less severe CHD compared to the Western population (Khoo et al. 2000).

2 Methodology

Ethics Approval

Ethics approval was acquired from Universiti Teknologi MARA (UiTM) Ref No 600-RMI (5/1/6).

Selection of Subjects

Twelve subjects were recruited. Six of them were hypercholesterolemic with LDL levels above 4.1 mmol/L and another six served as controls with LDL levels below 3.0 mmol/L. The age of subjects ranged from 25 to 60 years old, consisting of both males and females. Their history of heart disease as well as their family history was obtained through questionnaires. Subjects who did not consent to be part of the research were excluded. Subjects with secondary hypercholesterolemia from underlying diseases such as renal, liver or thyroid disease were excluded from the research. Patients with triglyceride levels more than 4.0 mmol/L were also excluded from the study.

Blood Sampling and DNA Extraction

From each subject 4 ml of venous blood was drawn. For biochemical testing 2 ml was used. Another 2 ml was stored in citrate tube for DNA extraction using

Promega's Wizard ® Genomic DNA Purification Kit, according to manufacturer's protocol.

Genotyping of Exon 10 LDLR Gene

The sequence of exon 10 of *LDLR* was obtained from NCBI Gene Bank (NG_009060). Two specific primers were designed (5′ TGAGATGAGGG CTCC TGGC 3′ and 3′ CACTAACCAGTTCCTGAAGC 5′) to detect the presence of polymorphism(s). PCR amplification was performed in 25 μL reactions of 5 pmol/μL each of forward and reverse primers, 100 μM dNTPs, 1.5 mM $MgCl_2$, 1 X PCR buffer, 0.5 U *Taq Polymerase*, and 50 ng/μL of DNA samples.

DNA was denatured at 94 °C for 1 min, followed by 30 cycles of denaturation at 94 °C for 45 s, annealing at 61.9 °C for 45 s, and extension at 72 °C for 45 s, with a final extension step for 3 min at 72 °C. The PCR products were size-fractionated by electrophoresis in a 2 % agarose gel and stained with ethidium bromide (EtBr). Bands were visualized by the gel documentation system. PCR products were up-scaled to 100 μl, purified and sent for sequencing.

Analysis of Polymorphisms

The sequences of all 12 samples were analysed using ChromasPro software. Each DNA sequence was run in NCBI BLAST and aligned to the reference sequence (NG_009060). SNPs were then identified. Direct gene counting method was used to determine the allelic frequencies.

3 Results

Detection of 29326G > A and 20209 A > G Polymorphism

Two sites of polymorphism were detected: 29326 G > A (Fig. 1a) and 29209 A > G (Fig. 1b). Both polymorphisms do not cause amino acid changes. 29326 G > A resulted in Threonine-531-Threonine while 29209 A > G resulted in Argine-471-Arginie.

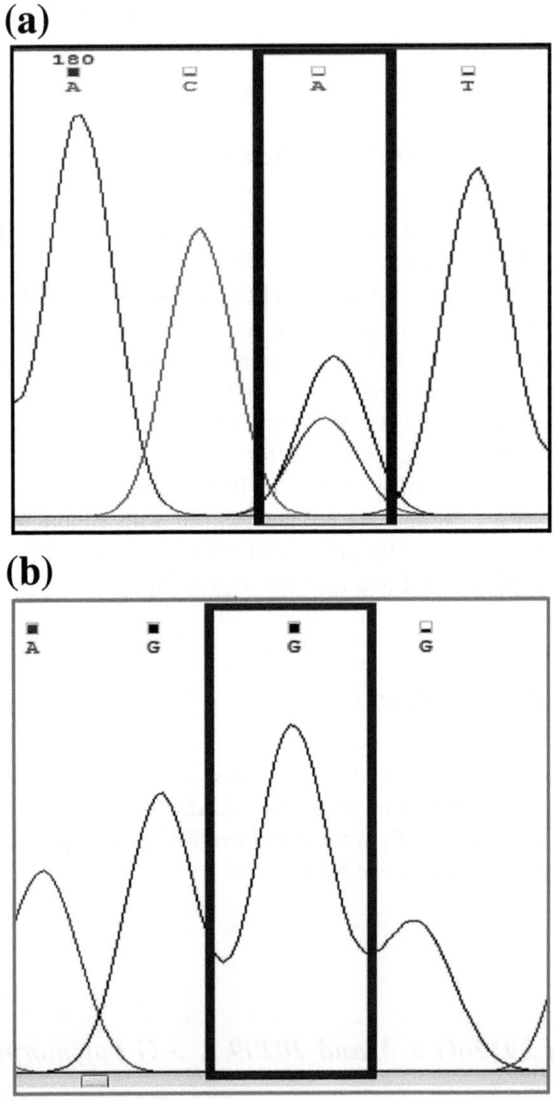

Fig. 1 **a** Chromatogram of a heterozygous SNP (29326 G > A). Heterozygous peaks are highlighted in *red/rectangular box*. **b** Chromatogram of homogyzous SNP. Homogygous peak of (29209 A > G) is highlighted in *red/rectangular box*

Allele Frequency for 29326 G > A

Out of 12 samples, 11 samples were found to have GG (wildtype) in which five are hypercholesterolemic and the remaining six normal. Only one sample was heterozygous mutant (GA) with hypercholesterolemic status, while none had AA. The frequencies of G and A were 0.958 and 0.042 respectively.

Allele Frequency for 29209 A > G

Out of 12 samples, 8 samples were found to have AA, with 4 of them being hypercholesterolemic and 4 normal. None of the samples were heterozygous, and four of them were homozygous mutants (GG), with two of them being hypercholesterolemic and two normal. The frequencies of A and G were 0.666 and 0.334, respectively.

4 Discussions

Pathological polymorphism that occurs within the *LDLR* gene would affect the ability of the receptor to function optimally. 29209 A > G polymorphism was reported to be one of the minor allele frequency (MAF) of major populations that are listed in HapMap. According to HapMap, the allele frequency for G allele is 0.84; whereas A allele is 0.16. The similar results were seen in the Han Chinese population in China which had higher frequencies of G alleles. Intriguingly, the allele frequencies for 29209 A > G were 0.666 and 0.334 for A and G respectively, in the present study. Malay Malaysians have higher wildtype allele frequencies as compared to Japanese population who have frequency of A allele and G allele of 0.447 and 0.553, respectively. However, the present findings are not conclusive as the sample size was small due to budget and time constraints. Another study done among the 154 Kelantanese reported that GG variants has the second highest frequency of 11 % (Al-Khateeb et al. 2011). Therefore, it is possible that this variant is commonly found among the Malaysian population. On the other hand, the amino acid of GG variants remains as arginine despite the change in nucleotide base from A/G. This arg471arg was reported in a study done in a Russian population (Tatishcheva et al. 2001). Hence, this SNP might be considered as a silent mutation that does not have any clinical significance in Malaysians too. However, further studies need to be done to identify the clinical significance of this variant.

The second SNP detected in this research was 29326G > A. Substitution of G/A does not result in change in amino acid in which it remains as threonine at position 531. This SNP was only detected in one out of the twelve samples. The significance of this SNP remains unknown as this is an unreported SNP. Hence, it could be a novel polymorphism. Similar to 29209A > G this variant maybe of non-clinical significance as there is no change in the resultant amino acid. Out of 11, 5 wildtype (AA) subjects were hypercholesterolemic. Since *LDLR* gene is a 45 kbp gene that consists of 18 exons. Exon 10 may not involve directly in upregulation of LDL levels in the studied subjects. Therefore, due to a wide range of possible points of mutation throughout *LDLR* gene, there is a possibility that polymorphisms had occurred in other exons which lead to increased level LDL in those 5 hypercholesterolemic wildtype samples.

The dysfunction of the LDLR may not be the only contributors of high LDL levels. DNA methylation, which involved in the inactivation of genes, may also lead to dysfunction of LDL regulation. DNA methylation is the addition of a methyl group to cytosine at 5′ position or to the adenine at the sixth nitrogen. The addition of this methyl group has been found to be related with low or no gene expression (Suzuki and Bird 2008). Therefore, there is also possibility that FH may occur due to DNA methylation present in the promoter region of the *LDLR* rendering them to express less receptors, hence increasing levels of LDL. A study done on epigenome-wide analysis of HDL levels showed that there is a positive correlation between DNA methylation and the variation in HDL metabolism (Guay et al. 2012). Hence, epigenomics may be a new platform for studying contributing genetic factors of FH.

5 Conclusion

Two SNPs were detected in exon 10 of *LDLR* gene in the present study. 29209G > A SNP was detected in 33 % of the samples. A probable novel SNP 29326A > G was detected in only 8 % of the samples. However, both SNPs are synonymous and they lack significant phenotypic changes in relation to the LDL levels in blood. Therefore, further studies need to be done on epigenetic aspects with regards to these two SNPs.

Acknowledgement This research is supported by RAGS-PSI grant (600-RMI/DANA 5/3/PSI (151/2013) awarded by Research Management Institute, Universiti Teknologi MARA, MALAYSIA.

References

Al-Khateeb A, Zahri M, Mohamed M, Sasongko T, Ibrahim S, Yusof Z, Zilfalil B (2011) Analysis of sequence variations in low-density lipoprotein receptor gene among Malaysian patients with familial hypercholesterolemia. BMC Med Genet 12:40

Austin MA, Hutter CM, Zimmern RL, Humphries SE (2004) Familial hypercholesterolemia and coronary heart disease: a HuGE association review. Am J Epidemiol 160:421–429

Goldberg AC, Hopkins PN, Toth PP, Ballantyne CM, Rader DJ, Robinson JG, Daniels SR, Gidding SS, De Ferranti SD, Ito MK, Mcgowan MP, Moriarty PM, Cromwell WC, Ross JL, Ziajka PE (2011) Familial Hypercholesterolemia: Screening, diagnosis and management of pediatric and adult patients: Clinical guidance from the National Lipid Association Expert Panel on Familial Hypercholesterolemia. J Clin Lipidol 5:133–140

Goldstein JL, Brown MS (2009) The LDL receptor. Arterioscler Thromb Vasc Biol 29:431–438

Guay SP, Voisin G, Brisson D, Munger J, Lamarche B, Gaudet D, Bouchard L (2012) Epigenome-wide analysis in familial hypercholesterolemia identified new loci associated with high-density lipoprotein cholesterol concentration. Epigenomics 4:623–639

Khoo KL, Van Acker P, Tan H, Deslypere JP (2000) Genetic causes of familial hypercholesterolaemia in a Malaysian population. Med J Malaysia 55(4):409–418

Livy A, Lye SH (2011) Familial hypercholesterolemia in Asia: a review. J OMICS Res 1:22–31

Suzuki MM, Bird A (2008) DNA methylation landscapes: provocative insights from epigenomics. Nat Rev Genet 9:465–476

Tatishcheva IA, Mandel'shtam MI, Golubkov VI, Lipovetskiĭ BM, Gaĭtskhoki VS (2001) Four new mutations and polymorphic variants of the low density lipoprotein receptor in patients with familial hypercholesterolemia in Saint Petersburg. Genetika 37:1290–1295

Chapter 55
Fruit Morphology Description of Seven Jackfruit Clones from Farmers Collection

Noor Baiti Abd Aziz, Abd Rahman Milan and Mohd Zaki Razali

Abstract Jackfruit scientifically known as *Artocarpus heterophyllus* and locally known as 'nangka' is a potential fruit for commercial production for local and export market. The fruit parts that can be consumed are only the flakes. The prospects of jackfruit in Malaysia are bright. It can be consumed either fresh or as minimally processed jackfruit. In response to clonal issue of commercial jackfruit, the performance of six jackfruit clones viz., Mastura, Mantin, Hong, Cj3, Cj6 and YL were evaluated from farmers' field with Tekam Yellow used as a check clone. The evaluation was done on the collected fruit that comes from the main states that produce this commodity such as Johor, Negeri Sembilan, Selangor and Pahang. Ten fruits each from seven clones were characterised for fruit quality characters namely fruit weight, fruit length, fruit width, fruit circumference, fruit rind thickness, number of flakes per fruit, number of kesep per fruit, flake thickness, 10 flakes weight, 10 flakes weight with seed, 10 seeds weight, fruit rind weight, rachis weight, flake TSS (% Brix), weight of one flake per fruit, total weight of flakes per fruit and recovery rate (%). Highly significant clonal differences in fruit characteristics such as fruit weight, fruit length, number of flakes per fruit, number of kesep per fruit, flake thickness, 10 flakes weight, 10 flakes weight with seed, fruit rind weight, rachis weight, flake TSS (% Brix), weight of one flake per fruit and total weight of flakes per fruit were obtained. Result also showed that there is no significant difference for the character of 10 seed weight. However, significant clonal differences were observed in fruit width, fruit circumference, fruit rind thickness and recovery rate. With desirable fruit quality traits, the mean of fruit weight was 16.9 kg ranging from 6.8 to 34.0 kg. The fruit length was 48.5 cm ranging from 20.5 to 73.0 cm and the flake TSS was 20.4 °Bx ranging from 11.0 to 38.8 °Bx. Mostly there is no wide variation of the fruit quality traits in clones as jackfruit is self-pollinated crops and all of them are monocrop plants. Fruit quality

N.B. Abd Aziz (✉) · M.Z. Razali
Horticulture Research Centre, MARDI, Kluang, Johor, Malaysia
e-mail: aabaiti.mardi@1govuc.gov.my

A.R. Milan
Horticulture Research Centre, MARDI, Serdang, Selangor, Malaysia
e-mail: armilan.mardi@1govuc.gov.my

N.A. Yacob et al. (eds.), *Regional Conference on Science, Technology and Social Sciences (RCSTSS 2014)*, DOI 10.1007/978-981-10-0534-3_55

can be used as a criterion in the selection of superior jackfruit clones. Therefore, this important information would be useful to select the best quality jackfruit to be recommended to the farmers or growers as high-quality clones.

Keywords Fruit morphology · Jackfruit clone · Farmers collection

1 Introduction

Jackfruit (*Artocarpus heterophyllus* L.) tree is important in tropical and subtropical regions such as Malaysia as a commercial edible fruit. It is a member of Moraceae family and relatively close to cempedak (*A. champeden*), breadfruit (*A. altilis*) and tarap (*A. adoratissimus*). The tree is a major component of subsistence and small farmers' farming systems and the fruit is often consumed fresh and also contributes to the country's agricultural industry. All the parts of the plant, including the fruit, contain white, sticky latex (Abd Rahman 2011). It is the largest edible fruit in the world up to 30–40 kg in weight. The edible, pulpy part represents the perianth (Goswami et al. 2011). When fully ripe, the unopened jackfruits emit a strong aroma. The exterior skin of the compound fruit is green or yellow when ripe depending on the cultivars. The interior part consists of large edible flakes of yellow, strong aroma flesh that encloses a smooth, oval and light brown seeds (Maimun and Abd Rahman 2005).

There has been an increased interest in jackfruit as a commercial crop and the plants establish well under Malaysian climate. In Malaysia, about 4,261 ha of area are cultivated with jackfruit and contributed 23,275 mt for RM 55,344,266 value production in 2011. In the past, the main jackfruit growing areas were in Kedah, Terengganu and Kelantan. However, it has gradually moved to Johor, Perak and Selangor because of the rich, well-drained and deep alluvial soil. In 2001–2007, Malaysian jackfruits are exported mainly to Singapore, United Kingdom, Hong Kong and Netherland, but we lost to Thailand in the early 2008 (Abu Kasim 2010). It was found that even though the areas of jackfruit cultivation increased from 3,123 ha (2006) to 3359 ha (2009), only 56–59 % of the area cultivated could yield. The jackfruit is mostly smallholder-owned and grown as monocrop by private farmers, TKPM projects and group projects of Department of Agriculture, SEDC, FELDA, FELCRA and LPP.

The Department of Agriculture (DOA) had registered more than 30 jackfruit clones. Out of these, two clones namely Tekam Yellow (J33) and Mastura (J35) are recommended for planting. More recently, two other clones namely Hong and YL are becoming popular amongst new jackfruit growers. These new clones have different characteristics to the other two recommended clones in which the fruits are much sweeter and good performance. The new clones are selected by growers using local variability. There are other clones selected by farmers and grown in small extent such as Mantin, Cj3 and Cj6. Nowadays, there are some private nurseries that still produce confirmed planting materials by Department of Agriculture for most clones including Cj3 and Cj6. Farmers or growers could procure the true-to-type grafts of certain jackfruit clone that is still available and recommended for current market of Malaysia.

Jackfruit is one of the fruit types that were given priority in commercial cultivation in Malaysia. The popularity of jackfruit as a commercial crop is very meagre due to wide variation in the fruit quality. Jackfruit comprises innumerable trees differing from each other in fruit characteristics such as shape, size and quality (Jagadeesh et al. 2010). These types may be further divided depending on the size of fruit, taste, odour of flesh, nature, shape and diversity of prickles on the rind for the maintenance of separate varieties (Singh 1995). There are many definitions and standards set by each industry but the simple definition of fruit quality is whatever the consumer desires (Britt 2001). The consumer initially judges the quality by the appearance of the fruit at the point of sale and then by the taste of the fruit (Maimun and Abd Rahman 2005). Attractiveness is usually based on fruit size and colour.

Jackfruit is a unique woody fruit species with a potential to produce fruits all year around. Similar to star fruit, the jackfruit is considered as a nonseasonal crop and this ability to consistently produce is an important economic characteristic, especially in its marketing. Yield of jackfruit is relatively high compared to other commonly grown woody fruit trees. Being an edible product and taken fresh, jackfruit flesh quality must both visually and organoleptically appealing to the average customer (Palaniappan and Hui 2001). It was mainly sold as whole fruits. However, apart from the sale of whole, fruit was also minimally processed to cater for the retail outlets mainly for supermarkets. This is because jackfruit is suitable for marketing in minimally processed form (Maimun and Siti Hawa 2008). In minimal processing, the flesh was extracted from the fruits and packed in polystyrene trays wrapped with transparent plastic.

This paper will report on the analysis of all fruit characters related to fruit qualities which contributes as (i) a guide to the farmers/growers to select the best clone for their cultivation based on these characters and (ii) important information for the consumers to differentiate jackfruit clones that exist in Malaysian market.

2 Materials and Methods

There are several steps in these fruit quality traits evaluation of selected jackfruit commercial clones in Malaysia.

Plant Materials

In the present study, seven jackfruit clones namely Mastura, Mantin, Hong, Cj3, Cj6, YL and Tekam Yellow as check clone selected from farmers' field in Selangor, Negeri Sembilan, Pahang and Johor were covered (Table 1). The plants used in this

Table 1 Name and origin of seven jackfruit commercial clones used in the study

Clone local name	DOA registration	Origin
Tekam Yellow	J33	Kota, Negeri Sembilan
Mastura	J35	Kota Tinggi, Johor
Mantin	J32	Kajang, Selangor
Hong	J34	Ayer Hitam, Johor
Cj3	J38	Kota, Negeri Sembilan
Cj6	Not registered	Bera, Pahang
YL	Not registered	Rawang, Selangor

study were recorded from the age of 8–10 years old trees and normally established as monocrop field by private farmers. They were planted by following the standard agronomic practices as recommended by Department of Agriculture with planting distance of individual trees done at 6 m × 6 m spacing. For this, mature unripe jackfruits were chosen and then collected to study the fruit quality traits at edible ripe stage. The collected fruits were brought to the nursery of MARDI Kluang and kept in ambient condition for ripening.

Fruit Quality Characterization

The research work was conducted at MARDI Kluang, Johor. Fruits were harvested at the stage of maturity, determined by the change in colour from pale green to greenish or brownish yellow. Ten fruits from each clone were randomly sampled and their fruit quality characteristics were recorded and evaluated viz. fruit weight, fruit length, fruit width, fruit circumference, fruit rind thickness, number of flakes per fruit, number of kesep per fruit, flake thickness, 10 flakes weight, 10 flakes weight with seed, 10 seeds weight, fruit rind weight, rachis weight, flake TSS (% Brix), weight of one flake per fruit, total weight of flakes per fruit and recovery rate (%).

Data Analysis

Data taken for those parameters were compiled and subjected to multivariate analysis among the characters using SAS for window software. All the fruit characteristics above were considered for variable studies through the statistical methods viz. standard deviation and coefficient of variation (CV). F test was also used to analyse the data.

3 Results and Discussion

A total of seven jackfruit commercial clones were successfully produced and the fruits were collected from the farmers' field (Table 1). It described the local name of clone, Department of Agriculture (DOA) registration number and the origin where the plant was collected. Research was studied to evaluate fruit quality traits for 17 characters as shown in Table 2. The resulting data showed an interesting output in parameters observed. Results showed that 12 characters are highly significant differences on fruit weight, fruit length, number of flakes per fruit, number of kesep per fruit, flake thickness, 10 flakes weight, 10 flakes weight with seed, fruit rind weight, rachis weight, flake TSS (% Brix), weight of one flake per fruit and total weight of flakes per fruit. The characters that showed significant differences are fruit width, fruit circumference, fruit rind thickness and recovery rate. There is no significant difference on character of 10 seeds weight per fruit (Table 3).

The characteristics that affect the fruit appearance are primarily fruit size while the most important aspect of fruit quality is taste (Maimun and Abd Rahman 2005). Average fruit weight is 16.9 kg, the fruit length is 48.5 cm and the flake TSS is 20.4 °Bx. In Table 3, wide range of fruit weight, fruit length and flake TSS value was observed. The range for fruit weight was 6.8–34.0 kg, while for fruit length, the data ranged from 20.5 to 73.0 cm and flake TSS, the data ranged from 11.0 to 38.8 °Bx. Most of the parameters studied showed that small variations exist as shown by coefficient of variation (CV) value less than 30 % because of bring planted as monocrop field. A high coefficient of variation was observed for number of kesep per fruit (101.5), flake thickness (41.9 cm) and 10 seeds weight per fruit

Table 2 Fruit quality characters and its description used in the study

Character	Description
Fruit weight	kg/fruit
Fruit length	cm
Fruit width	cm
Fruit circumference	cm
Fruit rind thickness	cm
Number of flakes/fruit	–
Number of kesep/fruit	–
Flake thickness	cm
10 flakes weight (with seeds)	g
10 flakes weight	g
10 seeds weight	g
Fruit rind weight	kg
Rachis weight	kg
Flake TSS	°Bx
Weight of one flake/fruit	g
Total weight of flakes/fruit	kg
Recovery rate	%

Table 3 Mean, standard deviation, coefficient of variation (CV), range of fruit quality traits and F value of seven jackfruit commercial clones used in the study

No	Character	Mean	Standard deviation	Coefficient of variation	Range	F value
1.	Fruit weight (kg)	16.9	4.0	23.3	6.8–34.0	18.31**
2.	Fruit length (cm)	48.5	7.1	14.6	20.5–73.0	10.28**
3.	Fruit width (cm)	27.2	5.3	19.3	11.5–42.0	2.30*
4.	Fruit circumference (cm)	82.2	11.2	13.6	34.3–100.8	3.64*
5.	Fruit rind thickness (cm)	1.4	0.5	34.5	0.7–4.3	5.47*
6.	Number of flakes/fruit	190.1	70.0	36.8	62–541	6.98**
7.	Number of kesep/fruit	4.2	4.3	101.5	0–23	7.49**
8.	Flake thickness (cm)	0.8	0.3	41.9	0.3–1.7	8.32**
9.	10 flakes weight with seeds (g)	506.6	132.8	26.2	240–980	11.47**
10.	10 flakes weight (g)	420.8	129.0	30.7	140–915	14.48**
11.	10 seeds weight (g)	86.8	41.9	48.2	30–140	1.94ns
12.	Fruit rind weight (kg)	6.2	1.4	21.9	2.8–13.6	27.40**
13.	Rachis weight (kg)	1.8	0.6	32.3	0.3–6.0	14.76**
14.	Flake TSS (% Brix)	20.4	3.7	17.9	11.0–38.8	23.59**
15.	Weight of one flake/fruit (g)	41.9	12.9	30.7	14.0–91.5	14.48**
16.	Total weight of flakes/fruit (kg)	7.3	2.2	30.6	2.5–14.6	13.03**
17.	Recovery rate (%)	42.6	8.8	20.7	18.9–71.6	2.37*

Note ns—not significant
*significant at $p \leq 0.05$
**significant at $p \leq 0.01$

(48.2 g). However, parameters for fruit rind thickness (34.5 cm) and number of flakes per fruit (36.8) exhibited a wide variation. This result will help a lot in selecting high-quality clone of jackfruit.

4 Conclusion

This study showed that there is significant difference in fruit quality among jackfruit clones. The fruit characters studied were very important for the consumer. Based on these characters consumer can select high-quality clone especially for farmers or growers. Selection of recommended clone with good quality traits will help them fetch a higher price of the products.

Acknowledgements The authors are grateful to those who have directly and indirectly contributed to this study especially to all jackfruit staffs at MARDI Kluang for their help in fruit collection and data recording. This study was supported by MARDI Development fund

(P-RH134-0108). We also would like to thank to Ms. Johana Ahamad and Mr. Johafizal Ahamad for their technical assistance. Appreciation is extended to Deputy Director and Director of Horticulture Research Centre for permission to present this paper.

References

Abu Kasim A (2010). Potensi dan strategi pemasaran buah-buahan tempatan. MARDI

Abd Rahman M (2011). Jackfruit in Malaysia. Studium Press LLC, pp 445–462

Britt BH (2001) Apple quality for consumer. Int Dwarf Fruit Tree Assoc 34:54–56

Goswami C, Hossain MA, Kader HA, Islam R (2011) Assessment of physicochemical properties of jackfruits' (*Artocarpus heterophyllus* Lam) pulps. J Hortic For Biotechnol 15(3):26–31

Jagadeesh SL, Reddy BS, Basavaraj N, Swamy GSK, Hegde L (2010) Varibility studies in physico-chemical qualities of jackfruit (*Artocarpus heterophyllus* Lam.) of coastal zone of Karnataka. *Karnataka*. J Agric Sci 23(2):293–297

Maimun T, Abd Rahman M (2005) Fruit quality evaluation of jackfruit at MARDI Kluang, Johor. In: Paper presented at Mesyuarat Teknikal Pusat HR, Penang, pp 59–67

Maimun T, Siti Hawa J (2008) Jackfruit. Breeding horticulture crop. MARDI, pp 119–132

Palaniappan S, Hui TF (2001) An insight into FASSB's experience on the commercial growing of jackfruits. In: Seminar on potentials of the fruit industry in Sarawak, pp 1–6

Singh A (1995) Jackfruit, fruit physiology and production, 4th edn. Kalyani Publishers, New Delhi, pp 364–367

Chapter 56
Diversity of Dragonfly Communities at Two Habitats in Negeri Sembilan

Amira Md. Zaliyati and Syazuani Mohd Shariff

Abstract Kuala Pilah and Batu kikir are located in Negeri Sembilan, Malaysia. The changes of flora and fauna community have always been related with the anthropogenic activities. This study was conducted in order to determine the species diversity of dragonfly within two different habitats in Negeri Sembilan. The sampling was carried out from August 2013 to October 2013 at two habitats comprising of UiTM Forest Reserve (Kuala Pilah) and Kampung Lonek Paddy Field (Batu Kikir). A total of 164 individuals of dragonflies were collected, which comprise of 14 species. For Kuala Pilah, a total of 11 species with 62 individuals were collected while for Batu kikir, a total of 9 species with 102 individuals were collected. Family Libellulidae was the dominant family that indicates 99 % of the total collected followed by one percent of total collected by Family Gomphidae. The highest total number of dragonfly individuals was shown by *Potamarcha congener* with 52 individuals that indicate 32 % of total collected while the least number of individuals was shown by the species of *Orthetrum chrysis*, *Tholymis tillarga,* and *Ictinogomphus decoratus* with one individual that indicate only one percent, respectively. Thus, there was high species diversity ($H' = 1.9$), high species richness ($R = 2.423$), and high species evenness of dragonfly ($E = 0.792$) in UiTM Forest Reserve as compared to Kampung Lonek Paddy Field.

Keywords Dragonfly · Diversity · Negeri Sembilan · UiTM Forest Reserve

A. Md. Zaliyati · S. Mohd Shariff (✉)
Faculty of Applied Sciences, Universiti Teknologi MARA,
Kuala Pilah, Negeri Sembilan, Malaysia
e-mail: syazuani@ns.uitm.edu.my

A. Md. Zaliyati
e-mail: amamira3@gmail.com

N.A. Yacob et al. (eds.), *Regional Conference on Science, Technology and Social Sciences (RCSTSS 2014)*, DOI 10.1007/978-981-10-0534-3_56

1 Introduction

Dragonflies can be found almost at all places in Earth. According to Kalkman et al. (2008), dragonflies with about 5,680 species form a relatively small insects group where most species are found in the tropics. Research conducted by Orr et al. (2004), about 342 species of Malaysian Odonata fauna are found where 181 of them were dragonflies from five families. In addition, Norma Rashid et al. (2001) claim that there are approximately 293 species of dragonflies found at Sabah, Sarawak, and Brunei while other 226 species are reported from Peninsular Malaysia together with Singapore. Major family of dragonflies found in various ecosystems of Peninsular Malaysia is Libellulidae. *Orchithemis pulcherrima* and *Neurothemis fluctuans* are the dominant species of dragonflies found in Malaysia.

Population of dragonflies was discovered to be threatened at near 10 % (IUCN 2007). This is due to the anthropogenic threat to the species such as extensive logging of lowland forest on many islands, climate change and its effect on freshwater systems (Riservato et al. 2009). Main threats to Odonata in Indo-Burma at both aquatic and the terrestrial environments are habitat loss, degradation, or modification. Both habitat loss for commercial development, logging, agriculture, and aquaculture can be estimated as major threats to dragonfly (Harabis and Dolny 2012; Reels et al. 2012). These activities will give effect as they are sensitive to pollution and alteration of nutrients resulting from habitat disturbance. According to Che Salmah et al. (2005), dragonfly can be found at riverine ecosystem included paddy field, freshwater swamp, and peat swamp. All types of aquatic habitats are suitable but microhabitat will become primary factor in determining their diversity and distribution.

Hence, it is our priority to take basic studies on dragonfly numbers and populations which can lead to create the awareness programme, since lack of its information within Negeri Sembilan. Therefore, this paper reveals the comparison on diversity of dragonfly between undisturbed habitat (UiTM Forest Reserve) and disturbed habitat (Paddy Field of Kampung Lonek) as basic data for future studies in Negeri Sembilan.

2 Methodology

Sampling was carried out at undisturbed habitat in Fig. 1 (UiTM Forest Reserve, Negeri Sembilan) and undisturbed habitat in Fig. 2 (Paddy field of Kampung Lonek) are located at N 02° 47.710′ E 102° 13.112′ and N 02° 49.050′ E 102° 20.639′, respectively. The sample collection was carried out using the insect net from end of August 2013 until end of October 2013. During these months, the total amounts of rainfall recorded were 218, 130, and 357 mm for august, september, and october 2013, respectively. All samples were identified using book, "A Guide to the Dragonflies of Borneo" (Orr 2003) and "Dragonflies of Peninsular Malaysia and

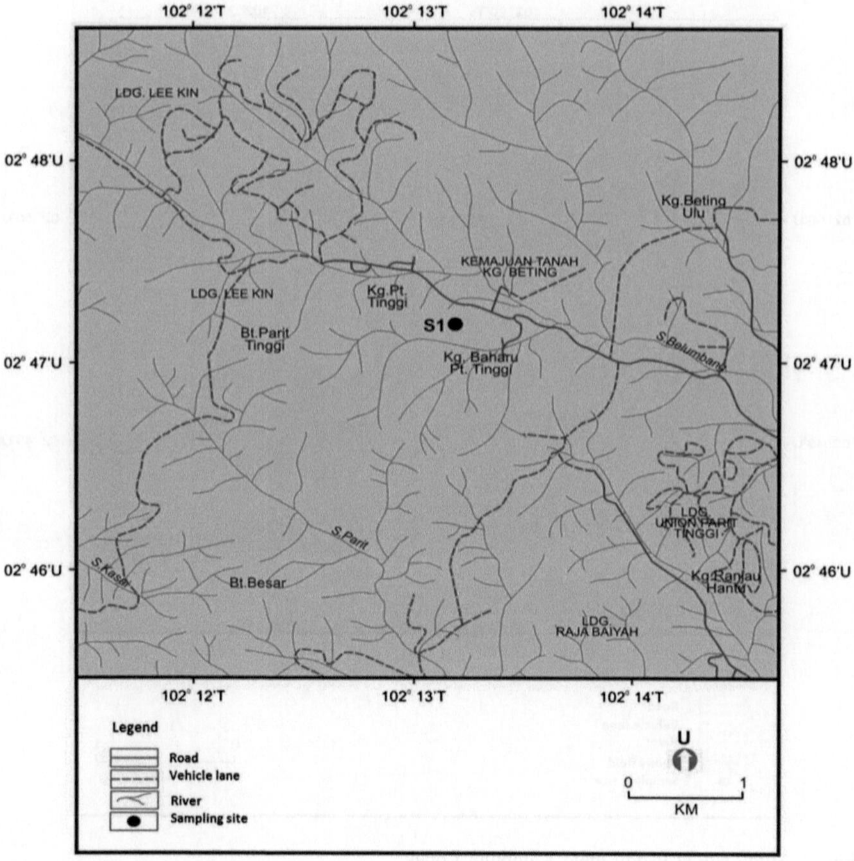

Fig. 1 Study site at UiTM Forest Reserve, Negeri Sembilan

Singapore" (Orr 2005). The data were analyzed by using various statistical methods including Shahnon-Weiner Index (H'), Margaleft (Richness) Index (R), and Evenness Index (E). The Shahnon-Weiner Index (H') was used to measure the degree of species composition or species diversity for each site. The formula being used is $H' = -\sum[(\mathrm{ni}/N)\ln(\mathrm{ni}/N)]$, where N is the total of individuals of population sample and ni is the total individuals belonging to the i species. The Margaleft (Richness) Index (R) being used to measured the number of species in a sample or abundance of the species per unit area. The formula being used is $R = \mathrm{S} - 1/\ln(N)$, where S is total of species and N is total of individuals sample. The evenness Index (E) was used to count the homogeneity or pattern of distribution of species in relation to other species in a sampled per unit area. This index range lies between 0 to 1, in which 0 showed no evenness and 1 signifies complete evenness. The formula is $E = H'/H'$ max, where H' is the Shahnon-Weiner Index and H' max is the diversity index observed to maximum diversity.

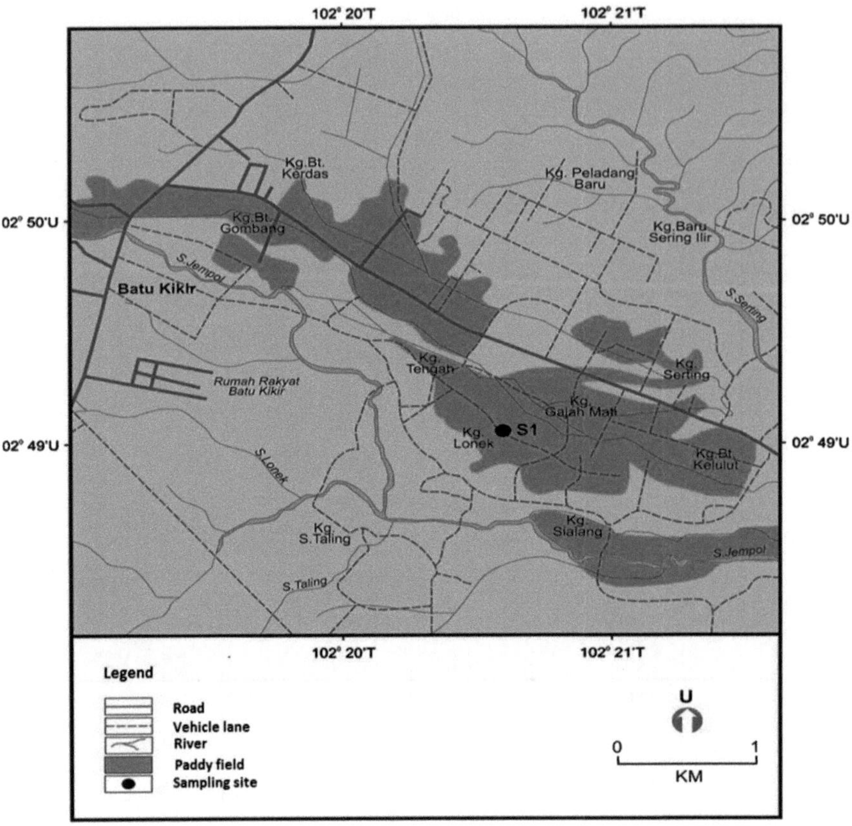

Fig. 2 Study site at Paddy Field Kampung Lonek

3 Results and Discussions

This study used a total collection of 164 dragonflies, representing two families with 14 species of dragonflies from both study sites (Table 1). The two families collected come from Family Libellulidae and Gomphidae. Libellulidae was the abundant family in both sites that makes up 99 % of the total individuals collected while Gomphidae constituted 1 %. Based on the data obtained, Libellulidae was the major family with 13 species followed by Gomphidae with one species. These families were dominant in Anisoptera families, recorded at Rantau Abang Recreational Forest, Terengganu with 73 and 9 %, respectively. They have also been found at Padang Nenas, Terengganu where Libellulidae was recorded as the major family with 85 % of the total individuals collected (Nurul Nadia 2008). Both Rantau Abang Recreational Forest and Padang Nenas were classified as disturbed areas in Terengganu and this means Libellulids can adapt to many environment conditions including disturbed habitat. This was closely related to the alterations of the

Table 1 Total collection of dragonfly and their percentage at UiTM Forest Reserve and Paddy Field of Kampung Lonek, Negeri Sembilan

Family	Species	Number of individuals	
		UiTM Forest Reserve	Paddy Field of Kampung Lonek
Libellulidae	Neurothemis fluctuans	25 (40 %)	0
	Acisoma panorpoides	10 (16 %)	3 (2.9 %)
	Trithemis aurora	6 (10 %)	1 (1.0 %)
	Rhyothemis triangularis	5 (8 %)	0
	Crocothemis servilia	4 (6 %)	13 (12.7 %)
	Rhyothemis phyllis	4 (6 %)	0
	Orthetrum sabina	3 (5 %)	9 (8.8 %)
	Brachydiplax chalybea	2 (3 %)	4 (3.9 %)
	Brachythemis contaminata	1 (2 %)	1 (1 %)
	Orthetrum chrysis	1 (2 %)	0
	Tholymis tillarga	1 (2 %)	0
	Potamarcha congener	0	52 (51 %)
	Pantala flavescens	0	18 (17.6 %)
Gomphidae	Ictinogomphus decoratus	0	1 (1.0 %)
	Total number of individuals	62	102

landscape disturbance that occur in some area at Rantau Abang Forest Reserve and Padang Nenas, Terengganu. Earlier studies had reported the dominance of Libellulidae such as seven species of Libellulidae from 12 species in an irrigated paddy field of Madurai, Tamil Nadu, India (Kandibane et al. 2005). Furthermore, from 162 species recorded, there were 42 species of Libellulidae from Western Himalaya, India (Prasad 2002).

There was 62 individuals from 11 species of dragonflies were collected at UiTM Forest Reserve. *Neurothemis fluctuans* was the species with the highest number of individuals collected with 25 individuals (40 %) followed by *Acisoma panorpoides* with 10 individuals (16 %), and *Trithemis aurora* with 6 individuals (10 %). Study done by Amiruddin et al. (2011), revealed that *Neurothemis fluctuans* was the very common species, it breeds at open area and disturbed habitats. These species are also widely distributed and ubiquitous together with *Orthetrum sabina* and *Rhyothemis phyllis*. Least number of individuals are represented by the species of *Brachythemis contaminata*, *Orthetrum chrysis, and Tholymis tillarga* with one individual (2 %), respectively. While, at the Paddy Field of Kampung Lonek, Negeri Sembilan, a total of 102 individuals with 9 species were collected. *Potamarcha congener* shows the highest number of individuals collected with 52 individuals (51 %) followed by *Pantala flavescens* with 18 individuals (17.6 %) and *Crocothemis servilia* with 13 individuals (12.7 %). Study done by Talmale and

Kulkarni (2003), *Potamarcha congener* was recognized in a wide range close to standing water bodies such as puddles. According to Sharma and Joshi (2007), *Pantala flavescens* was a migratory species which was found very dominant in case of number of individuals with 607 individuals (17.12 %) of the total from Dholbaha Dam in Punjab Shivalik, India. It was the usually found breed in streams of agricultural and take to flock from September to December (Kulkarni et al. 2004). *Brachythemis contaminata*, *Trithemis aurora*, and *Ictinogomphus decoratus* were the species with least number of individual with one individual (1 %), respectively. From this study, *Brachythemis contaminata* can be found at both habitats. The presence of this species indicates that the habitat was contaminated as it acts as an indicator for the polluted habitats (Manwar et al. 2012). This is due to the anthropogenic activity such as forest edge, fragmentation of forest, and herbicides used during paddy plantation flowing into water which serve as habitat for eggs and larvae of dragonfly.

The value for diversity of dragonflies in UiTM Forest Reserve and Paddy Field of Kampung Lonek, Negeri Sembilan has been calculated using Shannon Weiner Index (H′), Richness Index (R), and Evenness Index (E) (Table 2). The value for H′ in UiTM Forest Reserve (H′ = 1.9) and Paddy Field of Kampung Lonek (H′ = 1.493) show the UiTM Forest Reserve has high diversity of dragonflies species than Paddy Field of Kampung Lonek as their range lies between 0 and 4.6, where the value near 0 means every species in the sample is equal, while value 4.6 indicates that the number of individuals are evenly distributed between all the species. According to Sahlen (1999) in undisturbed forests the highest diversity of dragonflies was found unless there were changes in the wetlands. Anbalagan et al. (2013) claim that diversity of Odonata including dragonfly was lower in paddy fields compared to vegetable fields. It occurs due to type of ecosystems and climate that can influence the diversity.

Meanwhile, value for richness in UiTM Forest Reserve (R = 2.423) and Paddy Field Kampung Lonek (R = 1.73) shows the UiTM Forest Reserve has high richness than Paddy Field Kampung Lonek. According to Jennings et al. (2008), this indicates that there was high species richness at UiTM Forest Reserve than species richness at the Paddy Field of Kampung Lonek due to the disturbance that occurs. As stated by John et al. (1996), the species richness is proportionate with disturbance, which means the UiTM Forest Reserve is in early disturbance. As compared to a study done by Buczynski and Tonczyk (2013) high species richness of dragonfly was found in Tuchola Forests, northern Poland due to the high forest

Table 2 The ecological index of dragonfly population at UiTM Forest Reserve and Paddy Field of Kampung Lonek, Negeri Sembilan

Ecological index	UiTM Forest Reserve	Paddy Field of Kampung Lonek
Shahnon-Weiner index (H′)	1.9	1.493
Richness Index (*R*)	2.423	1.73
Evenness Index (*E*)	0.792	0.68

coverage. Clausnizer et al. (2012) claim that there was an enhancement of dragonflies richness within tropical forest matrix at Africa.

At the meantime, Evenness Index of both sites seen varies slightly at UiTM Forest Reserve, (E = 0.792) and Paddy Field of Kampung Lonek (E = 0.68). These results indicate that the UiTM Forest Reserve has the higher number of evenness index and no dominant species, in contrast to Evenness Index of dragonfly at Rantau Abang Recreational Forest (E = 0.76) and Padang Nenas (H = 0.78) (Nurul Nadia 2008). It also indicates that there was also no dominant species of dragonfly at both habitats.

4 Conclusion

In this study, UiTM Forest Reserve recorded 10 species with 62 individuals being caught, while the Paddy field shows 9 species with 102 individuals recorded. These collections comprise of two families which are Libellulidae and Gomphidae. Study shows the Shahnon Weiner index for UiTM Forest Reserve and Paddy Field with $H' = 1.9$ and $H' = 1.493$, respectively. While for richness index, UiTM Forest Reserve recorded 2.423 and the Paddy Field showed 1.73. As for Evenness Index, UiTM Forest Reserve showed 0.792 while the Paddy Field shows 0.68. The high number in term of ecological indices of UiTM Forest Reserve is higher than Paddy field, which shows that habitat condition will have effect toward dragonfly species. As there are species that exist in disturbed habitat that have been caught in UiTM Forest Reserve, it is an expectation for future development to take full concern regarding on the habitat changes as it affects the dragonfly species.

Acknowledgements We would like to thank the Faculty of Applied Science, Universiti Teknologi MARA (Negeri Sembilan) for research facilities provided.

References

Amirrudin A, Wahizatul Afzan A, Johari MN, Nurul Nadia R, Siti Aminah MS (2011) Dragonfly of Jambu Bongkok Forest Reserve, Terengganu, Peninsular Malaysia. In: Jamilah MS, Faridah M (eds) A biological assessment of Jambu Bongkok forest reserve, Terengganu and Nearby ecosystem. Department of Biological Sciences, Faculty of Science and Technology Universiti Malaysia Terengganu, Terengganu, pp 45–60

Anbalagan V, Paulraj MG, Ignacimuthu S (2013) Odonata diversity (Insecta: Arthropoda) in rice and vegetable fields in a north-eastern district of Tamil Nadu, India. J Res Biol 3(4):977–983

Buczynski P, Tonczyk G (2013) Dragonflies (Odonata) of Tuchola Forests (Northern Poland). Odonatrix 1:20–21

Che Salmah MR, Abu Hassan A, Wahizatul Afzan A (2005) Preliminary study on the composition and distribution of Odonata in Perlis State Park. Malayan Nat J 57(3):317–326

Clausnitzer V, Dijkstra K-DB, Koch R, Boudot J-P, Darwall WRT, Kipping J, Samraoui B, Samways MJ, Simaika JP, Suhling F (2012) Focus on Africa freshwaters: hotspots of dragonfly diversity and conservation concern. Front Ecol Environ 10(3):129–134

Harabis F, Dolny A (2012) Human altered ecosystem: suitable habitats as well as ecological traps for dragonflies (Odonata): the matter of scale. J Insect Conserv 16:121–130

IUCN (2007) IUCN Red List of threatened species. http://www.iucnredlist.org. Accessed 20 Nov 2012

Jennings MD, Hoekstra J, Higgins J, Boucher T (2008) A comparative measure of biodiversity based on species composition. Biodivers Conserv 17(4):833–840

John V, Douglas B, Ivette P, de la Inigo Granzow C (1996) A theory of disturbance and species diversity: evidence from Nicaragua after Hurricane Joan. Biotropica 28(4):600–613

Kalkman VJ, Clausnitzer V, Dijkstra K-DB, Orr AG, Paulson DR, Tol JV (2008) Global diversity of dragonflies (Odonata) in freshwater. Hydrobiologia 593:351–363

Kandibane M, Raguraman S, Ganapathy N (2005) Relative abundance and diversity of Odonata in an irrigated rice field of Madurai, Tamil Nadu. Zoos' Print J 20(11):2051–2052

Kulkarni PP, Prasad M, Talmale SS (2004) Insecta: Odonata in Fauna of Pench National Park (Maharashtra). In: Conservation area series 20. Zoological Survey of India, Kolkata, pp 175–206

Manwar NA, Rathod PP, Ahmad Raja I (2012) Diversity and abundance of dragonflies and damselflies of Chatri Lake region, in Pohara—Malkhed reserve forest, Amravati, Maharashtra (India). Int J Eng Res Appl (IJERA) 2(5):521–523

Norma Rashid Y, Mohd Sofian A, Zakaria Ismail M (2001) Diversity and distribution of Odonata (dragonflies and damselflies) in the freshwater swamp lake Tasek Bera, Malaysia. Hydrobiologia 459:135–146

Nurul Nadia R (2008) "Diversity and species composition of dragonflies at two disturbed areas in Terengganu". Undergraduate thesis, Bachelor of Applied Science, Faculty of Science and Technology, Universiti Malaysia Terengganu

Orr AG, Butler SG, Hamalainen M, Kemp RG (2004) Insect: Odonata. In: Yule CM, Yong HS (ed) Freshwater invertebrates of the Malaysian region. Academy of Sciences Malaysia, Kuala Lumpur, Malaysia, pp 409–442

Orr AG (2003). A guide to the dragonfly of Borneo: their identification and biology. Natural History Publication (Borneo) Sdn. Bhd, Kota Kinabalu, Sabah, Malaysia, p 195

Prasad M (2002). Odonata diversity in Western Himalaya, India. In: Kumar A. (ed) Current trends in odonatology. Daya Publishing House, Delhi, pp 221–254

Riservato E, Boudot J-P, Ferreira S, Jovic M, Kalkman VJ, Schneider W, Samraoui B, and Cuttelod A (2009) The status and distribution of dragonflies of the Mediterranean Basin. IUCN, Gland, Switzerland and Malaga, Spain, p 33

Reels G, Dow R, Hamalainen M, Cuong DM (2012) The status and distribution of dragonflies and damselflies (Odonata) in Indo-Burma. In: Allen DJ, Smith KG, Darwall WRT (ed) The status and distribution of freshwater biodiversity in Indo-Burma. IUCN, Cambridge, UK and Gland, Switzerland, pp 90–101

Sahlen G (1999) The impact of forestry on dragonfly diversity in Central Sweden. Int J Odonatol 2 (2):177–186

Sharma G, Joshi PC (2007) Diversity of Odonata (Insecta) from Dholbaha Dam (Distt. Hoshiarpur) in Punjab Shivalik, India. J Asia Pac Entomol 10(2): 177–180

Talmale SS, Kulkarni PP (2003) Odonata in paddy fields of Bhandara district, Maharashtra. Bionotes 5(3):67–68

Chapter 57
Study on Population Size of *Hirundo tahitica* in UiTM Negeri Sembilan, Kuala Pilah Campus

Ahmad Zaimi Mohd Zawawi, Siti Nabilah Ishak, Izzati Adilah Azmir, Nursyazni Abdul Rahim and Amirah Sharif

Abstract Phenomenon of bird droppings by *Hirundo sp.* in UiTM Negeri Sembilan has now become a crucial issue as their droppings not only caused unsightly messes and property damages but also possed a serious threat to students' and staff's health. In order to overcome the problem, the actual number of the birds, especially *Hirundo tahitica,* needs to be assessed so that a suitable way to control the population size of the birds can be implemented. Therefore, a study on the population size of *H. tahitica* had been conducted by using a multiple captures of Jolly-Seber method, as this method allows natality, mortality, immigration, and emigration are suitable for assessment of open population. A variety of trapping methods were used to maximize the number of birds captured including mist nets, sweep nets, and fish nets and they were released back to population after marked. Analyses of data on 656 individual of *H. tahitica* captured during eight sampling periods between November 2012 and February 2013 showed that the total population size is 23, 157 birds. Five morphological traits of the species also were recorded namely wing length, tail length, body length, wing span, and body weight as they are important in sex and age determination of the birds. Among the birds captured, 255 of them are male and 401 are female. Morphological traits measurement recorded showed that only tail length have significant difference between male and female of *H. tahitica* (male-7.53 ± 1.26 g, female-4.96 ± 0.70 g) whereas other traits are slightly different.

A.Z. Mohd Zawawi (✉) · S.N. Ishak · I.A. Azmir · N. Abdul Rahim · A. Sharif
Faculty of Applied Sciences, Universiti Teknologi MARA, Kuala Pilah,
Negeri Sembilan, Malaysia
e-mail: ahmadzaimi@ns.uitm.edu.my

S.N. Ishak
e-mail: izzati_adilah@ns.uitm.edu.my

I.A. Azmir
e-mail: syaznirahim@ns.uitm.edu.my

N. Abdul Rahim
e-mail: amirahsharif@ns.uitm.edu.my

A. Sharif
e-mail: biela_229@yahoo.com

N.A. Yacob et al. (eds.), *Regional Conference on Science, Technology and Social Sciences (RCSTSS 2014)*, DOI 10.1007/978-981-10-0534-3_57

Keywords Bird population · *Hirundo sp.* · Jolly-Seber · Migration bird · Population size

1 Introduction

Migration is a fundamental characteristic of the life history of many organisms from monarch butterflies to marine mammals and also one of the most fascinating of all behaviours (Moore et al. 2005). Nearly all animals group capable of movements undertakes some kind of daily displacement or seasonal migration (Pulido 2007). Bird migration is one of the most fascinating of all life history traits in wildlife. Birds are the most mobile of vertebrates, whether they fly, swim, or run, their ability to cover a great distances quickly enables many species to exploit different areas at different seasons in the annual cycle and at different stages in their life history (Cox 2010).

According to Huppop (2011), migrants stay from hours to weeks, depending on species, age, sex, body condition, season, weather, food availability, intraspecific and interspecific competition, and predation pressure. The direction of departure also appears to be influenced by body and weather conditions. Migration is not, however, of a unitary nature. Migration varies among species, populations, age groups, and sexes, and may also vary between years within individuals. Differences occur with respect to the distances migrant birds travel, the routes they follow, the timing of departure and arrival and the behaviour en route. Birds can thereby occupy widely separated areas at different seasons, returning repeatedly to the same localities from year to year, and adopting an itinerant lifestyle of a kind not open to less mobile creatures (Newton 2008).

Information provided by Nurfamieza (2012) through her research studies on diversity of birds in UiTM Negeri Sembilan stated that birds from genus *Hirundo* are the most abundant at the administration buildings which are the barn swallow (*Hirundo rustica*) and pacific swallow (*Hirundo tahitica*). In addition, Halmos et al. (2010) claimed that both of these species were categorized as long-distance migratory birds and widely distributed in both temporal and tropical habitats. These birds are highly aerial and exclusive insectivores (Balakrishnan 2010).

The barn swallow (*H. rustica*) is a familiar and a popular bird in the world as it is one of the most widely disturbed migratory bird species, breeding in North America, Europe, Asia, and North Africa and wintering in South America, southern Africa, southern Asia, and even northern Australia. Previous studies by Dor et al. (2010) claimed that this abundant species usually builds its nest in human-made structures such as barns, bridges, and buildings. Meanwhile, Pacific swallow (*H. tahitica*) mostly found along coast and rivers in Andaman and Sri Lankan hills. It also occurs in Southeast Asia, Australia, and Pacific Islands. Pacific swallow is resident where it inhabits open country in hills, often around houses and plantations. It feeds on aerial insects and flight less sweeping than barn swallow.

Currently, the population of the *Hirundo sp.* at the area of UiTM Kuala Pilah has tremendously created serious problems not just because of their droppings but also has caused great damage to university structures, crops, and machinery. Previous research claimed that most swallows are known to use architectures structure for roosting and nesting, and this feature has given new opportunities for population expansion (Balakrishnan 2010). Studies by Sakraoui et al. (2005) also showed swallows prefer buildings for nesting. Besides, Higuchi (2011) stated in his research, migratory birds establish a link not only among natural environments in different countries along their migration routes but also among people in different countries and encounter many problems during their migration. Therefore, studies about these migratory birds should be urgently conducted in order to minimize the problems created by these birds.

2 Materials and Methods

The study took place at UiTM Negeri Sembilan (Kuala Pilah Campus) with the estimated area of campus 400 acres which includes the Pelangai Forest Reserve (Mazhani et al. 2010). This campus is located 7 km from main road. Three villages are located nearby which are Kampung Parit Tinggi, Desa Parit Tinggi, and Kampung Beting Ulu (Mazhani et al. 2010).

Sample Collection

For sample collection, mist net was used as it is a popular and important tool to capture birds for monitoring species diversity, relative abundance, population size, and demography. The mist nets used vary in sizes and depended on the species. The mesh size should be ideally suited with target species. The size of mist nets for *H. tahitica* is very small because the body size of the species is small. For the majority of species, the best times for mist netting are either at dawn or at dusk, when birds tend to move between roosting and feeding areas at these times.

However, in this experiment, the mist nets were operated only at dusk because during that time the species were at their peak activity and most abundant in the university's area. Four mist nets were set up randomly around the administration and faculty buildings for each sampling. At each location, the mist nests were opened from 7.00 pm to 12.00 am in each sampling time with 3 days interval gap between another subsequent sampling within 1 month. The nets were checked every 30 min. These mist nets were operated simultaneously at each site. It was started immediately after sunset as it is the period of greatest flight. The sampling was made two times per week for 1 month which means there are only eight samplings as weather conditions also play vital role in mist nets trapping.

Sweep net was also used as alternative tools to maximize number of birds other than mist nets. Its common function is to collect insect on grass and brush. However, it also can be used to catch birds if it is used in a correct way. This is because if the sweep nets were not handled properly, not even a single bird would be trapped in the nets.

Therefore, careful and proper handling is required in order to catch the birds. This net was used to catch birds that perch at the edge of the university's buildings. Also, the net were used during midnight which is around 12.00 am to 3.00 am. This is because at that time all the birds perch, rest, and sleep all around the buildings. Therefore no movements of flight by the birds which makes it easier to catch them. A method other than mist nets and sweep nets used to catch the birds in this research project was fish net. Primary function of this net is to capture small fish; however, it can also be used to catch birds if used in appropriate way. But, using this fish nets is not effective compared to mist nets and sweep nets. This is because the numbers of individual birds trapping is less compared to mist and sweep nets due to its small size of net trap. Thus, it can only catch one or two individual birds per activity. However, it is still a useful tool to maximize capturing birds and contribute to the numbers of trapping birds.

Morphometric Measurement

The morphometric measurements were conducted which included bird's wing length, wing span, tails length, and body length by using a ruler while body weight of each individual birds were measured by a weight scale. These morphometric measurements are important in order to determine the sex of *H. tahitica* in UiTM Negeri Sembilan.

Tagging Process

After all the measurement data had been recorded, the tagging process was carried out where individual birds were marked with numbers, various colours of leg bands that functioned to differentiate between each sampling occasions. Then, marked individual birds were released back to population. Lastly, the recapture process was carried out again after 3 days to collect the data.

3 Results and Discussion

In this study, a total of 656 individual of *H. tahitica* were captured over eight sampling periods and 13 individual birds were recaptured again in the subsequent sampling. Among them, 255 are male and 401 are female (Table 1). However, one of the birds caught in mist nets was found dead. According to Lettink and Armstrong (2003) a few considerations need to be taken during mark, capture, and

Table 1 Total number of individuals of *H. tahitica* according to sex captured at UiTM Negeri Sembilan

Species		Total number of individual birds
H. tahitica		
	Female	401
	Male	255
	Total	656

recapture process which includes potential impacts of capture and marking methods, size and location of study area, and factors that influence capture rates.

In order to get the population size of this bird, Jolly-Seber method was used as it allowed natality, mortality, immigration, and emigration (Orell and Belda 2002). After calculating using a spreadsheet, the total population size of *H. tahitica* in UiTM Negeri Sembilan was 23,157 birds (Table 2).

Based on the data recorded shown in Fig. 1, the total number of individual birds captured decreased in each sampling occasions from the first sampling to the eighth sampling. Two reasons for this phenomenon to occur are the use of mist nets as the main tool to capture the birds and the weather conditions. During the second sampling, the birds manage to avoid and acquaint with the location of mist nets. While observing these species, instead of passing through the net, the birds flew above and below it as if they could sense and were familiar the invisible mist nets. A previous study conducted by Zakaria and Nawaz (2010) showed that the nets were placed for 3 days at the same location before they were moved to other locations and 3 days of netting was sufficient to capture most of the birds. After 3 days, birds may become familiar with the mist nets. A study conducted by Lettink and Armstrong (2003) showed that passerines birds such as swallows have the net avoidance ability especially in sunny days where greater visibility in sunlight may have caused the swallows to avoid the nets.

Table 2 Total number of capture, mark, and recapture of *H. tahitica* in UiTMNegeri Sembilan by using Jolly-Seber method

	Time of capture							
	1	2	3	4	5	6	7	8
Time of last capture								
1		3	2	2	1	1	0	0
2			2	0	0	0	0	0
3				1	1	0	0	0
4					0	0	0	0
5						0	0	0
6							0	0
7								0
Total marked	0	3	4	3	2	1	0	0
Total unmarked	143	133	123	95	67	40	23	19
Total caught	143	136	127	98	69	41	23	19
Total release	143	135	127	98	69	41	23	19

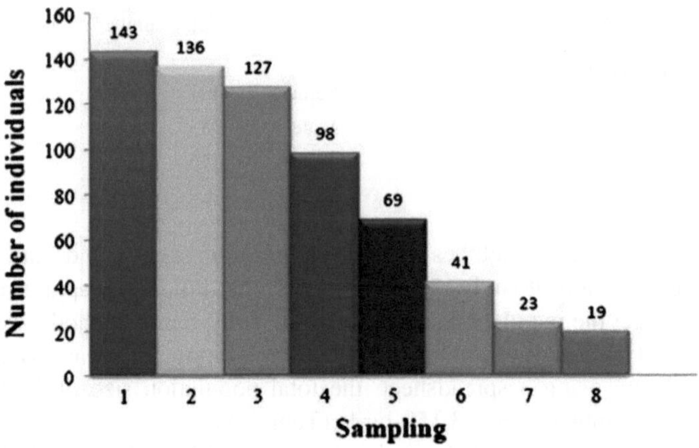

Fig. 1 The total number of *H. tahitica* captured in each sampling at UiTM Negeri Sembilan

Besides that, two weather conditions which adversely affect the mist nets are winds and rain. When conducting this research on February 2013, almost every week, the weather conditions in UiTM Negeri Sembilan were not good and thus, affected the rate of bird capture. A research conducted by Muhammad-Saiful et al. (2011) also experienced the same situation where both of weather conditions affect the rate of capture.

Estimated population size of these birds in UiTM Negeri Sembilan is important in order to have successful bird control programme. This results supported by previous studies where Bijlsma and Brink (2005) stated that migratory birds such as swallow migrates in such a large flocks up to several millions. Also, by migrating and roosting in huge numbers, swallow could reduce predation risks.

4 Conclusion

In conclusion, multiple capture and mark recapture as in Jolly-Seber method is applicable and best used to estimate the population size of *Hirundo tahitica* at UiTM Negeri Sembilan. Analyses of the data from Jolly-Seber method showed the total population size of *H. tahitica* at UiTM Negeri Sembilan is 23,157 birds. Morphometric data of these species help to determine the proportion of both sexes where females have higher proportion compared to male. *H. tahitica* is categorized under sexual dimorphic and thus it is easy to distinguish between male and female especially through body colour and tail length.

Acknowledgement This research financially supported by the DANA KECEMERLANGAN UiTM (600-RMU/ST/DANA 5/3/Dst (9/2012) is gratefully acknowledged.

References

Balakrishnan P (2010) Breeding biology of the Hill Swallow *Hirundo domicola* in Western Ghats, India. J Bombay Nat Hist Soc 107(2):109–115

Bijlsma RG, van den Brink B (2005) A Barn Swallow *Hirundo rustica* roost under attack: timing and risks in the presence of African hobbies Falco cuvieri. Ardea 93(1):37–48

Cox WG (2010) Bird migration and global change. Island Press

Dor R, Safran RJ, Sheldon FH, Winkler DW, Lovette IJ (2010) Phylogeny of the genus Hirundo and Barn Swallow subspecies complex. Mol Phylogenet Evol 56:409–418

Halmos G, Karcza Zs, Nemeth A, Csorgo T (2010) The migratory fattening of the Barn Swallow *Hirundo rustica* in Hungary. Acta Zool. Hung 56(1):73–87

Higuchi H (2011) Bird migration and the conservation of the global environment. J Ornithol 153(1):3–14

Huppop O (2011) Bird migration on helgoland: the yield from 100 years of research. J Ornithol 152(1):25–40

Lettink M, Armstrong DP (2003) An introduction to using mark recapture analysis for monitoring threatened species. Dept Conserv Tech Ser 28:5–32

Mazhani M, Norfazlina B, Raja NurAteeka RO (2010) A preliminary checklist of flora in UiTM forest reserve, Negeri Sembilan. UiTM Shah Alam, Shah Alam

Moore FR, Woodrey MS, Buler JJ, Woltmann S, Simons TR (2005) Understanding the stopover of migratory birds: a scale dependent approach. Forest Serv Gen Tech Rep 191:684–689

Muhammad-Saiful M, Shahrul-Anuar MS, Lim CK, Mustafa AR (2011) Bird species diversity in the Padawan Limestone area, Sarawak. Trop Life Sci Res 22(2):65–80

Newton I (2008) The migration ecology of birds, 1st edn. Academic Press, London, pp 3–8

Nurfamieza M (2012) A preliminary study of birds diversity in UiTM Negeri Sembilan, Kampus Kuala Pilah (Beting). Undergraduate thesis, Bachelor of Science (Hons) Biology, Faculty of Applied Sciences, Universiti Teknologi MARA

Orell M, Belda JE (2002) Delayed cost of reproduction and senescence in the willow *Tit Parusmontanus*. J Anim Ecol, Br Ecol Soc 71:55–64

Pulido F (2007) The genetics and evolution of Avian migration. Bioscience 57(2):165–174

Sakraoui R, Dadci W, Chabi Y, Banbura J (2005) Breeding biology of Barn Swallows Hirundorustica in Algeria, North Africa. OrnisFennica 82:33–42

Zakaria M, Nawaz MR (2010) Bird species composition and feeding guilds based on point count and Mist Netting methods at the Paya Indah Wetland reserve, Penisular Malaysia. Trop Life Sci Res 21(2):7–26

Part XI
Environmental Science and Management

Chapter 58
Evaluation of Vertical Accuracy of Airborne IFSAR and Open-Source Digital Elevation Models (DEMs) for Flood Inundation Mapping

Suhaila Hashim, Wan Mohd Naim Wan Mohd, Nor Aizam Adnan and Eran Sadek Said Md Sadek

Abstract Acquiring an accurate elevation model for flood inundation mapping is essential and can be considered as an expensive task. One possible solution is to acquire readily available digital elevation model (DEM) or open-source DEMs such as ASTER GDEM and SRTM. The issue is how accurate are these datasets and can it be used for flood inundation mapping. The objective of this paper is to evaluate the vertical accuracy of DEMs acquired from different data sources. Results presented in this paper are part of an undergoing research to evaluate the use of global DEMs for flood inundation mapping. For the evaluation of vertical accuracy, DEMs acquired from Airborne IFSAR, ASTER and SRTM Void Fill are used to generate height points. The results obtained are compared with height from GPS observation of two different test sites located in Alor Setar and Kuala Nerang, Kedah. The terrain characteristics of Alor Setar and Kuala Nerang test sites are flat and variable terrain respectively. The overall vertical accuracies of IFSAR DTM, IFSAR DSM, ASTER, and SRTM Void Fill for Alor Setar are ±0.497, ±1.529, ±5.848, and ±4.268 m, respectively. For the undulating terrain, the vertical accuracies of these DEMs are slightly lower, i.e., ±0.841, ±2.092 and ±5.300 m. The height accuracy of ASTER GDEM in undulating area is found to be higher than relatively flat area, i.e., ±3.278 m. Findings from this preliminary test have shown that the vertical accuracy of airborne IFSAR DTM is quite high and can potentially be used in flood management.

S. Hashim (✉)
Faculty of Architecture, Planning & Surveying,
Universiti Teknologi MARA, Arau, Perlis, Malaysia
e-mail: suhailahashim@perlis.uitm.edu.my

W.M.N. Wan Mohd · N.A. Adnan · E.S.S. Md Sadek
Faculty of Architecture, Planning & Surveying, Universiti Teknologi MARA,
Shah Alam, Selangor, Malaysia
e-mail: wmn@salam.uitm.edu.my

N.A. Yacob et al. (eds.), *Regional Conference on Science, Technology and Social Sciences (RCSTSS 2014)*, DOI 10.1007/978-981-10-0534-3_58

Keywords Airborne IFSAR · ASTER · SRTM void fill · Digital elevation model · Digital terrain model

1 Introduction

Flood inundation mapping is an important tool for engineers, planners, and government agencies for municipal and urban growth planning, emergency action plans, flood insurance rates and ecological studies. Flood inundation maps have been prepared from satellite data for more than a decade by hydrologist all over the world. Recently, digital elevation model (DEM) and geographical information system (GIS) have became practical tools for development of cost effective methods for flood inundation mapping in gauged and ungauged river basins. Furthermore, DEMs can be utilized to derive the flow direction and the computational sequence for flow routing for each of the discretized cells of the catchment represented as a proper hydrologic cascading system (Basarudin and Adnan 2014).

Digital elevation model (DEM) is a quantitative representation of terrain and is important for earth science and hydrological applications. The accuracy of hydraulic and hydrologic modeling results largely depend on resolution and accuracy of DEM. DEM can be generated from different techniques such as photogrammetric method using stereo data (Hohle 2009), interferometry (Kervyn 2001) and airborne laser scanning (Favey et al. 1999). ASTER is an international collaboration project between the Ministry of Economy, Trade and Industry of Japan (METI) and the United States National Aeronautics and Space Administration (NASA). The DEM covers 99 % of the earth's land mass. Near-infrared stereo imagery is collected simultaneously at both nadir and off nadir angles with along-track alignment. This stereo imagery is then utilized to develop a DEM through stereo correlation technique. As reported by Hirano et al. (2003), the vertical accuracy of ASTER DEMs is in the range of 7–15 m. The most complete DEM available to the public was the Shuttle Radar Topography Mission (SRTM) dataset. It was acquired as a joint mission by NASA, German Aerospace Center, and the National Geospatial-Intelligence Agency. The SRTM data were created using interferometric processing of L-band synthetic aperture radar (SAR) data. Airborne INTERMap IFSAR provides three main products, viz., digital surface models (DSM), digital terrain models (DTM), and orthorectified radar imagery (ORI). The vertical accuracy of 0.5–1.0 m of both the airborne IFSAR DSM and DTM can be achieved by the airborne Intermap mapping system (Wei and Coyne 2008).

In order to have an accurate map, the DEM need to be evaluated. Various techniques have been used by different authors to evaluate the accuracy of different DEM data. Zhou et al. (2012), Jarvis et al. (2004) and Hall and Tragheim (2010) generated elevation profile to compare the differences between DEMs while Nikolakopoulos et al. (2006) carried out correlation analysis to compare the difference in DEM accuracy. Another method of assessing the DEM accuracy is by comparing the relationship between topographic characteristics such as slope and aspect (Gorokhovich and Voustianiouk 2006). Yang et al. (2011) used matching contour method to evaluate the accuracy of ASTER GDEM elevation. In a study by Kuuskivi and Li (2006) the accuracy performance of DEM products from airborne and spaceborne IFSAR are compared with high-accuracy ground control points (GCPs) and higher accuracy DEM.

The cost of acquiring DEM is largely influenced by the technique and accuracy required. Although Airborne LiDAR is considered as the most expensive technique, it has become preferred technology for digital elevation data acquisition in a wide range of applications (Liu 2011). Other freely available open-source DEM products are Shutter Radar Topography Mission version 3 (STRM3) with 90 m DEM and ASTER GDEM (Farr and Kobrick 2000; Farr et al. 2007) and can be acquired via the United States Geological Survey (USGS) website whilst airborne NEXTMap IFSAR is cost effective DEM for large coverage applications. The objective of this paper is to evaluate the accuracy of IFSAR DTM, IFSAR DSM, ASTER GDEM and SRTM compared GPS observation height.

2 Study Area

The study areas are located within the District of Kota Setar and Padang Terap (i.e., near the city of Alor Setar and Kuala Nerang town). The area is selected as the study area due to the availability of GPS observation data, NEXTMap IFSAR, ASTER GDEM, SRTM, and variable terrain characteristics. The height range of this area is from 1 m up to 140 m. The lower part of the study areas is situated within the District of Kota Setar and mainly covered by residential area, commercial area and paddy field. The higher part of the study area is situated within District of Padang Terap and mainly covered by forest and agricultural areas. The geographical location of the study area is shown in Fig. 1. The extent of small study area is shown in Fig. 2a, b.

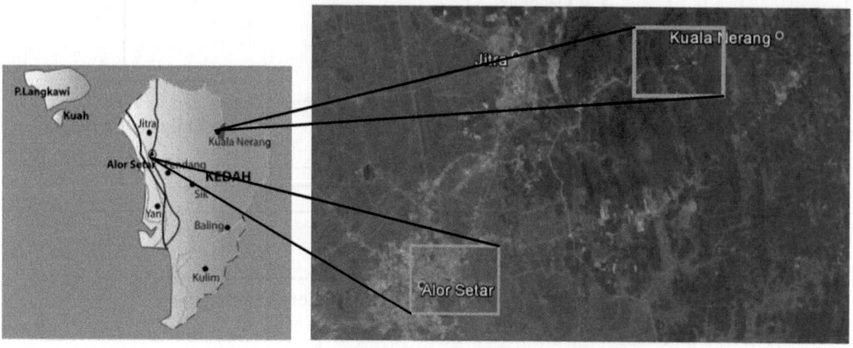

Fig. 1 Location of study area (Adopted from Google Earth, 2014)

3 Methodology

The methodology adopted for this study is divided into three main stages: (i) data acquisition, (ii) data processing, and (iii) data analysis. Figure 3 shows the general methodology adopted for this study.

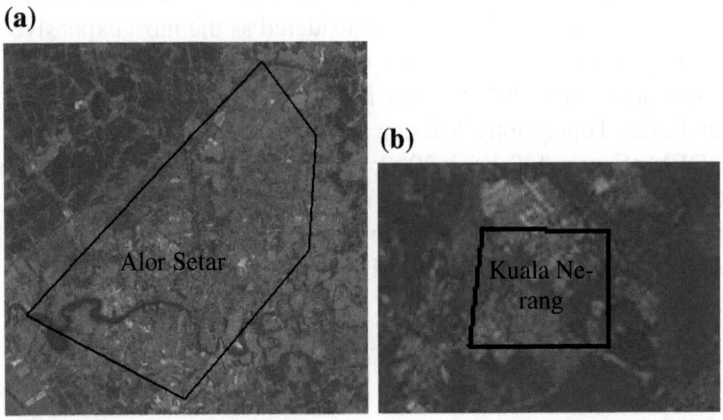

Fig. 2 Location of study areas **a** Test site 1 (Alor Setar) **b** Test site 2 (Kuala Nerang) (Adopted from Google Earth, 2014)

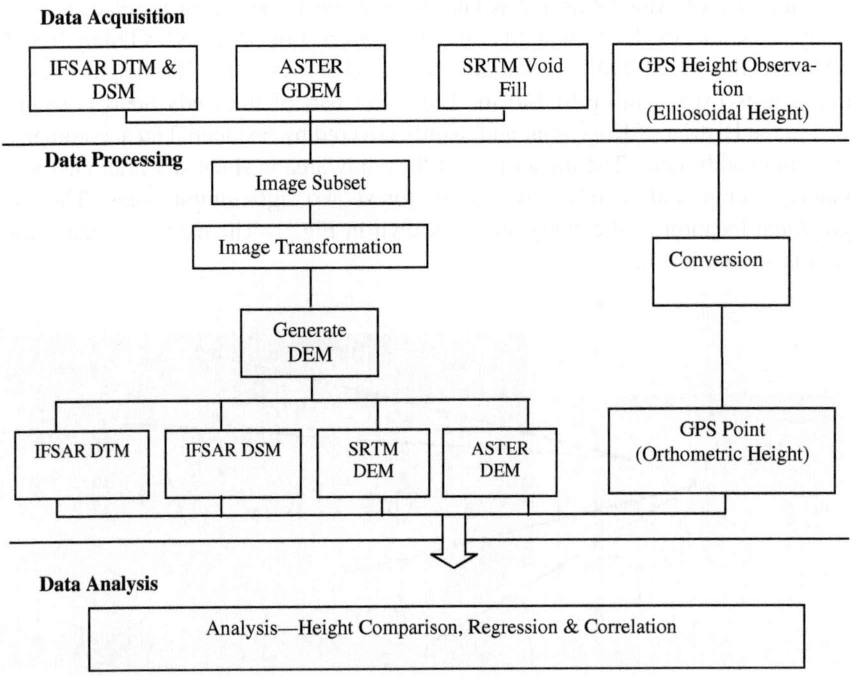

Fig. 3 Flowchart of methodology

Data Acquisition

The data used in this study are elevation of selected points observed with global positioning system, open-source DSM (ASTER GDEM and SRTM) of 2011 and NEXTMap Airborne IFSAR DEM. ASTER GDEM and SRTM Void Filled DEM data are downloaded from the USGS website (earthexplorer.usgs.gov) while the IFSAR dataset are acquired from Intermap Technologies Malaysia. The three types of IFSAR data are the DSM, DTM, and ORI. The NEXTMAP DSM represents the earth's surface and includes all features such as buildings and trees on it while DTM is a bare-earth model of the terrain. A total number of 30 static observations using dual frequency GPS receiver were observed within the two test sites.

Data Processing

As the ASTER GDEM and SRTM data downloaded from the USGS website cover a large area, image subset is carried out to clip data according to the coverage of the IFSAR DTM and IFSAR DSM. All these datasets are later transformed into Malayan Rectified Skew Orthomorphic (MRSO) projection in the ArcGIS software. Spatial Analyst tool in the ArcGIS software is used to generate the DEMs. The output of the data processing steps are four different DEMs (IFSAR DTM, IFSAR DSM, ASTER, SRTM).

A total of 30 GPS points are measured within two different test sites (refer Fig. 4). The relative positioning technique whereby one base station was selected from myRTKnet station. For this project, Tokai station (located in Kedah) was selected as the base station. This base station was used with rover station to complete the

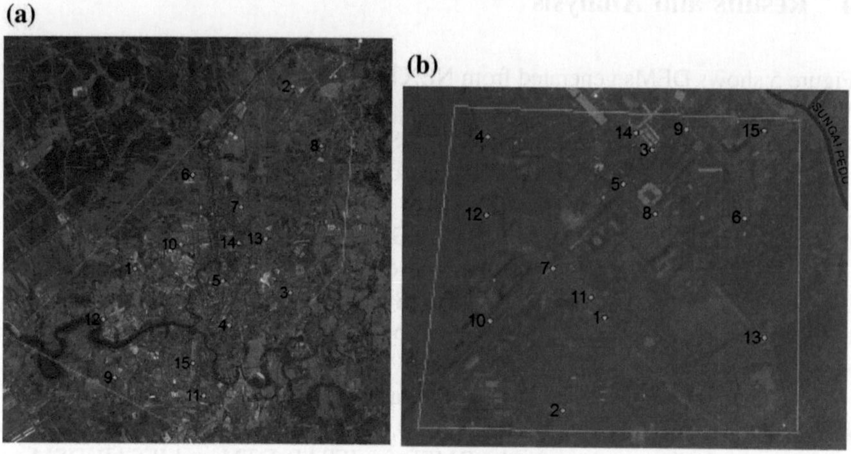

Fig. 4 Distribution of GPS point a Alor Setar b Kuala Nerang

baseline processing. An appropriate rover stations were chosen from suitable location based on Google Map. The observed ellipsoidal heights of the rover stations are later converter into orthometric heights using geoid-ellipsoid separation for Peninsular Malaysia Software.

Data Analysis

Further processing involved the measuring of selected height points and then determining the vertical accuracy of the observed height points. To determine the accuracy of different DEMs, the extracted DEM are overlaid onto the GPS orthometric heights points in the ArcGIS software. The minimum error, maximum error and the root mean square error (RMSE) are computed based on Eqs. (1)–(3), respectively. In order to determine the degree of relation between the different DEMs and GPS height, spatial correlation is computed.

$$\text{Minimum error} = \min(|Z_{\text{obs}} - Z_{\text{ref}}|) \tag{1}$$

$$\text{Maximum error} = \max(|Z_{\text{obs}} - Z_{\text{ref}}|) \tag{2}$$

$$RMSE = \sqrt{\frac{\sum_{i=1}^{n} (Z_{\text{obs},i} - Z_{\text{ref},i})^2}{n}} \tag{3}$$

where Z_{obs} is the observed heights in different DEMs, Z_{ref} is the observed heights in reference DEM and n is the total number of observations.

4 Results and Analysis

Figure 5 shows DEMs generated from NEXTMap IFSAR, ASTER GDEM, SRTM. Visual inspection on these figures shows that the generated DEMs are almost similar except for DEM generated from ASTER GDEM (especially in low-lying areas, i.e., elevation less than 50 m). This could be due to the courser DEM grid resolution (i.e., 30 m) as compared to the 5 m resolution of the IFSAR data or inaccurate height generated from ASTER GDEM.

Table 1 shows the elevation point observed from different DEMs based on location of GPS observation points in Test Site 1 and 2. The descriptive statistics of the accuracy of DEMs based on the height points measured within the study area is summarized in Table 2.

For the Test Site 1 (relatively flat area), the RMSE for IFSAR DTM, IFSAR DSM, ASTER and SRTM are ±0.497, ±1.529, ±5.848, and ±4.268 m, respectively. In the terrain area, the RMSE for IFSAR DTM and IFSAR DSM are

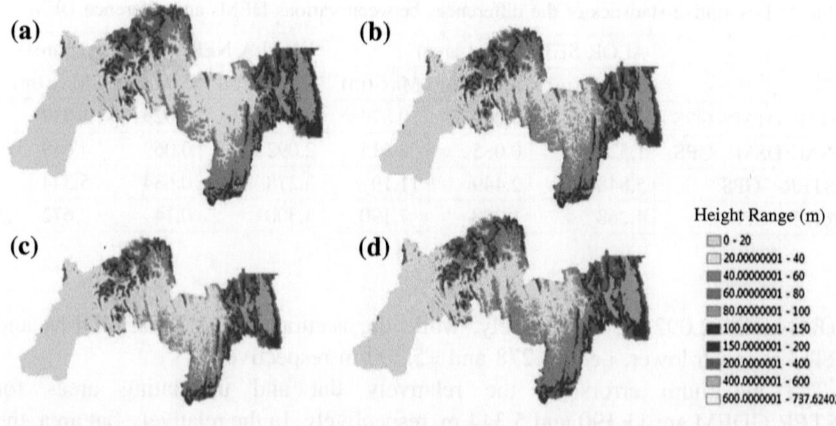

Fig. 5 Height range maps generated from **a** IFSAR DTM **b** IFSAR DSM **c** ASTER GDEM **d** SRTM

Table 1 Comparison between height points measured with GPS, NEXTMap IFSAR DTM, NEXTMap IFSAR DSM, ASTER GDEM, and SRTM

PT NO	Test site 1—Alor Setar (flat area)					Test site 2—Kuala Nerang (undulating area)				
	GPS	NEXTMap IFSAR		ASTER	SRTM	GPS	NEXTMap IFSAR		ASTER	SRTM
	(m)	DTM (m)	DSM (m)	(m)	(m)	(m)	DTM (m)	DSM (m)	(m)	(m)
1	2.982	2.469	2.897	11	6	22.328	22.193	24.84	26	31
2	4.223	5.297	5.438	10	4	22.419	22.391	23.593	26	21
3	3.14	3.047	3.495	7	5	20.366	20.638	19.951	21	26
4	1.691	2.092	6.206	8	6	19.717	19.61	24.366	18	21
5	1.821	2.298	3.928	7	7	18.656	19.556	20.24	24	27
6	3.298	3.837	3.667	7	2	28.041	25.134	27.972	29	31
7	3.517	3.819	2.78	8	8	19.3	20.025	19.735	17	26
8	3.551	3.822	3.425	6	5	19.982	19.349	20.069	23	26
9	1.657	2.528	1.03	8	3	22.897	22.599	27.066	28	27
10	2.81	3.689	3.183	14	10	20.19	20.219	22.118	23	26
11	2.482	3.039	3.65	7	7	21.145	21.194	20.893	18	24
12	2.263	2.345	1.738	6	6	19.14	17.361	19.872	18	19
13	2.742	4.378	3.438	9	6	41.192	41.088	41.631	40	42
14	3.012	3.061	3.475	8	10	21.809	21.74	21.94	17	25
15	1.886	3.167	2.97	8	6	21.927	20.09	21.481	17	33

Table 2 Descriptive statistics of the differences between various DEMs and reference DEM

	ALOR SETAR (flat area)			KUALA NERANG (terrain area)		
	RMSE (m)	Min (m)	Max (m)	RMSE (m)	Min (m)	Max (m)
IFSAR DTM—GPS	0.497	0.049	0.879	0.841	0.029	1.837
IFSAR DSM—GPS	1.529	0.085	4.515	2.092	0.069	4.649
ASTER—GPS	5.848	2.449	11.19	3.278	0.634	5.344
SRTM—GPS	4.268	1.298	7.190	5.300	0.14	8.672

±0.841 and ±2.092 m, respectively, while the accuracy of ASTER GDEM and SRTM is much lower, i.e., ±3.278 and ±5.300 m respectively.

The maximum errors in the relatively flat and undulating areas for ASTER GDEM are 11.190 and 5.344 m, respectively. In the relatively flat area, the minimum and maximum height difference IFSAR DTM and GPS heights are 0.049 and 0.879 m, respectively. The minimum and maximum errors of IFSAR DSM as compared to GPS observation in the relatively flat area are 0.085 and 4.515 m and 0.069 and 4.649 m undulating area respectively. The accuracies for IFSAR DTM, IFSAR DSM, and SRTM DEM are lower in the terrain area (refer to Table 2). On the other hand, the accuracy for ASTER GDEM is much better for undulating area. The correlation between the elevations obtained from different DEMs and the reference DEM are graphically shown in Figs. 6 and 7. Based on these figures and Table 3, it is evident that the correlation between DEMs and reference DEMs is highest in IFSAR DTM (i.e., 99.40 %) followed by IFSAR DSM (i.e., 98.30 %). The figure also shows strong correlation in ASTER GDEM and SRTM (i.e., 89.40 and 93.20 % respectively).

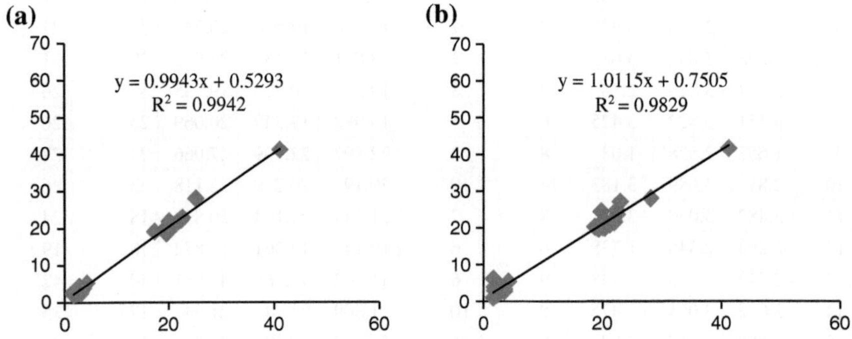

Fig. 6 Correlation plot of **a** IFSAR DTM versus GPS observation **b** IFSAR DSM verses GPS observation

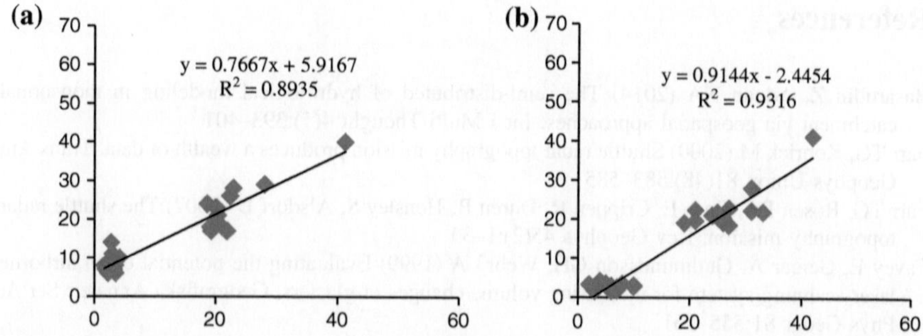

Fig. 7 Correlation plot of **a** ASTER GDEM versus GPS observation **b** SRTM verses GPS observation

Table 3 Correlation and regression coefficients between IFSAR DTM, IFSAR DSM, ASTER GDEM, and SRTM with GPS height

	Correlation coefficient (R^2)	Gradient (m)	Intercept
IFSAR DTM—GPS	0.994	0.994	0.529
IFSAR DSM—GPS	0.983	1.012	0.751
ASTER—GPS	0.893	0.767	0.894
SRTM—GPS	0.932	0.914	2.445

5 Conclusion

The vertical accuracies of IFSAR DTM, IFSAR DSM, ASTER GDEM, and SRTM are evaluated in the present study. The accuracy evaluation of these datasets is performed based on GPS height observation. Initial findings have indicated the potential use of IFSAR DTM products for generating accurate flood inundation map. Although DEM accuracies of SRTM and ASTER are much lower as compared to that of Airborne IFSAR, it could still be used to generate the DEM infill in hilly areas as the IFSAR DTM is very expensive. As this study is part of a more comprehensive research to evaluate the suitability of using different DEMs including open-source Global DEMs for flood inundation mapping, a detail study to evaluate the effects of using different DEMs on the accuracy of the generated flood inundation maps is needed.

Acknowledgements The authors would like thank the Department of Survey and Mapping Malaysia (JUPEM), Intermap Malaysia Inc., Malaysian Remote Sensing Agency and the Department Public Works Malaysia, for providing necessary data for this study. Many thanks go to the Universiti Teknologi MARA for sponsoring my study and the Ministry of Education, Malaysia for funding this research under the Long Term Research Grant Scheme (LRGS/b-u/2012/UUM/Teknologi Komunikasi dan Informasi).

References

Basarudin Z, Adnan NA (2014) The semi-distributed of hydrological modeling in monsoonal catchment via geospatial approaches. Int J Multi Thought 4(1):393–401

Farr TG, Kobrick M (2000) Shuttle radar topography mission produces a wealth of data. Trans Am Geophys Union 81(48):583–585

Farr TG, Rosen PA, Caro E, Crippen R, Duren R, Hensley S, Alsdorf D (2007) The shuttle radar topography mission. Rev Geophys 45(2):1–33

Favey E, Geiger A, Gudmundsson GH, Wehr3 A (1999) Evaluating the potential of an airborne laser-scanning system for measuring volume changes of glaciers. Geografiska Annaler: Ser A, Phys Geogr 81:555–561

Gorokhovich Y, Voustianiouk A (2006) Accuracy assessment of the processed STRM-based elevation data by CGIAR using field data from USA and Thailand and its relation to terrain characteristics. Remote Sens Environ 104:409–415

Hall M, Tragheim DG (2010) The accuracy of ASTER digital elevation models: a comparison with NEXTMap Britain. In: Fleming C, Marsh SH, Giles JRA (eds) Elevation models for geoscience. Geological Society of London, Special Publications, vol 345, pp 43–53

Höhle J (2009) DEM generation using a digital large format frame camera. Photogram Eng Remote Sens 75(1):87–93

Hirano A, Welch R, Lang H (2003) Mapping from ASTER stereo image data: DEM validation and accuracy assessment. ISPRS J Photogram Remote Sens 57(5):356–370

Jarvis A, Rubiano J, Nelson A, Farrow A, Mulligan M (2004) Practical use of SRTM data in the tropics—comparisons with digital elevation models generated from cartographic data. CIAT working document No. 198 http://ciatlibrary.ciat.cgiar.org/articulos_ciat/Jarvis4.pdf. Accessed 20 June 2015

Kuuskivi T, Li X (2006) Three-dimensional mapping using both airborne and spaceborne IFSAR technologies. In: American society of photogrammetry and remote sensing conference 2006

Kervyn F (2001) Modelling topography with SAR interferometry: illustrations of a favourable and less favourable environment. Comput Geosci 27(9):1039–1050

Liu X (2011) Accuracy assessment of LiDAR elevation data using survey marks. Survey Rev 43 (319):80–93

Nikolakopoulos K, Kamaratakis E, Chrysoulakis N (2006) SRTM vs ASTER elevation products–comparison for two regions in crete, Greece. Int J Remote Sens 27:4819–4838

Wei M, Coyne T (2008) Integrated airborne IFSAR mapping system. Int Arch Photogram, Remote Sens Spat Inf Sci 37:367–372

Yang X, Zhang W, Zhu S (2011) Accuracy assessment of ASTER GDEM in North Shaanxi of China. Adv Cartography GISci 2:371–382

Zhou H, Zhang J, Gong L, Shang X (2012) Comparison and validation of different DEM data derived from InSAR. Procedia Environ Sci 12:590–597

Part XII
Food Science

Chapter 59
Quality Attributes of Different Purple Sweet Potato Variety and Sensory Evaluation of Purple Sweet Potato Straight Drink

Nur Izalin Mohamad Zahari, Jeeven Karuppan, Erwan Shah Shaari, Kasmah Mohamad, Rosnah Othman and Yusnita Yaacob

Abstract Sweet potato (*Ipomoea batatas* L.) is an extremely important crop in many parts of the world, being cultivated in more than 100 countries. In Malaysia, sweet potato is the second crop after cassava (*Manihot esculenta crantz*). In this study, the effect of several Malaysian purple sweet potato varieties (K2, K5 and K6) on physical and chemical characteristics and sensory evaluation of purple sweet potato straight drink were evaluated. These samples were evaluated in the form of fresh flesh and puree before being processed into straight drink. The total anthocyanin content (TAC) of the purees ranging from 25.71 to 426.67 mg/L. TAC difference between varieties was statistically significant ($p > 0.05$). All varieties contain high minerals in the range of 70.1–92.0 mg/100 g for phosphorus, 45.3–69.1 mg/100 g for calcium, 435–518 mg/100 g for potassium, 28.9–34.7 mg/100 g for magnesium and 5.4–13.1 mg/100 g for sodium. Vitamin C was 11.7–29.9 mg/100 g and vitamin E was 11.2–3.8 mg/100 g. For sensory evaluation, there were no significant differences ($p > 0.05$) in terms of aroma and

N.I. Mohamad Zahari (✉) · J. Karuppan · K. Mohamad · R. Othman · Y. Yaacob
Food Technology Research Centre, Malaysian Agricultural Research & Development Institute (MARDI), Serdang, Selangor, Malaysia
e-mail: izalin@mardi.gov.my

J. Karuppan
e-mail: jeeven@mardi.gov.my

K. Mohamad
e-mail: kasmah@mardi.gov.my

R. Othman
e-mail: rosnah@mardi.gov.my

Y. Yaacob
e-mail: eyta@mardi.gov.my

E.S. Shaari
Rice & Industrial Crops Centre, Malaysian Agricultural Research & Development Institute (MARDI), Serdang, Selangor, Malaysia
e-mail: erwanss@mardi.gov.my

N.A. Yacob et al. (eds.), *Regional Conference on Science, Technology and Social Sciences (RCSTSS 2014)*, DOI 10.1007/978-981-10-0534-3_59

sweetness of sweet potato straight drink with different percentage of K2 puree (T1, T2, T3). Thus, K2 was found suitable for various processing, which contains high total anthocyanin content and could be a high-phytonutrient source in human nutrition.

Keywords Anthocyanin · Purple sweet potato · Quality evaluation · Sensory · Straight drink

1 Introduction

Sweet potatoes (*Ipomoea batatas* L.) are known to have high energy, dietary fibre, minerals, vitamins and antioxidants. The colourful orange and purple fleshed sweet potatoes are normally rich with antioxidants which are β-carotene and anthocyanin, respectively. Anthocyanin are members of the flavonoid group of phytochemicals, which peonidin and cyanidin act as the major components in red and purple fleshed sweet potatoes (Lila 2004; Wang et al. 2006). Previous studies showed that some of the purple sweet potato (PSP) cultivars have high content of anthocyanin pigments (Ray and Tomlins 2010). Wrolstad (2000) reported that most fruits and vegetables with red and purple colour contain anthocyanin ranged from 20 to 6000 mg/kg while Teow et al. (2007) reported that total anthocyanin contents of 19 genotypes of PSP varied in the range of 17–531 mg/kg roots.

In terms of its role in health benefits, anthocyanin was found to have free radical scavenging and antioxidant capacities (Wang and Jiao 2000; Tsuda et al. 2002, 2003), to enhance memory function, to decrease blood sugar (Ray and Tomlins 2010) and to inhibit cancer cell growth (Wang et al. 2006). Purple sweet potato straight drink is a type of beverages which contains not less than 5 % of fruit, with or without permitted preservative, colourant, flavour and other food conditioner, and ready to be consumed without any dilution (Malaysian Food Act 1983). The use of purple sweet potato tubers for straight drink would further enhance nutrition of the consumers. Consequently, the objectives of this study are to characterize the quality of different variety of local purple sweet potato and to evaluate the organoleptic characteristics of purple sweet potato straight drink.

2 Methodology

Three purple sweet potato accessions were collected from a farmer's field for planting material. These local-purple sweet potato accessions were named Kedudut 2, Kedudut 5 and Kedudut 6 and have been used by farmers in Bachok area, Kelantan for traditional food. The tubers were harvested at 16 weeks and sent to lab for further processing.

Preparation of Purple Sweet Potato Puree

Purple sweet potato puree was prepared from the fresh K2, K5 and K6 tubers. The mature, undamaged and disease-free tubers were obtained from the farm 1 day before processing. The selected tubers were washed and peeled. The cleaned tubers was then cut into 1–1.2 in. and steamed in the steamer for 15–20 min. The steamed flesh was meshed and packed into plastic packaging material. The puree was stored in freezer (−4 °C) prior to analysis and straight drink processing.

Production of Sweet Potato Straight Drink

Straight drink with puree concentrations of 5, 10 and 15 % were prepared from purple sweet potato. The sweet potato puree was thawed and used in the processing according to the formulation. The total soluble solid (TSS) and acidity of the puree were determined and maintained as 14°Brix and 0.3 %, respectively, in all formulation. Purple sweet potato straight drink was prepared using the puree, distilled water, sugar and citric acid. The ingredients were mixed together and heated at 85 °C for 20 min for pasteurization. The drink was hot filled into 200 ml glass bottles, capped with crown cork immediately and cooled. Nutritional analysis, TAC and sensory analyses were carried out for each sample.

Nutritional Analysis

The purple sweet potato puree was analysed for proximate and minerals as well as vitamin C and vitamin E while purple sweet potato straight drink were analysed for their moisture, protein, fat, ash and dietary fibre according to the methods of AOAC (1995). Carbohydrate was calculated by subtracting the sum of the moisture, protein, fat and ash from 100. Factors 4, 4, 9 and 2 are used for calculating energy values from protein, fat, dietary fibre and carbohydrate, respectively.

Total Anthocyanin Content (TAC)

Measurement of total anthocyanin content was done using pH differential method. The content of anthocyanin was expressed as mg/l.

Sensory Evaluation

Sensory evaluation was conducted for the purple sweet potato straight drink with different concentration level of K2 puree. Twenty five panels from Malaysian Agricultural Research & Development Institute (MARDI) with some experience in sensory evaluation were requested to analyse the sensory qualities using 9-point hedonic rating scale as follows: (1) dislike extremely; (2) dislike very much; (3) dislike moderately; (4) dislike slightly; (5) neither like nor dislike; (6) like slightly; (7) like moderately; (8) like very much and (9) like extremely.

Statistical Analysis

Statistical calculation was carried out by Minitab Version 14 software (Minitab Inc., State College, PA, USA). Analysis of variance (ANOVA) was applied for determining significant difference at $p < 0.05$.

3 Results and Discussion

The composition of K2, K5 and K6 purees is shown in Figs. 1 and 2. Results showed that there are significant differences between different sweet potato variety in terms of proximate, vitamins and minerals. Among the varieties, K2 contains higher total

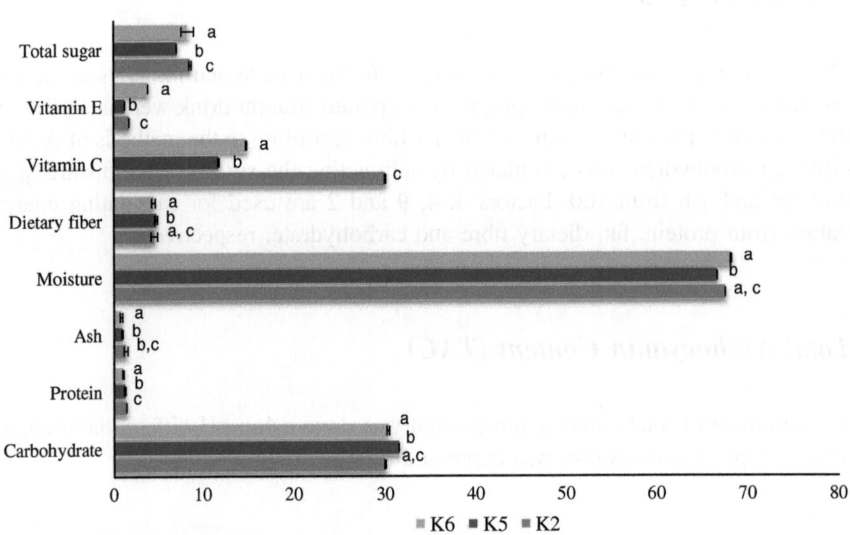

Fig. 1 Proximate, vitamin C and vitamin E results of different purple sweet potato variety. Column with the same letter do not show statistical differences at $p < 0.05$

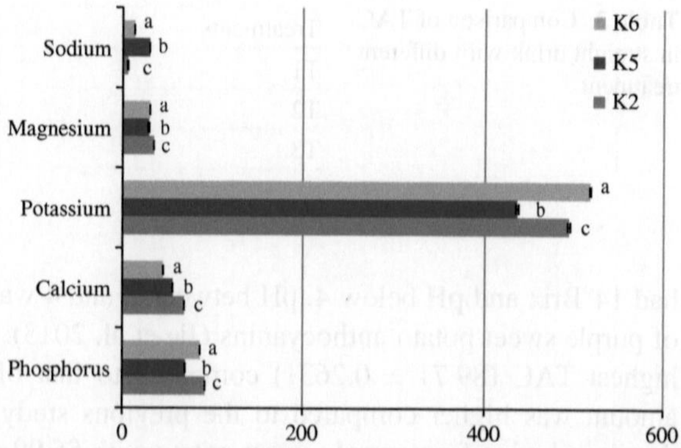

Fig. 2 Mineral content in different purple sweet potato varieties. Column with the same letter do not show statistical differences at $p < 0.05$

sugar, vitamin C, ash and protein. The purees also contain dietary fibre and vitamin E. Potassium is the highest mineral present among the minerals in sweet potato puree (435–518 mg/100 g), followed by phosphorus (70.1–92.0 mg/100 g), calcium (45.3–69.1 mg/100 g), magnesium (28.9–34.7 mg/100 g) and sodium (5.4–13.1 mg/100 g). Since potassium causes sodium to be excreted, it is effective for lowering blood pressure.

The total anthocyanin content (TAC) of the purple sweet potato purees ranged from 25.71 to 426.67 mg/l (Table 1), which were statistically significant ($p < 0.05$). It was found that K2 variety contain 24.32 mg/l total anthocyanin in raw and increased to 426.67 mg/l during heat treatment. This could be due to the cell membrane of the sweet potato which was destroyed during the heat treatment and caused the pigment to increase. Thus, steaming process can greatly influence the anthocyanin content to increase in the puree. However, this finding was contradictory to Turfan et al. (2011) which reported that heating caused 8–14 and 9–13 % loss of anthocyanins in pomegranate juice from the sac and the whole fruit, respectively.

For TAC and sensory evaluation, purple sweet potato straight drink was prepared according to the formulation in Table 2. The drinks with different treatments

Table 1 Total anthocyanin content (TAC) of purple sweet potato in raw and puree

Variety	TAC (mg/l)	
	In raw	In puree
K2	24.32	426.67
K5	ND	25.71
K6	ND	127.70

Table 2 Concentration of purple sweet potato puree in straight drink

Treatment	Purple sweet potato puree (%)
T1	05
T2	10
T3	15

Table 3 Comparison of TAC in straight drink with different treatment

Treatments	TAC (mg/l)
T1	40.57 ± 0.02205
T2	63.97 ± 0.6547
T3	89.71 ± 0.2631

had 14°Brix and pH below 4. pH between 3 and 4 was found good for stabilization of purple sweet potato anthocyanins (Jie et al. 2013). T3 straight drink exhibits the highest TAC (89.71 ± 0.2631) compared to that of T1 and T2 (Table 3). This amount was higher compared to the previous study by Ray et al. (2012) which developed wine from purple sweet potato with 55.09 mg/l TAC. Skibola and Smith (2000) reported that estimated daily intake of anthocyanin is from 500 mg to 1 g. Thus, straight drink developed in this study was high in TAC and safe to be consumed for health benefit.

Sensory attributes of purple sweet potato straight drink are shown in Table 4. The drink was dark purple in colour and acidic in taste. All three treatments gave high sensory scores which implied higher preference of the samples. T3 showed no significant differences ($p < 0.05$) compared to T2 and T3, especially in terms of aroma and sweetness scores. However, in the case of colour, there were significant differences between samples (T1 and T2, T1 and T3), which changed with an increase in the puree concentration from 5 to 15 %. T3 have darker purple colour and the results from this study showed that consumers can accept the natural strong colour from the sweet potato. T3 drink also showed better acceptance than that of T2 and T3 drinks as indicated by its higher mean value of overall acceptability score.

Table 4 Sensory evaluation results of purple sweet potato (K2) straight drink with different treatments

Treatments	Sensory attributes					
	Colour	Aroma	Consistency	Sweetness	Taste	Overall acceptability
T1	6.52 ± 1.22^{a}	6.2 ± 0.95^{abc}	6.4 ± 1.11^{ab}	6.44 ± 1.12^{abc}	6.24 ± 0.96^{ab}	6.40 ± 1.00^{ab}
T2	7.28 ± 0.93^{bc}	6.56 ± 1.04^{abc}	7.04 ± 0.67^{abc}	6.92 ± 0.81^{abc}	6.88 ± 0.78^{abc}	6.92 ± 0.70^{abc}
T3	7.5 ± 0.87^{bc}	6.84 ± 0.80^{abc}	7.24 ± 0.83^{bc}	6.96 ± 0.97^{abc}	7.12 ± 0.97^{bc}	7.08 ± 0.81^{bc}

All the given values are means of three determinations ± standard deviation. Means within the columns with different letters are significantly different at $p < 0.05$

4 Conclusion

This research was designed to utilize the purple sweet potato local variety, which is still low in terms of product variation. The findings of the study showed that the straight drink prepared using 15 % of K2 variety was selected as the most preferred based on the nutritional and organoleptic point of view. Purple sweet potato straight drink formulated with 15 % of puree is found to be rich in TAC (89.71 mg/l) and is acceptable among the consumers. The formulation was economical, using feasible method and enables the consumers to purchase good-quality products for their health.

Acknowledgements This study was funded by the Ministry of Agricultural (MOA) under Mega Project Grant (21100030002450001).

References

Association of Official Analytical Chemists (1995) Official methods of analysis of the association of official analytical chemists, 16th edn. Methods 984.27. Association of Official Analytical Chemists, Washington DC

Jie L, Xiao-ding L, Yun Z, Zheng-dong Z, Zhi-ya Q, Meng L, Shao-hua Z, Shuo L, Meng W, Lu Q (2013) Identification and thermal stability of purple-fleshed sweet potato anthocyanins in aqueous solutions with various pH values and fruit juices. Food Chem 136:1429–1434

Lila MA (2004) Anthocyanins and human health: an in vitro investigation approach. J Biomed Biotechnol 5:306–313

Malaysian Food Act—Act 281 (1983) Act 352 Fruit drink, p 230–231

Ray RC, Tomlins KI (2010) Sweet potato: post harvest aspects in food, feed and industry. Nova Science Publishers Inc, New York

Ray RC, Panda SK, Swain MR, Sivakumar PS (2012) Proximate composition and sensory evaluation of antocyanin-rich purple sweet potato (*Ipomoea batatas* L.) wine. Int J Food Sci Technol 47(3):452–458

Skibola C, Smith M (2000) Potential health impacts of excessive flavonoid intake. Free Radical Biol Med 29:375–383

Teow CC, Truong V, McFeeters RF, Thompson RL, Pecota KV, Yencho GC (2007) Antioxidant activities, phenolic and β-carotene contents of sweet potato genotypes with varying flesh colours. Food Chem 103:829–838

Tsuda T, Horio F, Osawa T (2002) Cyanidin 3-o-beta-D-glucoside suppresses nitric oxide production during a zymosan treatment in rats. J Nutr Sci Vitaminol 48(4):305–310

Tsuda T, Horio F, Uchida K, Aoki H, Osawa T (2003) Dietary cyanidin 3-O-beta-D-glucoside-rich purple corn colour prevents obesity and ameliorates hyperglycemia in mice. J Nutr 133 (7):2125–2130

Turfan O, Turkyilmaz M, Yemis O, Ozkan M (2011) Anthocyanin and colour changes during processing of pomegranate (*Punica granatum* L., cv. *Hicaznar*) juice from sacs and whole fruit. Food Chem 129:1644–1651

Wang GL, Yue J, Su DX, Fang HJ (2006) Study on the antioxidant activity of sweet potato and its inhibiting effect on growth cancer S810. Acta Nutrimenta Sinica 28:71–74

Wang S, Jiao H (2000) Scavenging capacity of berry crops on superoxide radicals, hydrogen peroxide, hydroxyl radical and singlet oxygen. J Agricu Food Chem 48(11):5677–5684

Wrolstad RE (2000) Anthocyanins. In: Lauro GJ, Francis FJ (eds) Natural food colorants. Marcel Dekker, New York, pp 237–252

Chapter 60
Quantitative Analysis of Hydrophilic and Lipophilic Antioxidant Components in Palm Puree

Haswani Maisarah Mustafa, Noriham Abdullah and Zainon Mohd. Noor

Abstract Palm oil is an edible vegetable oil derived from the mesocarp (reddish pulp) of the fruit of the oil palms. The leftover mesocarp, after which the oil is extracted out, is normally used as animal feed. Hence, the aim of this study is to develop value-added food product namely palm puree (PP) derived from mesocarp and crude palm oil (CPO). Palm puree (PP) was developed from different breeds of oil palm fruit (*Elaeis Guineensis*) namely PP24 and PP99. Both PPs were labelled as PP24A, PP24B, PP99A and PP99B, where A consisted of 2 % mesocarp fibre (MF) with 98 % CPO, while B consisted of 5 % MF with 95 % CPO. Analysis on antioxidant activity using total phenolic content (TPC), total flavonoid content (TFC), ferric reducing antioxidant power (FRAP) and oxygen radical absorption capacity (ORAC) was conducted. Identification and quantification of individual hydrophilic and lipophilic antioxidants was also performed using reversed-phase high-performance liquid chromatography (HPLC). TPC of all samples varied from 486.33 to 778.29 mg GAE/100 g EW. The amount of TFC ranged between 30.08 and 52.01 mg CE/100 g EW. The FRAP value varied from 585.58 to 1234.06 mM TE/g EW. While the ORAC ranged between 209.60 ± 2.0 and 214.40 ± 3.00 mmol TE/g sample. The quantitative analysis had shown the presence of seven different phenolic acids and four vitamin E derivatives where syringic acid and γ-Tocotrienol (γ-T_3) were found to be the most dominant hydrophilic and lipophilic antioxidant in PP, respectively.

Keywords Antioxidant capacity · Hydrophilic antioxidant · Lipophilic antioxidant · Palm puree · Syringic acid

---

H.M. Mustafa (✉)
Section of Food Engineering Technology, Universiti Kuala Lumpur
Malaysian Institute of Chemical & Bioengineering Technology (UniKL MICET), Melaka, Malaysia
e-mail: haswanimaisarah@gmail.com

N. Abdullah · Z.M. Noor
Faculty of Applied Sciences, Universiti Teknologi MARA, Shah Alam, Selangor, Malaysia
e-mail: noriham985@salam.uitm.edu.my

N.A. Yacob et al. (eds.), *Regional Conference on Science, Technology and Social Sciences (RCSTSS 2014)*, DOI 10.1007/978-981-10-0534-3_60

1 Introduction

Oil palm fruit (*Elaeis guineensis* Jacq.) being introduced in 1870s and since grown in Malaysia as an ornamental plant belongs to the family Palmae. It is a very unique plant that can bear two types of oil which are the palm oil that comes from the palm mesocarp and palm kernel oil from the palm kernel (MPOC 2008). Palm oil has neutral effect on the blood cholesterol elevation makes the belief of palm oil nutritionally contain saturated fats untrue. Cholesterol lowering effect found to be important health benefits among Malaysian that uses palm oil in their daily intake diet (Ong and Goh 2002). Sundram et al. (2003) reported that rich amounts of β-carotene (precursor of vitamin A), vitamin E (tocopherol and tocotrienol) as well as water-soluble phenolic-flavonoid-rich antioxidant complex abundantly present in the palm oil exhibit the potent antioxidant properties coupled with other beneficial effects such as skin and breast cancers. In recent years, there has been much demand on the use of natural antioxidant instead of synthetic antioxidant. Natural antioxidant known to be safe and can protect human body against free radicals and from any progress of chronic diseases. Antioxidant may act in different ways in human body depending on its mode of action. It may act primarily as a chain-breaking antioxidant that react with lipid radical to form stable compound and such antioxidant act as a free radical interceptors. Antioxidant also acts secondarily as preventive antioxidant that can reduce the rate of chain initiation by many ways which include metal inactivators, hydroperoxide decomposers, oxygen scavengers, and synergists. Thus, the aim of this work was to evaluate the hydrophilic and lipophilic antioxidants presence in palm puree samples that were developed from Tenera breeds using various assays such as total phenolic content (TPC) and total flavonoid content (TFC), oxygen radical absorption capacity (ORAC) and ferric reducing antioxidant power (FRAP). The amount of individual phenolic acid compounds and tocol derivatives in PP samples were carried out using high-performance liquid chromatography (HPLC) assay.

2 Materials and Methods

Chemicals and Reagents

Ethanol and methanol were purchased from J.Kollin, UK. Folin–Ciocalteu reagent, sodium carbonate (Na_2CO_3) and ferric chloride hexahydrate ($FeCl_3.6H_2O$) were purchased from Merck, Germany. 2,4,6-tripyridyl-*s*-triazine (TPTZ), sodium acetate trihydrate, (+) catechin hydrate, gallic acid and Trolox (6-hydroxy-2,5,7,8-tetramethylchromane-2-carboxylic acid, 97 %) standards were purchased from Sigma Chemical Co. (St Louis, MO, USA). Glacial acetic acid and hexanes were purchased from J.T. Baker. Hydrochloric acid (HCl) was purchased from R&M chemicals. All reagents used were of analytical grade unless otherwise stated.

Acetonitrile and methanol used for HPLC analyses were of HPLC grade purchased from Merck. Tocol standards were obtained from Sime Darby Jomalina Teluk Panglima Garang, Banting, Selangor. Phenolic acid standards such as gallic acid ($C_7H_6O_5$, MW = 170.12, >98 % pure), p-hydroxybenzoic acid ($C_7H_6O_3$, MW = 138.12, >99 % pure), caffeic acid ($C_9H_8O_4$, MW = 180.16, >98 %pure), vanillic acid ($C_8H_8O_4$, MW = 168.15, >97 % pure), syringic acid ($C_9H_{10}O_5$, MW = 198.17, >95 % pure), p-coumaric acid ($C_9H_8O_3$, MW = 164.16, >98 % pure), ferulic acid ($C_{10}H_{10}O_4$, MW = 194.18, 99 % pure) were purchased from Sigma Chemical Co. (St Louis, MO, USA). Bio ORAC antioxidant assay kit (Cat# AOX-2) was purchased from Zen-Bio Inc. (NC, USA).

Sample Preparation

Two Tenera breeds (T24 and T99) of *Elaies Guineensis* Jacq. were obtained from Sime Darby Plantation Sdn. Bhd Banting and processed to developPalm Puree (PP). Kernel of the palm fruits were separated from the mesocarp fibre (MF). The MF was pressed using screw pressed to obtain MF and crude palm oil (CPO). Both Tenerabreeds were further developed into four formulations namely PP24A, PP24B, PP99A and P99B where A is a combination of 2 % MF with 98 % CPO and B for combination of 5 % MF with 95 % CPO. All of them were packaged in cans and stored at room temperature until used for further analysis.

Sample Extraction

All PP samples that were used for hydrophilic antioxidant (TPC, TFC, FRAP, ORAC and quantification of hydrophilic antioxidants using HPLC) were subjected to *n*-hexane extraction by using Soxhlet technique until the oil is completely removed. Samples were then extracted according to method by Maisarah et al. (2013). A 200 mg of sample was extracted for 2 h with solvent mixture containing 2 ml of 80 % methanol containing 1 % hydrochloric acid at room temperature on incubator shaker set at 200 rpm. The mixture was centrifuged at 3500 rpm for 15 min and the supernatant was kept in amber bottle at −20 °C for further analysis.

Determination of Total Phenolic Content (TPC)

Total phenolic content assay of samples were determined using Folin–Ciocalteu assay as described by Maisarah et al. (2013). A 100 µl of supernatant was mixed with 0.5 ml Folin–Ciocalteu reagent (diluted 10 times with distilled water). Next, it was added with 7 ml of distilled water and allowed to stand at room temperature for

5 min. Then, 1.5 ml sodium bicarbonate (60 mg/ml) solution was added to the mixture and left at room temperature in dark place for 2 h. Absorbance was read at 725 nm against blank using UV–Visible spectrophotometer (Perkin Elmer Lambda 35, USA). A calibration curve was prepared using a standard solution of gallic acid ranged from 0.0 to 1.0 mg/ml. Results were expressed as gallic acid equivalents mg GAE/100 g extract.

Determination of Total Flavonoid Content (TFC)

Total flavonoid content assay of samples were determined as described by Iqbal et al. (2007). One millilitre of sample was added to 4 ml distilled water. Then, 0.3 ml sodium nitrite ($NaNO_2$) was added and left to stand in a dark place for 5 min. The solution was then, added with 0.3 ml aluminium chloride ($AlCl_3$) left at dark place for 6 min. Two millilitres sodium hydroxide was added to the solution and made to a final volume of 10 ml. Absorbance was read at 510 nm against blank using UV–Visible spectrophotometer (Perkin Elmer Lambda 35, USA). A calibration curve was prepared using a standard solution of catechin ranged from 0.0 to 1.0 mg/ml. Results were expressed as catechin equivalents mg CE/100 g extract.

Ferric Reducing Antioxidant Power (FRAP) Assay

The FRAP assay was carried out according to method described by Liu et al. (2009) with modifications. The FRAP reagent contain 20 mM $FeCl_3 \cdot 6H_2O$, 10 mM TPTZ (2,4,6-tripyridyl-s-triazine) solution in 40 mM HCl and 0.3 M acetate buffer, pH 3.6 and incubated at 37 °C for 10 min in an incubator (Memmert, Germany). The FRAP reagent was mixed in the ratio of 1:1:10. Aliquot of 100 µl sample was mixed with 2.9 ml of FRAP reagent. The absorbance of the reaction mixture was measured using a spectrophotometer model Perkin Elmer Lambda 35, USA at 593 nm after incubation at room temperature for 1 h. Trolox (1000 µM) was used for the calibration curve and the results were expressed as mM of Trolox equivalents per mg extract.

Oxygen Radical Absorption Capacity (ORAC) Assay

The Zen-Bio ORAC Antioxidant Assay Kit (Cat# AOX-2) was used to determine the antioxidant capacity against free radical. The preparation of ORAC assay was conducted as per instruction provided. The ORAC assay is measured by BioTek ELISA model Synergy™ HT Microplate Reader. The plate reader incubation chamber was warmed to 37 °C and was set up to perform a kinetic read with 1 min intervals for 30 min run. Excitation, emission and cut-off were set at 485, 538 and 530 nm, respectively. Fluorescein was prepared by mixing 16.8 ml of AOX assay buffer with 1.2 ml stock fluorescein to make a solution. Trolox standard was briefly spun down and mixed well by vortex. The mixture produced a diluted stock Trolox standard of 50 µM. The diluted stock Trolox standard was then serially diluted by pipetting 50 µl of AOX assay buffer into beakers. The 50 µM stock dilution serves as the highest concentration and the assay buffer serves as the zero standards. A volume of 150 µl of working fluorescein solution was added to each well of the assay plate provided. Then, 25 µl of samples or Trolox standards were added to individual wells, and 25 µl of assay buffer was also added to individual well as a negative control. The plate was placed at 37 °C. The AAPH working solution was prepared by adding 2.7 ml of AOX assay buffer to the AAPH tube provided while waiting for the plate assay to equilibrate to 37 °C. 25 µl of the AAPH working solution was added to each well containing standards and samples. The assay plate was placed in the plate reader and the kinetic fluorescence readings were noted.

Quantification of Hydrophilic Antioxidants by HPLC

The HPLC Agilent 1200 used for quantification of hydrophilic antioxidants was from Hewlett-Packard (Agilent Technologies, Baudrats, Germany) equipped with Zorbax Eclipse SB-C18 (i.d 4.6 × 250 mm, 5 µm) column. All phenolic acid compounds (gallic acid, 2,4-dihydroxybenzoic acid, caffeic acid, vanillic acid, syringic acid, *p*-coumaric acid and ferulic acid) were prepared in various standard concentrations and serial dilutions of standard concentration were diluted with methanol before injected into HPLC system to establish standard calibration curve (0.0–0.5 mg/ml). Three replicate injections were performed for each concentration. The slope and R^2 were determined by the least square linear regression analysis method. Phenolic acids in PP samples were analysed according to Maisarah et al. (2013). A mobile phase was prepared which consist of ratio acidified water to acetonitrile of 90:10 (v/v) containing 2 % acetic acid, at a flow rate of 1 ml/min. Detection of phenolic acids using HPLC were coupled using a diode array detector (DAD) at 295 nm. All volume samples were injected at 20 µl. Identification of phenolic acids was carried out by comparing the retention times and peak spectrums. Quantification in each gram of sample was carried out using external standard. The amount of each phenolic acid was expressed as milligram per gram of extract weight (EW) of the deoiled PP (mg/g).

Quantification of Lipophilic Antioxidants by HPLC

The HPLC Agilent 1200 was used for quantification of lipophilic antioxidants was from Hewlett-Packard (Agilent Technologies, Baudrats, Germany) equipped with Zorbax Eclipse XDB-C18 (4.6 × 150 mm, 5 μm) column. All tocol derivatives also commonly known as vitamin E (α-tocopherol, α-tocotrienol, γ-tocotrienol and δ-tocotrienol) were prepared in various standard concentrations and serial dilutions of standard concentration were diluted methanol before injected into HPLC system to establish standard calibration curve (0.0–1.0 mg/ml). Three replicate injections were performed for each concentration. The slope and R^2 were determined by the least square linear regression analysis method. Tocol derivatives in PP samples were analysed according to Haswani (2014) by dissolving 0.05 g sample in 10 ml hexane and subsequently diluted ten times with methanol. A mobile phase with ratio of acetonitrile and methanol (95:5, v/v) at a flow rate of 1 ml/min was prepared. Detection of tocol derivatives were conducted using a DAD at 295 nm. All samples were injected at 20 μl. Identification of tocol derivatives were based on the comparison with the retention times and peak spectrums. Quantification in each gram of sample was carried out using external standard. The amount of each tocol derivatives were expressed as milligram per gram of fresh weight (FW) PP (mg/g).

Statistical Analysis

All data were analysed using SAS 9.0 software. Analysis of variance (ANOVA) and Duncan's multiple range method were used to compare any significant differences between samples. Values were expressed as means ± standard deviations. Differences were considered significant at $p < 0.05$. All analyses were carried out in triplicates.

3 Results and Discussion

TPC, TFC and Antioxidant Activities

Theresults of TPC, TFC and its respective antioxidant activities of PP samples are summarised in Table 1. The TPC and TFC of PP samples ranged from 486 to 778 mg GAE/100 g EW and 30.08–52.01 mg CE/100 g EW in the descending order ($p < 0.05$) of PP24A > PP24B > PP99A > PP99B. Different TPC and TFC results exhibited by PP samples were due to different breeds used (T24 and T99) and higher amount of phenolic compounds that were present and released from T24 puree matrix compared to T99. The Folin–Ciocalteu reagent is applied in order to estimate the presence of phenolic compound in an extract (Liu et al. 2009).

Table 1 Results of TPC, TFC and its respective antioxidant activities

Sample	Total phenolic content (TPC) (mg GAE/100 g EW)	Total flavonoid content (TFC) (mg GAE/100 g EW)	FRAP (mM TE/mg EW)	ORAC (mmol TE/g EW)
PP24A	778.29 ± 23.99[a]	52.01 ± 2.49[a]	1234.06 ± 17.50[a]	214.40 ± 3.00[a]
PP24B	709.97 ± 85.33[b]	49.85 ± 1.24[b]	1180.14 ± 27.97[b]	212.40 ± 1.45[a]
PP99A	537.97 ± 3.76[c]	34.55 ± 0.31[c]	646.24 ± 14.84[c]	209.60 ± 6.50[a]
PP99B	486.33 ± 17.36[d]	30.08 ± 2.62[d]	585.58 ± 28.54[d]	209.60 ± 2.00[a]

[abcd]Mean values in the same column followed by the same letter are not significantly different at $p < 0.05$ using DMRT

The assay does not focus on specific polyphenols and different phenolic compound respond differently to this assay (Huda-Faujan et al. 2007). The contents and distribution of phenolic acids also varies in different formulation of PP. This finding was supported by previous studies (Oliveira et al. 2011) who reported that sample with higher TPC value was capable to react better with Folin–Ciocalteu reagent. Therefore, PP24 samples react better than PP99 samples. It was reported that phenolic compound can be found in higher plants that contain diverse group of secondary metabolites and the bioactive polyphenol help in fighting from oxidative stress that can cause many diseases such as ageing, cancer and cardiovascular disease (Liu et al. 2009).

Previous studies reported that flavonoids structure which contains hydroxyl functional group and the heterocyclic rings substituent are the key factor for the scavenging activities. Apart from scavenging action, flavonoids may act by complexing with radical in order to stabilise them. By inhibiting enzyme action that is responsible for highly oxidising free radical production such as O_2•, RO• and HO•, it has fulfilled the criteria of the flavonoids to be considered as an antioxidant. Flavonoids were also important in human diet because of its role as radical scavengers, preventing heart disease and exhibited a potential as for the anti-immunodeficiency (Mohd et al. 2009).

Antioxidant activities by FRAP assay is based on the ability to reduce Fe^{3+} to form Fe^{2+} by antioxidant in the presence of TPTZ, forming Fe^{2+}-TPTZ, an intense blue colour complex with maximum absorption at 593 nm (Liu et al. 2009). All PP samples demonstrated high antioxidant activity up to 1234 mM TE/g EW. There were significant different among PP samples studied. PP24A shows the highest ability of reducing antioxidant power, followed by PP24B, PP99A and PP99B with 1234.06, 1180.14, 646.24 and 585.58 mM TE/g extract weight, respectively. FRAP value for PP samples were found higher than palm-pressed fibre phenolics (100–250 mM TE/g EW) as reported by Nang et al. (2007). FRAP is a method to screen the ability of cells and plant tissue for maintaining the redox status. It cannot act by hydrogen atom transfer for quenching radical particularly protein. FRAP assay is adjusted to pH 3.6 which is in acidic medium, able to maintain the solubility of the iron in the solution. Due to this medium, ionisation is decreased and hence drives the electron transfer and increases the redox potential, causing shift in dominant reaction mechanism.

ORAC assay is one the test that applied the area-under-curve technique in the process of quantitation. It measures the total antioxidant capacity from the decay of the fluorescein (FL) probe. Cao et al. (1993) has developed further the ORAC method by using FL as a probe protein replacing the B-phycoerythrin (B-PE) for its shortcomings. In this study, however, all PP samples show there is no significant different between samples. It can be indicated that the mechanism in PP sample does not contributed to the scavenging activities. Sanchez et al. (2007) reported that the ORAC values for olive oils in various test conducted ranging from 1.23 to 110×10^{-3} mmol TE/g. This indicates that other oil that was previously studied has lower ORAC values in contrast with PP samples.

Different antioxidant activities attributed to different mechanism of action. Sample may act as hydrogen atom transfer (HAT) or by electron transfer (ET). PP samples show it may act as HAT since FRAP results showed there were significant different between all PP samples. Several previous studies reported that polyphenolic compounds are correlated with antioxidant activities which are responsible for stabilising lipid oxidation (Mohd et al. 2009). Antioxidative activity of phenolic compounds was reported to be related to the ability to donate hydrogen atoms to free radicals. Balasundram et al. (2006) reported that antioxidant activity of phenolic acids depends on the number and positions of hydroxyl groups in relation to the carboxyl functional group. As the degree of hydroxylation in phenolic acids increase together with the position of the –OH groups in the phenolic acid structure, the antioxidant activity also increase, which confers higher stability to the radical by delocalization of electron. Therefore, high amount of TPC and TFC in PP samples could relate to good antioxidant activity of puree.

Hydrophilic Antioxidant (Phenolic Acid)

Phenolic acid compounds found in all PP samples are summarised in Table 2. Results shown that there were seven phenolic acids detected using HPLC in PP samples namely gallic acid (GA), *p*-hydroxybenzoic acid (HBA), caffeic acid (CA), vanillic acid (VA), syringic acid (SA), *p*-coumaric acid and ferulic acid (FA). The chromatogram of mixed phenolic acids identified in PP samples is shown in Fig. 1. Based on the total amount of phenolic acids in all PP samples, there were four major phenolic acids abundantly present in Tenera breeds which were syringic acid (218.64 mg/g EW), vanillic acid (161.05 mg/g EW), gallic acid (58.15 mg/g EW) and *p*-hydroxybenzoic acid (56.33 mg/g EW). From the study, there were no significant different of syringic acid and vanillic acid in PP samples that comes from the same Tenera breeds. All of the seven phenolic acids identified are known to be the hydrophilic antioxidant.

Results in this study were in agreement with Neo et al. (2010) who stated that *p*-hydroxybenzoic acid is one of the phenolic compound that significantly present in the mesocarp of the oil palm fruit (*E. guineensis*). Different amount of phenolic acids in different breeds could be due to the environmental conditions such as

Table 2 Total amount of phenolic acids in PP samples as determined by HPLC

Compound	Sample (mg/g EW)				
	PP24A	PP24B	PP99A	PP99B	Total**
Gallic acid	15.81 ± 0.55^{a}	16.45 ± 0.21^{a}	13.23 ± 0.18^{b}	12.66 ± 0.36^{b}	58.15 ± 1.30^{C}
p-hydroxybenzoic acid	8.37 ± 0.45^{c}	7.98 ± 1.38^{c}	17.55 ± 0.49^{b}	22.43 ± 2.98^{a}	56.33 ± 5.30^{C}
Caffeic acid	4.73 ± 0.31^{a}	4.39 ± 0.33^{a}	3.92 ± 0.08^{b}	2.74 ± 0.01^{c}	15.78 ± 0.73^{D}
Vanillic acid	46.91 ± 2.43^{a}	45.76 ± 1.13^{a}	35.40 ± 1.51^{b}	32.98 ± 0.70^{c}	161.05 ± 5.77^{B}
Syringic acid	66.65 ± 1.82^{a}	65.05 ± 1.49^{a}	44.93 ± 1.50^{b}	42.01 ± 1.60^{b}	218.64 ± 6.41^{A}
p-coumaric acid	4.26 ± 0.21^{a}	4.24 ± 0.28^{a}	3.48 ± 0.25^{b}	3.50 ± 0.02^{b}	15.48 ± 0.76^{D}
Ferulic acid	9.83 ± 2.57^{a}	6.30 ± 1.97^{a}	Trace	Trace	16.13 ± 4.54^{D}
Total*	156.56 ± 8.54^{a}	150.17 ± 7.69^{a}	118.51 ± 4.61^{b}	116.32 ± 5.67b	

^{ABCD}Significant difference ($p < 0.05$) of one formulation over the other(s) if values in the same row carry different superscript

^{abc}Significant difference ($p < 0.05$) of one sample over the other(s) if values in the same row carry different superscripts

*Total is the sum of each phenolic compound identified in each PP sample

**Total is the sum of individual phenolic compound indentified all PP samples

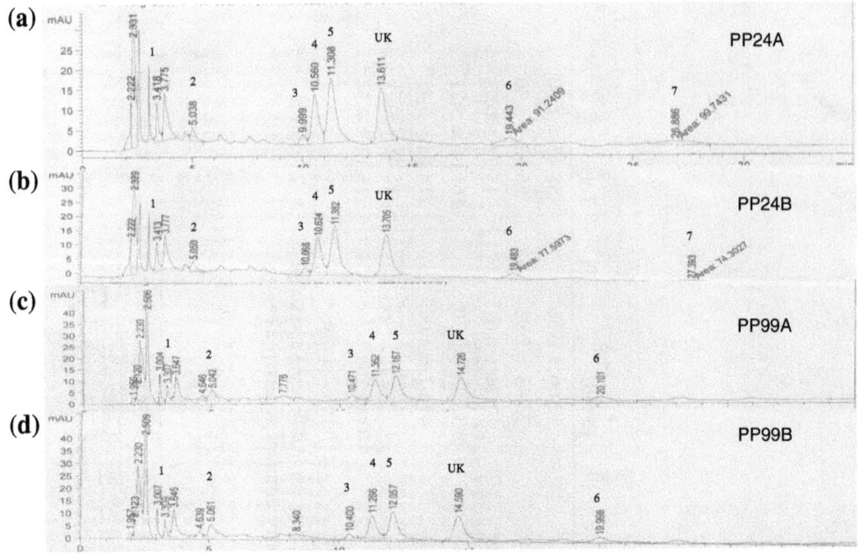

Fig. 1 **a** PP24A; **b** PP24B; **c** PP99A; **d** PP99B at 295 nm. Peak identification: (*1*) gallic acid; (*2*) *p*-hydroxybenzoic acid; (*3*) caffeic acid; (*4*) vanillic acid; (*5*) syringic acid; (*6*) *p*-coumaric acid; (*7*) ferulic acid; (*UK*) unknown peak

location or growing region, geographical factor, climate (temperature and humidity) and soil type (Vaher et al. 2010). Although PP samples have been subjected to various processing and heat treatments, it was found that it was still able to retain the phenolic acids content. This could be because phenolic acids occur in plants as metabolic intermediates and accumulated in the vacuoles. The action of thermal processing may release more bound phenolic acids from the cellular breakdown. The phenolics are divided into two which are the insoluble phenolics found in the cell wall components and soluble phenolics which are located within the plant cell vacuoles. Phenolics such as lignin and hydroxycinnamic acid are responsible for enhancing the mechanical strength of the cell wall, regulate the plant growth and morphogenesis. Major phenolic acids such as ferulic and *p*-coumaric are found in the form of dimers which esterified to pectins and arabinoxylans or cross-linked to cell wall polysaccharides (Naczk and Shahidi 2004).

Lipophilic Antioxidant (Vitamin E)

Vitamin E is one of the fat-soluble vitamins that is abundantly present in oil palm fruit especially in gamma-tocotrienol (γ-T_3) form. From previous study, the least amount vitamin E was found in delta-tocotrienol (δ-T_3) form. The amount of tocol derivatives (vitamin E) is shown in Table 3. Four vitamin E derivatives were

Table 3 Total amount of vitamin E in PP samples as determined by HPLC

Compound	Sample (mg/kg FW)			
	PP24A	PP24B	PP99A	PP99B
α-T_3	175.41 ± 19.55^{a}	154.60 ± 3.00^{a}	151.71 ± 0.01^{b}	151.65 ± 0.06^{b}
γ-T_3	227. 12 ± 7.65^{a}	214.44 ± 1.60^{a}	212.73 ± 0.04^{b}	212.64 ± 0.20^{b}
δ-T_3	82.04 ± 0.11^{a}	81.79 ± 0.05^{a}	81.64 ± 0.02^{b}	81.57 ± 0.09^{b}
α-T	122.19 ± 13.59^{a}	106.60 ± 14.94^{a}	74.94 ± 14.99^{b}	51.14 ± 6.96^{b}
Total	606.76 ± 42.5^{Aa}	556.89 ± 14.66^{Aa}	521.02 ± 17.21^{Bb}	497.00 ± 8.32^{Bb}

^{ABC}Significant difference ($p < 0.05$) of one formulation over the other(s) if values in the same row carry different superscript

^{abc}Significant difference ($p < 0.05$) of one sample over the other(s) if values in the same row carry different superscripts

identified in all PP samples which were α-T_3, γ-T_3, δ-T_3 and α-T. Based on the results, there were no significant different ($p > 0.05$) in total amount of vitamin E between PP formulations of the same Tenera. However, there was significant different between PP24 and PP99. PP24 exhibited significantly higher ($p < 0.05$) total amount of vitamin E as compared with PP99. The chromatograms of mixed vitamin E derivatives are identified in PP samples are shown in Fig. 2.

It was reported that vitamin E in palm oil possessed antioxidant activity that inhibits the lipid oxidation. In fats and oils, vitamin E is responsible in stabilising the hydroperoxy and free radical (O'Brien 2009). As an antioxidant, tocopherols react by transferring hydrogen atom from the hydroxyl on the chromanol ring

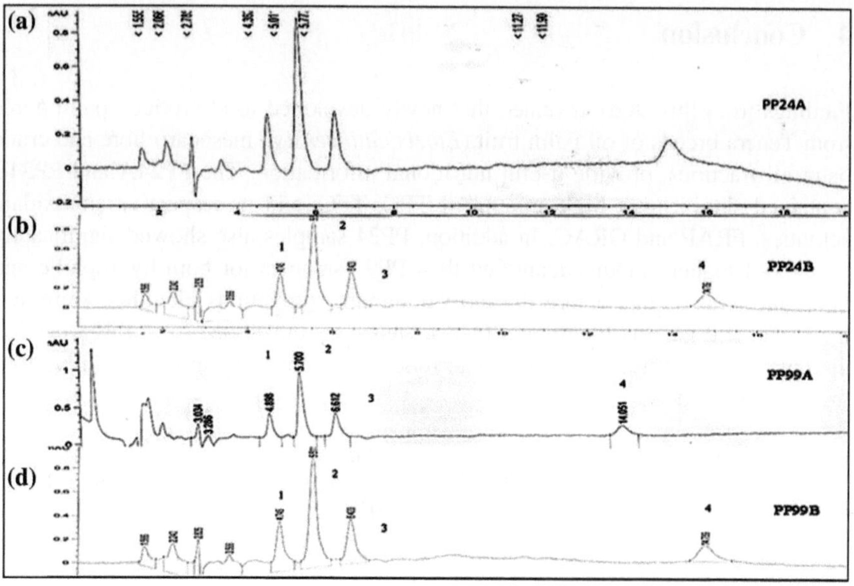

Fig. 2 Peak identification: (*1*) δ-T_3; (*2*) γ-T_3; (*3*) α-T_3d; (*4*) α-T

structure to a free radical. Tocopherols exhibit antioxidant activity 250 times greater than synthetic antioxidant, BHT due to the heterocyclic ring that serve as hydrogen donor with the chroman moiety that is also responsible for solubility of fats. Tocotrienols, on the other hand, has similar antioxidant activity as tocopherols as the structure also contain a hydroxyl group in the same position located in tocopherols. Antioxidant activity of tocotrienol was found to exert the same mechanism as tocopherols. It was reported that tocopherols act as pro-oxidant when used at higher concentration (more than 1000 ppm). Tocopherols also act as singlet oxygen scavenger when photosensitised initiation occurs in the presence of light.

Nesaretnam et al. (2012) reported that tocotrienols and tocopherol are the major fat-soluble antioxidant of vegetable oil. Palm tocotrienol was investigated to have the potential as anticancer properties. Crude palm oil was reported to be more effective than refined palm oil in treating rats. Vitamin E is identified as a potent antioxidant that serves to protect the cellular membranes from the attack of free radical. Palm oil was reported to promote antithrombotic effect by reducing the aggregation of platelet and modulating prostanoid synthesis. A new finding by Sen et al. (2007) reported that tocotrienols were not studied extensively as tocopherols even though it has been proven to have many advantages and beneficial to human health. Tocotrienols that are taken orally is found not only to reach the brain site but also to prevent stroke when taken in sufficient amount. Other benefits of consuming tocotrienols and tocopherols in the diet such as anticancer effect, neuroprotection effect, anticholesterol effect and hypercholesterolemic effect need to be extensively studied.

4 Conclusion

Findings from this study revealed that newly developed food product, palm puree from Tenera breeds of oil palm fruit (*Elaeis Guineensis*), mesocarp fibre and crude palm oil fractions, provide useful nutritional information. The PP24A and PP24B contained significantly high amount of TPC, TFC, and its respective antioxidant activities, FRAP and ORAC. In addition, PP24 samples also showed significantly ($p < 0.05$) higher amount quantified than PP99 samples for both hydrophilic and lipophilic antioxidants where the most dominant compounds identified were syringic acid and gamma-tocotrienol (γ-T_3). Hence PP developed from this work has the potential to be used as a functional food ingredient that can be used in various food applications.

Acknowledgements This research was made possible by funding assistance and laboratory facilities provided from UNIKL-MICET, Universiti Teknologi MARA Shah Alam and Sime Darby research grant (Grant No: 100-RMI.P 37/7/2005).

References

Balasundram N, Sundram K, Samman S (2006) Phenolic compounds in plants and agri-industrial by-products: antioxidant activity, occurrence, and potential uses. Food Chem 99:191–203

Cao G, Alessio HM, Cutler RG (1993) Oxygen radical absorbance capacity assay for antioxidants. Free Radical Biol Med 14:303–311

Haswani MM (2014) Development of palm puree from different Tenera breeds and its effect on the stability, physicochemical properties and acceptability of biscuit. pp 45–46. Unpublished Master Thesis, Universiti Teknologi MARA

Huda-Faujan N, Noriham A, Norakkiah AS, Babji AS (2007) Antioxidative activities of water extracts of some Malaysian herbs. ASEAN Food J 1:61–68

Iqbal S, Bhanger MI, Anwar F (2007) Antioxidant properties and components of bran extracts from selected wheat varieties commercially available in Pakistan. LWT 40:361–367

Liu L, Laura T, Liang X, Ye H, Zeng X (2009) Determination of polyphenolic content and antioxidant activity of kudingcha made from *Ilexkudingcha* C. J Tseng Food Chem 112:35–41

Maisarah MH, Noriham A, Zainon MN (2013) Quantification of polyphenolic acids and antioxidant capacity of Palm Puree from different Tenera breeds of *ElaeisGuineensis*. Int J Biosci, Biochem Bioinform 3:349–353

Malaysian Palm Oil Council (MPOC) (2008) Fact sheet: Malaysian palm oil, reprinted 2008. Kelana Jaya, p 7

Mohd FAB, Maryati M, Asmah R, Jeffery F (2009) Phytochemicals and antioxidant activity of different parts of bambangan (*Mangiferapajang*) and tarap (*Artocarpusodoratissimus*). Food Chem 113:479–483

Naczk M, Shahidi F (2004) Extraction and analysis of phenolic acids in food. J Chromatogr A 1054:95–111

Nang HLL, May CY, Ngan MA, Hock CC (2007) Extraction and identification of water-soluble compounds in palm-pressed fibre by SC-CO_2 and GC-MS. Am J Environ Sci 3:54–59

Neo YP, Azis A, Tan CP, Tan YA (2010) Phenolic acid analysis and antioxidant activity assessment of oil palm (*E. guineensis*) fruit extracts. Food Chem 122:353–359

Nesaretnam K, Meganathan P, Veerasenan SD, Selvaduray KR (2012) Tocotrienols and breast cancer: the evidence to date. Genes Nutr 7:3–9

O'Brien RD (2009) Fats and oils: formulating and processing for applications. CRC Press, Taylor and Francis Group, Boca Raton

Oliveira LS, Rufino MSM, Moura CFH, Cavalcanti FR, Alves RE, Miranda RA (2011) The influence of processing and long-term storage on the antioxidant metabolism of acerola (*Malpighiaemarginata*) purée. Braz Soc Plant Physiol 23:151–160

Ong AS, Goh SH (2002) Palm oil: a healthful and cost-effective dietary component. Food Nutr Bull 23:11–22

Sanchez CS, Gonzalez AMT, Garcia-Parilla MC, Granados JJQ, Garcia de la Serrana HL et al (2007) Different radical scavenging tests in virgin olive oil and their relation to the total phenol content. Analytica Chimica Acta 593:103–107

Sen CK, Khanna S, Roy S (2007) Tocotrienols in health and disease: the other half of the natural vitamin E family. Mol Aspects Med 28:692–728

Sundram K, Sambanthamurthi R, Tan YA (2003) Palm fruit chemistry and nutrition. Asia Pac J Clin Nutr 12:355–362

Vaher M, Matso K, Levandi T, Helmja K, Kaljurand M (2010) Phenolic compounds and the antioxidant activity of the bran, flour and whole grain of different wheat varieties. Procedia Chem 2:76–82

Part XIII
Sustainable Development

Chapter 61
Solid Waste Minimization Strategies: The First Step Towards Greening a University Campus

Noor Rizallinda Ishak and Siti Akhtar Mahayuddin

Abstract Tonnes of solid waste are generated daily in a university campus which need appropriate management. Problems in the management of solid waste are tremendously affecting the daily lives of people and give negative impact to the environment. Therefore, a comprehensive solid waste management system is one of the greatest challenges in order to attain campus sustainability. The least hierarchy in waste management practices is disposed. The disposal of waste can be reduced if other waste management practices such as reuse and recycle are increased. However, these practices can be increased by understanding the current profile of solid waste produced. For that reason, the current profile of solid waste generated need to be identified in the university campus. This paper presents the current status of solid waste generation and disposal in Universiti Teknologi MARA, Perak. Through a waste audit at four students' hostel, this study has quantified the amount and composition of solid waste generated. Thus, practical and economical strategies for waste minimization could be planned by the university's administration. To conclude, this paper will recommend waste minimization strategies, i.e., Environmental Management & Improvement (EMI) Community participation and social responsibility as the first step toward greening the university campus.

Keywords Greening campus · Solid waste management · UiTM Perak · Waste minimization · Waste generation

---

N.R. Ishak (✉)
Faculty of Architecture Planning and Surveying,
Universiti Teknologi MARA, Seri Iskandar, Perak, Malaysia
e-mail: noorr399@perak.uitm.edu.my

S.A. Mahayuddin
Faculty of Architecture Planning and Surveying,
Universiti Teknologi MARA, Kota Samarahan, Sarawak, Malaysia
e-mail: sitia880@sarawak.uitm.edu.my

N.A. Yacob et al. (eds.), *Regional Conference on Science, Technology and Social Sciences (RCSTSS 2014)*, DOI 10.1007/978-981-10-0534-3_61

1 Introduction

Greening influences every aspect of human culture which has impacts on the economic, the environment, and daily lives. With the increasing concerns on different environmental issues and a need to respond climate change, universities should create knowledge and integrate sustainability in educational and research programs, as well as promoting environmental issues to the society. In addition, a university campus can be considered as a small town because of its large size, population, and the various activities taking place in the campus (Alshuwaikhat and Abubakar 2008). Management of environmental resources in the university campus capable of giving long-term effects on the institutions themselves and society as a whole. Green university efforts have been engaged in many universities and made significant progress including improved environmental performance, enhance public awareness, and cost reduction on campus maintenance.

Universities would be expected to be a role model or a leader toward responsible waste management. In order to achieve sustainability in an institution, the greatest challenge is a comprehensive solid waste management (SWM) programs. Effective SWM programs require a complete understanding of the composition of the waste stream as well as the activities that determine its generation in the first place (Farmer et al. 1997). Examining waste by generation source is particularly important, as the characteristics and composition of solid waste vary according to its source (Tchobanoglous et al. 1996). Therefore, SWM programs that are based on source generation found to be far more successful than other programs (Armijo de Vega et al. 2008).

Another key element of comprehensive solid waste management is the understanding of the waste management hierarchy. The waste management hierarchy helps in the protection of the environment and conserve resources. This can be achieved through a priority approach established in waste policy, legislation, and sustainability (Hansen et al. 2002). The waste management hierarchy shows the order of preference or priority for action to reduce and manage waste. The hierarchy captures the progression of the material through successive stages of waste management and represents the latter part of the life-cycle for each material. (United Nations Environmental Program 2013). The aim of the waste management hierarchy is to exploit the maximum practical benefits from materials and to minimize the generation of waste. The proper application of the waste management hierarchy results in several benefits such as prevention of green gases emissions, reduction in pollutants, energy saving, conservation of resources, and stimulation of the development of green technologies.

A study on characterization of municipal waste provides local decision makers with a detailed understanding of the waste stream and enable waste management programs to be tailored to local needs (Chang and Davila 2008). Therefore, the waste characterization studies at colleges and universities give specific opportunities for waste reduction and recycling which is an essential step toward greening the campus (Keniry 1995; Creighton 1998). Waste characterization studies in the university campus are able to generate administrative support, cooperation among students, faculty, and staff and inspire further involvement in campus sustainability issues (Sharp 2002; Beringer et al. 2008). Moreover, proper waste management would bring benefits to the institutions such as financial reduction. Therefore, this paper presents the current status of solid waste generation and disposal in Universiti Teknologi MARA, Perak and also the waste minimization strategies toward green university campus.

2 Sampling and Characterization of Solid Waste

Presently, the management of solid waste in UiTM (Perak) is controlled and managed by the Department of Facilities Management of UiTM (Perak). Relevant information regarding current solid waste management were gathered from this department and the appointed waste contractor through an interview. Among the information gathered were on the method of waste collection, contractor involved, location and capacity of the bin, and the availability of the recycle bin. From this information, a waste audit was designed to sample and characterize the solid waste in the selected building of UiTM (Perak).

In this study, a waste audit was conducted at the hostel block only because major quantity of solid waste was produced by the students. The selected blocks were Angsana, Bidara, Cemara, and Damar which are located at Zone 3 (refer Fig. 1).

All the blocks' buildings are identical in terms of the floor level and layout. The Angsana and Bidara blocks are womens hostels. Meanwhile, Cemara and Damar

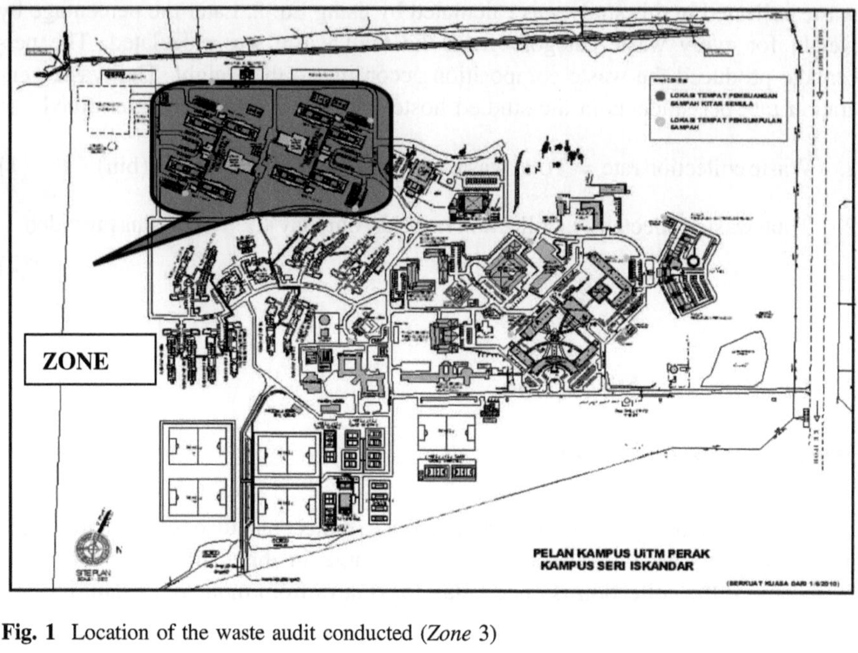

Fig. 1 Location of the waste audit conducted (*Zone* 3)

are mens hostels. Every block has four levels. There are approximately 13 bins located at each level. Ten of the bins are located along the corridor in front of the student's room, and three of them are located near the staircase for each level. There are a total of 415 existing bins for all four colleges Damar, Cemara, Bidara, and also Angsana. Meanwhile, the total number of rooms in the four blocks are 1,002 which accommodate 3,498 students.

During the waste audit, only two bins that are located near the stairs were selected for sampling at each level. For level 1, the ID sample was labeled as T1 and T2. For level 2, the ID sample was labeled as T3 and T4. Meanwhile for level 3, the ID sample was labeled as T5 and T6. Lastly, for level 4 the ID sample was labeled as T7 and T8. The waste audit for every sample at each block was conducted for six consecutive days (excluding Sunday) which align with the frequency of waste collection by the waste contractors. For sampling purposes, the solid waste was classified as paper, aluminum, glass, plastic, food waste, e-waste, clinical waste, and others and listed in a standard form. Waste from every sample was manually sorted according to its categories. All the sorted waste from every sample was weighed using a standard weighing scale with a sensitivity of 0.1 kg. Then, data was recorded in a standard form for analysis.

Data Collection, Recording, and Analysis

All the recorded data were computed in standard spreadsheet for waste analysis. The waste collection rate of each bin had been calculated using Eq. 1. Then the total waste collected in this study was calculated by using Eq. 2. Later the percentage by weight for every waste category from the total waste was calculated. The percentage produced the waste composition according to the weight. The waste generation rate per students in the studied hostel block also had been determined.

1. $\text{Waste collection rate} = \text{Total waste weighed (kg)}/\text{No. of sample (bin)}$ (1)

2. $\text{Total waste collected} = \text{Collection rate (kg/bin/day)} \times \text{No.of bins provided}$ (2)

3 Finding and Discussion

Waste Characterization

UiTM (Perak) has appointed three contractors to collect and dispose of all the waste in the campus. All the contractors are in charge in three different zones. The frequency of the collection is once a day for 6 days from monday to saturday. All the waste from the hostel blocks are then transferred to a temporary collection area nearby a parking lot before sending to a nearby municipal dumpsite. A one-tonne lorry and a five-tonne lorry are used to transport the collected waste.

The waste audit at four hostel blocks produced results as shown in Table 1. There were 192 samples from all the blocks with a total waste of 765.43 kg. The highest waste quantity was 245.95 kg recorded in Angsana Block. This was followed by Bidara and Damar blocks with waste quantity estimated at 232.49 kg and 172.30 kg, respectively. Meanwhile, the lowest waste quantity was recorded in Cemara block.

Angsana block was the last block being audited which was at the fourth week of the study. Coincidently, the audit at this block fell on the end of the semester. All the students need to clear their hostel before going off for semester holiday. This activity includes disposing of their unwanted belongings which contribute more waste at this block. The highest category of waste found at Angsana block was papers from students' projects such as assignments, notes, drawings, and models. There were also newspaper, corrugated cardboard, boxboard, and also, mixed paper found. The result also shows that mens hostel produced more waste than a female's hostel. Therefore, it can be concluded that males are the major contributors for waste disposal. The solid waste found at the males' hostel are papers, aluminum cans, and also glass drinking bottle.

Table 2 below shows the types and composition by weight of solid waste collected in Zone 3 hostel area. The highest composition of waste was food waste with a percentage of 48.30 %, followed by paper (30.90 %) and plastic (9.01 %).

Table 1 Waste sample for studied hostel block

Waste	Total sample (6 days)	Total waste (kg)
Angsana	48	245.95
Bidara	48	232.49
Cemara	48	114.69
Damar	48	172.30
Total	192	765.43

Table 2 Composition (% by wt) of solid waste generated in zone 3 hostel area

Types	Weight (kg/day)	Percentage (%)
Paper	236.54	30.90
Aluminum	25.59	3.34
Glass	9.11	1.19
Plastic	69.00	9.01
Food waste	369.70	48.30
e-waste	6.09	0.80
Clinical waste	4.90	0.64
Others	44.50	5.81
Total	765.43	100.00

Table 3 Total waste collected at zone 3 hostel area

Block	Total no of bins	Collection rate (kg/bin/day)	Total collection (kg/day)
Angsana	97	4	388
Bidara	104	4	416
Cemara	110	4	440
Damar	104	4	416
Total			1660

Meanwhile, the lowest composition of waste was clinical waste (0.64 %). Food waste was found to be the highest composition as the students having meals (packed food) in their hostels. They threw away the leftovers to the bins. The high composition of food waste is influenced by the student lifestyle of buying packed food instead of having their meals at the stall or restaurant. Normally, students buy packed food from the night market nearby UiTM (Perak), which is held every monday, tuesday, and wednesday. They prefer buying food from the night market as the food is cheaper and more varieties.

Waste Generation and Collection

The total amount of waste sampled in this study was 765.43 kg. By using Eq. 1, the collection rate for each bin provided in the hostel block found to be 4 kg/bin/day (refer Table 1). The total number of bins provided by the Department of Facilities Management of UiTM (Perak) for every block is shown in Table 3. Therefore, the estimated quantity of total waste collected at the four blocks was 1,660 kg per day. During this study, there were 3,498 students' reside in Zone 3 hostel. Thus, the waste generation rate for a hostel in Zone 3 was found to be 0.47 kg per student per day.

4 Waste Minimization Strategies

The findings from this study produced the necessary information to formulate practical and economical strategies for waste minimization. It was estimated that the UiTM (Perak) campus produces at least 1.6 tonne of solid waste per day. More than 44 % of these wastes are recyclables or potentially recyclable. This is shown in Table 1 that within the total waste generated in the hostel blocks, the category of paper represents the larger percentage (30.90 %) of recyclable waste in the hostel. Thus, a segregation and recycling program is feasible in this campus. Therefore, this paper proposes two solid waste minimization strategies that could be implemented in UiTM (Perak) as a first step toward greening the campus as shown in Fig. 2.

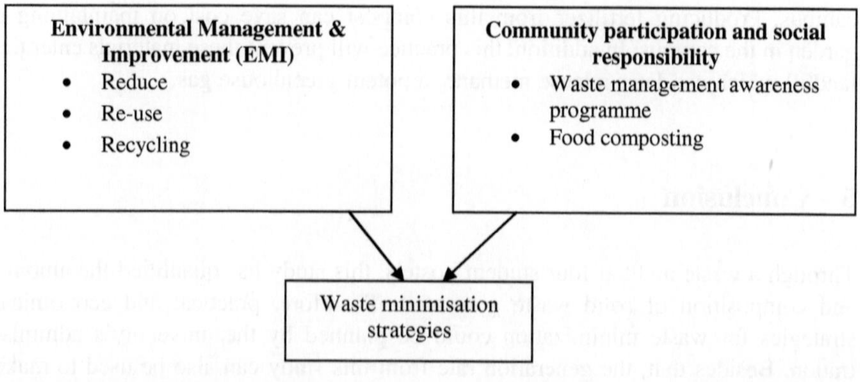

Fig. 2 Waste minimization strategies

The first strategy is environmental management and improvement (EMI) which will focus on the minimization of waste produced by the university's member. With regard to the waste management hierarchy, the most effective way to reduce waste is to not produce it in the first place. In fact, reduction and reuse are the top priorities to save natural resources, protect the environment, and save money. Recycling can also benefit the community and the environment as it is the process of collecting and processing materials into new products. Unfortunately, this study found that the provided recycle bins at every hostel blocks were underutilized as little quantity of recyclable waste was found in the bins. The location of the recycle bins near the staircase at the ground floor level of the hostel block is not easily reached by the students who stay at higher floor. This situation might hold back their intention to sort and send recyclable waste to the recycle bins. Therefore, the bins should be relocated to a more accessible area to facilitate and encourage students to sort and recycle waste.

The second strategy is community participation and social responsibility which focus on the waste management awareness programme and food composting. Full cooperation and participation from community in the campus are very important to make this strategy a success. There is a need to put more effort into raising community members' awareness. In addition, knowledge and awareness on waste sorting and recycling should be a major input in the awareness programme. This can be achieved by awareness campaign emphasized on the considerable change in the attitude and perception of the community in the campus toward solid waste management. Hopefully, with the raised awareness it will inevitably change their habit and also the behavior toward good solid waste management practices. This paper also recommends food waste composting to be conducted enormously at the hostel block as food waste accounted to 48.30 % which was the highest composition of waste collected. The compost from the organic material or food can be used as fertilizer in the campus. Currently, 20–30 % of waste are from food scraps and yard waste. Therefore, composting these waste should become a regular practice in the campus. Producing fertilizer from this compost can save cost on maintaining a garden in the campus. In addition, this practice will prevent these materials enter the landfill which can later release methane, a potent greenhouse gas.

5 Conclusion

Through a waste audit at four student hostels, this study has quantified the amount and composition of solid waste generated. Therefore, practical and economical strategies for waste minimization could be planned by the university's administration. Besides that, the generation rate from this study can also be used to make projections of total waste generated for the whole hostel area. Hence, it is suggested that the university must give encouragement, educate, and create awareness to its students in order to minimize waste and also improve recycling efforts. The two main waste minimization strategies proposed in this paper are essential in promoting sustainable behavior and attitude of the campus community. Hopefully, the outcomes and recommendations from this study would be beneficial to the university.

Acknowledgments We would like to thank Universiti Teknologi MARA (Perak), Excellence Fund for providing the research grant 600-KPK(PJI.5/2/2/4) (94)

References

Alshuwaikhat HM, Abubakar I (2008) An integrated approach to achieving campus sustainability: assessment of the current campus environmental management practices. J Clean Prod 16 (16):1777–1785

Armijo de Vega C, Ojeda-Benitez S, Ramirez-Barreto E (2008) Solid waste characterization and recycling potential for a university campus. Waste Manag 28:S21–S26

Beringer A, Wright T, Malone L (2008) Sustainability in higher education in Atlantic Canada. Int J Sustain High Educ 1(9):48–67

Chang N, Davila E (2008) Municipal solid waste characterization and management strategies for the Lower Rio Valley, Texas. Waste Manag 28:776–794

Creighton S (1998) Greening the Ivory Tower: improving the environmental track record of universities, colleges, and other institutions. The MIT Press, Cambridge, MA

Farmer G, Staniewicz N, Michael B, Wojcik A, Lim Y, Ivokovic D, Rajakulendran J (1997) Audit of waste collected over one week from ten dental practices: a pilot study. Aust Dent J 42(2):114–117

Hansen W, Christopher M, Verbuecheln M (2002) EU waste policies and challenges for local and regional authorities

Keniry JE (1995) Campus environmental stewardship at the turn of the 21st century. National Wildlife Federation, Washington, DC, USA

Sharp L (2002) Green campuses: the road from little victories to systematic transformation. Int J Sustain High Educ 3(2):128–145

Tchobanoglous G, Theisen H, Vigil S (1996) Integrated solid waste management. McGraw-Hill, New York, NY, USA

United Nations Environmental Program (2013) Guidelines for national waste management strategies moving from challenges to opportunities. ISBN 978-92-807-3333-4

Chapter 62
Reverse Logistics Network Design for Paper Recycling

Zurina Hanafi, Dong Li and Shen Cheng

Abstract Recycling becomes an important waste disposal tool nowadays due to the nature of value recovery and the benefits of greenhouse gas reduction. It is particularly suitable to be applied to products with short life cycles, such as paper. A paper recycling model is developed with the incorporation of critical design factors that contributes to the variation in the effectiveness, operations cost and environmental impact of the paper recycling model. The objective of the study is to see the impact of carbon emission policies on the current paper recycling network. The research is conducted in the United Kingdom (UK) with Liverpool as the collection point. The paper recycling network model is optimised in order to compare the overall cost which are carbon emissions and operations cost. The results reveal that carbon emission policies are currently ineffective in reducing the overall carbon emissions in the general society, and they will contribute to various short to long term negative effects to the strategic operation and break the initially designed counterbalance between the operations cost and environmental impact of the proposed model. In addition, the model suggests that the UK paper industry should reprocess as many recycled materials as possible so that it would reduce the additional cost and emissions associated with the import of finished paper products.

Keywords Carbon emission · Paper recycling · Reverse logistics

---

Z. Hanafi (✉)
School of Quantitative Sciences, Universiti Utara Malaysia,
Sintok, Kedah, Malaysia
e-mail: h.zurina@uum.edu.my

D. Li · S. Cheng
Management School, University of Liverpool, Liverpool, UK
e-mail: dongli@liv.ac.uk

S. Cheng
e-mail: chinatonewzealand@hotmail.com

N.A. Yacob et al. (eds.), *Regional Conference on Science, Technology and Social Sciences (RCSTSS 2014)*, DOI 10.1007/978-981-10-0534-3_62

1 Introduction

The growing awareness for environmental issues has already affected people's behaviours in many aspects and it has led to a range of government regulations and policies which are made with the purpose of achieving environmental sustainability. One of the major areas that are experiencing major culture change is the disposal of end-of-life products. Due to the massive amount of products being disposed every year, the traditional disposal methods, landfill and incineration, have resulted to accumulate a burden for both the economy and the environment. Reverse logistics refers to the sequence of activities required to collect the used product from the customers for the purpose of either reuse or repair or remanufacture or recycle or dispose of it (Agrawal et al. 2015). In United Kingdom (UK), it is estimated that recycling of waste products could save approximately 10 to 15 million tonnes of carbon emissions annually compared to the mix of landfill and incineration with energy recovery (CPI (1) 2009).

Paper is one of the most frequently used products with a very short life cycle. In the UK, about 14 million tonnes of paper products were consumed in 2006 (WRAP 2007), and the volume is expected to increase due to the boom in the UK paper manufacturing industry. Paper is a biodegradable product so that the disposal of it into landfill will create a powerful greenhouse gas p called methane, which contributes to severe climate change. Besides, paper is also a carbon-based product, resulting to a great amount of carbon emissions if it is incinerated (CPI (2) 2009). Therefore, the nature dictates that recycling is so far the most responsible and sustainable way to manage waste paper products. Moreover, the recycled paper can be used as a replacement for timber in paper production, so that the more paper being recycled, the less imported timber products that the country would be needed. Hence, in addition to the environmental benefits of recycling, it also makes sense in terms of the economy that recycling is the most preferable choice. The UK paper recycling supply chain is an integrated process containing collection, sorting, reprocessing and redistribution, and the logistics issue in connecting each stage within the network is important.

This paper is structured as follows: Sect. 2 provides a review of process in paper recycling network. Section 3 explains the methodology used in this paper. Section 4 explains the scenario analysis while Sect. 5 concludes this paper.

2 Paper Recycling Network

In 2005, the UK paper demand was estimated to be 12.5 million tonnes, of which only 39 % were supplied through domestic paper mills. The situation has been getting worse and worse year by year. In 2008, the UK paper manufacturing industry was only able to supply about 35 % of the total domestic market demand (CPI (3) 2009), and it had to import new paper products to fulfil the rest of its paper consumption. The reason for this significant imbalance in the market is obvious, that is, the country is seriously lacking of sufficient paper mill capacity. Despite that the UK is recognised to have the highest recovered paper utilisation rate, but at the end only 46 % of the recovered paper is recycled within the country. Therefore, while the government is attracting more investment to improve its paper manufacturing capacity, it is vital that the development of a paper recycling model is considered which enable efficient plan and manipulate the domestic recycling operation. Therefore, the best value and profit out of the limited 46 % of the recovered paper could be abstracted without causing serious environmental problems. In addition, the UK is currently in a rather embarrassing position with an increasing rate of waste paper collection but a shrinking level of domestic paper mill capacity. Although the officials believe that the current situation is sustainable, as long as there are expanding overseas markets which are able to absorb the majority of UK's collected paper (CPI (3) 2009), but due to the massive volume of recycled paper being exported, it suggests that a cost efficient and environmental responsible supply chain model for delivering the collected paper to overseas is called for. As a result, the UK paper recycling model should be developed with the attention of both domestic and international features of arrangement.

The framework of this paper recycling model is similar to any other reverse logistics models that are developed for managing other end-of-life products. The recycling network contains four fundamental stages of flow: collection, sorting, reprocessing and redistribution. Each stage of flow has a number of network characteristics which usually leads to various tradeoffs between social, economical and environmental considerations, and the valuation of these critical tradeoffs will affect the choice of such a network's design decisions.

Paper Collection

There are generally two collection systems that can be applied to retrieve products: industry-wide collection and company-specific collection (Barker and Zabinsky 2008). The industry-wide collection indicates that all products in the market are collected within the same system regardless of whom the original manufacturer is and it is frequently used for high-volume, commodity-type products. This type of system is able to cover a broader aspect of views and accumulate a much greater volume of collections, thus the benefit of economies of scale can be achieved. In addition, such system is a completely separated function and it is usually performed by a third party specialist, so the implementation will not complicate the manufacturer's forward supply chain activities. The company-specific collection on the other hand specifies that the manufacturer is responsible to collect its own products from its customers and it is suitable for high-value products. It helps the manufacturer strengthening the relationships with customers, and enhancing its marketing and sales efforts, however, the shattered location of its customers and the small amount of collecting volume from single source indicate that this method can be expensive and with more planning and operating complexity.

Paper is one of the typical low-value, high-volume commodity-type products and therefore, the industry-wide collection system tends to be used prevalently among paper recycling industry. The recycling vehicles are able to collect a substantial amount of paper along their travelling routes within the targeted areas, so they are likely to achieve a high level of utilisation and the overall efficiency of the recycling network can be maintained at a satisfied level.

Paper Sorting

Sorting can be classified into either centralised or decentralised manner. The former is used to deliver all collected wastes to a centralised facility to undertake further sorting and classification procedures whereas the latter is to detect and identify collected wastes in advance and transport the non-scrap and scrap materials to the sorting facility and the waste disposal in two separated streams. The process of sorting can give impact on the quality of the paper. If not properly separated, remaining stickies reduce the quality of the recovered paper or even disrupt the production process (Fugenschuh and Hayn 2014). Two sorting methods in fact represent a trade-off between cost and complexity. The centralised sorting method simplifies the recycling network as it only requires the operation of one single stream, but it contains the risk of incurring higher transportation cost as it ships unrecoverable products to the sorting facility first, rather than to waste disposal directly. The decentralised sorting on the contrast avoids the expense of unnecessary transportation cost, but it involves the collaboration of two separated streams which will create more administrational problems in the planning and the implementing phase of the recycling operation, and it also requires extra human resources and equipments.

Most paper in the UK is collected from householders through municipal waste stream where paper is usually mixed with other products such as glasses and plastic bottles. Sometimes paper is picked up and separated away from others upon collection, but further identification of whether the paper is recoverable or not can only take place once they are arrived at a single sorting facility. Therefore, the sorting stage in this paper recycling network generally follows a centralised principle.

Paper Reprocessing and Redistribution

Once the collected paper is detected and classified by grades, they are sent to paper mills to convert into pulp, which is an essential raw material for paper manufacturing. The alternative choices in our network's design decisions are that paper can be sent to either UK paper mills or overseas paper mills. The decision is rather flexible and the final choice depends on various decisive factors such as paper mill's capacity, availability, transportation cost and its associated environmental impact. However, if the recovered paper is destined to an overseas paper mill, the recycled paper products are most likely to be consumed in that foreign country, thus there will be no redistribution activities afterwards.

3 Methodology

Based on the UK waste management policy, the regional self-sufficiency and the proximity principle, the assumption is made where all paper collected would be delivered to a material recovery facility (MRF) within the Liverpool region to be processed. There are two MRFs operating within the Liverpool region. The first one is the Bidston MRF, which is operated by Mersey Waste Holdings Ltd, with an annual sorting capacity of 60,000 tonnes (MWDA 2009). The facility is using a combined system of labour and mechanical sorting techniques to separate paper from other household's wastes. The other one is the Gillmoss MRF with an annual capacity of up to 100,000 tonnes (MWDA 2009). Since these MRFs have almost the same characteristics, only Bidston MRF is used in the analysis.

The optimum travel route from Liverpool Centre to Biston MRF is measured to be 5.4 miles which is an approximate distance. Because of this inner-region delivery, the transportation of collected waste from Liverpool Centre to MRFs is usually taken by road. The vehicles used for kerbside sorting scheme is a heavy diesel truck with five compartments which can be classified as articulated lorry, and they collect wastes from householders every fortnight.

After screening and evaluating all the available UK paper reprocessors, this paper will only analyse the effect of delivering sorted paper to paper mills (PM) located in Darwen and Kent. They separately stand for an extreme but opposite case of the study, as the former represents the most adjacent paper mill from Liverpool region and the latter represents the most remote paper mill from Liverpool region, so the results gathered from the analysis will clearly indicate the difference between the best and the worst options in terms of carbon emissions if paper reprocessing takes place within the UK. Kent is located at the south part of UK and the distance is more than 400 km from Bidston and Gillmoss MRFs. Darwen on the other hand is located less than 80 km from both MRFs.

The sorted paper is usually transported in a containerised manner from MRFs to UK paper mills by either road or rail. The downside in setting up rail transport is that it often requires intermodal transportation with road to deliver the products to/from rail stations, and this handover procedure will further complicate the design and implementation of the recycling operation. However, in our case study Bidston MRF and paper mills (Kent and Darwen) are located very next to their local rail heads, so the little extra operating cost and carbon emissions resulting from the intermodal road transport at this point does not add any severe effects to our existing network. It is assumed that there is a door-to-door direct rail connection between MRF and paper mills.

The largest foreign buyer of UK's collected paper is China, who purchased 2.7 million tonnes of recovered paper in 2008, and now it is estimated that China is accounting for nearly 55 % of all UK's exports of recovered paper (Stones 2009). Therefore, this paper will take China as a representative for all international paper trading between UK MRFs and overseas paper mills. If any sorted paper is planned to go aboard, China will be their destination.

Recovered paper typically leaves UK from major ports such as Felixstowe or Southampton where all paper will be containerised and shipped directly to large Chinese ports. Only Southampton is used in this study. The typical maritime container used to store paper has a self-weight of 4 tonnes, with an inside capacity of 25 tonnes. The first step in planning such an international movement of goods is to transport sorted paper from MRFs to UK ports by either road or rail.

The next step is to ship the paper from UK ports to Chinese ports. The majority of paper mills in China are concentrated in Guangdong Province, which is located on the south coast of China. Thus, this paper will focus on measuring the amount of carbon emissions based on the assumption that ships leaving UK are destined to arrive in China via Shekou port, Guangdong Province.

The last segment of this paper recycling network is to bring reprocessed paper back from paper mills to redistribution centre (RC) located in Liverpool. For paper that has been exported to China, they will be reprocessed into new paper products and consumed in the Chinese market. In addition, it also reveals a fact that one of the underlying benefits of recycling which is to save UK raw materials for making new paper products is lost once the collected paper is exported to foreign countries.

4 Scenario Analysis

The recycling network is divided into three segments. The findings provide an idea of what the reverse logistics chains might be and how they could be undertaken within each segment of the network. Nevertheless, in order to design the entire recycling activities with an aim to minimise the overall environmental impact, the findings and measurements is combined together to create six different scenarios which represent and characterise the possible recycling supply chain arrangements that a Liverpool-based paper recycler is likely to adopt. The segmentation of the network is shown in Table 1.

Carbon emissions and operations cost is obtained using emission factors and the distance travelled with different transportation modes. By observing carbon emission and operations cost generated in Table 2, there are several critical decisions that can be concluded regarding to the design of such a UK paper recycling network. First of all, results have proven that the usefulness of regional self-sufficiency and the proximity principle can be clearly perceived in the analysis. Different scenarios reveal that the further away the paper mill is located from the source of

Table 1 Paper recycling network segmentation

Scenario	Segment 1	Segment 2	Segment 3
1	Liverpool—Bidston MRF	Bidston MRF—Kent PM by Road	Kent PM—Liverpool RC by Road
2	Liverpool—Bidston MRF	Bidston MRF—Kent PM by Rail	Kent PM—Liverpool RC by Rail
3	Liverpool—Bidston MRF	Bidston MRF—Darwen PM by Road	Darwen PM—Liverpool RC by Road
4	Liverpool—Bidston MRF	Bidston MRF—Darwen PM by Rail	Darwen PM—Liverpool RC by Rail
5	Liverpool—Bidston MRF	Bidston MRF—Southampton Port by Road	Southampton Port—Shekou Port by Sea
6	Liverpool—Bidston MRF	Bidston MRF—Southampton Port by Rail	Southampton Port—Shekou Port by Sea

Table 2 Carbon emissions and operations' cost

Scenario	Carbon emissions (kg)	Operations cost (£)
1	948	5445
2	724	3669
3	154	892
4	119	872
5	3509	5050
6	3410	4281

collection, the more severe negative impact that this supply chain will cause to the environment. Therefore, the design of any recycling network should always consider mileage saving in transportation as one of the primary objectives in order to reduce the level of carbon emissions.

Secondly, if the recycling network does not span over a large distance which implies that every station or facility in the network is located relatively close to each other and the delivery of products can be considered as short-haul freight, then there will be merely little variations in terms of environmental impact on whether to use road or rail as the primary mode of transport. Scenario 3 and 4 has backed up this opinion by having very similar results. However, the involvement of rail operation into the recycling network requires more careful coordination of personals and connecting activities due to its greater level of complexity and less versatile nature. Such negativities of rail transport cannot be strongly overweighed by its environmental-friendly advantage in short-haul freight, so it is comparatively more suitable to conduct road transport in a small-scaled recycling network. In contrast, if activities in the network do span broadly in a nation-wide scale and the delivery of goods is considered to be long-haul freight, then the carbon emissions gap between adopting different modes of transport can be noticeably reflected, as shown in Scenario 1 and 2. As a result, the environmental responsiveness of rail transport is obvious enough to cover its inconvenient nature in long-haul freight, and thus rail will become the dominating mode of transport if the recycling network covers a large geographical content.

At last, the analysis disclosed another decisive statement. This comparison implies that as long as there is vacant domestic paper mill capacity, all recovered paper should be reprocessed within the UK because it will not only increase the volume of domestic paper production so that the expenses on importing new paper products could be reduced, but it will also act in the best interest of the environment to enormously avoid the emission of carbons.

From the above analysis, the cost difference of transporting one container-load of paper throughout the entire recycling network between road and rail in long-haul freight is quite noticeable, as rail transport shown in scenario 2 has led to a nearly £2,000 cost-saving than road transport shown in scenario 1, so it is undoubted that rail should be used if this recycling supply chain has a large span. Nevertheless, if the network covers a small range of area, then the decision of either involving road or rail transport into the recycling network will not be affected by the cost factor, as scenario 3 and 4 show a very similar result. If the consideration is made for the difference in both modes' design and operating complexity, it is fair to come to a decision that road transport is a relatively better option in this case.

The current international cost for finished paper products comes to around £540 per tonne (World Scrap 2009), so let us assume that the sum of sorting and reprocessing cost in the UK is £500 per tonne of recycled paper. The total operating cost for scenario 1 will then increase to £17945, which has largely exceeded the cost of exporting the paper to China. However, because of the capacity insufficiency in UK's domestic paper production, it has to import new paper products to satisfy the rest of its market demand. Therefore, after the UK exports a container-load of paper to China, it has to spend £13500 to import the equivalent amount of new paper products back. In addition to that, it would incur an extra transportation cost for bringing the new paper into the UK market, which is assumed that the cost would be £4000 per trip. Taken into account of the profit resulting from the export of recovered paper, which is around £72 per tonne (McEntee 2005), the new overall cost for scenario 5 and 6 becomes £20750 and £19981, respectively. These figures clearly demonstrate the financial advantage of having sufficient domestic manufacturing capacity. It is much cheaper for the UK paper industry to manufacture new paper products itself rather than to import them from overseas, thus, the export of recycled paper should only be carried out in the case that the domestic paper mill capacity is fully saturated.

5 Conclusions

This paper examines the four fundamental stages of operation of the UK paper recycling market and analyses the potential elements that might alter the effectiveness as well as affecting the strategic design of each stage in the operation. It uses them as a basis to construct a conceptual recycling model taken into account of multiple design considerations and tradeoffs. Secondary data which are data published in white paper from Confederation of Paper Industries is used in the study. The model provides a general template for the chain of recycling activities and it can be adopted as a guideline to facilitate UK paper recyclers in most effective design and organise their operations. The overall level of carbon emissions and operating costs resulting from the mixture of different design decisions are evaluated, and an important conclusion has been brought out in the forefront, that is, the cost of sorting and reprocessing may not be an important decisive factor in the design of the network. Rather, the cost of transportation and its associated environmental impact, especially the carbon emissions, are the key driving forces for the choice of parameter design in such a reverse logistics supply chain. The results also reveal that the wider radius the network covers, due to the longer effort in transporting the collected paper, the more carbon emissions and operating costs the recycling operation would incur. Similarly, the analysis also implies that as long as there is domestic paper mill capacity available, it will always be a better option to produce paper products within the country rather than to import them from overseas. In addition, the choice of transportation mode significantly contributes to the variation in our results. Rail transport has been proven to have a substantial benefit in terms of the environmental impact as well as the operating cost. Therefore, the paper concludes that whereas it is possible, especially if the network has a large span, rail transport should be dominantly used in connecting each stage of the operation.

The impact of introducing carbon control policies might imply to the design and operation of the proposed model. Carbon emission policies, under its current implementation or the potential future application with an aim to reduce the use of road transport, will not only act ineffectively in terms of carbon emissions reduction, but it will eventually diminish the financial benefit of rail transport and leaves the network's strategic design with only two options: either maintain the use of road transport with a higher operating expense, or switch to the use of transport mode resulting to a compromise in more carbon being emitted.

References

Agrawal S, Singh RK, Murtaza Q (2015) A literature review and perspectives in reverse logistics. Resour Conserv Recycl 97:76–92

Barker TJ, Zabinsky ZB (2008) Reverse logistics network design: a conceptual framework for decision making. Int J Sustain Eng 1(4):250–260

CPI (1), Confederation of Paper Industries (2009) Recycling versus incineration. Fact sheet. http://www.paper.org.uk/information/factsheets/recycling_incineration.pdf. Accessed 18 June 2009

CPI (2), Confederation of Paper Industries (2009) Paper and cardboard recycling greenhouse gas benefits explained. Fact sheet. http://www.paper.org.uk/information/factsheets/greenhouse_gas.pdf. Accessed 20 June 2009

CPI (3), Confederation of Paper Industries (2009) UK paper recovery industry continues progress within Europe. http://www.paper.org.uk/news/2009/0107ukpaperrecovery.pdf. Accessed 20 July 2009

Fugenschuh A, Hayn C (2014) Mixed-integer linear methods for layout-optimization of screening systems in recovered paper production. Optim Eng 15:533–573

McEntee K (2005) Scrap paper prices look strong for 2005. The paper stock report. http://www.sssnews.com/directory/28-januaryfebruary-2005/99-scrap-paper-prices-look-strong-for-2005. Accessed 16 June 2009

MWDA. Merseyside Waste Disposal Authority (2009) Merseyside waste disposal authority selects Veolia environmental services as its preferred bidder for a new waste management and recycling contract. http://www.merseysidewda.gov.uk/media-news/current-newsDetail.asp?id=208&level=3&ParentID=1&MenuItemID=81&Current_ID=228. Accessed 5 June 2009

Stones M (2009) Food production daily. China to keep importing UK recovered paper and plastic. http://www.foodproductiondaily.com/Product-Categories/Packaging-Materials/China-to-keep-importing-UK-recovered-paper-and-plastic. Accessed 16 June 2009

World Scrap (2009) Paper price to increase by 4 percent. Paper and other materials. http://www.worldscrap.com/modules/news/article.php?aid=11402. Accessed 7 Jul 2009

WRAP (2007) Realising the value of recovered paper. Recycled paper. Market Situation Report. http://www.wrap.org.uk/downloads/Paper_report.1d437d40.4072.pdf. Accessed 1 June 2009

Part XIV
Physics

Chapter 63
Improvement of Insulated Wire Ball Bonding

Muhammad Faiz, Yap Boon Kar, Hung Yang Leong, Tan Chou Yong, Chin Teck Siong and Tan Lan C

Abstract Interconnect chip technology is getting smaller and the trend is continue until now. The semiconductor industry is seeing increasing demand for miniaturization which necessitates changes in chip design, including larger numbers of I/O per unit space, multitier devices, and stacked dies that increase wire density. Even the pitch—the horizontal distance between one wire and the next—is being reduced. This puts high demands on the stability of the wire bonding process, the bonding material properties, and their uniformity. The advancement of the IC chip technology must be follow by the wire technology too. Insulated wire in following the trend of the IC chip technology. Insulated wires would isolate one from the other, eliminating the need to avoid wires touching and creating short circuits. Relaxation of design rules would, thereby, simplify chip design and lower costs. However, there are many challenges to insulated wire bonding, one of which is the increased stress transmitted to the bond pad during ball bonding process. This high stress is not desirable as it leads to pad damage or cratering in the silicon under the pad. Another issue is pad splash where the pad material is squeezed outside the bonded area, which in severe cases can cause Al pad thinning and depletion. This paper compares the pad stress during wire bonding for insulated process. To study the root cause of the increased stress, ball bonding is performed with insulated wires using the different level of ultrasound (USG), bonding force (BF), scrub force, usg prebleed and capillary velocity (CV).

M. Faiz (✉) · T.C. Yong · C.T. Siong · T.L. C
Freescale Semiconductor (M) Sdn Bhd, Free Industrial Zone,
Petaling Jaya, Selangor, Malaysia
e-mail: faiz2036@gmail.com

Y.B. Kar · H.Y. Leong · T.L. C
Electronics and Communication Engineering Department,
Universiti Tenaga Nasional, Bandar Baru Bangi, Selangor, Malaysia
e-mail: Kbyap@uniten.edu.my

T.C. Yong · T.L. C
Department of Mechanical Engineering, Faculty of Engineering,
University of Malaya, Kuala Lumpur, Malaysia

N.A. Yacob et al. (eds.), *Regional Conference on Science, Technology and Social Sciences (RCSTSS 2014)*, DOI 10.1007/978-981-10-0534-3_63

Keywords Insulated · Wire bonding · Ball bonding

1 Introduction

The interconnect chip technology used in various devices is growing all over the world. It has become a trend whereby the smaller the device is, the more advanced the technology is. This trend pushes the semiconductor company to keep reducing the size of the IC chips for the usage of the smaller device. However, most of this cutting edge technology needs higher specifications for its parts so that it can cope with the high-end tasks for that particular device. Therefore, there is a need for the semiconductor company to continuously improve their current products for wider usage and better performance.

An increasing number of microelectronic packages are being bonded with Cu wire due to the skyrocketing prices of gold, coupled with the industries' continuous drive for cost reduction. In addition, Cu wire provides better thermal and electrical performance. Copper or Cu possesses about 30 % higher thermal and electrical conductivity compared to Au. This offers improved device thermal and electrical performance. However, there are a set of challenges coming with the reduced cost and performance, such as Cu wire and ball are easy oxidized in air, harder Cu ball can cause pad and under pad structure to crack, narrow second bond window and short tail problems.

This paper focuses on first bond issues with insulated wire to Al bond pad. It the root causes of Al splash and studies the influential parameters that affect the 1st bonding process. The samples are tested for reliability and bond ability by wire pull, wire peel, and ball shear tests.

2 Experimental Procedure

Experimental Study

The ball bonding is performed on Kulicke and Soffa (K&S) IConn automatic ball bonder using 0.8 mil insulated Cu wire and a commercial Cu wire with 4 N purity. The bonding is performed at a nominal heater plate temperature of 160 °C, resulting in actual bonding temperature of ≈138 °C measured at the pad surface. A copper kit with forming gas is used to prevent oxidation of copper during free air ball (FAB) formation. The forming gas used is a homogenous mixture of 95 % nitrogen and 5 % hydrogen. The flow rate of the forming gas is set to 0.62 l/min. The device is packaged BGA with a total of 700 wires/4 tiers. In this study, the root cause of the increased stress, ball bonding is performed with insulated wires using the different level of ultrasound (USG), bonding force (BF), and capillary velocity (CV).

Table 1 Screening parameter

Parameter
CV (capillary velocity)
USG prebleed
Scrub force
Force USG current
Force power

For early stage of development, screening characterization is used to determine the important parameter that affects ball bonding response. The parameter that affects ball bonding response is showed in Table 1:

Design of experiments (DOE) is a systematic, rigorous approach to engineering problem solving that applies principles and techniques at the data collection stage so as to ensure the generation of valid, defensible, and supportable engineering conclusions. In addition, all of this is carried out under the constraint of a minimal expenditure of engineering runs, time, and money (Table 2).

A total of 30 bonded ball size were measured. Al splash measurements were also carried out for both wires. Destructive testing (ball shear and wire pull) was performed to ensure good ball bond strength. The sample size was 48 readings per unit. No metal lift or ball lift was allowed. A set of 10 fully bonded samples for each pattern, were baked at 225 °C for 4.5 h to accelerate the IMC growth. The IMC coverage was inspected on the bond pad and calculated via the Imagen tool.

Table 2 Screening matrix result

Running order	Pattern	Capillary velocity	USG prebleed	Scrub force	Force USG current	Force power
1	0000	Nominal	Nominal	Nominal	Nominal	Nominal
2	−−−−+	Low	Low	Low	Low	High
3	+−−++	High	Low	Low	High	High
4	−+−+−	Low	High	Low	High	Low
5	+−+−−	High	Low	High	Low	Low
6	+++++	High	High	High	High	High
7	−−++−	Low	Low	High	High	Low
8	−++−+	Low	High	High	Low	High
9	++−−−	High	High	Low	Low	Low
10	00000	Nominal	Nominal	Nominal	Nominal	Nominal

3 Results and Discussion

Bonded Ball and Al Splash

The study focus on to control ball size and Al splash to avoid bond pad shortage after bonding process. The bonded ball and Al splash need to achieve standard specification. For bonded ball, the ball bond diameter (BBD) must be within 35–38 μm and the Al splash is not more 42 μm. The ball bond height (BBH) specification is 9–11 μm. Table 3 showed the comparison of ball bond diameter for every cell with different parameter.

Table 3 showed cell 1, cell 2, cell 3, cell 4, and cell 9 are meeting the requirement for bond ball diameter. By comparing the parameter that cell 1, cell 2, cell 3, cell 4, and cell 9 used, all 4 cells have the same scrub force parameter that have been set to "Low." The comparison showed the scrub force is affecting the ball bond diameter (BBD). If the parameter of scrub force is set to "High", the ball bond diameter (BBD) will increase same like cell 5, cell 6, and cell 7 that have bigger ball bond diameter (BBD) that low scrub force parameter. To meet the requirement, the good ball bond, measured the ball bond height (BBH) for all cell. Table 4 showed the result of bond bond height (BBH):

For the ball bond height (BBH), the requirement is 9–11 μm. In the Table 4 showed only cell 5 and cell 6 have low ball bond height. Al splash is an important thing to consider the good ball bond. The device used in this study have 43 μm bond pad opening. From the bond pad width, the Al splash must be below than

Table 3 Ball bond diameter (BBD) for all cell

BBD	Cell 1	Cell 2	Cell 3	Cell 4	Cell 5
Average	37.68	36.24	36.62	36.14	39.01
Minimum	37.25	35.4	35.35	35.5	37.9
Maximum	38.4	37.8	37.4	36.75	39.95
BBD	Cell 6	Cell 7	Cell 8	Cell 9	Cell 10
Average	39.02	38.92	39.38	36.18	37.96
Minimum	37.4	36.95	38.8	35	37.05
Maximum	40	39.95	39.95	37.35	38.4

Table 4 Ball bond height (BBH) for all cells

BBH	Cell 1	Cell 2	Cell 3	Cell 4	Cell 5
Average	9.88	10.5	10.69	10.38	8.63
Minimum	9	10	10	8	8
Maximum	10	11	11	11	10
BBH	Cell 6	Cell 7	Cell 8	Cell 9	Cell 10
Average	8.13	9.75	9.38	11.13	9.13
Minimum	8	9	9	11	8
Maximum	9	11	11	12	11

Table 5 Al splash diameter for all cell

Al splash	Cell 1	Cell 2	Cell 3	Cell 4	Cell 5
Average	41.44	40.2	43.91	38.61	42.08
Minimum	40	38.7	42.4	37	41.2
Maximum	43.6	41.6	45.6	39.9	43.9
Al splash	Cell 6	Cell 7	Cell 8	Cell 9	Cell 10
Average	44.93	44.5	42.24	40.18	41.39
Minimum	42.4	43	39.6	36.9	39
Maximum	47.6	45.9	43.9	42	44.2

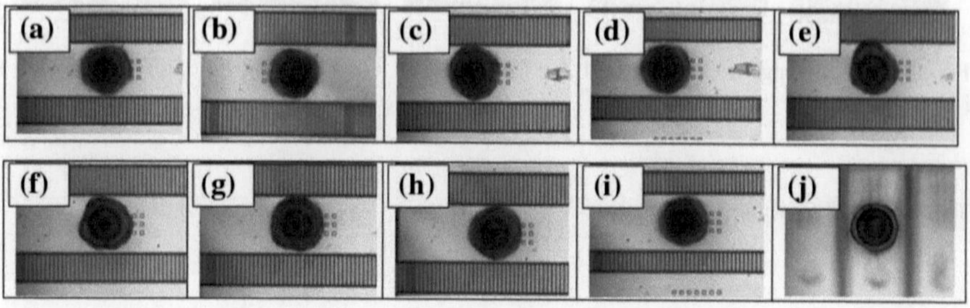

Fig. 1 Al splash characterization **a** Cell 1, **b** Cell 2, **c** Cell 3, **d** Cell 4, **e** Cell 5, **f** Cell 6, **g** Cell 7, **h** Cell 8, **i** Cell 9, **j** Cell 10

43 μm Al splash. If the Al splash is of same size or bigger than the bond pad opening, it will cause bond pad shortage after ball bond process. Table 5 shows the Al splash measurement for all cell.

In the Table 3, observed and focus on maximum Al splash diameter. The Al splash diameter target is between 40–41 μm. This range is considered safe from Al splash bond pad shortage. Cell 2, cell 4, and cell 9 are within the specification that have maximum Al splash diameter at 41 and 40 μm. The Al splash is inspecting via high power microscope to see the splash condition. Figure 1 shows the Al splash condition:

From Fig. 1, cell 2, cell 4, and cell 9 Al splash is within the bond pad opening and do not touch the edge of the bond pad. Other cells already touch the bond pad edge and one of them the Al splash already out from the bond pad.

IMC Study

The good ball bonded must have a good IMC (intermetallic) coverage. The good IMC coverage showed the strong bondability between ball bond and Al bond pad. Based on the IMC coverage test, all cell-bonded ball have >90 % IMC coverage. Figure 2 showed the IMC coverage for every cell (Table 6).

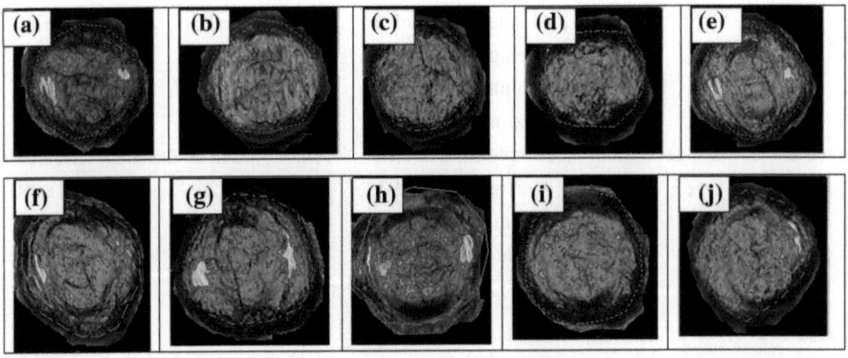

Fig. 2 IMC coverage **a** Cell 1, **b** Cell 2, **c** Cell 3, **d** Cell 4, **e** Cell 5, **f** Cell 6, **g** Cell 7, **h** Cell 8, **i** Cell 9, **j** Cell 10

Table 6 IMC Coverage

Cell	IMC coverage (%)
Cell 1	88.61
Cell 2	98.47
Cell 3	97.66
Cell 4	98.85
Cell 5	94.52
Cell 6	93.82
Cell 7	90.11
Cell 8	83.27
Cell 9	92.78
Cell 10	93.65

Destructive Test

Samples were built for T0 (room temperature) and 4.5 h thermal aging at 225 °C. Wire pull strength and ball shear strength as shown in Tables 7 and 8. Figure 3 showed the wire pull strength performance for all cell. Cell 9 has the lowest wire pull strength but no ball lift observed. For T0, only cell 8 has ball lift after wire pull test. For cell 8, ball bond parameter cannot be tested because cell 8 has failed the T0 wire pull test. There are no any metal peeling and ball lift at T0 for other group.

Figure 4 shows several of ball shear strength performance for every cell. The ball shear strength is affected by the size for Al splash. The bigger the Al splash, the high the ball shear strength. Cell 2, cell 4, cell 8, and cell 9 have the ball shear strength within spec 12–14 g that have been showed in Table 8:

Wire pull strength of insulated Cu-bonded samples were comparable for all cells after isothermal aging at 225 °C for 4.5 h. The wire pulls values for insulated Cu samples decreased by 1.6 and 5.3 %, respectively. This was due to the effect of high

Table 7 Wire Pull strength test performance

T0 wire pull	Cell 1	Cell 2	Cell 3	Cell 4	Cell 5
Average	8.04	8.06	8.01	7.97	7.954
Minimum	7.3	7.294	7.559	6.8	7.223
Maximum	8.62	8.823	8.667	8.635	8.449
Failure	0	0	0	0	0
T0 wire pull	Cell 6	Cell 7	Cell 8	Cell 9	Cell 10
Average	8.036	7.908	7.86	7.729	7.81
Minimum	7.37	7.162	5.554	6.765	6.83
Maximum	8.75	8.469	8.751	8.669	8.391
Failure	0	0	2/48 BL	0	0

Table 8 Ball shear strength test performance

T0 ball shear	Cell 1	Cell 2	Cell 3	Cell 4	Cell 5
Average	18.205	13.528	15.376	13.087	16.636
Minimum	14.646	12.042	14.252	10.285	16.386
Maximum	20.882	14.638	17.133	14.508	18.867
Failure	0	0	0	0	0
T0 ball shear	Cell 6	Cell 7	Cell 8	Cell 9	Cell 10
Average	16.449	15.302	13.256	12.614	14.313
Minimum	14.633	14.094	11.065	11.532	11.813
Maximum	17.857	16.98	15.154	13.886	15.526
Failure	0	0	0	0	0

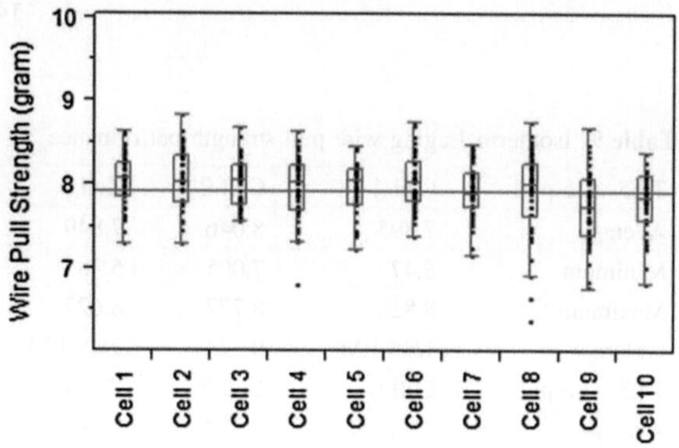

Fig. 3 Wire pull strength performance

temperature annealing of the wire. However, the wire pull break mode for both wires was break-at-neck for all the aging read points. The wire pull values have far exceeded the minimum requirement of 3.5 g. Observed ball lift and metal peeling at cell 1, cell 3, cell 4, cell 8, cell 9, and cell 10. However, cell 2, cell 5, cell 6, and cell 7 survive the isothermal aging. Figure 5 and Table 9 showed the difference.

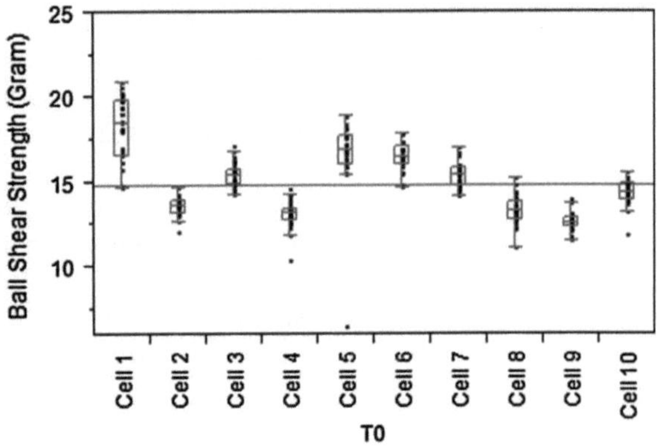

Fig. 4 Ball shears strength performance

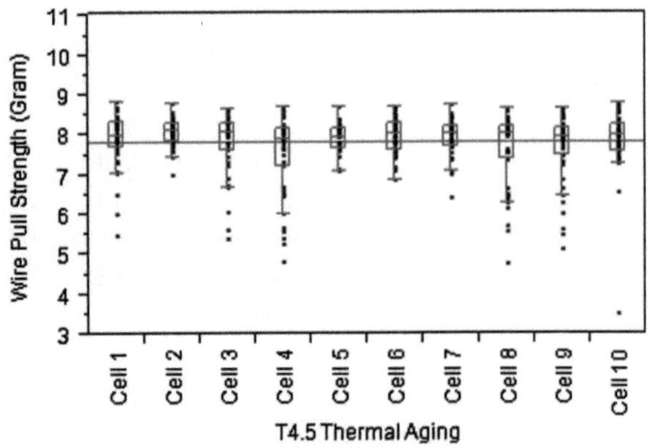

Fig. 5 Isothermal aging wire pull strength performance

Table 9 Isothermal aging wire pull strength performance

T4.5 wire pull	Cell 1	Cell 2	Cell 3	Cell 4	Cell 5
Average	7.895	8.046	7.849	7.492	7.903
Minimum	5.47	7.005	5.38	4.811	7.088
Maximum	8.828	8.777	8.627	8.708	8.712
Failure	1/48 LM	0	7/48 LM	14/48 LM	0
T4.5 wire pull	Cell 6	Cell 7	Cell 8	Cell 9	Cell 10
Average	7.943	7.932	7.595	7.676	7.851
Minimum	6.841	6.385	4.74	5.108	3.51
Maximum	8.706	8.737	8.647	8.666	8.785
Failure	0	0	6/48 BL	6/48 LM	1/48 BL

4 Conclusion

All cells perform well for all testing and study. The bonded ball have a good IMC coverage that >90 % that can be considered very good ball bonding. Same as the ball bond diameter and ball bond height. For Al splash, only cell 2, cell 4, and cell 9 have a good Al splash within the bond pad opening.

Other than these three cells, other cells have excessive Al splash which is not good for ball bonding. Cell 1 until cell 10 have nearly the same wire pull performance. However, detected ball lift at cell 8. The data showed the various ball shear strengths that have been affected by Al splash diameter. Bigger the Al splash can get higher ball shear strength. Isothermal aging showed increasing of number of failure cell. Only four cells survive that wire pull test without metal peeling and ball lift. From the above observations and data, cell 2 is the best choice for improved insulated wire ball bonding. From data above, cell 2 has a good BBD, BBH, Al splash within the specification, good intermetallic coverage, and also the ability to survive at isothermal aging wire pull test. Cell 2 showed that insulated ball bonding used the “Low” parameter to control the excessive Al splash without affecting the other responses.

Acknowledgment The authors would like to thank LC Tan, and TSO-PAE engineers from Freescale Semiconductor (M) Sdn. Bhd. for providing technical assistance, guidance, and supports. This project is funded by the (UMRG Grant RP024B-13AET).

Chapter 64
Eddy Current Thermography Testing on Lack of Fusion (LOF) Defect of Carbon Steel Welded Sample

Nurliyana Shamimie Rusli, Syamsyir Akmal Senawi, Sidek Abdul Aziz, Ilham Mukriz Zainal Abidin, Azhan Hashim, Azman Kassim, Henry Johann Ridzwan, Liyana Zolkarnain, Nurhana Ilmira Harun and Siti Sumaiyah Sheikh Abdul Aziz

Abstract Eddy current thermography (ECT) employs the combination of conventional and advanced nondestructive testing (NDT) technique. This technology is capable to detect defect in subsurface of carbon steel welded sample of dimension 200 × 150 × 20 mm. The optimum joule heating process via eddy current distribution is 300A with the heat diffusion around 50–500 kHz (which is relatively high-frequency electromagnetic wave). Along with the emerging development of eddy current thermography, the infrared camera permits its radiation range to visualize the real-time image of the defect via the application of the thermoimager TIM software. The quantitative information about the defect will be acquired by the camera which is sensitive to the transient signal of heat through material under inspection. The analysis due to the geometrical shape, position, and length of lack

N.S. Rusli · S.A. Senawi (✉) · A. Hashim · A. Kassim · H.J. Ridzwan · L. Zolkarnain
Faculty of Applied Sciences, Universiti Teknologi MARA, Jengka, Pahang, Malaysia
e-mail: syamsyir@pahang.uitm.edu.my

N.S. Rusli
e-mail: nurliyanashamimie@gmail.com

A. Hashim
e-mail: dazhan@pahang.uitm.edu.my

A. Kassim
e-mail: azman615@pahang.uitm.edu.my

H.J. Ridzwan
e-mail: hendrie@pahang.uitm.edu.my

L. Zolkarnain
e-mail: liyanazolkarnain90@gmail.com

N.S. Rusli · S. Abdul Aziz
Faculty of Sciences, Universiti Putra Malaysia, Serdang, Selangor, Malaysia
e-mail: sidek@upm.edu.my

N.A. Yacob et al. (eds.), *Regional Conference on Science, Technology and Social Sciences (RCSTSS 2014)*, DOI 10.1007/978-981-10-0534-3_64

of fusion (LOF) defect can be made through maximum amplitude temperature analysis. The investigation performs the temperature linescan graph of temperature amplitude via Matlab. The result obtained from ECT is compared to the error bar analysis in line with radiography technique. The studies demonstrate the effectiveness and reliability of the technique for the quantification analysis of lack of fusion (LOF) defect of welded sample.

Keywords Eddy current thermography · Lack of fusion (LOF) · Welded sample

1 Introduction

Advanced and efficient testing techniques are very crucial for every inspection process of engineering structures as well as complex components such as aircraft turbines blades to identify potential sites of failure (John et al. 2010; Pan et al. 2012). The continuous development of high speed train required this technique for safety maintenance and detection of rolling contact fatigue (RCF) (Wilson et al. 2011). The application of the technique covers the broad field of oil and gas pipelines industries, aircraft industry, electrical services, houses and buildings, nuclear industry, and also automotive industries (Simm et al. 2010; Yin et al. 2013). Eddy current thermography technique can be used for surface and subsurface defect detection of conductive parts (Yang et al. 2011). Joule heating process associated with the eddy current flow permits the detection of flaw in the material (Weekes et al. 2012). The method of inspection comes in package with the infrared technology to capture the distribution of heat on components being tested in the form of two-dimensional images and video images. Nondestructive testing by using thermography technique is the innovation of infrared camera that led to the improvement in spatial resolution, frame rate, and temperature sensitivity (Simm et al. 2010). This technique manages to provide quantitative information via the acquisition and images sequences analysis (Wilson et al. 2011).

I.M. Zainal Abidin
Non-Destructive Testing Group, Industrial Technology Division, Malaysian Nuclear Agency, Kuala Lumpur, Selangor, Malaysia
e-mail: mukriz@nuclearmalaysia.gov.my

N.I. Harun · S.S. Sheikh Abdul Aziz
Faculty of Applied Sciences, Universiti Teknologi Mara, Tapah, Perak, Malaysia
e-mail: nurhanailmira@perak.uitm.edu.my

S.S. Sheikh Abdul Aziz
e-mail: sumaiyahaziz@perak.uitm.edu.my

Information analysis from the detection in the form of defect size, location, and depth can be attained through features presented in frequency domain and period of transient signal (Abidin et al. 2010). Previously, fine defect such as surface, sub-surface, or within a complex geometry defect were detected and presented in signal forms by using conventional NDT method. The principal of defect detection using ECT techniques introduces the eddy current path that disrupted by the presence of defect (Yang et al. 2011) will be interpreted further in terms of signal due to the variation in current density (Weekes et al. 2012). The defect localized around the region consequent the increment and decrement of joule (Ohmic) heating process. Based on the principle of ECT, the process of eddy current induction and heat diffusion were integrated with the infrared radiation range from the IR camera. Thermographic data and images captured by the infrared camera can be assessed and interpret to provide both qualitative and quantitative information of the defects in the inspected samples (Abidin et al. 2012; Yin et al. 2013). This technique is able to detect hidden, subsurface defects even in complex geometry components (Abidin et al. 2010).

2 Methodology

The analysis is conducted over the carbon steel welded sample with LOF defect inside the joint of the sample. Analogically, defect usually can be detected by a direct observation due to the electromagnetic heating that visualizes the pattern of heat diffusion on the image pixels of IR camera (Yang et al. 2011). The ECT involves the application of high current and high-frequency electromagnetic wave (typically 50–500 kHz) to the sample being tested for a short period (typically 20 ms–1 s) (Wilson et al. 2011). The configuration of eddy current thermography system consists of an induction heating system at the range of high-frequency electromagnetic wave. The probe is connected to the work head and will induce the current toward the sample. The water flow inside the coil will regulate the direct heating of the coil (Yin et al. 2013). Figure 1 shows the setup for an eddy current thermography system.

The "thermoimager" software that links to the infrared camera will process the image captured by the camera. The IR camera has integrated link to the data stored on PC which is to interpret the signal of the presence of defect (Wilson et al. 2011) that could be predefined by visual analysis through the observation of different contrast on the processed image of the sample. The temperature linescans at the sample surface should be taken directly beneath the induction coil during the heating process for sample temperature profiles.

In providing distinctive investigation result, radiography analysis is conducted to scan the geometrical shape image of welded sample after the ECT analysis. Radiographic image in the form of films can be analyzed by visual inspection based on the particular criterions. The fast observation could be accomplished either by comparing defects with known defect templates or by thresholding the image

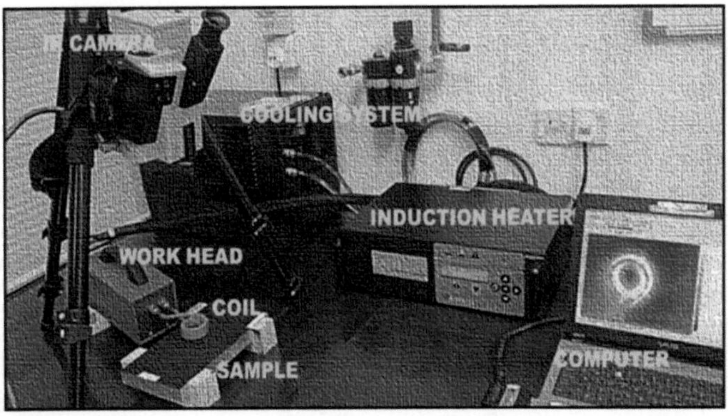

Fig. 1 Eddy current thermography system (Advanced NDT Lab, Malaysia Nuclear Agency)

(Wang and Warren 2002). NDT is known as comparative measurements before the acquired results can be quantified for a quantitative characterization of the investigated defect (Abidin et al. 2012). The acquired image is processed via Matlab to signify the temperature amplitude distribution at the respective position for every image of the sample. The position of defect produced different temperature amplitudes compared to the defect-free area. The infrared camera connected to the thermoimager software captured the image of the defect in a very fast period due to eddy current excitation. The temperature distribution at the surface of the samples will differ depending on the sensitivity of induced current toward the defect.

3 Results and Discussion

The carbon steel welded sample of dimension 200 × 150 × 20 mm has a defect located at the welded joint area which is occurred during the welding process as depicted in Fig. 2a. The defect was located at labeled B position whereas the positions labeled A and C are free from defects. Meanwhile, a radiography image obtained after the process of radiography testing by using X-rays or gamma rays imaging. The internal defect can be interpreted by the visibility of dark contrast that revealed the presence of defect projected on the radiographic film (Wang and Warren 2002). The results show temperature distribution captured by the infrared camera at different positions along the welded area of the sample, which are specified into areas A, B, and C. Figure 2a, b revealed the temperature profiles of the line scans graph as well as the amplitude of the temperature at defect and no-defect position. The optimum current applied at the setting of induction heater is 300 A while the heating time is 450 ms. The thermoimager TIM software that links to the infrared camera measured the surface temperature of the inspected material due to infrared energy emission. The radiometric image is analyzed through the

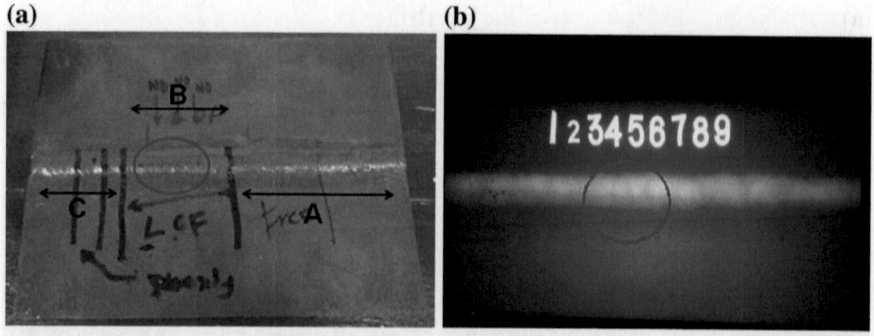

Fig. 2 **a** Welded sample test specimen having LOF type of defect labeled *A* (no-defect position), *B* (defect position-in circle), *C* (no-defect position) and **b** radiographic image of welded sample with *red circle* showing the defect position

software. The images during the maximum heating process were extracted into 15 processing line scan via Matlab at the preselected location nearest to the coil. These data were presented in pixels value corresponding to the respective temperature.

From Fig. 3a, the temperature profiles shows the amplitudes corresponding to the different positions along the welded area. The defect identification cannot be resolved clearly from the Fig. 2, therefore the amplitude from each linescans were extracted to provide the clear indication of defect or no-defect areas. Theoretically, defect area signifies the highest heat accumulation. This is due to less volume of metal occupied that makes heat diffuse faster on the LOF defect area. Error bar analysis on Fig. 3b is an alternative analysis to provide the scientific evidence for defect location. The error bar analysis represents the uncertainty and data validity.

The error bars plotted in Fig. 3b distinguish the presence and absence of defect. The defect is about 20 mm in length with referring to the error bar graph and validated by the radiography result. The highest temperature amplitudes which

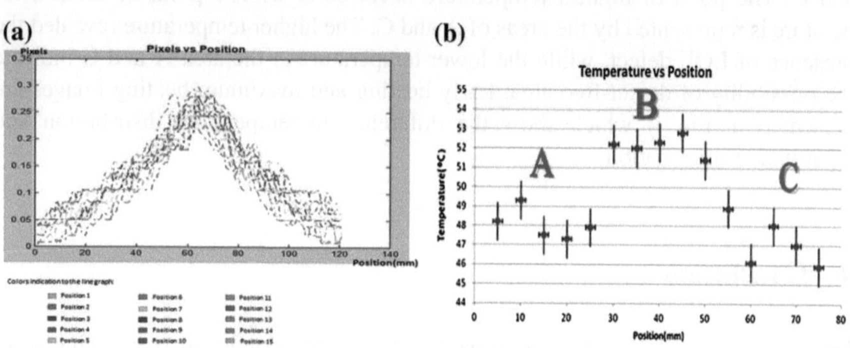

Fig. 3 **a** Temperature profiles of the linescans at every positions from area *A*, *B*, and *C* of the welded sample and **b** *error bar* shows the amplitude of the temperature at the position of defect (*B*) and without defect (*A*) and (*C*)

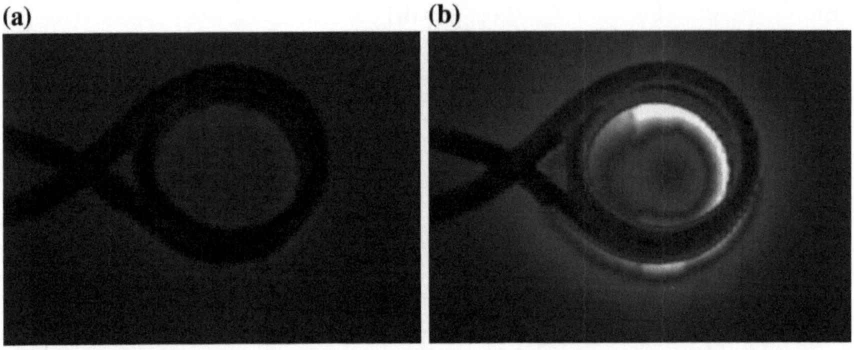

Fig. 4 Position of **a** early heating and **b** maximum heating snap by infrared camera

corresponds to the position of the LOF defect is located at the area B having the length of approximately 20 mm which starts at the point of 30 mm to the point of 50 mm. The lower temperature amplitude is shown at the position A and C. The maximum temperature at positions of A and C is approximately 49 °C whereas the maximum temperature at position B is approximately 53 °C. The highest maximum temperature at position B proved the presence of LOF defect because heat will accumulate faster at the position of less volume of metal. The location of the presence of LOF defect will generate higher temperature value compared to the defect free area. The heating distribution is higher at the LOF defect area but minimum at the defect-free area.

The area that possessed the same behavior is justified if the error bars overlapped each other in symmetrical line parallel to the horizontal axes. The error bars at positions A and C are considered to overlap each other. Analogically, if we start to draw a line at 0 mm until the end of 80 mm for area A and C on Fig. 3b, the line crosses through the width of leg at the point of the error bar respective to the both areas A and C. On the contrary, the error bar at B does not overlap with the point A and C. The point of highest temperature occurred at B. The point of lower temperature is represented by the areas of A and C. The higher temperature revealed the presence of LOF defect, while the lower temperature at the area A and C indicate the possibility of defect-free area. Early heating and maximum heating images are presented in Fig. 4 which show the difference in temperature distribution and quantitative information.

4 Conclusion

Eddy current thermography has proved as one of the nondestructive testing technique which can provide quantitative information of the inspected sample. By analysis, the position of the (LOF) defect is successfully detected and confirmed.

The relatively higher temperature amplitude acquired from the line scans of each position along the carbon steel welded sample shows the behavior of the presence of defect. Further research on eddy current thermography technique is recommended for modeling and experimental investigations of defect on subsurface of welded samples by using Comsol FEM Multiphysics. This would provide an alternative solution toward the testing and assessment of welding which is very crucial in the maintenance and overall structural integrity evaluation for industrial components.

Acknowledgment The authors would like to thank the Industrial Technology Division of Malaysian Nuclear Agency for the implementation of the lab work at the Advanced NDT Lab.

References

Abidin IZ, Umar MZ, Yusof MY, Ibrahim MM, Hamzah AR, Salleh MN (2012) Advantages and application of eddy current thermography testing for comprehensive and reliable defect asessment. In: Proceedings of the 18th world conference on nondestructive testing, 16–20 April, Durban, South Africa

John W, Tian GY, Abidin IZ, Yang S, Almond D (2010) Modelling and evaluation of eddy current stimulated thermography. Nondestr Test Eval 25(3):205–218

Pan M, He Y, Tian G, Chen D, Luo F (2012) Defect characterisation using pulsed eddy current thermography under transmission mode and NDT applications. NDT and E Int 52:28–36

Simm A, Abidin IZ, Tian GY, Woo WL (2010) Simulation and visualisation for electromagnetic nondestructive evaluation. In: Proceedings of the 2010 14th international conference information visualisation, pp 515–520

Wang G, Warren L (2002) Automatic identification of different types of welding defects in radiographic images. NDT and E Int 35(8):519–528

Weekes B, Almond DP, Cawley P, Barden T (2012) Eddy-current induced thermography—probability of detection study of small fatigue cracks in steel, titanium and nickel-based superalloy. NDT and E Int 49:47–56

Wilson J, Tian G, Mukriz I, Almond D (2011) PEC thermography for imaging multiple cracks from rolling contact fatigue. NDT and E Int 44(6):505–512

Yang S, Tian GY, Abidin IZ, Wilson J (2011) Simulation of edge cracks using pulsed eddy current stimulated thermography. J Dyn Syst Meas Contr 133(1):011008

Yin A, Gao B, Tian GY, Woo WL, Li K (2013) Physical interpretation and separation of eddy current pulsed thermography. J Appl Phys 113:064101

Zainal Abidin I, Yun Tian G, Wilson J, Yang S, Almond D (2010) Quantitative evaluation of angular defects by pulsed eddy current thermography. NDT and E Int 43(7):537–546

